Advanced Polymer Nanocomposites II

Advanced Polymer Nanocomposites II

Editors

Ting-Yu Liu
Yu-Wei Cheng

MDPI • Basel • Beijing • Wuhan • Barcelona • Belgrade • Manchester • Tokyo • Cluj • Tianjin

Editors
Ting-Yu Liu
Department of Materials Engineering
Ming Chi University of Technology
New Taipei City
Taiwan

Yu-Wei Cheng
Department of Chemical Engineering
Ming Chi University of Technology
New Taipei City
Taiwan

Editorial Office
MDPI
St. Alban-Anlage 66
4052 Basel, Switzerland

This is a reprint of articles from the Special Issue published online in the open access journal *Polymers* (ISSN 2073-4360) (available at: www.mdpi.com/journal/polymers/special_issues/Advanced_Polymer_NanocompositesII).

For citation purposes, cite each article independently as indicated on the article page online and as indicated below:

LastName, A.A.; LastName, B.B.; LastName, C.C. Article Title. *Journal Name* **Year**, *Volume Number*, Page Range.

ISBN 978-3-0365-6921-5 (Hbk)
ISBN 978-3-0365-6920-8 (PDF)

Contents

About the Editors

Ting-Yu Liu

Ting-Yu Liu received his PhD degree at the Department of Materials Science and Engineering, National Chiao Tung University, Taiwan, in 2008. He worked at the Department of Materials Science and Engineering, University of Pennsylvania, USA, for 1 year of research. After that, he was a post-doc fellow at the Institute of Atomic and Molecular Sciences, Academia Sinica, Taiwan, from 2009 to 2011, and a project assistant professor at the Institute of Polymer Science and Engineering, National Taiwan University, from 2011 to 2013. He is currently a distinguished professor at the Department of Materials Engineering, Ming Chi University of Technology, Taiwan. His research covers nanomaterials, biomaterials, polymer composites, and opto-electric (surface-enhanced Raman spectroscopy detection) and electrochemical sensing. He has published more than 100 SCI-indexed journal articles and has an h-index of 32 according to the citation report from Google Scholar.

Yu-Wei Cheng

Yu-Wei Cheng received his PhD degree from the Institute of Polymer Science and Engineering, National Taiwan University. He is currently an assistant professor at the Department of Chemical Engineering, Ming Chi University of Technology, Taiwan. His research interests included the synthesis of functionalized polymers and the fabrication of optimized Raman enhancing nanoparticle arrays for small-molecule detection. His recent research focuses on manipulated metal nanoparticle arrays grown on 2D and 3D substrates for multifunctional smart sensors and surface-enhanced Raman scattering detection, published in the highly ranking journals of chemistry and physical fields. He has published approximately 27 SCI-indexed journal articles and has an a h-index of 12 according to a citation report from Google Scholar.

Preface to "Advanced Polymer Nanocomposites II"

Polymer nanocomposites are currently of high industrial interest in the field of nanomaterials due to fact that they can improve the performance of polymeric matrixes and inorganic nanomaterials. They can be used to alter light/magnetic behaviors, electrical/thermal conductivity, toughness, stiffness and mechanical strength. The inorganic quantum dots/nanoparticles, nanorods/nanotubes, and 2D materials (such as graphene-based nanosheets) can be decorated in a polymer matrix through chemical synthesis or physical blending to improve its performance. Thus, how to fabricate a homogeneous dispersion of fillers in the polymer matrix has been a most crucial area of research in the nanomaterials field. This Special Issue reprint, "Advanced Polymer Nanocomposites II", collects high-quality original and review papers, focused on the scientific discussion and practical application in the field of functional polymer nanocomposites, including: (a) optoelectronic materials (papers 1-9); (b) biomedical materials (papers 10-14); and (c) other functional polymer nanocomposites (papers 15-23). We hope that this Special Issue reprint can promote academic research exchanges and identify and respond to tremendous challenges of research in this burgeoning field.

Ting-Yu Liu and Yu-Wei Cheng
Editors

polymers

MDPI

Article

Highly Thermal Stable Phenolic Resin Based on Double-Decker-Shaped POSS Nanocomposites for Supercapacitors

Wei-Cheng Chen [1], Yuan-Tzu Liu [1] and Shiao-Wei Kuo [1,2,*]

1 Department of Materials and Optoelectronic Science, Center of Crystal Research, National Sun Yat-Sen University, Kaohsiung 80424, Taiwan; chwei566@gmail.com (W.-C.C.); ruby860420@gmail.com (Y.-T.L.)
2 Department of Medicinal and Applied Chemistry, Kaohsiung Medical University, Kaohsiung 807, Taiwan
* Correspondence: kuosw@faculty.nsysu.edu.tw; Tel.: +886-7-525-4099

Received: 25 August 2020; Accepted: 19 September 2020; Published: 21 September 2020

Abstract: In this study we incorporated various amounts of a double-decker silsesquioxane (DDSQ) into phenolic/DDSQ hybrids, which we prepared from a bifunctionalized phenolic DDSQ derivative (DDSQ-4OH), phenol, and CH_2O under basic conditions (with DDSQ-4OH itself prepared through hydrosilylation of nadic anhydride with DDSQ and subsequent reaction with 4-aminophenol). We characterized these phenolic/DDSQ hybrids using Fourier transform infrared spectroscopy; ^{1}H, ^{13}C, and ^{29}Si nuclear magnetic resonance spectroscopy; X-ray photoelectron spectroscopy (XPS); and thermogravimetric analysis. The thermal decomposition temperature and char yield both increased significantly upon increasing the DDSQ content, with the DDSQ units providing an inorganic protection layer on the phenolic surface, as confirmed through XPS analyses. We obtained carbon/DDSQ hybrids from the phenolic/DDSQ hybrids after thermal curing and calcination at 900 °C; these carbon/DDSQ hybrids displayed electrochemical properties superior to those of previously reported counterparts.

Keywords: POSS; phenolic resin; hydrogen bonding; supercapacitors

1. Introduction

Compared with inorganic materials, organic polymers usually possess lower moduli and poorer thermal stabilities and mechanical properties. The incorporation of inorganic nanoparticles (NPs) nanotubes, or nanosheets (e.g., clay [1,2]), graphene [3–5], carbon nanotubes [6–8], or polyhedral oligomeric silsesquioxanes (POSS) [9–13] into a polymer matrix is, however, a relatively simple approach toward improving its mechanical and thermal properties.

Phenolic resin is a typical thermosetting resin that has many applications (e.g., in laboratory countertops, adhesives, and coatings) because of its structural integrity and solvent-resistance. Nevertheless, phenolic resin has been neglected as a material for use in polymer nanocomposites because it has a three-dimensional structure even when it is not crosslinked. Several approaches have been developed, however, to overcome this problem to enhance thermal stability [14–25]. For example, phenolic nanocomposites have been synthesized through intercalative polymerization in organic-modified clays [14] and through sol–gel reactions [15,16]. In addition, the ability of POSS to improve the thermal properties of phenolic resin have been discussed widely [17–25]. Indeed, we have prepared several phenolic/POSS nanocomposites incorporating various functionalized (e.g., isobutyl, acrylate, and acetoxystyrene) POSS NPs, stabilized through hydrogen bonding [17–21]. Zhang et al. incorporated multi-functionalized amine POSS NPs into a phenolic resin, stabilized through hydrogen bonding and covalent crosslinking, to enhance its mechanical properties [22]. Zheng and Liu et al. used a multi-functionalized epoxy POSS as a crosslinking agent within phenolic resin, thereby improving its thermal stability [23,24]. Furthermore, we synthesized a multi-functionalized phenol POSS derivative and reacted it with phenol and CH_2O to obtain various phenolic/POSS hybrids, which also exhibited high thermal stability and low surface energies [25].

Unfortunately, these phenolic/POSS nanocomposites prepared using multi-functionalized POSS NPs have led to insoluble cross-linked phenolic resins [22–25], restricting their applicability. When the POSS NPs presented more than two functional groups, the polymer/POSS nanocomposites formed crosslinked structures, even in the presence of only a small amount of the POSS NPs. Therefore, bifunctionalized POSS NPs, including double-decker silsesquioxane (DDSQ), that can form main-chain-type POSS units have allowed new approaches for the synthesis of novel polymer/POSS hybrids from, for example, polyurethane [26,27], polybenzoxazine [28–30], block copolymer [31,32], and polyimide [33,34] systems.

In this study, we synthesized the bifunctionalized phenolic DDSQ derivative DDSQ-4OH (Scheme 1e) through the hydrosilylation of nadic anhydride (ND) with DDSQ (itself obtained from phenyltrimethylsilane (Scheme 1a) and DD-Na (Scheme 1b), giving DDSQ-ND (Scheme 1d), and subsequent reaction with 4-aminophenol. We then prepared phenolic/DDSQ (PDDSQ) hybrids incorporating various amounts of DDSQ through the reactions of DDSQ-4OH with phenol and CH_2O under basic conditions (Scheme 1f). Furthermore, we obtained carbon/DDSQ hybrids after thermal curing and calcination of these PDDSQ hybrids; these carbon materials displayed electrochemical properties superior to those of related non-carbonized and even carbonized materials.

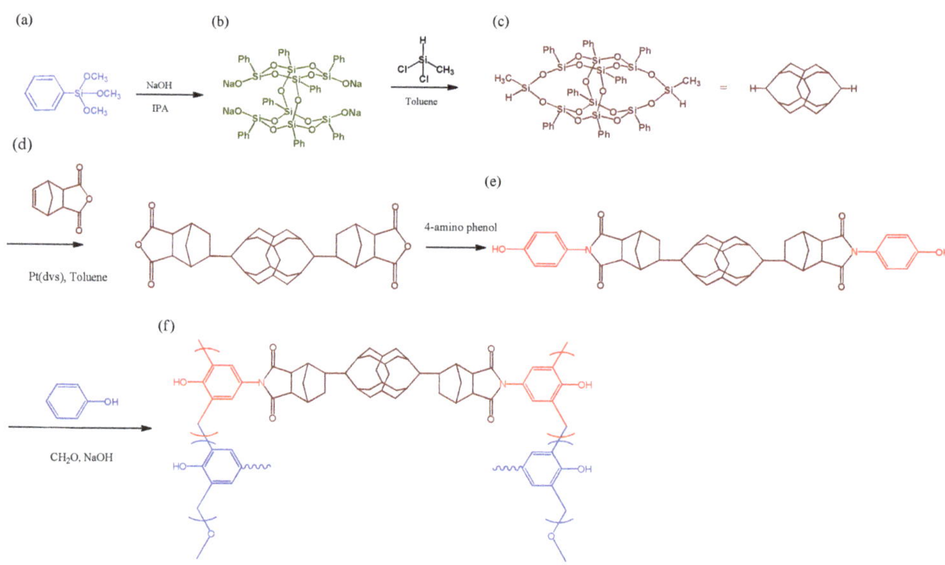

Scheme 1. Chemical structures of (**a**) phenyltrimethoxysilane, (**b**) DD-Na, (**c**) double-decker silsesquioxane (DDSQ), (**d**) DDSQ-nadic anhydride (ND), and (**e**) DDSQ-4OH and (**f**) the preparation of phenolic/DDSQ (PDDSQ) hybrids from DDSQ-4OH and phenol under basic conditions.

2. Materials and Methods

2.1. Materials

Phenyltrimethoxysilane (Scheme 1a), sodium hydroxide (NaOH), methyl dichlorosilane (CH_3Cl_2SiH), charcoal, acetonitrile, ND, paraformaldehyde (CH_2O), platinum divinyltetramethyldisiloxane complex

(Pt (dvs)), 4-aminophenol, and phenol were purchased from Sigma–Aldrich (St. Louis, MO, USA). DDSQ-4OH (Scheme 1e) was synthesized according to a method described previously [18,35].

2.2. PDDSQ hybrids

Mixtures of phenol and various amounts of DDSQ-4OH in dioxane and aqueous NaOH solution were stirred for 10 min; formaldehyde in water was added and then the mixtures were stirred for 10 min before heating at 70 °C for 1 h. All compositions of these PDDSQ hybrids used in this study were summarized in Table 1. The mixture was cooled to room temperature and treated with dilute HCl solution to pH 7.0. The water content in the viscous liquid was decreased through distillation under reduced pressure for 2 h at 45 °C. The resulting PDDSQ hybrids were dissolved in tetrahydrofuran (THF) and stirred vigorously; the precipitated NaCl was removed through centrifugation.

Table 1. Characterization data for the PDDSQ hybrids synthesized in this study.

Name	Monomer Feed (g)				DDSQ-4OH in hybrids (wt.%)	T_d (°C)	Char Yield (wt.%)	M_n (g/mol)
	Phenol	DDSQ-4OH	CH_2O	NaOH	FTIR			
Pure Phenolic	10.0	-	17.25	4.25	0	364	41.6	5440
PDDSQ-20	9.0	1.0	15.72	3.87	20.2	389	47.7	4670
PDDSQ-30	4.0	1.0	7.09	1.75	32.3	403	49.7	4320
PDDSQ-45	2.3	1.0	4.22	1.04	44.9	411	55.5	4580
PDDSQ-50	1.5	1.0	2.78	0.68	50.7	438	58.2	4160
PDDSQ-80	1.0	1.0	1.92	0.47	80.5	443	63.8	4460
Pure PDDSQ	-	1.0	0.19	0.04	100	532	70.4	4500

2.3. Carbon/DDSQ Hybrids

A PDDSQ hybrid was stirred in THF until it had dissolved completely. The solution was poured into a Teflon pan and then the solvent was evaporated at room temperature over 2 days. The resulting solid was heated at 150 °C for 24 h to ensure complete thermal curing. Finally, the PDDSQ hybrid was subjected to thermal treatment at 900 °C for 6 h under a N_2 atmosphere in a tubular furnace to give a carbon/DDSQ hybrid.

3. Results and Discussion

3.1. Synthesis of the Pure PDDSQ Hybrid

Scheme 1 displays our approach for the synthesis of PDDSQ hybrids featuring various contents of DDSQ; we confirmed their structures using gel permeation chromatography apparatus (GPC, Waters 510, Labx, Midland, ON, Canada) and Fourier transform infrared (FTIR, Nicolet iS50, Thermo Fisher Scientific, Waltham MA, USA) and nuclear magnetic resonance (NMR, Bruker 500, Billerica, MA, USA) spectroscopy. First, we prepared the PDDSQ hybrid from DDSQ-4OH, synthesized as presented in Scheme 1e,f, in the absence of phenol. Figure 1 provides the ^{1}H NMR spectra of DDSQ-4OH and its corresponding pure PDDSQ. The spectrum of DDSQ-4OH (Figure 1a) featured a signal for the $SiCH_3$ unit at 0.36 ppm and signals located between 7.14 and 7.50 ppm representing the aromatic protons of the DDSQ unit. Furthermore, signals from the aliphatic protons of the ND units appeared between 3.25 and 0.83 ppm, a broad signal appeared at 5.33 ppm for the OH units, and signals for the aromatic protons derived from 4-aminophenol appeared at 6.24 (*l*) and 6.80 (*k*) ppm, confirming the formation of DDSQ-4OH. The ^{1}H NMR spectrum of the pure PDDSQ (Figure 1b) featured small signals at 5.30 and 9.73 ppm, representing its aliphatic and phenolic OH units, respectively. The spectrum of the pure PDDSQ also contained signals from various kinds of CH_2 units: $ArC\underline{H}_2Ar$ at 1.83 ppm (peak a), methylene ether ($PhC\underline{H}_2OR$) at 3.73 ppm (peak b), and methylol ($OC\underline{H}_2OH$) (peak c) at 4.82 ppm. Furthermore, the intensity of the signal of the aromatic protons from the 4-aminophenol unit at 6.24 ppm (*l*) had decreased significantly (Figure S1b), consistent with the formation of those bridging aromatic rings, ether, and methylol units.

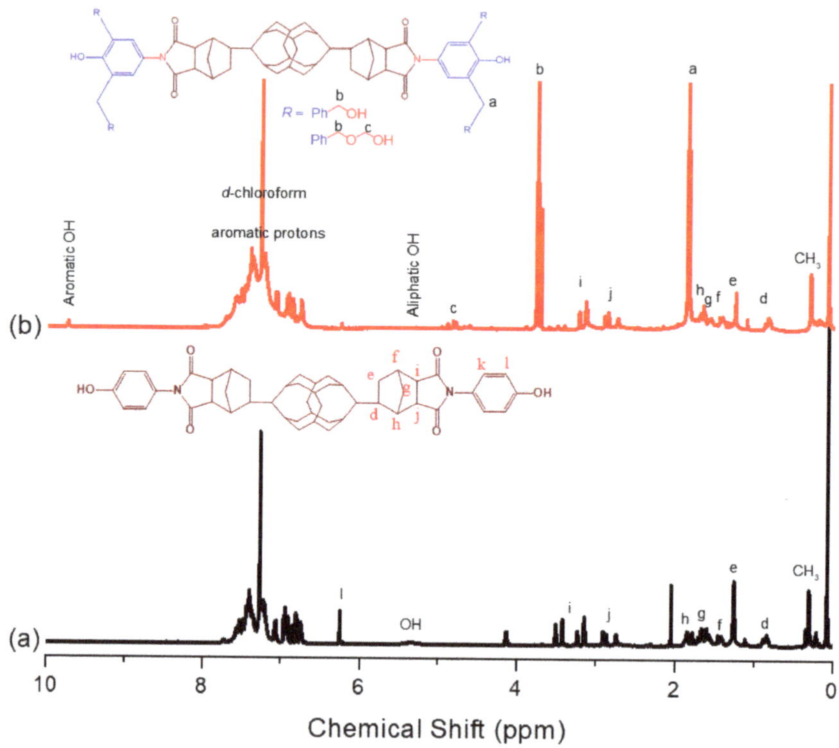

Figure 1. ^{1}H NMR spectra of (**a**) DDSQ-4OH and (**b**) pure PDDSQ.

Figure 2 displays the ^{13}C NMR spectra of DDSQ-4OH and its corresponding pure PDDSQ. The spectrum of DDSQ-4OH (Figure 2a) featured a signal for the Si$\underline{C}H_3$ unit at 0.81 ppm, with signals located between 116.3 and 133.0 ppm representing aromatic carbon nuclei. The signals of the aliphatic carbon nuclei of the ND units appeared between 24.5 and 51.3 ppm and those of the aromatic C–OH (peak *a*) and C=O (peak *b*) units appeared at 157.3 and 178.8 ppm, confirming the formation of DDSQ-4OH. In the spectrum of the pure PDDSQ (Figure 2b), additional peaks appeared at 25.5 ppm for the Ar$\underline{C}H_2$Ar (peak c) units and at 68.1 ppm for the methylene ether (peak d) and methylol (peak e) units, along with the other signals from the DDSQ-4OH moieties, confirming the synthesis of the pure PDDSQ.

Figure S1 presents the FTIR spectra of DDSQ-4OH and its corresponding pure PDDSQ. The spectrum of DDSQ-4OH featured two strong signals at 1090 and 1130 cm^{-1}, corresponding to the Si–O–Si and C–O units, as well as a signal at 1265 cm^{-1} representing the Si–CH$_3$ groups. The spectrum of the pure PDDSQ did not feature any obvious additional peaks when compared with that of DDSQ-4OH, with only the intensity of the signal representing the OH groups having increased in intensity after the reaction of CH$_2$O to form the pure PDDSQ. Figure 3 displays the ^{29}Si NMR spectra of DDSQ-4OH and the pure PDDSQ. The spectrum of DDSQ-4OH featured three major peaks at –79.05, –64.17, and –21.34 ppm representing the SiO$_3$R (T$_3$), SiO$_2$R$_2$ (T$_2$), and Si-C units, respectively (Figure 3a). In the spectrum of the pure PDDSQ (Figure 3b), two peaks at −79.05 and −64.17 ppm were remained, suggesting that the cage structure had not been destroyed; nevertheless, the intensities of these three peaks had decreased, as expected because the concentration of DDSQ units had decreased after the formation of the pure PDDSQ. The molecular weight of the pure PDDSQ (ca. 4500 g mol^{-1}) was determined through GPC analysis, with a polydispersity index of 1.10 (Figure S2).

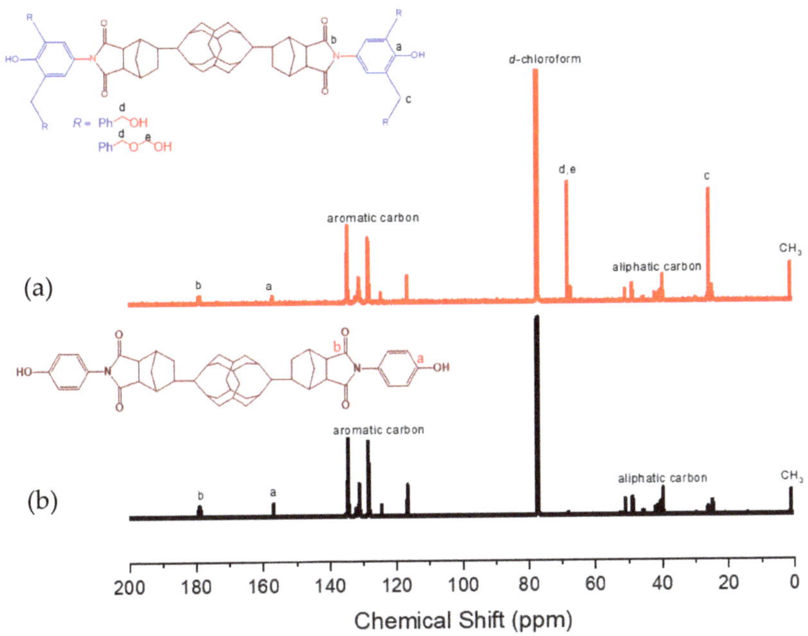

Figure 2. ^{13}C NMR spectra of (**a**) DDSQ-4OH and (**b**) pure PDDSQ.

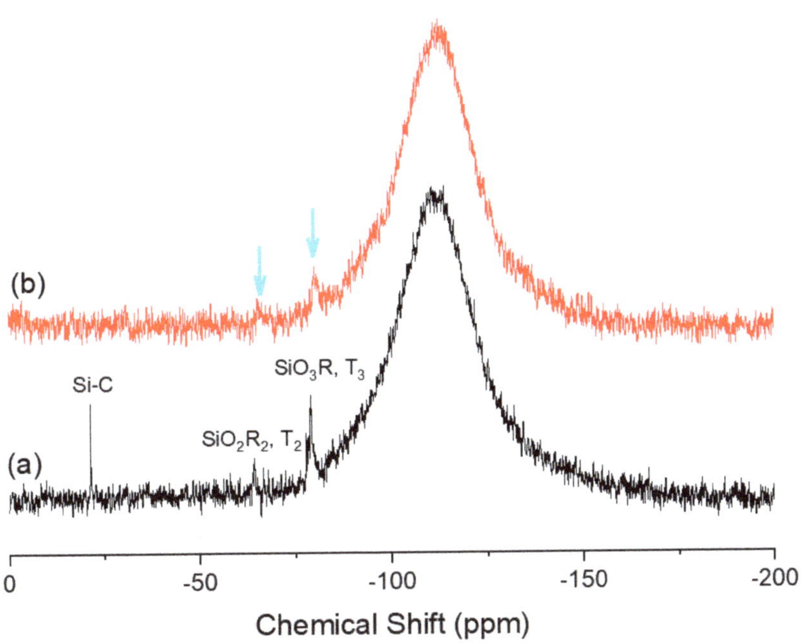

Figure 3. ^{29}Si NMR spectra of (**a**) DDSQ-4OH and (**b**) pure PDDSQ.

3.2. Synthesis of PDDSQ Hybrids

Having successfully synthesized the pure PDDSQ, we prepared PDDSQ hybrids incorporating various amounts of DDSQ, as displayed in Scheme 1f. Figure 4 presents FTIR spectra of these PDDSQ hybrids, recorded at room temperature. As mentioned above, the spectrum of the pure PDDSQ featured two strong signals at 1090 and 1265 cm^{-1} representing the Si–O–Si and Si–CH_3 units of the DDSQ cage structure; in addition, two other obvious peaks appeared at 1702 and 1510 cm^{-1}, representing the C=O groups and aromatic units of the ND and phenolic moieties, respectively. Here, we maintained the signal for the aromatic unit of the phenolic unit at 1510 cm^{-1}, with the intensity of the signal at 1702 cm^{-1} for the C=O groups increasing upon increasing the concentration of DDSQ units in the PDDSQ hybrids. Thus, the composition of DDSQ could be determined readily by the peak ratio A_{1702}/A_{1510} (Table 1). Furthermore, the spectrum of the pure phenolic resin displayed two major signals at 3375 and 3525 cm^{-1}, corresponding to self-associated and free OH units, respectively. The intensity of the signal for the free OH units at 3525 cm^{-1} decreased upon increasing the DDSQ content in the PDDSQ hybrids from 20 to 100 wt %. The signal for the self-association of the phenolic OH units shifted to higher frequency upon increasing the concentration of DDSQ units in the PDDSQ hybrids, with the peak shifting to 3435 cm^{-1} as a result of the self-associative OH···OH bonding transforming into inter-associative OH···siloxane bonding. The frequency difference ($\Delta\nu$) between the signals for the free and hydrogen-bonded OH groups can give a rough estimation of the average hydrogen bonding strength [36]. The average strength of the phenolic OH groups interacting with the DDSQ siloxane moieties ($\Delta\nu = 95$ cm^{-1}) appeared to be weaker than that for the self-associative OH···OH bonding of the pure phenolic resin ($\Delta\nu = 150$ cm^{-1}). This result is consistent with our previous finding that the inter-association equilibrium constant of phenolic/POSS ($K_A = 38.7$) is smaller than the self-association equilibrium constant of pure phenolic ($K_B = 52.3$), based on the Painter–Coleman association model [17,18]. Furthermore, Figure S3 presents ^{1}H spectra of pure phenolic resin, PDDSQ-50 and pure PDDSQ recorded at room temperature. The peak assignment is similar with Figure 1 and we could observe that the PDDSQ hybrids are also synthesized successfully based on NMR and FTIR spectra analyses.

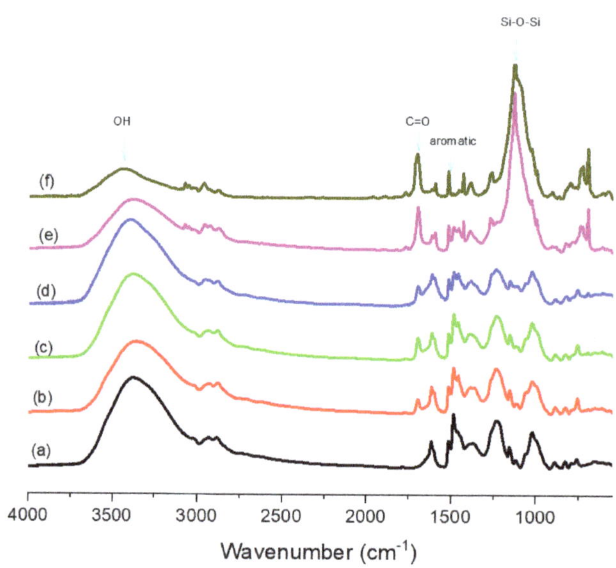

Figure 4. FTIR spectra of (**a**) pure phenolic, (**b**) PDDSQ-30, (**c**) PDDSQ-45, (**d**) PDDSQ-50, (**e**) PDDSQ-80, and (**f**) pure PDDSQ.

Although we had detected only weak intermolecular interactions between the phenolic OH groups and the DDSQ siloxane units, these interactions did still enhance the thermal stability of the PDDSQ hybrids, as determined using thermal gravimetric analyzer (TGA, Q-50, TA Instruments, New Castle, DE, USA). The TGA trace (Figure 5) of the pure phenolic resin revealed three thermal degradation steps, typical of the degradation of phenolic resin [37] and the derivative curve was displayed in Figure S4. The first step ca. 250 °C from the degradation of CH_2 bridges, the second step at ca. 450 °C is due the broken of crosslinking network structure, and the third at ca. 500 °C comes from the C–OH units in phenol were broken and then form H_2 gas. Table 1 also summarizes the thermal degradation temperatures (T_{d10}) and char yields at 800 °C of these PDDSQ hybrids. The char yields and values of T_d increased significantly upon increasing the DDSQ concentration, as expected. For example, the char yield and value of T_d increased by 12.2 wt% and 79 °C, respectively, for the PDDSQ-80 hybrid when compared with those of the pure phenolic resin. This behavior can be explained by considering the nano-reinforcement effect of incorporating inorganic DDSQ NPs into a phenolic matrix. The DDSQ units were presumably dispersed well within the phenolic resin at the nanoscale, stabilized through hydrogen bonding and covalent bonding, and, thus, they could enhance its initial decomposition temperature. Furthermore, the DDSQ units could retard the thermal motion of the polymer through the covalent bonding, and could also increase the chain spacing, giving rise to the lower thermal conductivity. It has been proposed that DDSQ units prefer to be oriented toward the air side in related nanostructures, because the low surface free energy of the siloxane units would screen out the polar functional units (e.g., carboxyl, OH, and urethane) [38,39]. As a result, we suspect that the DDSQ NPs in our PDDSQ hybrids would also have preferred to be oriented toward the air side, thereby forming an inorganic silica protection layer on the phenolic surface, resulting in higher thermal stability. To confirm this hypothesis, we used X-ray photoelectron spectroscopy (XPS) to examine the surface behavior of our PDDSQ hybrids (Figure 6). Figure 6a reveals the presence of three major elements: C atoms at 285.1 eV (Figure 6b), Si atoms at 103.6 eV (Figure 6c), and O atoms at 533.2 eV (Figure 6d). Large amounts of Si (20.0 wt %) and O (33.8 wt %) atoms were present at the surface, indicating that the DDSQ units had to migrated to the surface of the phenolic resin.

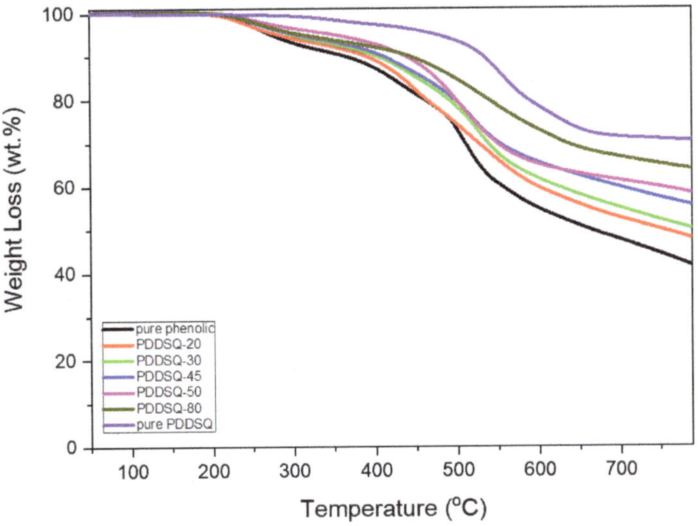

Figure 5. TGA analyses of PDDSQ hybrids containing various DDSQ contents.

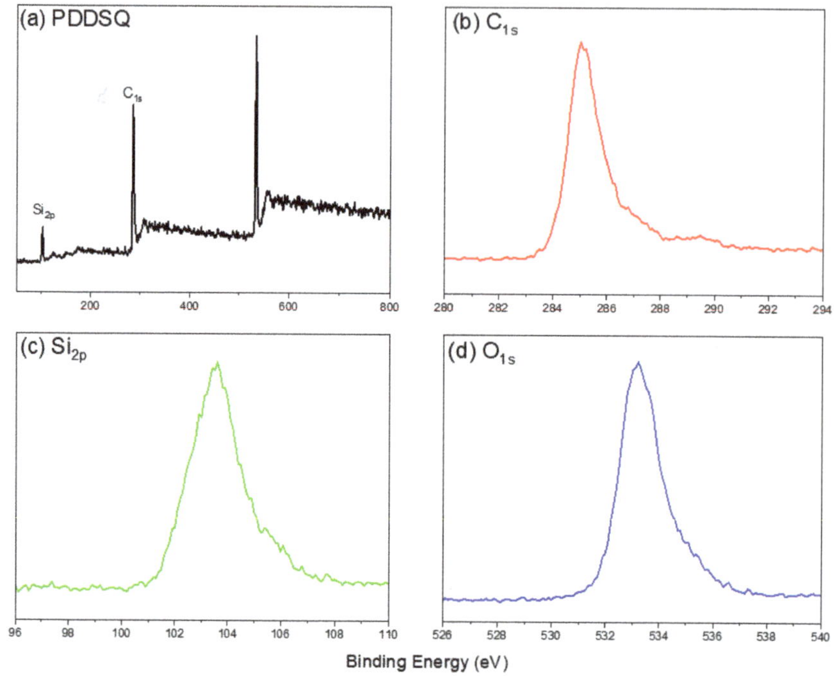

Figure 6. XPS analysis of (**a**) pure PDDSQ and its (**b**) C 1s, (**c**) Si 2p, and (**d**) O 1s atoms.

3.3. Synthesis of Carbon/DDSQ Hybrids for Electrochemical Applications

To obtain carbon/DDSQ hybrids with potential electrochemical applications, we heated the PDDSQ hybrids at 150 °C for 24 h to complete their thermal curing, and then subjected them to further heating at 900 or 1000 °C for 6 h under a N_2 atmosphere. Figure 7a displays FTIR spectra of the pure PDDSQ before thermal curing, after thermal curing at 150 °C, and after heating at 900 °C to give its carbon/DDSQ hybrid. We have already discussed the spectrum of the pure PDDSQ (Figure S1); the spectrum of the sample obtained after thermal curing at 150 °C did not exhibit any major changes in its absorption peak. After thermal heating at 900 °C, however, the signals at 1701 cm^{-1} for the C=O groups and at 1432 cm^{-1} for the aliphatic CH_2 units from the ND moieties both disappeared, and new peaks appeared at 1385 and 1612 cm^{-1} representing the carbon structure. Most importantly, the signals at 1090 cm^{-1} for the Si–O–Si units and at 1265 cm^{-1} for the Si–CH_3 units were still present after thermal heating at 900 °C, suggesting that the DDSQ units remained dispersed within the carbon matrix in the carbon/DDSQ hybrids. Raman spectroscopy (HORIBA Jobin–yvon, T6400, Edison, NJ, USA) revealed the local graphitic structure of this carbon/DDSQ hybrid (Figure 7b), with two major carbonized structures identified by G- and D-bands representing sp^2- and sp^3-hybridized carbon atoms at 1592 and 1340 cm^{-1}, respectively [40,41]. The G- and D-bands correspond to first- and second-order Raman scattering, respectively; this G-band had undergone a major shift from the wavenumber of a typical G-band (ca. 1582 cm^{-1}), suggesting the presence of oxidized graphene after thermal treatment at 900 °C (1592 cm^{-1}) or 1000 °C (1620 cm^{-1}). Furthermore, the intensity ratio I_D/I_G can be used to roughly characterize the degree of graphitization. The I_D/I_G ratio (1.35 after treatment at 900 °C) decreased significantly to 1.01 after thermal treatment at 1000 °C, indicating a highly defected carbon structure containing Si, O, and N atoms in the former system, as confirmed by the Transmission electron microscope (TEM, JEOL, JEM-2100, Akishima, Japan) images in Figure 7c,d. The carbon/DDSQ hybrid after thermal treatment at 900 °C exhibited an irregular structure (Figure 7c), but its interlayer spacing was approximately 0.38 nm (based on the (002) plane of the carbon structure) in Figure 7d, confirmed by its WAXD pattern [42]. After thermal treatment at 1000 °C, the high value of q of 16.52 nm^{-1} (d = 0.38 nm) was due to the (002) plane of the carbon structure (Figure 7e). Therefore, the carbon/DDSQ hybrids had highly defected structures after thermal treatment at 900 °C; we selected three carbon materials—those prepared from the pure phenolic resin, PDDSQ-50, and the pure PDDSQ—to test their suitability for use as electrode materials in supercapacitors.

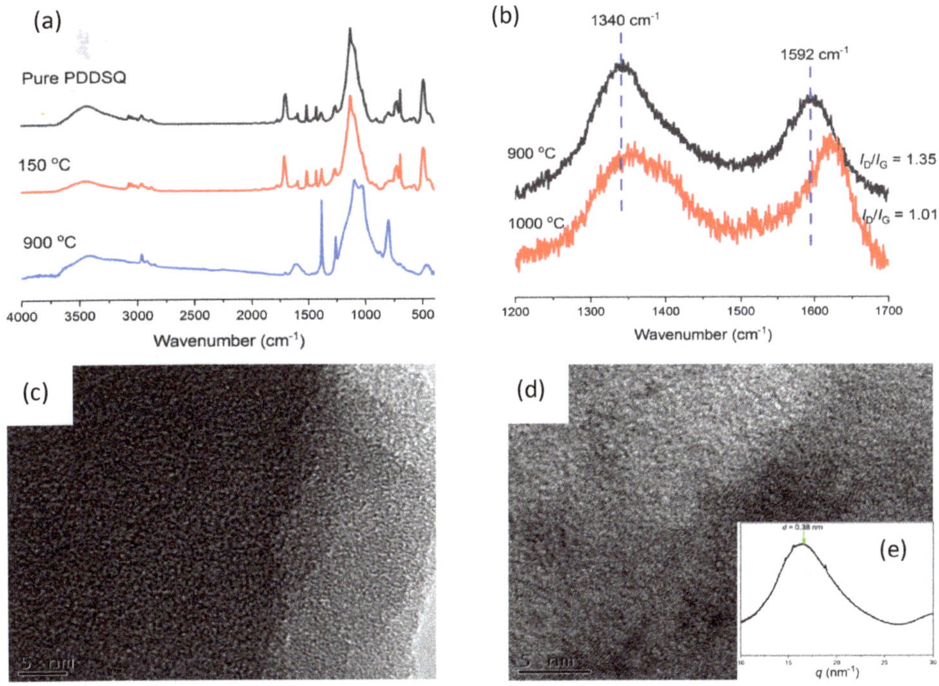

Figure 7. (**a**) FTIR spectra of pure PDDSQ before thermal treatment and after thermal curing at 150 and 900 °C. (**b**) Raman spectra of pure PDDSQ after thermal treatment at 900 and 1000 °C. (**c**,**d**) TEM images of pure PDDSQ after thermal treatment at (**c**) 900 and (**d**) 1000 °C. (**e**) corresponding WAXD pattern after thermal treatment at 1000 °C.

We used cyclic voltammetry (CV) and the galvanostatic charge/discharge (GCD) method in 1 M KOH as the aqueous electrolyte in a three-electrode system to investigate the electrochemical performance of our PDDSQ hybrids after thermal treatment at 900 °C. Figure 8a–c reveal quite different CV curves for the carbon/DDSQ hybrids derived from the pure phenolic, PDDSQ-50, and the pure PDDSQ, measured at sweep rates from 5 to 200 mV s^{-1} in a potential window from +0.35 to −1.00 V. The most distinct differences appeared at the lowest tested sweep rate of 5 mV s^{-1} (Figure 8d–f). The CV curve of the carbon material derived from pure phenolic (Figure 8d) had a rectangle-like shape, implying that its capacitive response resulted from electric double-layer capacitance (EDLC) [43,44]. The CV curve of the carbon material derived from PDDSQ (Figure 8f) had a quasi-rectangular shape featuring a hump, indicating that its capacitive response originated from pseudocapacitance, due to doping by heteroatoms (e.g., N and and O atoms), allowing a reversible redox process at the relatively low sweep rate (5 mV s^{-1}) [45,46]. A similar phenomenon appeared in the CV curve of the carbon material derived from PDDSQ-50 (Figure 8e), suggesting a combination of pseudocapacitance and EDLC. Because of the many electroactive centers (e.g., N and O atoms and electron-rich phenyl rings) in the carbon material derived from the pure PDDSQ, reversible radical redox procedures occurred during the charging and discharging processes [47]. Furthermore, the current density of the carbon material derived from pure PDDSQ increased dramatically upon increasing the sweep rate, while the shape of its CV curve was retained (Figure 8c), indicating good rate capability and facile kinetics [48,49].

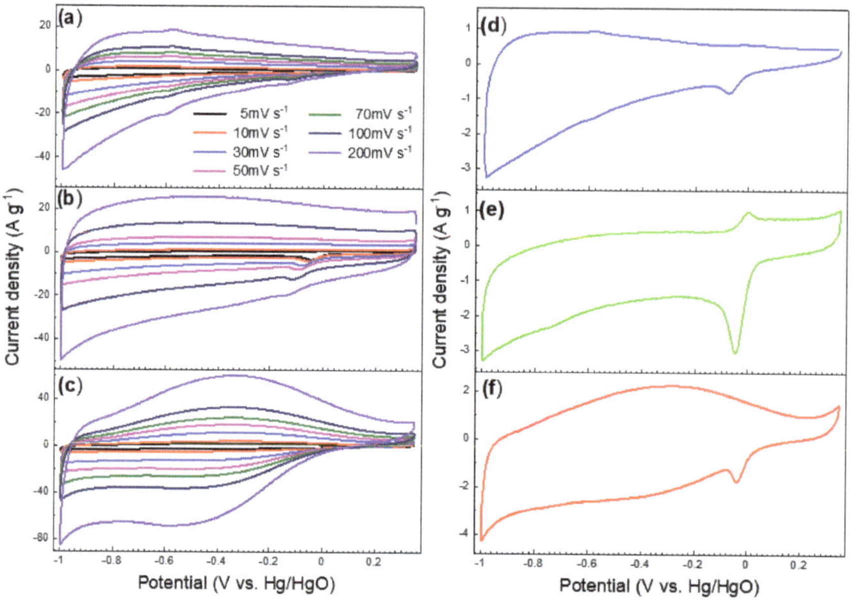

Figure 8. Cyclic voltammetry (CV) curves of carbon/DDSQ hybrids derived from (**a**,**d**) pure phenolic, (**b**,**e**) PDDSQ-50, and (**c**,**f**) pure PDDSQ, measured at (**a**–**c**) various scan rates and (**d**–**f**) a scan rate of 5 mV s^{-1} in 1 M KOH.

Figure 9 displays the GCD curves of the carbon/DDSQ hybrids derived from pure phenolic, PDDSQ-50, and pure PDDSQ, measured at current densities from 0.5 to 20 A g^{-1}. The GCD curves of the carbon material derived from pure phenolic were triangular in shape, suggesting EDLC characteristics, while those of the carbon materials derived from PDDSQ-50 and pure PDDSQ were triangular with a slight bend, implying the presence of both pseudocapacitance and EDLC [49]. The discharging time of the carbon material derived from pure PDDSQ (Figure 9c) was much longer than that of the carbon material derived from pure phenolic (Figure 9a), suggesting that the former capacitance (pure PDDSQ) was larger than that of the latter.

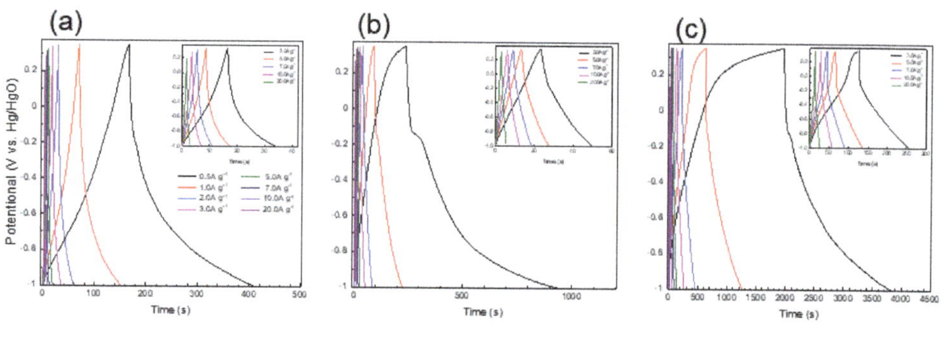

Figure 9. Specific capacitances of carbon/DDSQ hybrids derived from (**a**) pure phenolic, (**b**) PDDSQ-50, and (**c**) pure PDDSQ, measured at various current densities.

We used Equation S1 to calculate the specific capacitances from the GCD curves (Figure 10). At a current density of 0.5 A g^{-1}, the specific capacitance of the carbon material derived from pure PDDSQ (689.5 F g^{-1}) was much larger than those of the carbon materials derived from PDDQ-50 (258.8 F g^{-1}) and pure phenolic (89.9 F g^{-1}). The excellent performance of the carbon material derived from pure PDDSQ was due to its many electroactive centers (e.g., N and O atoms and electron-rich phenyl rings) making it easier for the electrolyte to access the electrode surface [50]. In addition, the cage structure of DDSQ provided a large number of fine and interlaced channels for the free passage of electrolyte ions. After cycling over 2000 times at 10 A g^{-1}, all of the carbon/DDSQ hybrids displayed high cycling stability, retaining over 90% of their original capacitances (Figure 10b). Although the carbon material derived from pure PDDSQ allowed the electrolyte ions to pass freely, a current density that is too large may cause many thermal effects, resulting in poorer durability. Interestingly, even though the specific capacitance of the carbon material derived from pure phenolic was relatively low, its stability was better than that of the carbon material derived from pure PDDSQ. After cycling the carbon material derived from PDDSQ-50 over 2000 times at 10 A g^{-1}, it displayed close to 99.9% retention; thus, by adjusting the ratios of phenol and DDSQ-4OH during the copolymerization reaction we could balance the electrode materials to obtain both a large specific capacitance and high stability.

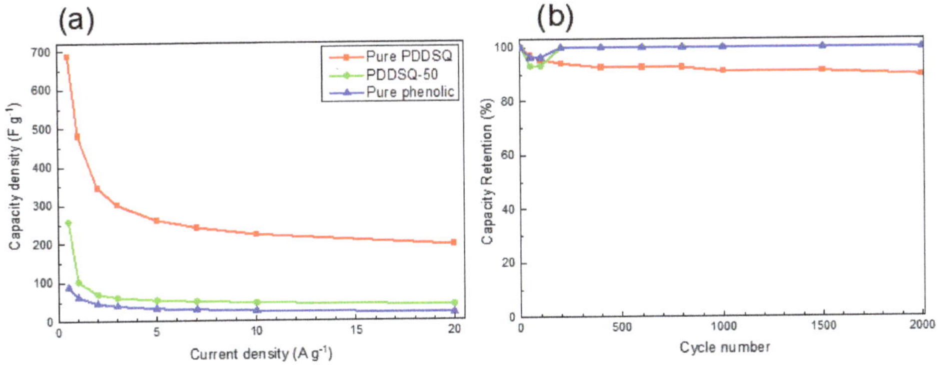

Figure 10. (**a**) specific capacitances measured at current densities from 0.5 to 10 A g^{-1} and (**b**) cycling performance measured at a current density of 10 A g^{-1} after 2000 cycles for the carbon/DDSQ hybrids derived from pure phenolic, PDDSQ-50, and pure PDDSQ.

Figure 11 displays the specific energy plotted with respect to the specific power. The carbon material derived from pure PDDSQ exhibited a maximum energy density of 174.5 W h kg^{-1} at a power density of 337.5 W kg^{-1}; its energy density was maintained at 49.5 W h kg^{-1} even at a high power density of 13.5 kW kg^{-1}. Compared with other materials prepared in previous studies, our present carbon/DDSQ hybrids displayed excellent properties and were much simpler to prepare, suggesting that they have potential for use as future electrode materials [50–61].

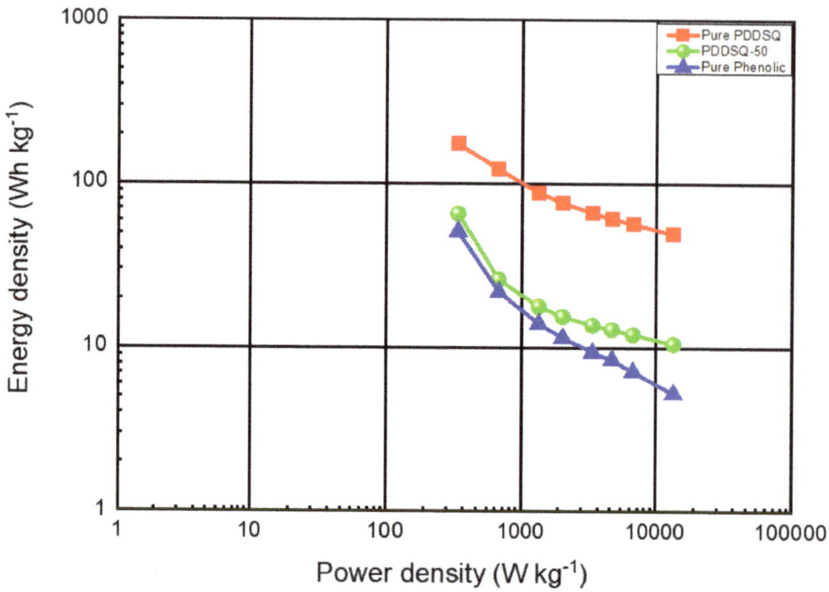

Figure 11. Ragone plots of the energy density and power density of the carbon/DDSQ hybrids derived from pure phenolic, PDDSQ-50, and pure PDDSQ.

4. Conclusions

We have synthesized PDDSQ hybrids incorporating various amounts of DDSQ and confirmed their chemical structures using FTIR and NMR spectroscopy. All of these PDDSQ hybrids had char yields and thermal stabilities higher than those of the pure phenolic resin, and they increased upon increasing the DDSQ content in the hybrid systems. The DDSQ units provided an inorganic protection layer on the phenolic surface, as confirmed through XPS analysis. We obtained carbon/DDSQ hybrids after thermal curing and calcination of our PDDSQ hybrids; these carbon materials displayed electrochemical properties superior to those of previously reported counterparts.

Supplementary Materials: The following are available online at http://www.mdpi.com/2073-4360/12/9/2151/s1, Table S1. Comparison between the energy density and power density data /specific capacitance of phenolic/DDSQ hybrids with different materials for supercapacitor application; Figure S1: FTIR spectra of (a) DDSQ-4OH and (b) pure PDDSQ; Figure S2: GPC analysis of pure PDDSQ; Figure S3: ^{1}H NMR spectra of (a) pure phenolic, (b) PDDSQ-50 and (c) pure PDDSQ; Figure S4: TGA analyses of pure phenolic resin and its corresponding derivative curve.

Author Contributions: Data curation, W.-C.C. and Y.-T.L.; Supervision, S.-W.K.; Writing–original draft, S.-W.K.; Writing–review & editing, S.-W.K. All authors have read and agreed to the published version of the manuscript.

Funding: This research was funded by the Ministry of Science and Technology, Taiwan, under contracts MOST 109-2221-E-110-067-MY3.

Conflicts of Interest: The authors declare no conflict of interest.

References

1. Chee, S.S.; Jawaid, M. The Effect of Bi-Functionalized MMT on Morphology, Thermal Stability, Dynamic Mechanical, and Tensile Properties of Epoxy/Organoclay Nanocomposites. *Polymers* **2019**, *11*, 2012. [CrossRef]
2. Ayat, M.; Rahmouni, A.; Belbachir, M.; Bensaada, N.; Baghdadli, M.C.; Meghabar, R. Thermoplastic block copolymer: Alpha-MethylStyrene and vinyl acetate catalyzed by clay layered called Maghnite-Na^{+} (Algerian MMT). *J. Polym. Res.* **2019**, *26*, 230. [CrossRef]

3. He, J.; Xiao, P.; Liu, W.; Shi, J.; Zhang, L.; Liang, Y.; Pan, C.; Kuo, S.W.; Chen, T. A Universal high accuracy wearable pulse monitoring system via high sensitivity and large linearity graphene pressure sensor. *Nano Energy* **2019**, *59*, 422–433. [CrossRef]
4. Diez-Pascual, A.M.; Sanchez, J.A.L.; Capilla, R.P.; Diaz, P.G. Recent Developments in Graphene/Polymer Nanocomposites for Application in Polymer Solar Cells. *Polymers* **2019**, *10*, 217. [CrossRef] [PubMed]
5. He, J.; Xiao, P.; Shi, J.; Liang, Y.; Lu, W.; Chen, Y.; Wang, W.; Theato, P.; Kuo, S.W.; Chen, T. High Performance Humidity Fluctuation Sensor for Wearable Devices via a Bioinspired Atomic-Precise Tunable Graphene-Polymer Heterogeneous Sensing Junction. *Chem. Mater.* **2018**, *30*, 4343–4354. [CrossRef]
6. Mohamed, M.G.; Kuo, S.W. Functional Silica and Carbon Nanocomposites Based on Polybenzoxazines. *Macromol. Chem. Phys.* **2019**, *220*, 1800306. [CrossRef]
7. Kanimozhi, C.; Shea, M.J.; Ko, J.; Wei, W.; Huang, P.S.; Arnold, M.S.; Gopalan, P. Removable Nonconjugated Polymers To Debundle and Disperse Carbon Nanotubes. *Macromolecules* **2019**, *52*, 4278–4286. [CrossRef]
8. Zhao, G.X.; Huang, X.B.; Tang, Z.W.; Huang, Q.F.; Niu, F.L.; Wang, X.K. Polymer-based nanocomposites for heavy metal ions removal from aqueous solution: A review. *Polym. Chem.* **2018**, *9*, 3562–3582. [CrossRef]
9. Mohamed, M.G.; Kuo, S.W. Polybenzoxazine/Polyhedral Oligomeric Silsesquioxane (POSS) Nanocomposites. *Polymers* **2016**, *8*, 225. [CrossRef]
10. Mohamed, M.G.; Kuo, S.W. Functional Polyimide/Polyhedral Oligomeric Silsesquioxane Nanocomposites. *Polymers* **2019**, *11*, 26. [CrossRef]
11. Zhang, W.C.; Camino, G.; Yang, R. Polymer/polyhedral oligomeric silsesquioxane (POSS) nanocomposites: An overview of fire retardance. *Prog. Polym. Sci.* **2017**, *67*, 77–125. [CrossRef]
12. Chen, W.C.; Lin, R.C.; Tseng, S.M.; Kuo, S.W. Minimizing the Strong Screening Effect of Polyhedral Oligomeric Silsesquioxane Nanoparticles in Hydrogen-Bonded Random Copolymers. *Polymers* **2018**, *10*, 303. [CrossRef]
13. Ullah, A.; Shah, S.M.; Hussain, H. Amphiphilic tadpole-shaped POSS-poly(glycerol methacrylate) hybrid polymers: Synthesis and self-assembly. *J. Polym. Res.* **2019**, *26*, 4. [CrossRef]
14. Byun, H.Y.; Choi, M.H.; Chung, J. Synthesis and characterization of resol type phenolic resin/layered silicate nanocomposites. *Chem. Mater.* **2001**, *13*, 4221–4226. [CrossRef]
15. Ma, C.C.C.; Sung, S.C.; Wang, F.Y.; Chiang, L.Y.; Wang, L.Y.; Chiang, C.L. Thermal, mechanical, and morphological properties of novolac-type phenolic resin blended with fullerenol polyurethane and linear polyurethane. *J. Polym. Sci. Polym. Phys.* **2001**, *39*, 2436–2443. [CrossRef]
16. Wei, J.; Wang, G.; Chen, F.; Bai, M.; Liang, Y.; Wang, H.; Zhao, D.; Zhao, Y. Sol–Gel Synthesis of Metal–Phenolic Coordination Spheres and Their Derived Carbon Composites. *Angew Chem. Int. Ed.* **2018**, *130*, 9986–9991. [CrossRef]
17. Lee, Y.J.; Kuo, S.W.; Huang, W.J.; Lee, H.Y.; Chang, F.C. Miscibility, specific interactions, and self-assembly behavior of phenolic/polyhedral oligomeric silsesquioxane hybrids. *J. Polym. Sci. Polym. Phys.* **2004**, *42*, 1127–1136. [CrossRef]
18. Kuo, S.W.; Lin, H.C.; Huang, W.J.; Huang, C.F.; Chang, F.C. Hydrogen bonding interactions and miscibility between phenolic resin and octa (acetoxystyryl) polyhedral oligomeric silsesquioxane (AS-POSS) nanocomposites. *J. Polym. Sci. Polym. Phys.* **2006**, *44*, 673–686. [CrossRef]
19. Chiou, C.W.; Lin, Y.C.; Wang, L.; Hirano, C.; Suzuki, Y.; Hayakawa, T.; Kuo, S.W. Strong Screening Effect of Polyhedral Oligomeric Silsesquioxanes (POSS) Nanoparticles on Hydrogen Bonded Polymer Blends. *Polymers* **2014**, *6*, 926–948. [CrossRef]
20. Chiou, C.W.; Lin, Y.C.; Wang, L.; Maeda, R.; Hayakawa, T.; Kuo, S.W. Hydrogen Bond Interactions Mediate Hierarchical Self-Assembly of POSS-Containing Block Copolymers Blended with Phenolic Resin. *Macromolecules* **2014**, *47*, 8709–8721. [CrossRef]
21. Yu, C.Y.; Kuo, S.W. Phenolic Functionality of Polyhedral Oligomeric Silsesquioxane Nanoparticles Affects Self-Assembly Supramolecular Structures of Block Copolymer Hybrid Complexes. *Ind. Eng. Chem. Res.* **2018**, *57*, 2546–2559. [CrossRef]
22. Zhang, Y.; Lee, S.H.; Mitra, Y.; Kaiwen, L.; Pittman, U., Jr. Phenolic resin–trisilanolphenyl polyhedral oligomeric silsesquioxane POSS hybrid nanocomposites: Structure and properties. *Polymer* **2006**, *47*, 2984–2996. [CrossRef]
23. Liu, Y.H.; Zeng, K.; Zheng, S. Organic–inorganic hybrid nanocomposites involving novolac resin and polyhedral oligomeric Silsesquioxane. *React. Funct. Polym.* **2007**, *67*, 627–635. [CrossRef]
24. Lei, Z.; Ji, J.; Wu, Q.; Zhang, J.; Wang, Y.; Jing, X.; Liu, Y. Curing behavior and microstructure of epoxy-POSS modified novolac phenolic resin with different substitution degree. *Polymer* **2019**, *178*, 121587. [CrossRef]

25. Lin, H.C.; Kuo, S.W.; Huang, C.F.; Chang, F.C. Thermal and surface properties of phenolic nanocomposites containing octaphenol polyhedral oligomeric silsesquioxane. *Macromol. Rapid Commun.* **2006**, *27*, 537–541. [CrossRef]
26. Wei, K.; Wang, L.; Zheng, S. Organic–inorganic polyurethanes with 3,13-dihydroxypropyloctaphenyl double-decker silsesquioxane chain extender. *Polym. Chem.* **2013**, *4*, 1491–1501. [CrossRef]
27. Zhao, B.; Wei, K.; Wang, L.; Zheng, S. Poly(hydroxyl urethane)s with Double Decker Silsesquioxanes in the Main Chains: Synthesis, Shape Recovery, and Reprocessing Properties. *Macromolecules* **2020**, *53*, 434–444. [CrossRef]
28. Liu, N.; Li, L.; Wang, L.; Zheng, S. Organic-inorganic polybenzoxazine copolymers with double decker silsesquioxanes in the main chains: Synthesis and thermally activated ring-opening polymerization behavior. *Polymer* **2017**, *109*, 254–265. [CrossRef]
29. Liao, Y.T.; Lin, Y.C.; Kuo, S.W. Highly Thermally Stable, Transparent, and Flexible Polybenzoxazine Nanocomposites by Combination of Double-Decker-Shaped Polyhedral Silsesquioxanes and Polydimethylsiloxane. *Macromolecules* **2017**, *50*, 5739–5747. [CrossRef]
30. Chen, W.C.; Kuo, S.W. Ortho-Imide and Allyl Groups Effect on Highly Thermally Stable Polybenzoxazine/ Double-Decker-Shaped Polyhedral Silsesquioxane Hybrids. *Macromolecules* **2018**, *51*, 9602–9612. [CrossRef]
31. Zhao, B.; Mei, H.; Liu, N.; Zheng, S. Organic–Inorganic Polycyclooctadienes with Double-Decker Silsesquioxanes in the Main Chains: Synthesis, Self-Healing, and Shape Memory Properties Regulated with Quadruple Hydrogen Bonds. *Macromolecules* **2020**, *53*, 7119–7131. [CrossRef]
32. Chen, W.C.; Tsao, Y.H.; Wang, C.F.; Huang, C.F.; Dai, L.; Chen, T.; Kuo, S.W. Main Chain–Type Block Copolymers through Atom Transfer Radical Polymerization from Double-Decker–Shaped Polyhedral Oligomeric Silsesquioxane Hybrids. *Polymers* **2020**, *12*, 465. [CrossRef] [PubMed]
33. Wu, S.; Hayakawa, T.; Kikuchi, R.; Grunzinger, S.J.; Kakimoto, M.; Oikawa, H. Synthesis and characterization of semiaromatic polyimides containing POSS in main chain derived from double-decker-shaped silsesquioxane. *Macromolecules* **2007**, *40*, 5698–5705. [CrossRef]
34. Wu, S.; Hayakawa, T.; Kakimoto, M.; Oikawa, H. Synthesis and Characterization of Organosoluble Aromatic Polyimides Containing POSS in Main Chain Derived from Double Decker Shaped Silsesquioxane. *Macromolecules* **2008**, *41*, 3481–3487. [CrossRef]
35. Chen, W.C.; Ahmed, M.M.M.; Wang, C.F.; Huang, C.F.; Kuo, S.W. Highly thermally stable mesoporous Poly (cyanate ester) featuring double-decker–shaped polyhedral silsesquioxane framework. *Polymer* **2019**, *185*, 121940. [CrossRef]
36. Wang, C.F.; Ejeta, D.D.; Wu, J.Y.; Kuo, S.W.; Lin, C.H.; Lai, J.Y. Tuning the Wettability and Surface Free Energy of Poly(vinylphenol) Thin Films by Modulating Hydrogen-Bonding Interactions. *Polymers* **2020**, *12*, 523. [CrossRef]
37. Wang, F.Y.; Ma, C.C.M.; Wu, W.J. Kinetic Parameters of Thermal Degradation of Polyethylene Glycol-Toughened Novolac-Type Phenolic Resin. *J. Appl. Polym. Sci.* **2001**, *80*, 188–196. [CrossRef]
38. Turri, S.; Levi, M. Structure, dynamic properties, and surface behavior of nanostructured ionomeric polyurethanes from reactive polyhedral oligomeric silsesquioxanes. *Macromolecules* **2005**, *38*, 5569–5574. [CrossRef]
39. Turri, S.; Levi, M. Wettability of polyhedral oligomeric silsesquioxane nanostructured polymer surfaces. *Macromol. Rapid Commun.* **2005**, *26*, 1233–1236. [CrossRef]
40. Mohamed, G.M.; Hung, W.S.; EL-Mahdy, A.F.M.; Ahmed, M.M.M.; Dai, L.; Chen, T.; Kuo, S.W. High-Molecular-Weight PLA-b-PEO-b-PLA Triblock Copolymer Templated Large Mesoporous Carbons for Supercapacitors and CO2 Capture. *Polymers* **2020**, *12*, 1193. [CrossRef]
41. EL-Mahdy, A.F.M.; Liu, T.E.; Kuo, S.W. Direct synthesis of nitrogen-doped mesoporous carbons from triazine-functionalized resol for CO2 uptake and highly efficient removal of dyes. *J. Hazard. Mater.* **2020**, *391*, 122163. [CrossRef] [PubMed]
42. Qu, B.; Li, C.; Zhu, C.; Wang, S.; Zhang, X.; Chen, Y. Growth of MoSe2 nanosheets with small size and expanded spaces of (002) plane on the surfaces of porous N-doped carbon nanotubes for hydrogen production. *Nanoscale* **2016**, *8*, 16886–16893. [CrossRef] [PubMed]
43. EL-Mahdy, A.F.M.; Kuo, C.H.; Alshehri, A.A.; Kim, J.; Young, C.; Yamauchi, Y.; Kuo, S.W. Strategic design of triphenylamine-and triphenyltriazine-based two-dimensional covalent organic frameworks for CO_2 uptake and energy storage. *J. Mater. Chem. A* **2018**, *6*, 19532–19541. [CrossRef]

44. EL-Mahdy, A.F.M.; Young, C.; Kim, J.; You, J.; Yamauchi, Y.; Kuo, S.W. Hollow Microspherical and Microtubular [3+3] Carbazole-Based Covalent Organic Frameworks and Their Gas and Energy Storage Applications. *ACS Appl. Mater. Interfaces* **2019**, *11*, 9343–9354. [CrossRef]
45. Conway, B. *Electrochemical Supercapacitors: Scientific Principles and Technological Application*; Kluwer Academic/Plenum: New York, NY, USA, 1999.
46. Simon, P.; Gogotsi, Y. Materials for electrochemical capacitors. *Nat. Mater.* **2008**, *7*, 845–854. [CrossRef]
47. Su, C.; He, H.; Xu, L.; Zhao, K.; Zheng, C.; Zhang, C. A mesoporous conjugated polymer based on a high free radical density polytriphenylamine derivative: Its preparation and electrochemical performance as a cathode material for Li-ion batteries. *J. Mater. Chem. A* **2017**, *5*, 2701–2709. [CrossRef]
48. Hu, F.; Wang, J.; Hu, S.; Li, L.; Shao, W.; Qiu, J.; Lei, Z.; Deng, W.; Jiang, X. Engineered Fabrication of Hierarchical Frameworks with Tuned Pore Structure and N,O-Co-Doping for High-Performance Supercapacitors. *ACS Appl. Mater. Interfaces* **2017**, *9*, 31940–31949. [CrossRef]
49. Khattak, A.M.; Ghazi, Z.A.; Liang, B.; Khan, N.A.; Iqbal, A.; Li, L.; Tang, Z. A redox-active 2D covalent organic framework with pyridine moieties capable of faradaic energy storage. *J. Mater. Chem. A* **2016**, *4*, 16312–16317. [CrossRef]
50. EL-Mahdy, A.F.M.; Hung, Y.H.; Mansoure, T.H.; Yu, H.H.; Chen, T.; Kuo, S.W. A Hollow Microtubular Triazine-and Benzobisoxazole-Based Covalent Organic Framework Presenting Sponge-Like Shells That Functions as a High-Performance Supercapacitor. *Chem. Asian J.* **2019**, *14*, 1429–1435. [CrossRef]
51. Hong, M.S.; Lee, S.H.; Kim, S.W. Use of KCl Aqueous Electrolyte for 2V Manganese Oxide/Activated Carbon Hybrid Capacitor. *Electrochem. Solid-State Lett.* **2002**, *5*, A227–A230. [CrossRef]
52. Choi, B.G.; Chang, S.-J.; Kang, H.-W.; Park, C.P.; Kim, H.J.; Hong, W.H.; Lee, S.; Huh, Y.S. High Performance of a Solid-State Flexible Asymmetric Supercapacitor based on Graphene Films. *Nanoscale* **2012**, *4*, 4983–4988. [CrossRef] [PubMed]
53. Su, F.; Miao, M. Asymmetric Carbon Nanotube–MnO_2 Two-Ply Yarn Supercapacitors for Wearable Electronics. *Nanotechnology* **2014**, *25*, 135401. [CrossRef]
54. Tang, J.; Salunkhe, R.R.; Liu, J.; Torad, N.L.; Imura, M.; Furukawa, S.; Yamauchi, Y. Thermal Conversion of Core–Shell Metal–Organic Frameworks: A New Method for Selectively. *J. Am. Chem. Soc.* **2015**, *137*, 1572–1580. [CrossRef] [PubMed]
55. Chaudhary, M.; Nayak, A.K.; Muhammad, R.; Pradhan, D.; Mohanty, P. Nitrogen-enriched nanoporous polytriazine for high-performance supercapacitor application. *ACS Sustainable Chem. Eng.* **2018**, *6*, 5895–5902. [CrossRef]
56. Ai, W.; Zhou, W.; Du, Z.; Du, Y.; Zhang, H.; Jia, X.; Xie, L.; Yi, M.; Yu, T.; Huang, W. Benzoxazole and benzimidazole heterocycle-grafted graphene for high-performance supercapacitor electrodes. *J. Mater. Chem.* **2012**, *22*, 23439–23446. [CrossRef]
57. Puthusseri, D.; Aravindan, V.; Mahdavi, S.; Ogale, S. 3D micro-porous conducting carbon beehive by single step polymer carbonization for high performance supercapacitors: The magic of in situ porogen formation. *Energy. Environ. Sci.* **2014**, *7*, 728–735. [CrossRef]
58. Mohamed, G.M.; EL-Mahdy, A.F.M.; Takashi, Y.; Kuo, S.W. Ultrastable conductive microporous covalent triazine frameworks based on pyrene moieties provide high-performance CO_2 uptake and supercapacitance. *New J. Chem.* **2020**, *44*, 8241–8253. [CrossRef]
59. Samy, M.M.; Mohamed, M.G.; Kuo, S.W. Pyrene-functionalized tetraphenylethylene polybenzoxazine for dispersing single-walled carbon nanotubes and energy storage. *Compos. Sci. Tech.* **2020**, *199*, 108360. [CrossRef]
60. EL-Mahdy, A.F.M.; Hung, Y.H.; Mansoure, T.H.; Yu, H.H.; Hsu, Y.S.; Wu, K.C.W.; Kuo, S.W. Synthesis of [3+3] bketoenamine-tethered covalent organic frameworks (COFs) for high-performance supercapacitance and CO_2 storage. *J. Taiwan Inst. Chem. Eng.* **2019**, *103*, 199–208. [CrossRef]
61. Laing, Z.; Liu, H.; Zeng, J.; Zhou, J.; Li, H.; Xia, H. Facile Synthesis of Nitrogen-Doped Microporous Carbon Spheres for High Performance Symmetric Supercapacitors. *Nanoscale Res. Lett.* **2018**, *13*, 314. [CrossRef]

polymers

MDPI

Article

Polyimide-Derived Carbon-Coated $Li_4Ti_5O_{12}$ as High-Rate Anode Materials for Lithium Ion Batteries

Shih-Chieh Hsu [1], Tzu-Ten Huang [2], Yen-Ju Wu [3], Cheng-Zhang Lu [4], Huei Chu Weng [5,*], Jen-Hsien Huang [6], Cai-Wan Chang-Jian [7,*] and Ting-Yu Liu [8,*]

1 Department of Chemical and Materials Engineering, Tamkang University, No. 151, Yingzhuan Road, Tamsui District, New Taipei City 25137, Taiwan; roysos@mail.tku.edu.tw
2 National Synchrotron Radiation Research Center, 101 Hsin-Ann Road, Hsinchu Science Park, Hsinchu 30076, Taiwan; huang.ty@nsrrc.org.tw
3 International Center for Young Scientists (ICYS), National Institute for Materials Science (NIMS), 1-2-1 Sengen, Tsukuba, Ibaraki 305-0047, Japan; WU.YenJu@nims.go.jp
4 Material and Chemical Research Laboratories, Industrial Technology Research Institute, No. 195, Chung Hsing Road, Chutung, Hsinchu 31040, Taiwan; chengzhanglu@itri.org.tw
5 Department of Mechanical Engineering, Chung Yuan Christian University, No. 200, Chungpei Road, Chungli District, Taoyuan City 32023, Taiwan
6 Department of Green Material Technology, Green Technology Research Institute, CPC Corporation, No. 2, Zuonan Rd., Nanzi District, Kaohsiung City 81126, Taiwan; 295604@cpc.com.tw
7 Department of Mechanical and Automation Engineering, I-Shou University, No.1, Sec. 1, Syuecheng Rd., Dashu District, Kaohsiung City 84001, Taiwan
8 Department of Materials Engineering, Ming Chi University of Technology, 84 Gungjuan Road, Taishan District, New Taipei City 24301, Taiwan
* Correspondence: hcweng@cycu.edu.tw (H.C.W.); cwchangjian@isu.edu.tw (C.-W.C.-J.); tyliu0322@gmail.com (T.-Y.L.); Tel.: +886-3-2654311 (H.C.W.); +886-7-6577711 (ext. 3231) (C.-W.C.-J.); +886-2-29089899 (ext. 4456) (T.-Y.L.)

Citation: Hsu, S.-C.; Huang, T.-T.; Wu, Y.-J.; Lu, C.-Z.; Weng, H.C.; Huang, J.-H.; Chang-Jian, C.-W.; Liu, T.-Y. Polyimide-Derived Carbon-Coated $Li_4Ti_5O_{12}$ as High-Rate Anode Materials for Lithium Ion Batteries. *Polymers* **2021**, *13*, 1672. https://doi.org/10.3390/polym13111672

Academic Editor: Vito Di Noto

Received: 25 April 2021
Accepted: 19 May 2021
Published: 21 May 2021

Publisher's Note: MDPI stays neutral with regard to jurisdictional claims in published maps and institutional affiliations.

Abstract: Carbon-coated $Li_4Ti_5O_{12}$ (LTO) has been prepared using polyimide (PI) as a carbon source via the thermal imidization of polyamic acid (PAA) followed by a carbonization process. In this study, the PI with different structures based on pyromellitic dianhydride (PMDA), 4,4′-oxydianiline (ODA), and *p*-phenylenediamine (*p*-PDA) moieties have been synthesized. The effect of the PI structure on the electrochemical performance of the carbon-coated LTO has been investigated. The results indicate that the molecular arrangement of PI can be improved when the rigid *p*-PDA units are introduced into the PI backbone. The carbons derived from the *p*-PDA-based PI show a more regular graphite structure with fewer defects and higher conductivity. As a result, the carbon-coated LTO exhibits a better rate performance with a discharge capacity of 137.5 mAh/g at 20 C, which is almost 1.5 times larger than that of bare LTO (94.4 mAh/g).

Keywords: $Li_4Ti_5O_{12}$; polyimide; carbon coating; lithium ion battery; rate performance

1. Introduction

With the continuous depletion of fossil fuels and associated increasing air pollution, it is essential to raise the proportion of renewable energy supplies. However, renewable energy sources such as wind and solar are intermittent and cannot be stockpiled without energy storage systems (ESSs). ESSs based on lithium ion batteries (LIBs) are emerging as one of the key solutions to effectively integrate high shares of variable renewable energy due to their high energy density, zero memory effect and long lifespan [1]. By introducing an EES into the power generation system, it can smooth the output of wind or solar power generation, and reduce the impact on the power grid. Unfortunately, most of the anode and cathode materials suffer from low electronic conductivity and poor ionic diffusivity in the lattice resulting in poor rate capability. Therefore, the rate performance of the active materials must be further improved for use in voltage regulation and frequency modulation.

To address the above issue, alien ion doping [2], the building of a nanoporous structure [3,4] and surface coating [5–7] have proven to be effective for improving the rate performance by narrowing the band gap, shortening migration paths for Li^+ ions and reduction of interfacial resistance, respectively. Among these approaches, the surface coating is a feasible strategy to improve the LIB performance due to the multi-functional advantages such as enhancement of electric conductivity [8], structural stability [9] and offering a physical protection to avoid side reactions with the electrolyte [10]. Several compounds have been proposed to be an efficient coating layer such as carbon-based materials [11,12], metal oxides/hydroxides [13,14], phosphide-based materials [15] and glass-based materials [16]. Among these candidates, carbon coating is one of the most effective and facile ways to improve the electrochemical performance of LIB materials due to its excellent electron conductivity, low cost, and superior chemical/electrochemical stability. Compared with the metal oxide coating, carbon coating can easily form a smooth and uniform thin layer with high surface coverage on the active materials. Moreover, the carbon coating layer can also serve as a buffer layer to accommodate the dimensional variation of the active material during the lithiation and delithiation process leading to improved structural stability [17].

Recently, polyimide (PI) has been found to be a high-quality carbon source for the synthesis of graphene [18,19] and highly conductive carbon [20], because of the abundant hexagonal crystalline carbon within the imide structures. The obtained PI derived graphene reveals remarkable conductivity and electrochemical properties, which are useful for various applications e.g., supercapacitors [21], sensors [22], electrocatalysts [23] and electrothermal heaters [24]. Therefore, polyimide is expected to be suitable for forming the conductive carbon layer of LIB materials. It has been reported that the carbon films derived from PI with different structures can show quite different physical properties and graphite microstructure. However, the effect of carbon coating derived from different PIs on the LIB performance is still unclear.

In this study, two different PIs have been prepared from pyromellitic dianhydride (PMDA), 4,4′-oxydianiline (ODA) and *p*-phenylenediamine (*p*-PDA). By introducing the rigid planar moiety (*p*-PDA) into the main chain of classical PMDA/ODA PI (PO-PI), the PMDA/ODA/*p*-PDA PI (POP-PI) shows an increased crystallinity and orientation degree. Here, we used the PO-PI and POP-PI as carbon sources to modify the $Li_4Ti_5O_{12}$ (LTO) anode material, followed by thermal treatment to obtain the carbon-coated LTO materials. The results indicate that the uniformly coated carbon layer on LTO can reduce its resistance and polarization leading to better rate performance and the corresponding electrochemical properties. Moreover, with incorporation of the rigid *p*-PDA segment, the carbon derived from the POP-PI exhibits better molecular packing and less defect. As a result, the carbon-coated LTO prepared from the POP-PI can display an even better kinetic performance than that obtained from the PO-PI.

2. Experimental Section

2.1. Material

The monomers ODA (97%), *p*-PDA (98%) and PMDA (97%) were purchased from Jinyu Co., Ltd. (Kaohsiung, Taiwan). The TiO_2 (85%, anatase, Hombikat 8602) and Li_2CO_3 (≥99%, Aldrich) used for the synthesis of LTO were purchased from World Chem Industries Co., Ltd. (Taipei, Taiwan) and Sigma Aldrich (St. Louis, MI, USA), respectively.

2.2. Preparation of Pyromellitic Dianhydride/4,4′-Oxydianiline Polyimide (PMDA/ODA PI) and PMDA/ODA/p-Phenylenediamine (p-PDA) PI

In this study, the PI precursor, poly(amic acid) (PAA) was obtained through the two-steps synthesis method from its monomers. First, the ODA monomer or mixed ODA/*p*-PDA (1:1) were dissolved in *N*-methyl-2-pyrrolidone (NMP). Then, equimolar PMDA was added in the diamine solution under continuous stirring at 25 °C with a solid content of 15 wt% to produce the PAA solution. Here, the PAAs prepared from PMDA/ODA and PMDA/ODA/*p*-PDA are denoted as PO-PAA and POP-PAA, respectively. The PAA was then cast on glass substrate by doctor blade coating, s followed by soft baking at 80 °C for 60 min. The soft-baked precursor films were thermally imidized (150 °C for 30 min, 250 °C for 30 min, 350 °C for 60 min and 400 °C for 30 min) under N_2 atmosphere to obtain the PMDA/ODA PI and PMDA/ODA/*p*-PDA PI denoted herein as PO-PI and POP-PI, respectively.

2.3. Preparation of Carbon-Coated $Li_4Ti_5O_{12}$ (LTO)

The carbon-coated LTO powders were prepared by spray drying precursor solution of lithium titanium peroxide, followed by solid-state calcination. Firstly, the TiO_2 and Li_2CO_3 were dispersed in de-ionized water with a Li:Ti molar ratio 4:5 and a solid content of 15 wt%. The solution was ball milled for 12 h to produce the homogenous slurry, which was fed to a pilot spray dryer (OHKAWARA KAKOHKI, model L-8i, No. 145874). The spherical LTO precursor obtained was added to the PMDA/ODA or PMDA/ODA/*p*-PDA PAA solution, which was then stirred for 60 min and subsequently centrifuged. The centrifuged powders were dried in vacuum at 80 °C. Finally, the samples were thermally annealed via a stepwise process (150 °C for 30 min, 250 °C for 30 min, 350 °C for 60 min, 400 °C for 30 min, 500 °C for 60 min and 800 °C for 120 min). The modified LTO prepared from PO- and POP-PI are denoted herein as PO-LTO and POP-LTO, respectively. For the preparation of the bare LTO sample, the spray-dried precursor was directly annealed with the same heating program.

2.4. Characterization

The crystal structure of the sample was characterized by X-ray powder diffraction (XRD, Philips X'Pert/MPD instrument, El Dorado County, CA, USA). The morphologies were monitored by scanning electron microscopy (SEM, JEOL JSM-6701F, Tokyo, Japan) and transmission electron microscopy (TEM, JEOL 2010, Tokyo, Japan). The absorption measurement was taken using a Cintra 2020 (GBC scientific equipment, Australia) spectrophotometer. The differential scanning calorimetry (DSC) was measured on DSC 2500 (TA instrument, Lukens Drive, New Castle, DE, USA). Thermal gravimetric analysis (TGA) was carried out using a TGA 8000 (PerkinElmer, Boston, MS, USA). Functional group and chemical composition were characterized by using Fourier transform infrared (FTIR, PerkinElmer, Boston, MS, USA) spectroscopy and X-ray photoelectron spectroscopy (XPS, ULVAC-PHI, Tokyo, Japan). Raman spectrum of the as-prepared samples was measured using a Raman microscope (HR800, HORIBA, Tokyo, Japan).

2.5. Electrochemical Analysis

The energy storage performance of the samples was evaluated by fabricating the coin cells in an argon filled glovebox. The working electrode was prepared by mixing active material, polyvinylidene fluoride (PVdF), KS4 and Super P with a ratio of 8:0.5:0.5:1 in *N*-methyl-2-pyrrolidone (NMP). The as-prepared slurry was cast onto a Al foil and then dried in a vacuum oven at 120 °C overnight. The electrolyte comprises 1.0 M $LiPF_6$ dissolved in a mixture of ethylene carbonate and dimethyl carbonate at a volumetric ratio of 1:1. Cyclic voltammetry (CV) and electrochemical impedance spectroscopy (EIS) were carried out on a potentiostat (PARSTAT 4000 Potentiostat Galvanostat). The galvanostatic charge/discharge performance was examined between 1 and 2.5 V (vs. Li^+/Li) at various C rates (1 C = 175 mA/g).

3. Results and Discussion

In general, the molecular packing of PI can be improved by using rigid *p*-PDA units to replace the rotatable ODA moiety. It has been reported that the regular PI with higher orientation degree of molecular chains can produce better graphite structure after a carbonization process [25]. Therefore, in this study, we prepared the carbon-coated LTO powders using the less-ordered PO-PI and more ordered POP-PI as carbon sources. PO- and POP-PI were synthesized through a conventional two-step polycondensation of PMDA dianhydride with ODA and *p*-PDA diamines and the reaction process is shown in Figure 1a. The ultraviolet–visible (UV-vis) absorption spectra of the PO- and POP-PAA solutions are shown in Figure 1b. The inset in Figure 1b presents the digital photograph of the PO- and POP-PAA solutions. The PO-PAA solution exhibits strong absorption in the UV region with an absorption edge of 410 nm. With the incorporation of the PDA moiety into the polymer backbone, the absorption profile of POP-PAA tends to be shifted toward the long wavelength region. The red-shift of the absorption spectrum of POP-PAA can be explained by the formation of a charge transfer complex (CTC) between the alternating electron-acceptor (dianhydride) and electron-donor (diamine) moieties [26]. The C–O–C bond in ODA units can separate the chromaphoric centers and cut the electronic conjugations [27]. As a result, the PO-PAA shows a lighter color compared with that of POP-PAA due to its reduced intra-/intermolecular CTC formation. These results indicate that the PDA diamine units are successfully introduced into the copolymer. To study the effect of the *p*-PDA moiety on the imidization process, the enthalpy of imidization for the PO-PAA and POP-PAA was recorded by DSC measurement. As shown in Figure 1c, both the PAA samples exhibit a distinctive endothermic peak centered at around 175.5~179.8 °C in the first run which originated from the imidization reaction. In the second run, the DSC curves show a smooth profile without any peak, indicating the imidization process is complete. In addition, the enthalpy of imidization integrated from the DSC curve (first run) was calculated to be 228.6 and 235.6 J/g for PO-PAA and POP-PAA, respectively. The similar peak temperature and enthalpy of imidization for the two PAAs suggest that the incorporation of the *p*-PDA unit cannot alter the imidization behavior and the corresponding chemical conversion.

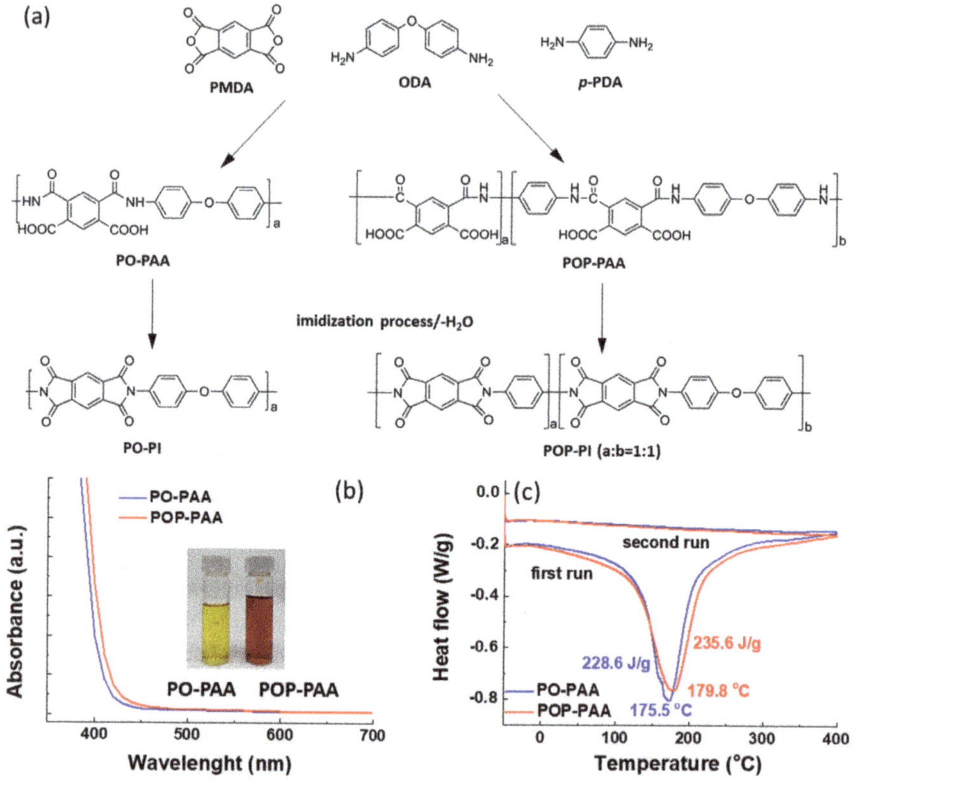

Figure 1. (**a**) The process of preparation of PMDA (pyromellitic dianhydride) and ODA (4,4′-oxydianiline) (PO-) and PMDA/ODA/*p*-phenylenediamine (*p*-PDA) polyimide (POP-PI) with PMDA, ODA, and *p*-PDA; (**b**) the absorbance spectrum of PO- and POP-PAA solution and (**c**) the differential scanning calorimetry (DSC) profile of the PO- and POP-PAA (poly(amic acid)).

The FTIR spectra of the as-prepared PAA and its corresponding PI are shown in Figure 2a. Both the two PAAs exhibit a broad band located at 1630 cm^{-1}, which can be assigned to amide C=O stretching mode [28,29]. After thermal treatment, this peak disappeared in the spectra of PO-PI and POP-PI (Figure 2b). Moreover, a new peak due to the C–N–C stretching vibration appears at 1365 cm^{-1} [30]. Meanwhile, the C=C double bond in the benzene ring at around 1500 cm^{-1} remains unchanged after the imidization process. These results indicate that the two PAAs were successfully converted into PO-PI and POP-PI after the imidization process. The peak at 1234 cm^{-1} is characteristic of C–O–C stretching vibration belonging to the ODA moiety [25]. It is worth noting that the POP-PAA and POP-PI reveal a much weaker intensity of C–O–C bond than their counterparts, confirming the successful incorporation of PDA moiety into the polymer backbone. The XPS O 1s and N 1s spectra of PO-PI are also shown in Figure 2c,d. After deconvolution, the O 1s XPS spectrum of the PO-PI can be fitted with two peak components assigned to C=O and C–O–C species at 531.9 and 533.2 eV, respectively [31]. For the N 1s spectrum of the PO-PI, only one characteristic peak contributed from O=C–N specie at 399.7 eV can be obtained after deconvolution [32]. These results indicate that the PAA has been successfully converted into the PI structure after the thermal treatment.

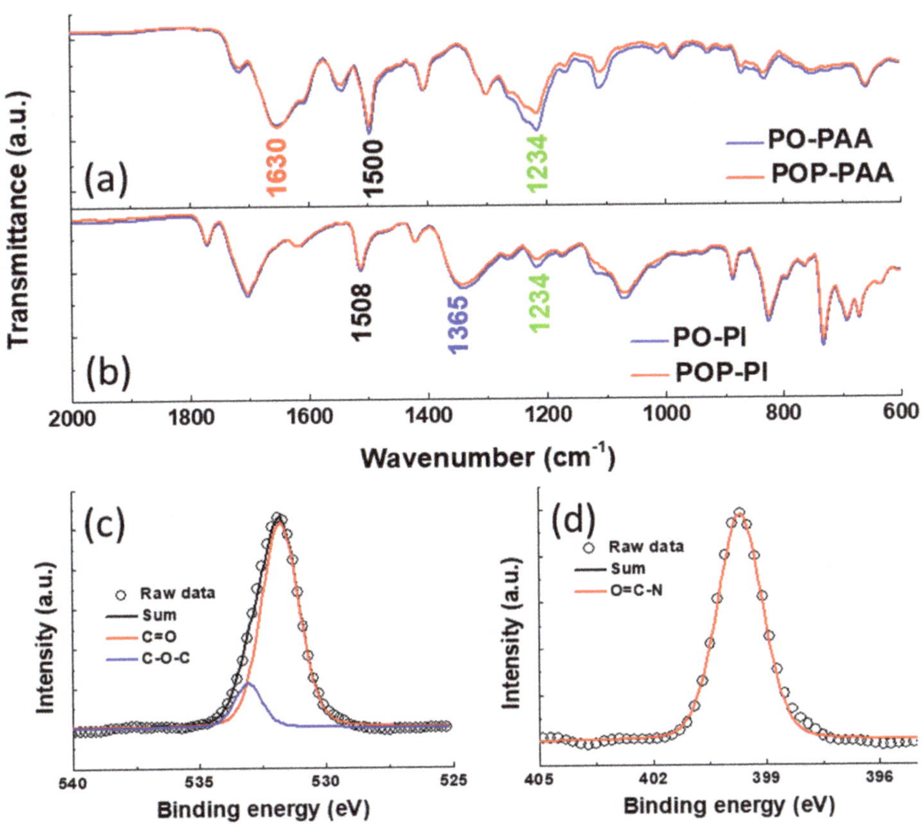

Figure 2. The Fourier transform infrared (FTIR) spectra of (**a**) PO-PAA and POP-PAA and (**b**) PO-PI and POP-PI; the peak deconvolution of the (**c**) O 1s and (**d**) N 1s X-ray photoelectron spectroscopy (XPS) spectra of the as-prepared PO-PI.

The microscopic molecular packing state of the PO- and POP-PI films was investigated by XRD measurement. As shown in Figure 3a, the XRD pattern of PO-PI is blunt without obvious diffraction signals, suggesting its amorphous nature. The amorphous state of

PO-PI originated from the flexible ether linkage within the ODA unit which loosens the chain packing of the PO-PI. In contrast, incorporation of the *p*-PDA unit in the polymer backbone can significantly enhance the polymer chain stacking. The regularly ordered structure of POP-PI is attributed to the rigid and planar skeletal structure of *p*-PDA leading to the better crystallinity [33]. Figure 3b compares the TGA curves of the PIs to investigate the thermal stability. Both the PI samples exhibit a major thermal decomposition ranging from 530 to 720 °C. The 5 wt% thermal decomposition temperature (T_d) of the POP-PI is found to be 584 °C, which is higher than that of PO-PI (576 °C), suggesting its better thermal stability of POP-PI. The improved thermal stability of POP-PI can be interpreted by the presence of the *p*-PDA group, which increases the intra- and interpolymer chain interactions, resulting in tight polymer chain packaging.

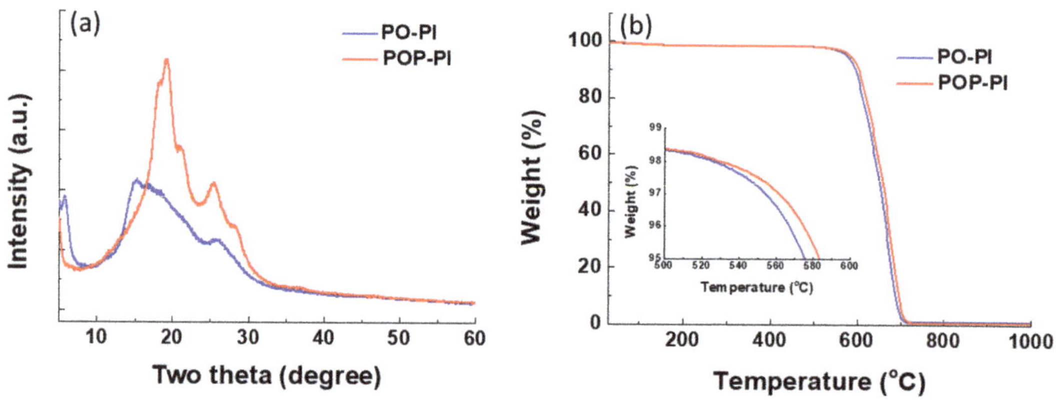

Figure 3. (**a**) The XRD patterns of PO-PI and POP-PI; (**b**) the TGA profile of the PO-PI and POP-PI.

Figure 4a displays the XRD pattern of the carbon materials derived from PO- and POP-PI. The resultant carbon materials reveal two peaks located at 23.5 and 43.8°, indicating the development of (002) and (100) planes, respectively [34]. These results suggest that both the two PI samples can be converted into the hexagonal structures of the carbonized materials after the thermal treatment. Moreover, the POP-PI derived carbon exhibits much stronger XRD peaks than that derived from PO-PI, indicating its higher degree of graphitization. Figure 4b shows the Raman spectra of the two carbonized materials to further provide atomic-scale structural information. The Raman spectrum of the two samples shows the characteristic D band (1342 cm^{-1}) and G band (1582 cm^{-1}), which correspond to the sp^3 and sp^2 carbon, respectively [35,36]. In general, the intensity ratio between the D band and G band (I_D/I_G) can be used to determine the graphitization degree of the carbonized materials. The I_D/I_G values of the PO- and POP-PI derived carbons are calculated to be 1.04 and 0.75, respectively. The lower I_D/I_G ratio of POP-PI derived carbon also suggests its more ordered structure and the increased sp^2 content. The C 1s XPS spectra of the PO- and POP-PI derived carbons are shown in Figure 4c,d to further characterize their chemical composition. Both the samples show four peaks with binding energies of 284.5, 285.4, 286.5 and 288.0 eV, which correspond to the functional groups of C=C, C–C, C–O and C=O, respectively. Based on the XPS analysis, the proportions of C=C for PO- and POP-PI derived carbons are 63.3% and 66.4%, respectively. Conversely, the proportion of oxygen-containing functional groups (C–O and C=O) of POP-PI based carbon (6.98%) is lower than that of the PO-PI based one (10.77%). In addition, the electronic conductivity of the PI derived carbons was directly measured with a four-point probe. The values of PO-PI and POP-PI derived carbons are 89.5 and 100.1 S/cm, respectively. These results further confirm that it is crucial to ensure the increased level of the structural arrangement of the initial PI precursor to prepare high-quality carbon materials with graphite-like structure [37].

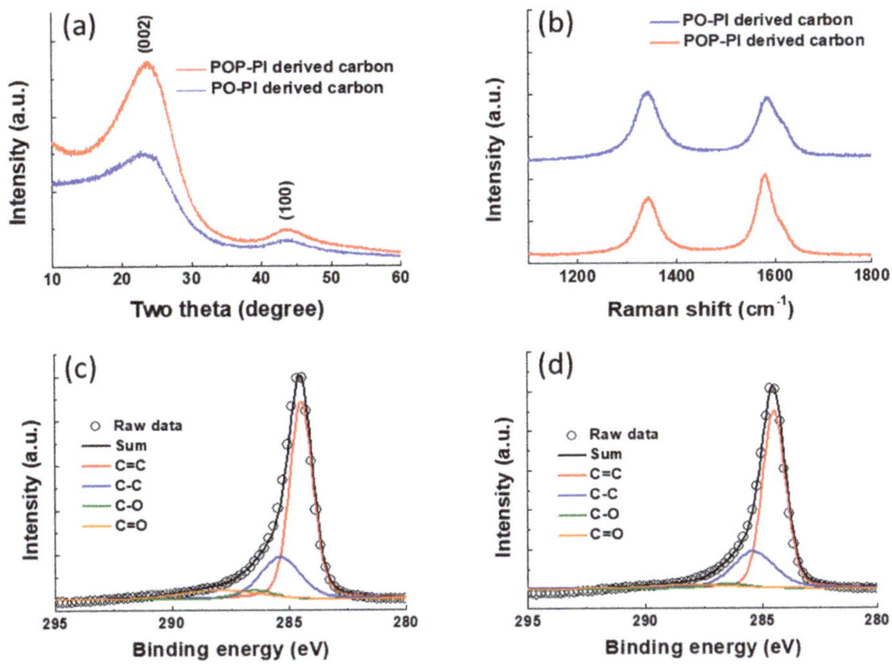

Figure 4. (**a**) The X-ray diffraction (XRD) pattern of the carbon materials derived from PO- and POP-PI; (**b**) Raman spectra of the carbon materials derived from PO- and POP-PI; peak deconvolution of the C 1s XPS spectra of (**c**) PO-PI derived carbon and (**d**) POP-PI derived carbon.

The XRD pattern of the spray-dried LTO powders is exhibited in Figure 5a. All the peaks between 5 and 70° can be assigned to the spinel LTO structure without any impurity phases. The particle size distribution of the as-prepared LTO is shown in Figure 5b. The particle size of the LTO powders ranges between 1.88 and 27.4 μm with a mean particle size of 10.7 μm (D_{50}). The micro-sized particles can be favorable for the powder packing during the electrode fabrication leading to better energy density. The surface area of the LTO was also evaluated by the Brunauer–Emmett–Teller (BET) analysis as shown in Figure 5c. Based on the nitrogen adsorption/desorption isotherms, the surface area of the LTO was calculated to be 11.3 m^2/g. Figure 5d displays the Ti 2p core level XPS spectrum of the as-prepared LTO samples. There are two pairs of Ti 2p peaks observed at 464.3, 458.5 and 462.2, 456.4 eV for Ti^{4+} $2p_{1/2}$, Ti^{4+} $2p_{3/2}$ and Ti^{3+} $2p_{1/2}$, Ti^{3+} $2p_{3/2}$, respectively. The partial reduction of the Ti ions from Ti^{4+} to Ti^{3+} originates from the generation of oxygen vacancies during the thermal annealing in N_2 ambience [38]. The microstructure of the LTO was investigated by SEM observation. The low-magnification SEM examination as shown in Figure 5e reveals that the morphology of LTO sample is perfectly preserved as highly uniform microspheres. In addition, the enlarged SEM image (Figure 5f) shows that the surface of the microspheres is composed of a primary nanoparticle with an average size of around 80 nm leading to a porous surface. The porous structure can facilitate the ionic transport during the charge/discharge process.

The morphologies of the pristine and carbon-coated LTO are monitored by SEM and TEM as shown in Figure 6. It can be observed that the whole sample reveals similar SEM morphology, indicating the spherical structure can be maintained after the carbon coating process. Moreover, the high-resolution TEM images of the PO-LTO and POP-LTO show a carbon layer with a thickness of around 2 nm was uniformly deposited on the LTO surface. In contrast, no carbon layer can be observed for bare LTO. It has been reported that the carbon layer can offer a conductive pathway for the electron transport. In addition, the randomly distributed defects and vacancies within the carbon layer also can improve the Li^+ ion migration [6]. As a result, the kinetic balance between the electronic and ionic transport can be established leading to better rate performance. The XPS survey of the POP-LTO and Raman spectrum of the three samples are also provided in the Supporting Information.

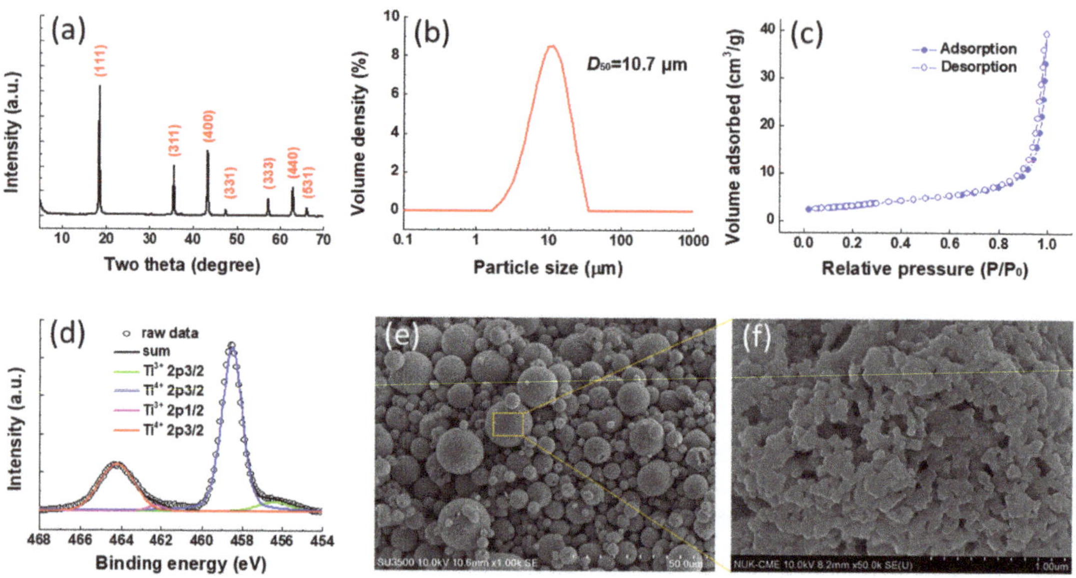

Figure 5. (**a**) The XRD pattern of the spray-dried $Li_4Ti_5O_{12}$ (LTO) powder; (**b**) the particle size distribution of the LTO; (**c**) the nitrogen adsorption/desorption isotherms of the LTO; (**d**) peak deconvolution of the Ti 2p XPS spectrum of LTO powder; (**e**,**f**) scanning electron microscopy (SEM) images of the LTO with different magnifications.

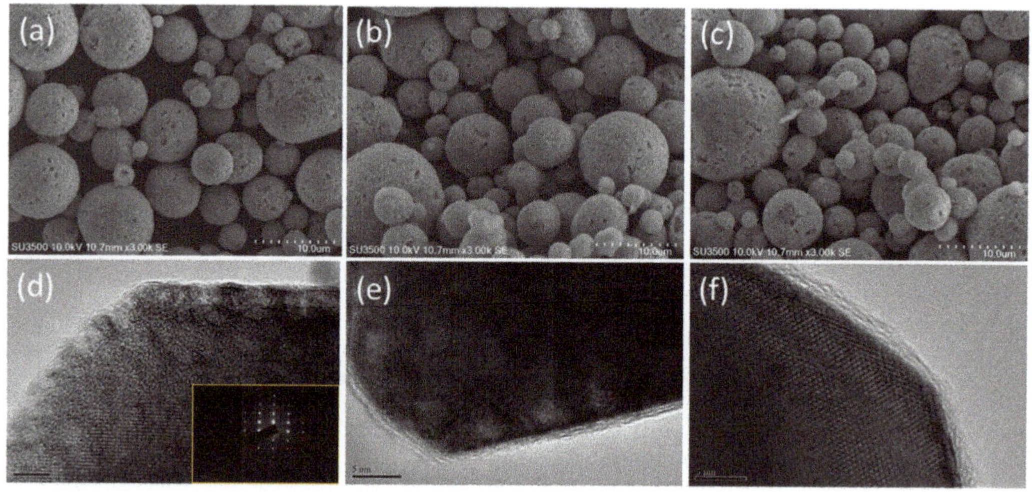

Figure 6. The SEM morphology of (**a**) bare LTO; (**b**) PO-LTO and (**c**) POP-LTO; the transmission electron microscopy (TEM) image of (**d**) bare LTO; (**e**) PO-LTO and (**f**) POP-LTO.

The electrochemical properties of the bare and carbon-coated LTO samples were studied by measuring their CV profiles with a scan rate of 1 mV/s between 1.0 and 2.5 V as shown in Figure 7a. All the samples reveal a pair of sharp redox couple which corresponds to the transition between $Li_4Ti_5O_{12}$ and $Li_7Ti_5O_{12}$. The separation between the anodic and cathodic peaks (ΔE) of the LTO would be dramatically reduced with the incorporation of carbon coating, indicating the reduction of polarization. The ΔE of LTO, PO-LTO and POP-LTO is found to be 0.7, 0.63 and 0.52 V, respectively. The lower polarization is mainly attributed to the enhancement of electric conductivity by the interfacial carbon modification. As shown in Figure 4, the POP based carbon layer shows a more regular graphite structure with a higher conductivity (100.1 S/cm) than that of PO based carbon layer (89.5 S/cm). As a result, the POP-LTO reveals a much lower polarization (0.52 V) due than that of PO-LTO (0.63 V). Furthermore, the EIS spectra of the three electrodes were taken to investigate the interfacial impedance in the electrodes. Figure 7b shows the Nyquist plots of LTO, PO-LTO and POP-LTO electrodes with a frequency range from 10^5 Hz to 10^{-2} Hz at an amplitude of 10 mV. All the plots exhibit a semicircle with a sloping line. The charge transfer resistance (*R*ct) determined from the size of the semicircle in the high frequency is around 117, 43.7 and 21.5 Ω for LTO, PO-LTO and POP-LTO, respectively. The *R*ct of POP-LTO is much lower than that of the bare one due to the deposition of the conductive carbon layer. The rate performance of the three electrodes evaluated at various C rates is displayed in Figure 7c. As expected, the POP-LTO with the modification of the POP-PI derived carbon layer can deliver the best rate capability due to its low polarization and *R*ct. The POP-LTO shows a high capacity retention of 83.2% (137.5 mAh/g) at a high rate of 20 C, which is higher than 70.7% (115.3 mAh/g) and 58.5% (94.4 mAh/g) for PO-LTO and pristine LTO, respectively. The corresponding charge/discharge profiles of the POP-LTO with different C rates are also provided in Figure 7d. The comparison of LTO performance with different carbon coating is also summarized in the Supporting Information.

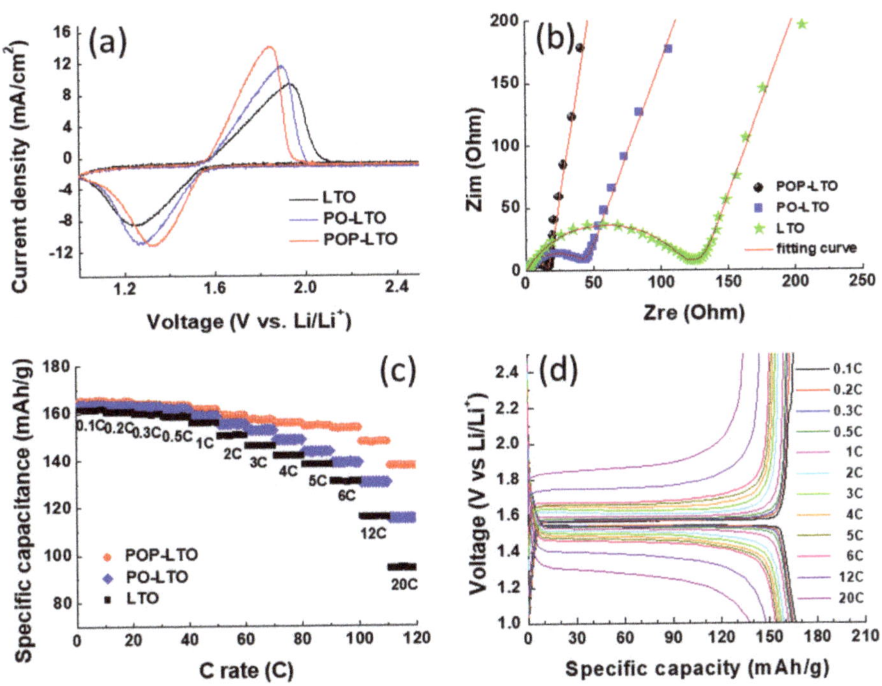

Figure 7. (**a**) The cyclic voltammetry (CV) curves of bare LTO, PO-LTO and POP-LTO stepped between 1.0 and 2.5 V with a scan rate of 1 mV/s; (**b**) Nyquist plots of bare LTO, PO-LTO and POP-LTO in the frequency range of 10^5 Hz to 10^{-2} Hz; (**c**) rate capability of the as-prepared samples at C rates between 0.1 C and 20 C; (**d**) the corresponding charge/discharge profile of the POP-LTO with various C rates.

4. Conclusions

In summary, a facile and scalable method has been developed for synthesizing carbon-coated LTO, using PI as the carbon source. After carbonization treatment at 800 °C in N_2 ambience, the PI coating can be transferred into the high-conductive carbon layer. The nano-carbon layer on LTO can improve both the electronic and ionic conductivities. As compared to pristine LTO, results show that the POP-LTO exhibits substantially improved cell performances particularly in the rate capability. A high initial discharge capacity of 165.1 mAh/g is delivered at 0.1 C and the specific capacity still maintain 137.5 mAh/g even at 20 C. It is believed that the developed method also can be extended to improve the electrochemical performances of other alternative LIB materials with low intrinsic electrical conductivity and poor Li^+ diffusion coefficient.

Supplementary Materials: The following are available online at https://www.mdpi.com/article/10.3390/polym13111672/s1, Figure S1: The XPS survey of the POP-LTO and the corresponding atomic percentages, Figure S2: Raman spectrum of the pristine LTO, PO-LTO and POP-LTO, Table S1: The comparison of LTO performance with different carbon coating.

Author Contributions: Writing—original draft preparation, S.-C.H.; writing—review and editing, T.-T.H. and T.-Y.L.; resources, Y.-J.W.; resources, C.-Z.L.; supervision, H.C.W.; conceptualization, J.-H.H.; investigation, C.-W.C.-J. and T.-Y.L.; funding acquisition, T.-Y.L. All authors have read and agreed to the published version of the manuscript.

Funding: This research received no external funding.

Acknowledgments: We are grateful to the Ministry of Science and Technology (MOST 104-2113-M-152-001-MY2, MOST 105-2320-B-038-014-) and CPC Corporation (105-3011). for financial support. We would also like to thank the research funding from Taipei Medical University (TMU102-AE1-B02, TMUTOP103004-2). The research endeavors at Ming Chi University of Technology were supported in part by the Ministry of Science and Technology (MOST 103-2221-E-131-002-MY2), by the Chang Gung Memorial Hospital, Linkou, Taiwan (CMRPG3E2091), and by the Academia Sinica Research Project on Thematic Project (AS-104-TP-A11).

Conflicts of Interest: The authors declare no conflict of interest.

References

1. Ding, Y.; Cano, Z.P.; Yu, A.; Lu, J.; Chen, Z. Automotive Li-ion batteries: Current status and future perspectives. *Electrochem. Energy Rev.* **2019**, *2*, 1–28. [CrossRef]
2. Zhang, Q.; Li, X. Recent developments in the doped-$Li_4Ti_5O_{12}$ anode materials of lithium-ion batteries for improving the rate capability. *Int. J. Electrochem. Sci.* **2013**, *8*, 6449–6456.
3. Wu, L.; Leng, X.; Liu, Y.; Wei, S.; Li, C.; Wang, G.; Lian, J.; Jiang, Q.; Nie, A.; Zhang, T.Y. A strategy for synthesis of nanosheets consisting of alternating spinel $Li_4Ti_5O_{12}$ and rutile TiO_2 lamellas for high-rate anodes of lithium-ion batteries. *ACS Appl. Mater. Interfaces* **2017**, *9*, 4649–4657. [CrossRef] [PubMed]
4. Jiang, Y.M.; Wang, K.X.; Wu, X.Y.; Zhang, H.J.; Bartlett, B.M.; Chen, J.S. $Li_4Ti_5O_{12}/TiO_2$ hollow spheres composed nanoflakes with preferentially exposed $Li_4Ti_5O_{12}$ (011) facets for high-rate lithium ion batteries. *ACS Appl. Mater. Interfaces* **2014**, *6*, 19791–19796. [CrossRef]
5. Park, K.S.; Benayad, A.; Kang, D.J.; Doo, S.G. Nitridation-driven conductive $Li_4Ti_5O_{12}$ for lithium ion batteries. *J. Am. Chem. Soc.* **2008**, *130*, 14930–14931. [CrossRef] [PubMed]
6. Long, D.H.; Jeong, M.G.; Lee, Y.S.; Choi, W.; Lee, J.K.; Oh, I.H.; Jung, H.G. Coating lithium titanate with nitrogen-doped carbon by simple refluxing for high-power lithium-ion batteries. *ACS Appl. Mater. Interfaces* **2015**, *7*, 10250–10257. [CrossRef] [PubMed]
7. Jo, M.R.; Nam, K.M.; Lee, Y.; Song, K.; Park, J.T.; Kang, Y.M. Phosphidation of $Li_4Ti_5O_{12}$ nanoparticles and their electrochemical and biocompatible superiority for lithium rechargeable batteries. *Chem. Commun.* **2011**, *47*, 11474–11476. [CrossRef] [PubMed]
8. Cao, L.; Kou, L.; Li, J.; Huang, J.; Yang, J.; Wang, Y. Nitrogen-doped carbon-coated V_2O_5 nanocomposite as cathode materials for lithium-ion battery. *J. Mater. Sci.* **2018**, *53*, 10270–10279. [CrossRef]
9. Kim, T.; Kim, H.; You, T.S.; Kim, J. Carbon-coated V_2O_5 nanoparticles derived from metal-organic frameworks as a cathode material for rechargeable lithium batteries. *J. Alloys Compd.* **2017**, *727*, 522–530. [CrossRef]
10. Lu, Q.; Jianhua Fang, J.; Yang, J.; Feng, X.; Wang, J.; Nuli, Y. A polyimide ion-conductive protection layer to suppress side reactions on $Li_4Ti_5O_{12}$ electrodes at elevated temperature. *RSC Adv.* **2014**, *4*, 10280–10283. [CrossRef]
11. Belharouak, I.; Amine, C.J.K. Synthesis and electrochemical analysis of vapor-deposited carbon-coated $LiFePO_4$. *Electrochem. Commun.* **2005**, *7*, 983–988. [CrossRef]

12. Chen, Z.; Dahn, J.R. Reducing carbon in $LiFePO_4/C$ composite electrodes to maximize specific energy, volumetric energy, and tap density. *J. Electrochem. Soc.* **2002**, *149*, A1184–A1189. [CrossRef]
13. Cho, J.; Kim, Y.J.; Park, B. Novel $LiCoO_2$ cathode material with Al_2O_3 coating for a Li ion cell. *Chem. Mater.* **2000**, *12*, 3788–3791. [CrossRef]
14. Jang, S.B.; Kang, S.H.; Amine, K.; Bae, Y.C.; Sun, Y.K. Synthesis and improved electrochemical performance of $Al(OH)_3$-coated $Li[Ni_{1/3}Mn_{1/3}Co_{1/3}]O_2$ cathode materials at elevated temperature. *Electrochim. Acta* **2005**, *50*, 4168–4173. [CrossRef]
15. Appapillai, A.T.; Mansour, A.N.; Cho, J.; Shao-Horn, Y. Microstructure of $LiCoO_2$ with and without "$AlPO_4$" nanoparticle coating: Combined STEM and XPS studies. *Chem. Mater.* **2007**, *19*, 5748–5757. [CrossRef]
16. Ying, J.; Wan, C.; Jiang, C. Surface treatment of $LiNi_{0.8}Co_{0.2}O_2$ cathode material for lithium secondary batteries. *J. Power Sources* **2001**, *102*, 162–166. [CrossRef]
17. Li, H.; Zhou, H. Enhancing the performances of Li-ion batteries by carbon-coating: Present and future. *Chem. Commun.* **2012**, *48*, 1201–1217. [CrossRef]
18. Lin, J.; Peng, Z.; Liu, Y.; Ruiz-Zepeda, F.; Ye, R.; Samuel, E.L.G.; Yacaman, M.J.; Yakobson, B.I.; Tour, J.M. Laser-induced porous graphene films from commercial polymers. *Nat. Commun.* **2014**, *5*, 5714. [CrossRef]
19. Ye, R.; James, D.K.; Tour, J.M. Laser-induced graphene. *Acc. Chem. Res.* **2018**, *51*, 1609–1620. [CrossRef]
20. Inagaki, M.; Harada, S.; Sato, T.; Nakajima, T.; Horino, Y.; Morita, K. Carbonization of polyimide film "Kapton". *Carbon* **1989**, *27*, 253–257. [CrossRef]
21. Peng, Z.; Ye, R.; Mann, J.A.; Zakhidov, D.; Li, Y.; Smalley, P.R.; Lin, J.; Tour, J.M. Flexible boron-doped laser induced graphene microsupercapacitors. *ACS Nano* **2015**, *9*, 5868–5875. [CrossRef] [PubMed]
22. Carvalho, A.F.; Fernandes, A.J.S.; Leitão, C.; Deuermeier, J.; Marques, A.C.; Martins, R.; Fortunato, E.; Costa, F.M. Laser-induced graphene strain sensors produced by ultraviolet irradiation of polyimide. *Adv. Funct. Mater.* **2018**, *28*, 1805271. [CrossRef]
23. Zhang, J.; Ren, M.; Wang, L.; Li, Y.; Yakobson, B.I.; Tour, J.M. Oxidized laser-induced graphene for efficient oxygen electrocatalysis. *Adv. Mater.* **2018**, *30*, 1707319. [CrossRef] [PubMed]
24. Bobinger, M.R.; Romero, F.J.; Salinas-Castillo, A.; Becherer, M.; Lugli, P.; Morales, D.P.; Rodríguez, N.; Rivadeneyra, A. Flexible and robust laser-induced graphene heaters photothermally scribed on bare polyimide substrates. *Carbon* **2019**, *144*, 116–126. [CrossRef]
25. Cao, L.; Zhang, M.; Niu, H.; Chang, J.; Liu, W.; Yang, H.; Cao, W.; Wu, D. Structural relationship between random copolyimides and their carbon fibers. *J. Mater. Sci.* **2017**, *52*, 1883–1897. [CrossRef]
26. Ando, S.; Matsuura, T.; Sasaki, S. Coloration of aromatic polyimides and electronic properties of their source materials. *Polym. J.* **1997**, *29*, 69–76. [CrossRef]
27. Matsuura, T.; Hasuda, Y.; Nishi, S.; Yamadat, N. Polyimide derived from 2,2′-Bis(trifluoromethyl)-4,4′-diaminobiphenyl.1. Synthesis and characterization of polyimides prepared with 2,2-Bis(3,4-dicarboxyphenyl)hexafluoropropane dianhydride or pyromellitic dianhydride. *Macromolecules* **1991**, *24*, 5001–5005. [CrossRef]
28. Kim, S.; Jang, K.S.; Choi, H.D.; Choi, S.H.; Kwon, S.J.; Kim, I.D.; Lim, J.A.; Hong, J.M. Porous polyimide membranes prepared by wet phase inversion for use in low dielectric applications. *Int. J. Mol. Sci.* **2013**, *24*, 8698–8707. [CrossRef] [PubMed]
29. Jeong, K.M.; Li, Y.; Yoo, D.G.; Lee, N.K.; Lee, H.G.; Ando, S.; Ha, C.S. Effects of crosslinking agents on the physical properties of polyimide/amino-functionalized graphene oxide hybrid films. *Polym. Int.* **2018**, *67*, 588–597. [CrossRef]
30. Ji, D.; Li, T.; Zou, Y.; Chu, M.; Zhou, K.; Liu, J.; Tian, G.; Zhang, Z.; Zhang, X.; Li, L.; et al. Copolymer dielectrics with balanced chain-packing density and surface polarity for high-performance flexible organic electronics. *Nat. Commun.* **2018**, *9*, 2339. [CrossRef] [PubMed]
31. Ektessabi, A.N.; Hakamata, S. XPS study of ion beam modified polyimide films. *Thin Solid Film.* **2000**, *377–378*, 621–625. [CrossRef]
32. Liu, J.N.; Sil, M.C.; Cheng, R.; Feng, S.P.; Chen, C.M. Surface silanization of polyimide for autocatalytic metallization. *JOM* **2020**, *72*, 3529–3537. [CrossRef]
33. Wang, J.; Zhao, X.; Berda, E.B.; Chen, C.; Wang, K.; Chen, S.; Zou, B.; Liu, B.; Zhou, Q.; Li, F.; et al. The elastic properties and piezochromism of polyimide films under high pressure. *Polymer* **2016**, *90*, 1–8. [CrossRef]
34. Zhang, M.Y.; Niu, H.Q.; Qi, S.L.; Tian, G.F.; Wang, X.D.; Wu, D.Z. Structure evolutions involved in the carbonization of polyimide fibers with different chemical constitution. *Mater. Today Commun.* **2014**, *1*, 1–8. [CrossRef]
35. Ferrari, A.C.; Robertson, J. Interpretation of Raman spectra of disordered and amorphous carbon. *Phys. Rev. B* **2000**, *61*, 14095. [CrossRef]
36. Sarno, M.; Baldino, L.; Scudieri, C.; Cardea, S.; Ciambelli, P.; Reverchon, E. SC-CO_2 assisted process for high energy density aerogel supercapacitor: The effect of GO loading. *Nanotechnology* **2017**, *28*, 204001. [CrossRef] [PubMed]
37. Smirnova, V.E.; Gofman, I.V.; Maritcheva, T.A.; Yudin, V.E.; Eto, K.; Takeichi, T.; Kaburagi, Y.; Hishiyama, Y. The effect of different orientations in rigid rod polyimide films on the graphitized products. *Carbon* **2007**, *45*, 839–846. [CrossRef]
38. Song, H.; Jeong, T.G.; Moon, Y.H.; Chun, H.H.; Chung, K.Y.; Kim, H.S.; Cho, B.W.; Kim, Y.T. Stabilization of oxygen-deficient structure for conducting $Li_4Ti_5O_{12-\delta}$ by molybdenum doping in a reducing atmosphere. *Sci. Rep.* **2014**, *4*, 4350. [CrossRef]

polymers

MDPI

Article

Superparamagnetic, High Magnetic α-Fe & α''-$Fe_{16}N_2$ Mixture Prepared from Inverse Suspension-Polymerized Fe_3O_4@polyaniline Composite

Yen-Zen Wang [1], Yu-Wei Cheng [2], Lin-Chia Ho [3], Wen-Yao Huang [4,*,†], Ko-Shan Ho [5,*,†] and Yu-Ting Syu [5]

1 Department of Chemical and Materials Engineering, National Yu-Lin University of Science & Technology, Yun-Lin 640301, Taiwan; wangzen@yuntech.edu.tw
2 Department of Chemical Engineering, Ming Chi University of Technology, New Taipei City 24301, Taiwan; louischengblue@gmail.com
3 National Defense Medical Center, 161 Sec. 6 Minquan E. Rd., Neihu District, Taipei City 11490, Taiwan; yms40426@gmail.com
4 Department of Photonics, National Sun Yat-sen University, 70 Lienhai Rd., Kaohsiung 80424, Taiwan
5 Department of Chemical and Materials Engineering, National Kaohsiung University of Science & Technology, 415 Chien-Kuo Rd., Kaohsiung 80782, Taiwan; 1104311103@nkust.edu.tw
* Correspondence: wyhuang@faculty.nsysu.edu.tw (W.-Y.H.); hks@nkust.edu.tw (K.-S.H.)
† Wen-Yao Huang and Ko-Shan Ho equally contributed to this work.

Abstract: Oleic acid (OA)-modified Fe_3O_4 nanoparticles were successfully covered with polyanilines (PANIs) via inverse suspension polymerization in accordance with SEM and TEM micrographs. The obtained nanoparticles were able to develop into a ferrite (α-Fe) and α''-$Fe_{16}N_2$ mixture with a superparamagnetic property and high saturated magnetization (SM) of 245 emu g^{-1} at 950 °C calcination under the protection of carbonization materials (calcined PANI) and other iron-compounds (α''-$Fe_{16}N_2$). The SM of the calcined iron-composites slightly decreases to 232 emu g^{-1} after staying in the open air for 3 months. The calcined mixture composite can be ground into homogeneous powders without the segregation of the iron and carbon phases in the mortar without significantly losing magnetic activities.

Keywords: polyaniline; ferrite; α''-$Fe_{16}N_2$; superparamagnetic; inverse suspension polymerization

Citation: Wang, Y.-Z.; Cheng, Y.-W.; Ho, L.-C.; Huang, W.-Y.; Ho, K.-S.; Syu, Y.-T. Superparamagnetic, High Magnetic α-Fe & α''-$Fe_{16}N_2$ Mixture Prepared from Inverse Suspension-Polymerized Fe_3O_4@polyaniline Composite. *Polymers* **2021**, *13*, 2380. https://doi.org/10.3390/polym13142380

Academic Editor: Patrick Ilg

Received: 11 June 2021
Accepted: 15 July 2021
Published: 20 July 2021

1. Introduction

Recently, with the advance of technology, the requirement for fast and wireless charging becomes more and more urgent. In particular, 3C mobile electronic products need to charge within several hours using portable power sources that are usually heavy and take one or two to recharge. Therefore, we also need a fast wireless power source to solve the problem. The effective charging distance of wireless charging [1] and large-scale charging development [2], thus, becomes very important. Wireless charging can be carried out by absorbing electromagnetic (EM) waves that emit from the power generators. Additionally, we also require a stable EM wave absorbing material to transform it into power, which can be fulfilled by superparamagnetic materials with high magnetization. There are other interesting application fields related to the calcined carbon (polyaniline) coated magnetic particles [3–6] such as nanoelectronics, catalysis, optical application, biosensors, environmental remediation, energy, hydrogen storage, drug transport, magnetic resonance imaging and cancer diagnosis [7]. Metamaterial theory is also used for the applications of the magnetic materials [8].

The regular material that meets these requirements is magnetite (Fe_3O_4), which can be obtained from the sol-gel method [9] conducted at room temperature. Although it demonstrates a superparamagnetic property, its low magnetization (around 100 emu g^{-1}) shortens the wireless charging distance. Other possible candidates with a higher magnetization than Fe_3O_4 are cementite (Fe_3C) [10], ferric nitride (FexNy) [11,12] and ferrite

(α-Fe) [13]; they all demonstrate saturated magnetization over 150 emu g^{-1} and even over 200 emu g^{-1} for α-Fe and α''-$Fe_{16}N_2$. However, α-Fe, which is the pure Fe element, is vulnerable to the O_2 in the open air to become iron oxide.

Consequently, we need to provide some solid protection on the α-Fe, and α''-$Fe_{16}N_2$ itself can also provide additional protection to keep it from oxidation and maintain its high magnetization in the air.

Usually, PANIs that provide carbon and nitrogen sources can be prepared in affluent water [14–18] easily using water-soluble anilinium monomers. The Fe_3O_4 particles prepared from the sol-gel method are usually hydrophilic materials with some –OH groups on the surfaces. If it stays with anilinium monomers in the water, the prepared PANI molecules cannot cover most of the Fe_3O_4 particles that become very mobile in the water [19–26] due to the hydrophilicity. The obtained PANIs are not able to protect Fe_3O_4 particles during calcination, since most of them will stay on the surface of PANI molecules, not inside.

Inverse suspension polymerization (ISP) is usually applied to cover functional particles with polymers that can be polymerized in the water phase [27–31]. In this study, we are trying to prepare the protecting Fe_3O_4 particles by the ISP of PANIs. The sol-gel-prepared Fe_3O_4 particles usually own surface –OH, which can be esterificated with OA to attach some hydrophobic tails to the particles [32,33] and allow them to stably stay inside the micelles in the toluene system. Moreover, the additional aliphatic tails that come from the attaching OA can also stabilize the micelles or polymer droplets during the polymerization of anilines. In other words, hydrophilic OA-modified Fe_3O_4 (Fe_3O_4(OA)) was first mixed with the anilinium monomers in the water and became micelles in the hydrophobic toluene solvents after stirring. Eventually, the Fe_3O_4(OA) particles can be surrounded with long polyaniline molecules after water soluble initiators such as APS (ammonium persulfate) are introduced. The composites prepared via ISP can then be subject to calcination in the argon to transform into other iron-compounds with high magnetization.

2. Experimental

2.1. Preparation

2.1.1. Synthesis of Fe_3O_4(OA)

In a beaker, 7.08 g of ferric chloride hexahydrate ($FeCl_3 \cdot 6H_2O$, J. T. Baker, NJ, USA) and 2.58 g of ferrous chloride tetrahydrate ($FeCl_2 \cdot 4H_2O$, J. T. Baker, NJ, USA) were mixed with 40 mL of deionized water by a magnetic stirrer. The homogenized solution was transferred to a round bottom three-necked flask equipped with a water condenser in one of the mouths. One of the two remaining mouths was purged with high-purity nitrogen to prevent the oxidation of the reaction mixture at a high temperature, the other one behaved as the exhaust release outlet. The temperature of the reaction solution was ramped up to 80 °C in a silicone oil bath and kept with purging nitrogen for ten minutes. Then, 2 mL of OA (Hitachi Astemo Ltd, Tokyo, Japan) and some ammonia water (Fisher Sci., Bridgewater, NJ, USA) was introduced to tune the solution to become alkaline and then the reaction was started. The reaction continued for 30 min and was finished by stopping the stirring of the magnetic stirrer, followed by attaching a powerful magnet on the bottom of the reactor to separate the magnetic precipitate. The precipitate was washed several times with deionized water, and the clear, upper layer of the solution was discarded. The isolated black precipitate was placed in an ultrasonic oscillator for 20 min and then dried in an oven for 12 h at 60 °C, the Fe_3O_4(OA) was available.

2.1.2. Synthesis of PANI/Fe_3O_4(OA) Nanocomposite

An amount of 3 g (0.091 mol) of n-dodecylbenzenesulfonic acid (DBSA: Tokyo Kasei Kogyo Co., Tokyo, Japan) was dissolved in 50 mL of de-ionized water, the mixture was slowly stirred until a homogeneous solution was obtained, followed by the addition of 9 g (0.0968 mol) of aniline monomer (Tokyo Kasei Kogyo Co., Tokyo, Japan) and the solution was stirred to be clear. Eventually, the Fe_3O_4(OA) obtained from the previous experiment was added and the mixture was stirred again to become homogeneous [32,33].

A comparison emulsion polymerization prepared PANI(EB)/Fe_3O_4(OA) [29–31] was obtained in water in the absence of toluene. The resultant composite was named as PANI(EB)-Em/Fe_3O_4(OA).

2.1.3. Calcination of PANI Nanocomposites

PANI(EB)/Fe_3O_4(OA) prepared in Section 2.1.2 was calcined in a tube furnace, ramping up from RT to 600–950 °C at most and staying for 30 min in the argon atmosphere. The obtained N, C-doped iron composites are named as FeNCs. When samples were kept at room temperature for two or three months in the air, they are named as FeNC-2 and FeNC-3, respectively. A sample prepared at 600 °C is named FeNC-600, etc.

2.2. *Characterization*

2.2.1. Fourier Transform Infrared Spectroscopy (FTIR)

The main functional groups of neat Fe_3O_4(OA) and calcinated FeNC were assigned in accordance with the FTIR spectra recorded on an IFS3000 v/s FTIR spectrometer (Bruker, Ettlingen, Germany) at room temperature with a resolution of 4 cm^{-1} and 16 scanning steps.

2.2.2. Ultraviolet and Visible, Near-IR Spectroscopy (UV–Vis–NIR)

The UV–Vis–NIR spectra of the PANI(ES) (PANI without dedoping by $NH_4OH_{(aq)}$) in the PANI(ES)/Fe_3O_4(OA) nanocomposites were obtained from a Hitachi U-2001 and DTS-1700 NIR Spectrometer (Nicosia, Cyprus). The wavelength ranged from 300 to 1600 nm.

2.2.3. TGA (Thermogravimetric Analysis)

The mass loss percentages of neat PANI(EB) and PANI(EB)/Fe_3O_4(OA) upon calcination (thermal degradation) were monitored and recorded using TGA (TA SDT-2960, New Castle, DE, USA) thermograms.

2.2.4. Scanning Electron Microscopy (SEM)

The sizes and morphologies of neat Fe_3O_4(OA), non-calcinated PANI(EB)/Fe_3O_4(OA), and calcinated FeNCs were characterized using SEM (field emission gun scanning electron microscope, AURIGAFE, Zeiss, Oberkochen, Germany).

2.2.5. Transmission Electron Microscopy (TEM)

Samples, of which photos were taken using the field emission transmission electron microscope, HR-AEM (HITACHI FE-2000, Hitachi, Tokyo, Japan), were first dispersed in acetone and put on carbonic-coated copper grids dropwise before subjecting to the emission.

2.2.6. Raman Spectroscopy

The Raman spectra of calcinated PANI(EB)s and FeNCs treated at different temperatures were obtained from a Raman spectrometer (TRIAX 320, HOBRIA, Kyoto, Japan).

2.2.7. Powder X-ray Diffraction (Powder XRD)

A copper target (Cu-Kα) Rigaku X-ray source (Rigaku, Tokyo, Japan) generating X-ray with a wavelength of 1.5402 Å after electron bombarding was used to create the diffraction patterns of neat Fe_3O_4(OA) and FeNCs. The scanning angle (2θ) that ranged from 10° to 70° with a voltage of 40 kV and a current of 30 mA, operated at 1° min^{-1}.

2.2.8. X-ray Photoelectron Spectroscopy (XPS)

The binding energy spectra of Fe 2p of FeNCs treated at different temperatures were used to characterize the characteristic crystallization planes of α–Fe, Fe_3C, FeNx and, Fe_3O_4 after calcination, and were obtained from an XPS instrument of Fison (VG)-Escalab 210 (Fison, Glasgow, UK) using an Al Ka X-ray source at 1486.6 eV. The pressure in the chamber maintained was under 10^{-6} Pa or lower during performance. Tablet samples were prepared by pressing in a stapler with a ring mold.

2.2.9. Superconductor Quantum Interference Device (SQUID)

The paramagnetic properties of neat Fe_3O_4(OA) and various FeNCs were measured from a SQUID of Quantum Design MPMS-XL7 (San Diego, CA, USA)

3. Results and Discussion

3.1. FTIR Spectra

The hydrophilicity of Fe_3O_4(OA) nanoparticles comes from the hydroxylated surfaces created during the sol-gel process, assigned at ~3300 cm^{-1} according to Figure 1. Some of the hydroxyl groups are still present after esterification with OA, as described in Figure 1 as well, indicating that some of the –OH groups remained intact after esterification and polymerization. The remaining hydrophilicity of the nanocomposite makes it still dispersible in micelles before the polymerization mixing with anilinium monomers or staying in polymer droplets after polymerization, which was randomly dispersed in the hydrophobic toluene matrix. The polymerization of anilinium monomers after the addition of water-soluble initiator APS in the presence of Fe_3O_4(OA) did not destroy the carbonyl groups of the ester that link the OA onto the Fe_3O_4 surface of the nanoparticles either, illustrating that the long hydrocarbon tails of the OA are still firmly connected to the Fe_3O_4 nanoparticle's surface. It provided the nanocomposites with some hydrophobicity and affinity to the toluene and the large polymer droplets were still able to suspend in the solvent after polymerization. The polymer droplets need to de-emulsify with acetone before collecting the nanocomposite products through filtration. The symmetric and asymmetric stretching mode of aliphatic methylene and the methyl groups of OA tails assign at 2920 and 2840 cm^{-1}, respectively. The additional peak around 587 cm^{-1} reveals the presence of the Fe–O bonding of both the neat Fe_3O_4(OA) and the nanocomposite, revealing that Fe_3O_4(OA) nanoparticles are staying inside the nanocomposite, even after de-emulsification and filtration. The assignments of the main functional groups of polyaniline and Fe_3O_4(OA) are listed in Table 1 and illustrated in Figure 1.

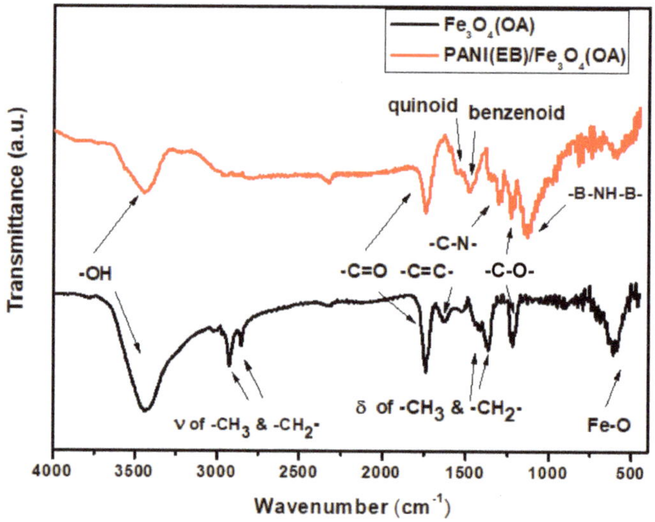

Figure 1. FTIR-spectra of neat Fe_3O_4(OA) and PANI(EB)/Fe_3O_4(OA) composite.

Table 1. Assignments of FTIR spectra.

$-OH$	$3400\ cm^{-1}$
$-CH_3$, $-CH_2$ (stretching)	$2900, 3000\ cm^{-1}$
$-C{=}O$	$1750\ cm^{-1}$
$-C{=}C-$	$1650\ cm^{-1}$
Quinoid ring	$1570\ cm^{-1}$
Benzoid ring	$1476,\ cm^{-1}$
$-CH_3$, $-CH_2$ (bending)	$1400, 1350\ cm^{-1}$
$-C-N-$	$1307\ cm^{-1}$
$-C-OH$	$1250\ cm^{-1}$
$-B-NH-B-$	$1135\ cm^{-1}$
Fe–O	$590\ cm^{-1}$

The obtained PANI(ES)/Fe_3O_4(OA) nanocomposites were then dedoped into PANI(EB)/Fe_3O_4(OA) in $NH_4OH_{(aq)}$ and its FTIR-spectrum is also demonstrated in Figure 1. The feature functional groups of PANI(EB) can be found for PANI(EB)/Fe_3O_4(OA) in Table 1 and Figure 1, revealing that the polymerization did not change the characteristic functional groups of PANI in the presence of Fe_3O_4(OA).

3.2. λ_{max} *of PANI in the Nanocomposite Obtained from UV–Vis–NIR Spectra*

If most of the Fe_3O_4(OA) in the nanocomposite is completely covered or embedded in the PANI matrix, it can also induce the shifting of the λ_{max} of PANI in the UV–Vis spectra due to the interaction. Figure 2 illustrates the flattening effect on the λ_{max} in the NIR region, the so-called free carrier-tail [14–18] in this region for the PANI(ES)-Em prepared via emulsion polymerization. It also represents a nanofibrous morphology that contributes to the carrier-tail due to the highly extended conjugation chain length. However, in accordance with Figure 2, the PANI(ES) prepared via ISP does not demonstrate any free carrier-tail, but the significant λ_{max} peak and the morphology is not fibrous either, which will be confirmed in the TEM micropicture. The λ_{max} of PANI(ES) prepared by ISP is around 860 nm, which is also in the near-IR region, indicating that its conjugation chain length is still long, and the chain is still extended. After the Fe_3O_4(OA) nanoparticles were introduced before the beginning of the polymerization, the λ_{max} of PANI(ES) blue-shift from 860 to 800 nm with increasing Fe_3O_4(OA) nanoparticles and curve became more bent. The blue shift and the presence of the bended curves of the nanocomposites reflected the shortening of the conjugation of the PANI molecules whose extensions were slightly recoiled. Since the λ_{max} is still high at 810 nm, a random-coil morphology (λ_{max} = 780 nm) is not thought to be present. The attracting force that causes the recoiling of the PANI(ES) molecules is believed to stem from the formation of the H-bonding between the left –OH groups and the amino groups of PANI(ES). The blue-shifting is enhanced when more Fe_3O_4(OA) is introduced, referring to Figure 2. The ISP approach effectively polymerizes the anilinium monomers inside micelles where lots of Fe_3O_4(OA) nanoparticles are already present. It is believed that some of the H-bonds are already formed before the addition of water-soluble APS initiator and most of the Fe_3O_4(OA) nanoparticles inside micelles can be surrounded and protected by both monomers before polymerization and polymers after polymerization, which is described in Scheme 1. The presence of the long aliphatic/long tail of Fe_3O_4(OA) can improve the stability of the micelles in the toluene solvents as well and less CTAB is necessary to create the inverse micelle (W/O). Moreover, the hydrophobic tails of Fe_3O_4(OA) are extended to the toluene phase and can also immobilize the Fe_3O_4 inside water micelles, allowing the growing PANI molecules to entangle around them.

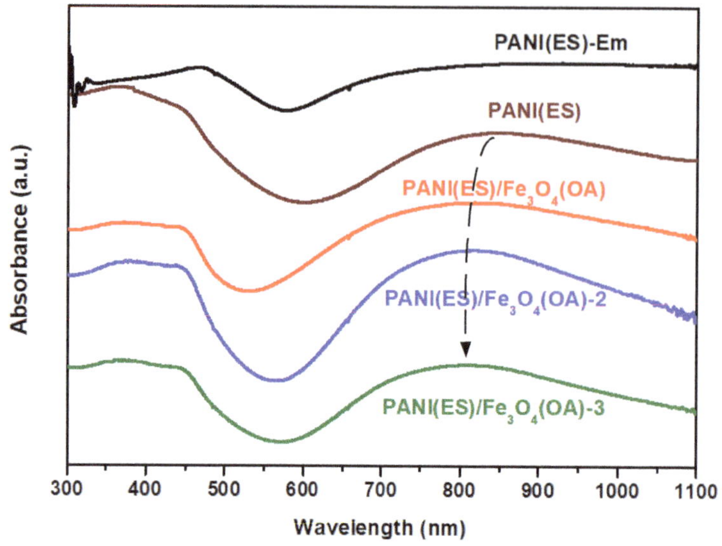

Figure 2. UV–Vis–NIR spectra of PANI(ES) and various PANI(ES)/Fe_3O_4(OA) composites.

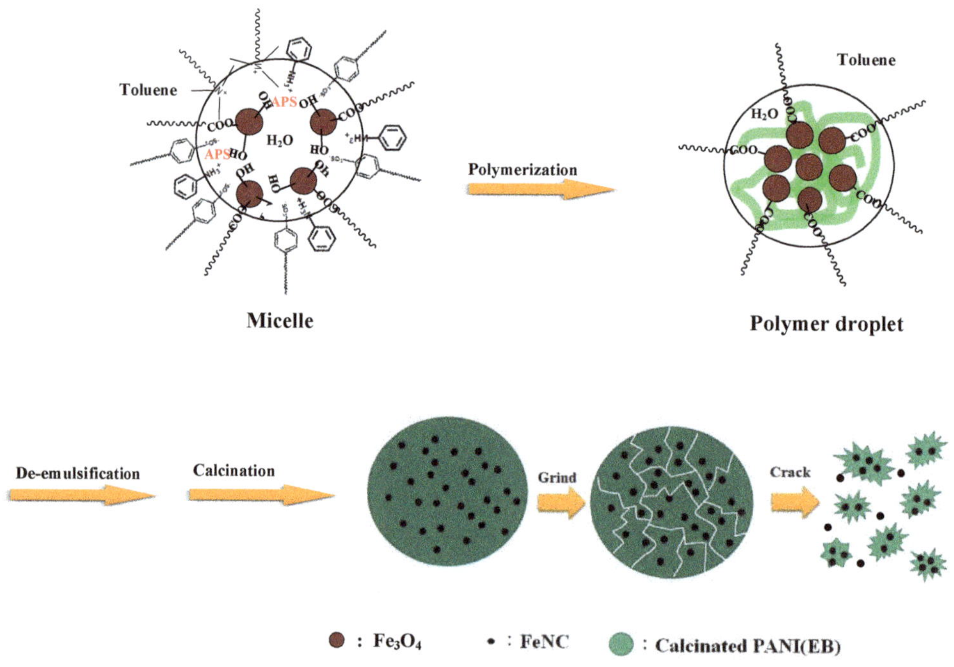

Scheme 1. Preparing diagram of FeNCs via inverse suspension polymerization and calcination.

When PANI is prepared via common emulsion polymerization, most of the Fe_3O_4(OA) particles will randomly distribute in the water and the fibrous PANI is created inside the micelles where less water is present. Eventually, the obtained nanofibrous PANI molecules can only attract lots of Fe_3O_4(OA) particles on the surface through H-bonding, which will be seen in the SEM micropicture. The Fe_3O_4(OA)-covered fibrous PANI surely cannot form a stable/high magnetic FeNC compound after calcination, since the formed high magnetic materials cannot be protected by the PANI during calcination.

3.3. TGA Thermogram

The mass loss and rate of the calcination at a high temperature can be monitored by the weights vs. temperatures in the thermogram demonstrated in Figure 3. The neat PANI(EB) experienced significant weight loss after 400 °C, which originated from the crosslinking of the neighboring molecules [18], and the weight loss continued gradually until 600 °C, when the crosslinking finished, and carbonization started. Almost no mass loss occurred after 600 °C for PANI(EB), which is the reason why the lowest calcination temperature chosen was 600 °C, after which the weight loss was entirely contributed from the PANI(EB)-covered iron-compounds not from PANI(EB) only, in accordance with Figure 3. If we check the difference of the residue weights of neat PANI(EB) after 600 °C and weight of PANI(EB)/Fe_3O_4(OA) at 600 °C from Figure 3 and Table 2, it is 22.4 wt% (35.9% − 13.5% = 22.4%). It means there is about 22.4 wt% of iron-compound in the composite, which is thermally stable until 600 °C. There is only 1.7 wt% loss (37.5% − 35.8% = 1.7%) after 600 °C for the composite, as seen in the inset of Figure 3, which is all contributed from the thermal degradation of the FeNCs from 600 to 950 °C. The actual degrading % for FeNC is around 1.7/22.4 ≈ 7.6 wt%. The composition variation and the types of atoms (Fe, N, C, or O) that are lost during the calcination of the FeNCs from 600 to 950 °C are strongly related to the magnetic activity, which will be characterized from the powder X-ray diffraction patterns, XPS, and SQUID spectra.

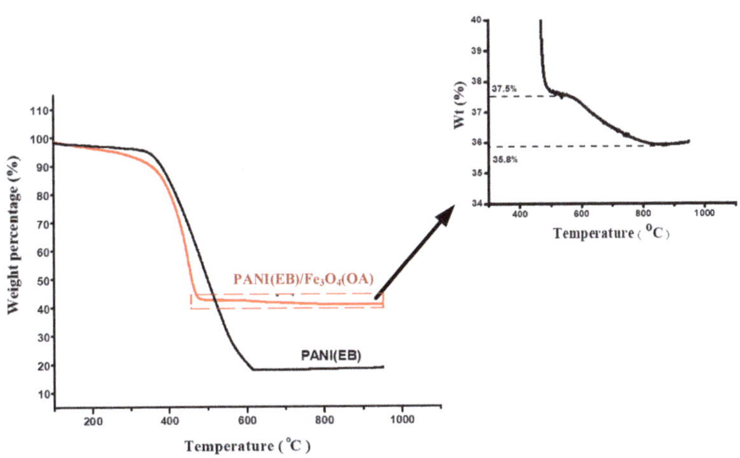

Figure 3. TGA thermograms of PANI(EB) and PANI(EB)/Fe_3O_4(OA).

Table 2. Various properties of FeNCs vs. temperature.

T (°C) \ Properties	Ms [(a)] (emu g^{-1})	Weight [(b)] (%) PANI(EB)	Weight [(b)] (%) FeNC	I_D/I_G [(c)] PANI(EB)	I_D/I_G [(c)] FeNC
Room Temp.	17.7	100	100 (PANI(EB)/Fe_3O_4(OA))	1.32	1.30 (PANI(EB)/Fe_3O_4(OA))
600	49.3	15.3	37.2	0.96	0.89
700	43.2	13.0	36.5	0.86	0.95
800	31.0	13.1	36.0	0.82	0.97
900	229.1	13.3	35.9	0.79	0.98
950	244.9	13.5	35.9	0.74	0.98

[(a)] Saturation magnetization at various temperatures obtained from SQUID; [(b)] Weight percentages at various temperatures obtained from TGA thermograms; [(c)] Intensity of D-band over G-band at various temperatures, obtained from Raman spectra.

3.4. SEM Micropicture

Unlike the neat Fe_3O_4 particles that demonstrate a pearl-like morphology, the Fe_3O_4(OA) particles are found to be able to coagulate by the interactive long, aliphatic chains of OA and become the huge cake-like morphology in Figure 4a after being treating with OA via esterification. The PANI(EB) prepared in the presence of Fe_3O_4(OA) particles via common emulsion polymerization demonstrates rib-like, juxtaposed nanofibers fully covered with lots of Fe_3O_4(OA) particles on their surfaces in Figure 4b, which is commonly seen in the traditional emulsion polymerization of PANI [31,34]. These exposed Fe_3O_4(OA) particles on the fibrous PANI(EB) certainly can transform or convert to other iron-compounds with higher magnetic activity. However, the formed iron-compounds would be directly exposed to the O_2 in the atmosphere at RT without any protection. Therefore, these iron-compounds would easily oxidize and recover to iron-oxide, whose magnetic force is far below that of the FeCx, FeNx, or α-Fe obtained after calcination. Consequently, an ISP system is designed to encapsulate the un-protected Fe_3O_4(OA) particles with PANI(EB) that can develop into a dense, strong protecting carbon layer after calcination at high temperature through crosslinking and carbonization.

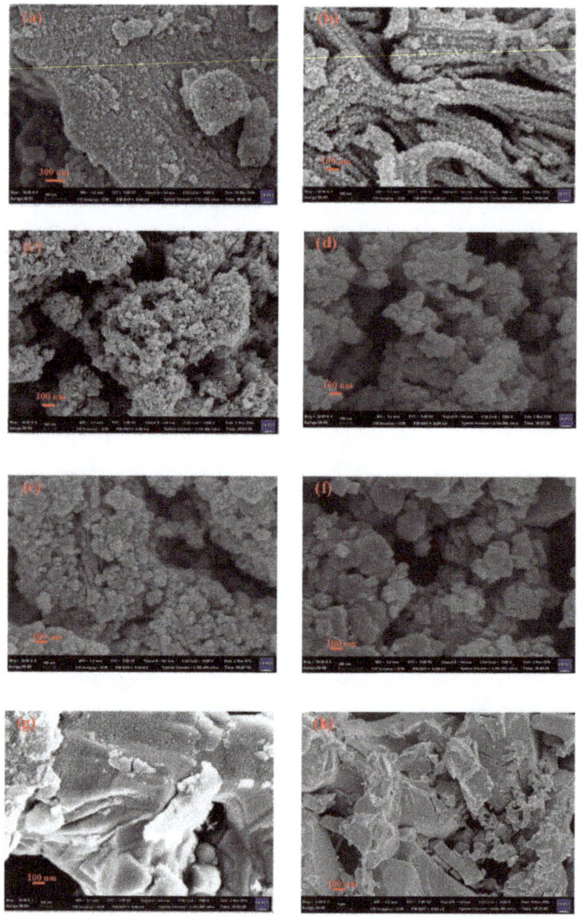

Figure 4. SEM micropictures of (**a**) Fe_3O_4(OA), (**b**) PANI(EB)-Em/Fe_3O_4(OA), (**c**) PANI(EB)/Fe_3O_4(OA), (**d**) FeNC-650, (**e**) FeNC-700, (**f**) FeNC-800, (**g**) FeNC-900, (**h**) FeNC-950.

The –OH groups of the sol-gel-prepared Fe_3O_4(OA) stay inside the micelles before polymerization. Additionally, after the polymerization in the inverse suspended system, most of the PANIs are formed in the micelles and become polymer droplets that then coalesce and develop the morphology seen in Figure 4c when the ISP system was eventually de-emulsified by the addition of acetone (the breaking of polymer droplets) and the fibrous morphology is never seen.

Upon calcination at high temperatures, the Fe_3O_4(OA) that covered with PANI(EB) would convert to other magnetic materials with a higher magnetic activity. Combining the increasing magnetic attractive forces and thermal energy provided by the high temperature calcination, these Fe_3O_4(OA) nanoparticles were able to transform and to merge into bigger particles due to the increasing magnetic attraction forces with increasing temperatures, as seen from Figure 4d–h. The sizes of the new particles are smaller when the calcination temperatures are below 700 °C. Bigger cake-like ensembles are perceivable when temperatures reach 800 °C. These cake-like ensembles even impinge further into huge slabs around 900 °C, revealing the occurrence of extremely high magnetic attractive forces after 900 °C. The types of magnetic materials created when Fe_3O_4(OA) are calcined inside of the PANI(EB) matrix at high temperatures can be studied by checking their X-ray diffraction patterns and XPS spectra.

3.5. TEM Micropicture

The pearl-like chain morphology of the TEM micropicture in Figure 5a expresses the neat Fe_3O_4(OA) particles that are connected to each other by the inter-entangled or inter-digitized long alkyl tails of the attached OA. Most of the Fe_3O_4(OA) particles were covered by the PANI(EB) after the ISP at RT, as seen in Figure 5b, where only some tiny ones can be seen in the margins of the big ensemble. The pretty dark and homogeneous color seen in Figure 5b reveals that Fe_3O_4(OA) particles are uniformly distributed in the PANI(EB) matrix. When the temperature reached 600 °C, most of the PANI(EB) were thermally degraded and only 13.5 wt% were left, according to Figure 3 and Table 2, the sample became more transparent in Figure 5c, and some tiny Fe_3O_4(OA) nanoparticles transformed and coalesced into bigger, dark particles after 600 °C. Whether they still remain in the form of Fe_3O_4 and what kind of new iron-compounds formed after 600 °C, can be analyzed using X-ray or XPS spectra. The particle-assembly phenomena were enhanced with the calcination temperature and huge grain-like particles developed from 600 to 950 °C, in accordance with Figure 5c–h. The growing size of the dark particles resulted from the increasing magnetic attractive forces with temperature, which originated from the formation of some iron-compounds with a high magnetic activity. There are N, C, and O atoms inside the nanocomposite, except Fe. Additionally, the FeNCs are all derivatives of Fe_3O_4(OA) via calcination. The increasing magnetic forces compared to the neat PANI(EB)/Fe_3O_4(OA) or the neat Fe_3O_4(OA) can be attributed to the newly formed FeCx, or α-Fe, even the α''-$Fe_{16}N_2$ compounds that own much higher magnetic forces than the neat Fe_3O_4(OA). Furthermore, the obtained Fe_3C (cementite) is the most stable FeCx compound. The variation % of these atoms and types of iron-compounds in the composite with increasing temperatures can be understood from either the XRD patterns or the XPS for each compound.

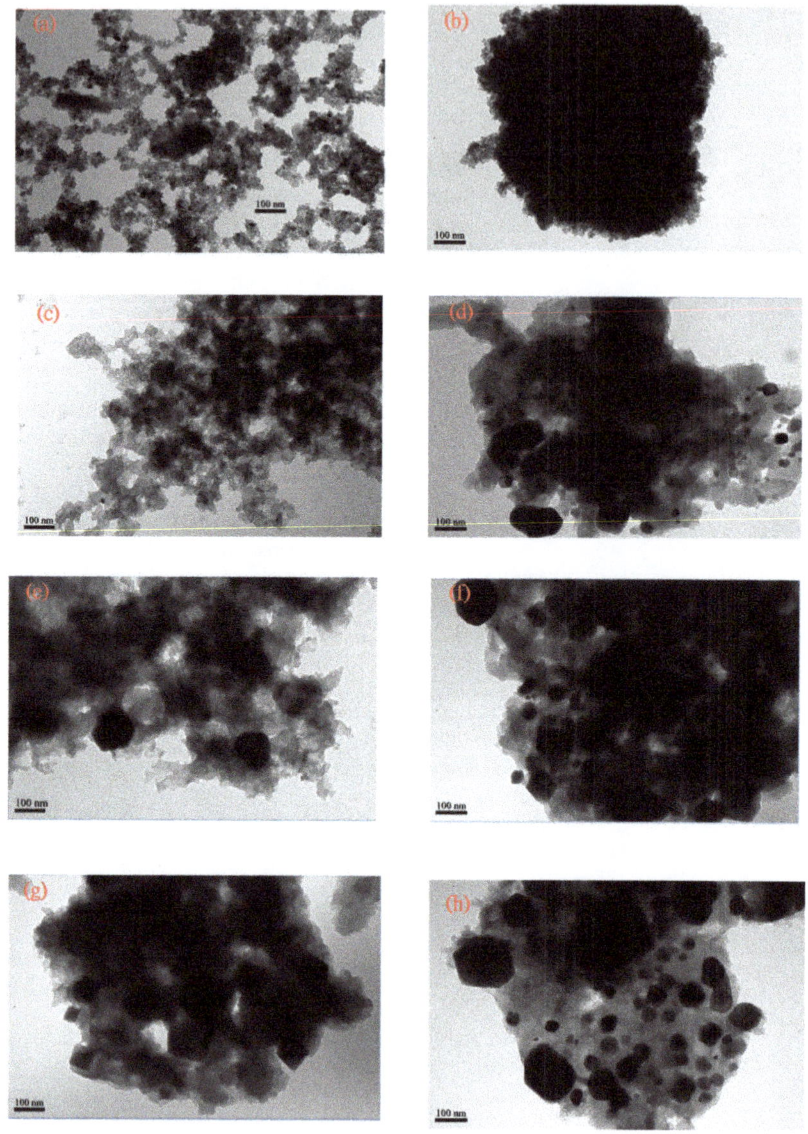

Figure 5. TEM micropictures of (**a**) Fe_3O_4(OA), (**b**) PANI(EB)/Fe_3O_4(OA), (**c**) FeNC-600, (**d**) FeNC-650, (**e**) FeNC-700, (**f**) FeNC-800, (**g**) FeNC-900, (**h**) FeNC-950.

The juxtaposed fibrous PANI(EB), fully covered with Fe_3O_4(OA) nanoparticles (PANI(EB)-Em/Fe_3O_4(OA)), can also be seen in Figure 6a. These fibers, which merged into a huge ensemble, gradually became hard stone in Figure 6b and cannot be easily broken into small pieces by simply grinding them in the mortar. It is believed that the surface Fe_3O_4(OA) nanoparticles developed into hard covering materials and the calcined PANI(EB) (950 °C) mostly remain inside the composites. In contrast, the ISP-prepared composite allowed the inclusive Fe_3O_4(OA) nanoparticles to develop into individual small magnetic particles inside the FeNCs, as seen in Figure 6c, which can be easily ground into tiny particles in the mortar, as depicted in Scheme 1. Amazingly, some highly crystallized stone powders can be released from the cracked particles after grinding, as seen in Figure 6d, which is also illustrated in Scheme 1. Actually, the breaching by grinding occurs following the carbonized PANI(EB) boundaries. It means we are able to fabricate any shapes of high magnetic stones by further sintering these magnetic powders in different shapes of molds at a temperature far below the melting points of the magnetic stones. Moreover, the magnetic powders can also be protected from oxidation during sintering by the carbonized coverings.

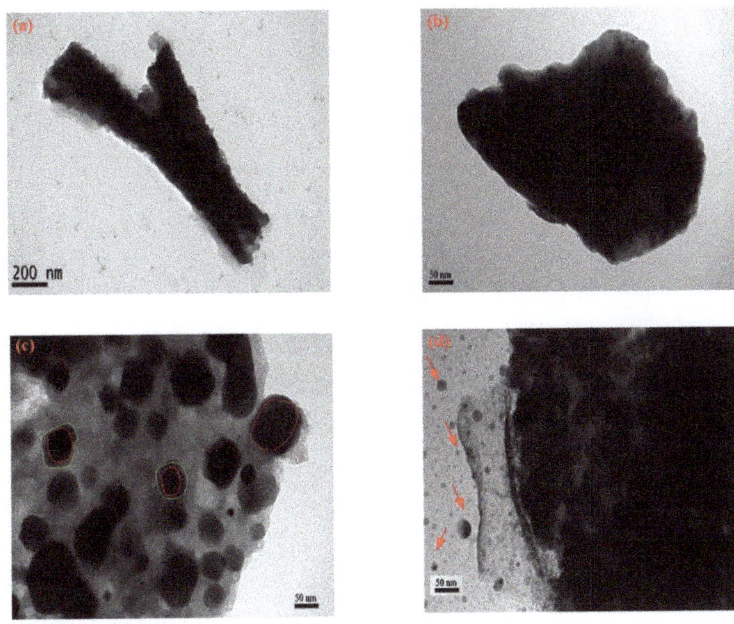

Figure 6. TEM micropictures of (**a**) PANI(EB)-Em/Fe_3O_4(OA), (**b**) PANI(EB)/Fe_3O_4(OA), (c) FeNC-950, (**d**) Ground FeNC-950.

3.6. *Raman Spectra*

The ratio (I_D/I_G) of –C–C– single bonds (sp^3, D-band) with –C=C– double bonds (sp^2, G-band) demonstrated in the Raman spectrum can be used to monitor the degree of the thermal degradation of organic compounds. When it increases with temperature (more G-bands thermally destroyed to become D-band carbons), it means a rougher surface structure is created after heating and vice versa. The I_D and I_G peaks assigned at 1348 and 1575 cm^{-1} in the Raman spectrum represent the D- and G-band of the covalent-bonded carbons, respectively. The surface structures that vary with the calcination temperatures for PANI(EB) and FeNCs are monitored using the Raman spectra in Figure 7 and their I_D/I_G values are listed in the last two columns of Table 2. The I_D/I_G of PANI(EB) in Figure 7a decrease with the calcination temperature due to the crosslinking and some degree of the ordering of the graphene lattice, referring to the PANI(EB) matrix structure of the FeNC composites that is actually becoming more and more smooth with the formation of plane sp^2 (C=C) bonding. However, the entire composite does not follow that trend with increasing temperatures when Fe_3O_4(OA) is present. The degradation of OA tails and the thermal transformation of the iron-compounds, which accompany the formation of bonds between Fe and N, or the C atoms of the PANI(EB), can destroy more C=C bonding too since it owns more active π-bonds. Moreover, the iron-compound itself would experience a crystallographic transformation at high temperatures in accordance with the Fe–C phase diagram after 900 °C. All these possible newly formed bonds and transformations at high temperatures play significant roles in the eventual structures of the FeNCs obtained at different temperature calcination, resulting in the increasing I_D/I_G with temperature in Figure 7b and Table 2. Certainly, the iron-compounds obtained at various temperatures all demonstrate much higher magnetic forces than the starting material PANI(EB)/Fe_3O_4(OA), which will be illustrated later.

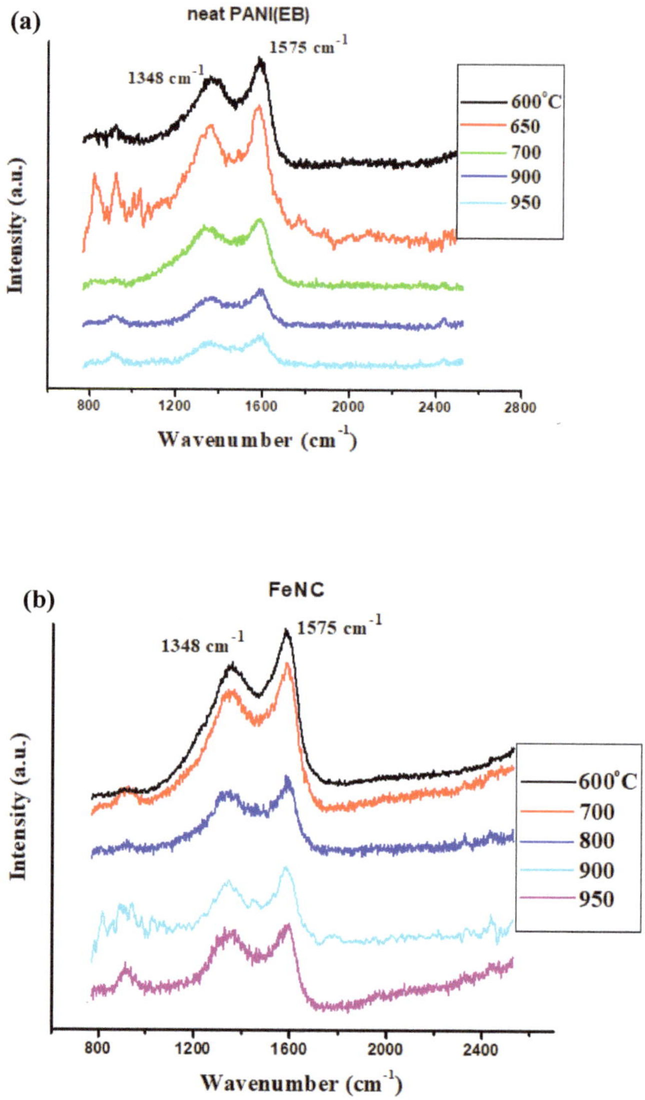

Figure 7. Raman-spectra of (**a**) neat PANI(EB) (**b**) FeNCs.

3.7. Powder XRD Patterns

The powder XRD patterns of neat Fe_3O_4 and FeNC treated at the various temperatures displayed in Figure 8a significantly demonstrates the variation of the characteristic crystalline plane-peaks from Fe_3O_4 to FeO, Fe_3C, Fe_3N, and α-Fe (α″-$Fe_{16}N_2$) with temperature. In particular, when the temperature was close to 900 °C, α-Fe and α″-$Fe_{16}N_2$ became the dominant core product due to the migration of O, N, and C atoms out of the core materials.

Other type of hard, stable iron-compounds such as Fe_3C or Fe_3N with lower magnetic activities formed the covering materials protecting the inner α-Fe from oxidation and losing magnetic activity in the atmosphere at RT. The presence of α''-$Fe_{16}N_2$ in the core also provided additional protection for the formed α-Fe. The characteristic diffraction patterns of α-Fe(110) and α''-$Fe_{16}N_2$(220) illustrated in Figure 8b remain almost intact after staying in the air for 3 months, which will otherwise become iron oxides in less than 1 month without any protection for neat α-Fe (ferrite). It again illustrates the necessity and importance of covering iron-materials with PANI(EB) via inverse suspension polymerization.

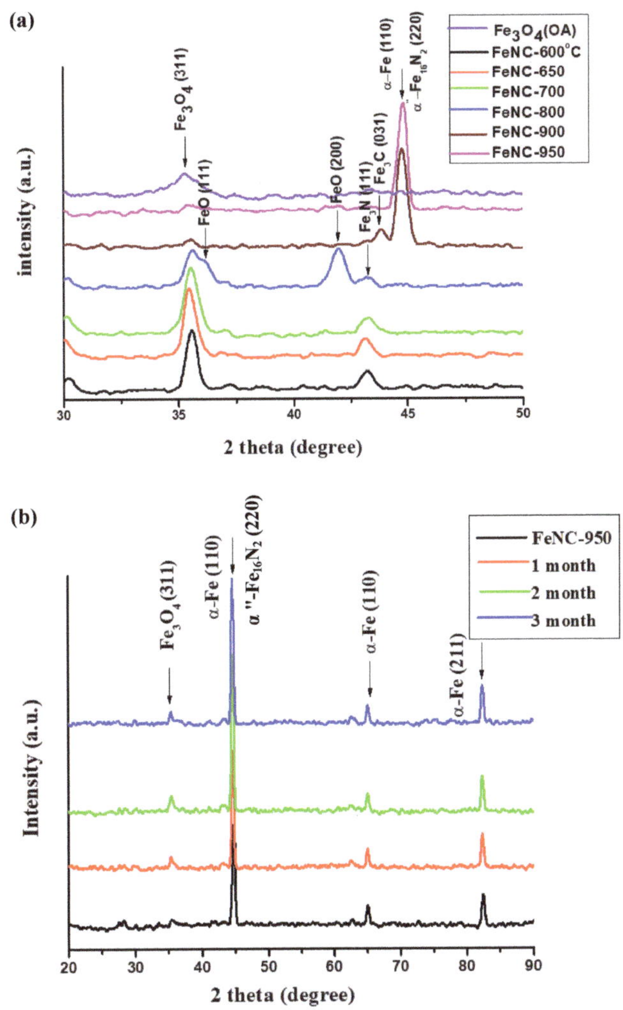

Figure 8. X-ray diffraction patterns of (**a**) FeNCs and (**b**) FeNC-950 staying in the open air for months.

According to FWHM, the crystallite sizes for FeNC-900 and -950 are 1.15 and 3.75 nm, respectively (calculated from the X-ray diffraction patterns of Figure 8b α-Fe(110) peak), which are much smaller than TEM (Figure 6c) The size of α-Fe (or α''-$Fe_{16}N_2$) crystallites also increased with temperature from 900 to 950 °C. The size of Fe_3C based on Fe_3C (031) plane in Figure 6c is 3.65 nm, which is also much smaller than the particles illustrated in TEM (Figure 6c).

3.8. XPS Spectra

Typical XPS spectra of FeNCs recording % of Fe, N, C, and O atoms in Figure 9a,b provide the % variation of each atom at different temperatures. Additionally, the Fe($2p_{3/2}$) of the composites calcined at various temperatures was revealed in Figure 9c. The characteristic peaks of α''-$Fe_{16}N_2$ (706.8 eV), α-Fe (707.7 eV), Fe_3C (708.5 eV), Fe_3O_4 (710.8 eV), and FeO (709.4 eV) [35–39], respectively, were illustrated in Figure 9c as well. Referring to the magnetic data obtained from Figure 10, we are able to construct Figure 9d, which illustrates how magnetic activity and each type of atom % varied with the calcination temperatures. The positions of the Fe($2p_{3/2}$) patterns, which ranged from 700 to 714 MeV, can be applied to understand what kind of iron-compounds the ISP-prepared PANI(EB)/Fe_3O_4(OA) composite became during calcination. Additionally, the satellite peaks that shift from ~715 up to ~725 eV when the thermal treatment temperature increases from 600 to 950 °C correspond to the ferrous compounds [40,41]. According to Figure 9c, α-Fe was not the dominant compound until the calcination temperature raised over 800 °C, but Fe_3C is already created above 700 °C by a reaction between the included Fe_3O_4(OA) nanoparticles and the surrounding PANI(EB) matrix seen in Figure 9d and exists for all calcination temperatures. In other words, Fe_3C, which is a very stable, hard magneton, dominates in the composite in the beginning of calcination and α-Fe formed later by driving some C atoms out of the core area at a temperature higher than 800 °C. Therefore, we can find a clear and sharp increase in the Fe % and a deep decease in the C % when the temperature was over 800 °C in Figure 9d, which was also accompanied with a sudden increase in magnetic forces. The formation of an α-Fe core and an Fe_3C shell provides a facile way to fabricate stable magnetons with ultrahigh SM values, which will be discussed in the following section.

3.9. SQUID Spectra

Ferrite (α-Fe) and α''-$Fe_{16}N_2$ are two of the magnetic materials with SM over 200 emu g^{-1}, α''-$Fe_{16}N_2$ even reaches 300 emu g^{-1}. Limited to its vulnerable structure, α-Fe easily fuses with O, C, or N atoms to become iron oxide, Fe_3C, and FeNx, respectively. Figure 10a clearly indicates that the SM of ISP-prepared PANI(EB)/Fe_3O_4(OA) varies with calcination temperatures from 600 to 950 °C. The SM increases from 10 to 249 emu g^{-1} when calcination increases from RT to 950 °C, in accordance with Figure 10a, due to the transformation from Fe_3O_4 to α-Fe and α''-$Fe_{16}N_2$, which has already been proven using the X-ray and XPS spectra discussed in the previous sections. However, the SM does not monotonously increase with temperature since different iron-compounds with a higher or lower SM formed at different calcination temperatures. The SM slightly increased from 10 to 67 emu g^{-1} at 650 °C and fell back to 27 emu g^{-1} after the temperature increased to 800 °C. Briefly, the SM does not exceed 70 emu g^{-1} if calcination maintains below 800 °C. The X-ray patterns in Figure 8a and the XPS spectra in Figure 9c demonstrated the presence of mixtures of Fe_3O_4, FeO, and Fe_3C and their SMs are well below 250 emu g^{-1} theoretically. Until 900 °C is reached, the SM abruptly raises to 230 emu g^{-1} and then 245 emu g^{-1} at 950 °C. Again, their X-ray patterns and XPS spectra illustrate the formation of affluent α-Fe and α''-$Fe_{16}N_2$ at this stage by driving other atoms out of the core area with the help of high thermal energy. Certainly, there might be the presence of the lattice transformation of the iron-compounds after 900 °C, which also propels N, C, and O atoms to the outer area of α-Fe and α''-$Fe_{16}N_2$ to become protecting shell materials. Moreover, α''-$Fe_{16}N_2$ is able to prevent α-Fe from oxidation as well. According to the common Fe–C diagram, BCC-Fe would convert to FCC-Fe after 912 °C. Therefore, it is very possible for the iron compounds to undergo a significant atom rearrangement when the temperature is over 900 °C.

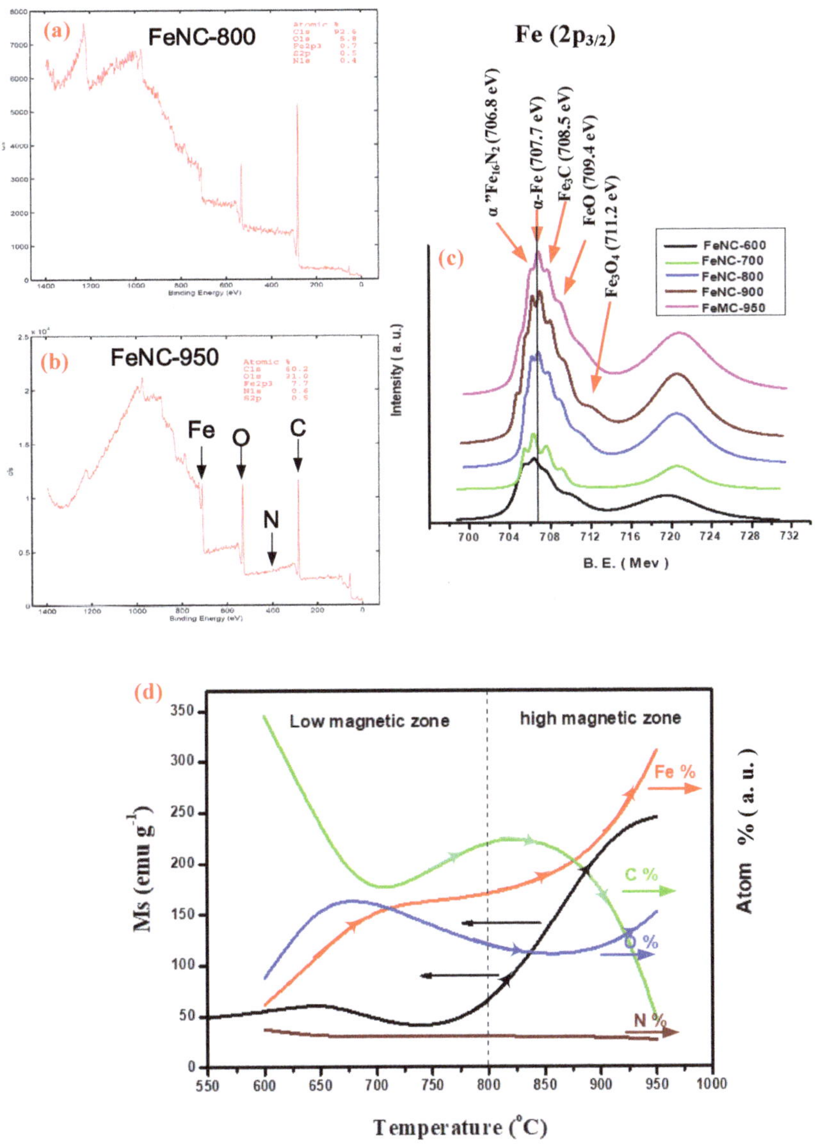

Figure 9. XPS spectra of (**a**) FeNC-800, (**b**) FeNC-950, (**c**) Fe ($2p_{3/2}$) of FeNCs, (**d**) specific magnetization (Ms) and various atom % of FeNCs at different temperatures.

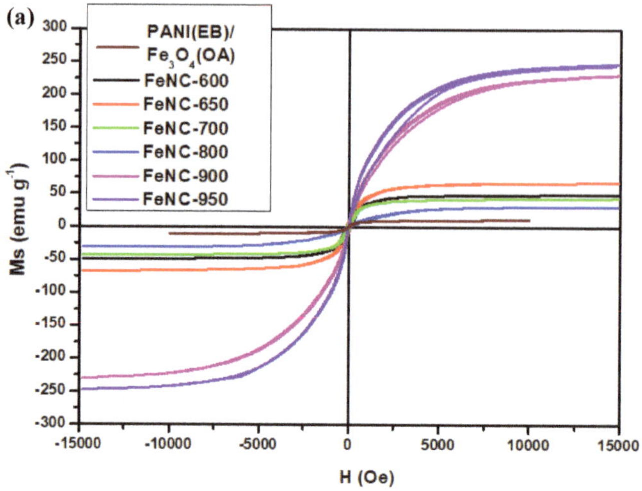

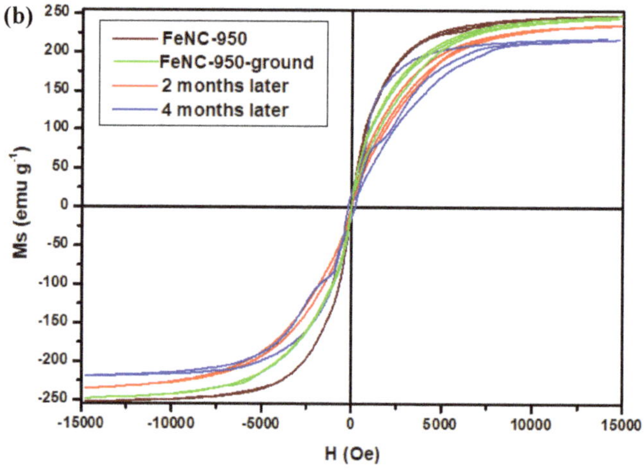

Figure 10. Saturated magnetization of (**a**) FeNCs and (**b**) various FeNC-950.

Under the protection of these stable iron-compounds, the SM of a mixture of α-Fe and α''-$Fe_{16}N_2$ decays only to 232 emu g^{-1} after staying in the air for 3 months, referring to Figure 10b. The slight decrease in the SM of FeNC-950 can be attributed to the possible surface oxidation of the powders during grinding, which largely created the surface area. All the FeNCs demonstrate superparamagnetic behaviors, referring to Figure 10a.

The ground FeNC-950 sample in the mortar retains the high SM of 245 emu g^{-1}, as shown in Figure 10a. The FeNC prepared from the calcination of ISP-prepared PANI(EB)/ Fe_3O_4(OA) is unlike other magnetic materials with a high SM, it can be easily broken into small pieces by a low degree of milling without losing the SM. The breaching points or parts of FeNC by grinding or milling are believed to follow the carbonized PANI(EB) boundaries that are affluent with carbon material, as illustrated in Scheme 1. In other words, FeNC is composed of small particles made of α-Fe and α″-$Fe_{16}N_2$ covered with carbon materials, which is very similar to the TEM micropicture in Figure 6c.

4. Conclusions

Fe_3O_4 nanoparticles were successfully and fully covered by polyanilines via inverse suspension polymerization in accordance with the SEM and TEM micrographs and the nanoparticles were able to develop into magnetons protected by the carbonization materials developed from polyaniline at different calcination temperatures. The saturated magnetization of the calcined iron-composites slightly increased from RT to 700 °C first and depressed continuously until 800 °C. A surprising jump of the saturated magnetization was found during 800~900 °C calcination. Based on the spectra of the X-ray diffraction and the XPS of iron-compounds calcined at temperatures higher than 900 °C, we understand a mixture of α-Fe and α″-$Fe_{16}N_2$ was formed in the core area and protected by the surrounding hard iron-compounds such as cementite (Fe_3C). The composite obtained from calcination at 950 °C slightly lost its saturated magnetization from 245 to 232 emu g^{-1} after staying for 3 months in the air. Moreover, unlike the common high magnetic materials, the calcined high magnetic product is not too hard to break by grinding or milling and does not cause the loss of saturated magnetization either. The easily broken properties do not originate from the presence of the cementite, since its hardness is also very high, but from the weak surrounding carbonized materials developed from polyaniline. For the time being, the application of inverse suspension polymerization to cover iron-magnetic materials with polymers before subjecting them to calcination to prepare the high magnetic materials proves to be very successful.

In the future, sintering approaches will be applied to shape the ground magnetic powders into various shapes of magnetons with high saturated magnetization in specific molds for different purposes.

Author Contributions: Conceptualization, Y.-Z.W. and W.-Y.H.; methodology, L.-C.H.; formal analysis, Y.-T.S.; writing—original draft preparation, Y.-W.C.; writing—review and editing, K.-S.H. All authors have read and agreed to the published version of the manuscript.

Funding: This research was funded by the Ministry of Science and Technology in Taiwan, grant number MOST 105-2622-E-151-012-CC3, MOST108-2221-E-992-037, and MOST 109-2221-E-992-083.

Acknowledgments: We appreciate the use of the soft matter-TEM equipment belonging to the Instrument Center of National Cheng Kung Univ. (NCKU), Ministry of Science and Technology in Taiwan, ROC.

Conflicts of Interest: The authors declare no conflict of interest.

References

1. Fuller, M. Wireless charging in California: Range, recharge, and vehicle electrification. *Transp. Res. Part C Emerg. Technol.* **2016**, *67*, 343–356. [CrossRef]
2. Lin, C.; Wu, G.; Obaidat, M.S.; Yu, C.W. Clustering and splitting charging algorithms for large scaled wireless rechargeable sensor networks. *J. Syst. Softw.* **2016**, *113*, 381–394. [CrossRef]
3. Niu, P.; Liang, X.; Lu, X.; Wang, S.; Li, Y.; Wang, L.; Guo, Y. Preparation of magnetic carbonized polyaniline nanotube and its adsorption behaviors of xanthene colorants in beverage and fish samples. *J. Chromatogr. A* **2019**, *1605*, 460369. [CrossRef]
4. Ghadimi, L.S.; Arsalani, N.; Tabrizi, A.G.; Mohammadi, A.; Ahadzadeh, I. Novel nanocomposite of MnFe2O4 and nitrogen-doped carbon from polyaniline carbonization as electrode material for symmetric ultra-stable supercapacitor. *Electrochim. Acta* **2018**, *282*, 116–127. [CrossRef]
5. Chen, X.; Zhang, Y.; Tao, L.; Nie, Q.; Meng, F.; Zhu, S.; Cui, L.; Huang, R. Ferromagnetic carbonized polyaniline/nanodiamond hybrids for ultrabroad-band electromagnetic absorption. *Carbon* **2020**, *164*, 224–234. [CrossRef]
6. Sim, B.; Chae, H.S.; Choi, H.J. Fabrication of polyaniline coated iron oxide hybrid particles and their dual stimuli-response under electric and magnetic fields. *Express Polym. Lett.* **2015**, *9*, 736–743. [CrossRef]
7. Siddiqui, M.T.H.; Nizamuddin, S.; Baloch, H.A.; Mubarak, N.M.; Dumbre, D.K.; Inamuddin; Asiri, A.M.; Bhutto, A.W.; Srinivasan, M.; Griffin, G.J. Synthesis of magnetic carbon nanocomposites by hydrothermal carbonization and pyrolysis. *Environ. Chem. Lett.* **2018**, *16*, 821–844. [CrossRef]

8. Prat-Camps, J.; Navau, C.; Sanchez, A. Quasistatic Metamaterials: Magnetic Coupling Enhancement by Effective Space Cancellation. *Adv. Mater.* **2016**, *28*, 4898–4903. [CrossRef] [PubMed]
9. Wang, H.; Wang, R.; Wang, L.; Tian, X. Preparation of multi-core/single-shell Fe_3O_4/PANI bifunctional nanoparticles via miniemulsion polymerization. *Colloids Surf. A Physicochem. Eng. Asp.* **2011**, *384*, 624–629. [CrossRef]
10. Zhang, B.; Du, Y.; Zhang, P.; Zhao, H.; Kang, L.; Han, X.; Xu, P. Microwave absorption enhancement of Fe_3O_4/polyaniline core/shell hybrid microspheres with controlled shell thickness. *J. Appl. Polym. Sci.* **2013**, *130*, 1909–1916. [CrossRef]
11. Yamanaka, K.; Onuma, Y.; Yamashita, S.; Masubuchi, Y.; Takeda, T.; Kikkawa, S. Humidity effects in $Fe_{16}N_2$ fine powder preparation by low-temperature nitridation. *J. Solid State Chem.* **2010**, *183*, 2236–2241. [CrossRef]
12. Li, X.; Xi, S.; Sun, L.; Dou, S.; Huang, Z.; Su, T.; Wang, X. Isolated FeN_4 Sites for Efficient Electrocatalytic CO_2 Reduction. *Adv. Sci.* **2020**, *7*, 2001545. [CrossRef]
13. Masubuchi, Y.; Yamashita, S.; Motohashi, T.; Kikkawa, S.; Niederberger, M. Magnetite/maghemite mixture prepared in benzyl alcohol for the preparation of α''-$Fe_{16}N_2$ with α-Fe. *J. Eur. Ceram. Soc.* **2011**, *31*, 2471–2474. [CrossRef]
14. Wang, Y.-Z.; Tsai, M.-J.; Hsieh, T.-H.; Tseng, P.-H.; Lu, C.-Y.; Ho, K.-S. Studies on one-dimensional polyanilines prepared with n-dodecylbenzenesulfonic and camphorsulfonic acids. *Polym. Int.* **2015**, *64*, 1568–1577. [CrossRef]
15. Cheng, Y.-W.; Chao, L.; Wang, Y.-M.; Ho, K.-S.; Shen, S.-Y.; Hsieh, T.-H.; Wang, Y.-Z. Branched and phenazinized polyaniline nanorod prepared in the presence of meta-phenylenediamine. *Synth. Met.* **2013**, *168*, 48–57. [CrossRef]
16. Wu, Y.-J.; Ho, K.-S.; Cheng, Y.-W.; Chao, L.; Wang, Y.-Z.; Hsieh, T.-H.; Ho, T.-H.; Han, Y.-K. Studies on the synthesis of low molecular weight, one-dimensional polyanilines prepared by fast emulsion polymerization using (n-dodecylbenzenesulfonic acid)/HCl emulsifiers. *Polym. Int.* **2013**, *62*, 581–590. [CrossRef]
17. Chao, L.; Ho, K.-S.; Shen, S.-Y.; Pu, H.-Y.; Hsieh, T.-H.; Kuo, C.-W.; Tseng, B.-H. Short polyaniline nanorod prepared in the presence of para-phenylenediamine. *J. Appl. Polym. Sci.* **2013**, *127*, 1853–1862. [CrossRef]
18. Ho, K.-S.; Han, Y.-K.; Tuan, Y.-T.; Huang, Y.-J.; Wang, Y.-Z.; Ho, T.-H.; Hsieh, T.-H.; Lin, J.-J.; Lin, S.-C. Formation and degradation mechanism of a novel nanofibrous polyaniline. *Synth. Met.* **2009**, *159*, 1202–1209. [CrossRef]
19. Basavaiah, K.; Kumar, Y.P.; Rao, A.V.P. A facile one-pot synthesis of polyaniline/magnetite nanocomposites by micelles-assisted method. *Appl. Nanosci.* **2013**, *3*, 409–415. [CrossRef]
20. Gabunada, J.C.; Vinothkannan, M.; Kim, D.H.; Kim, A.R.; Yoo, D.J. Magnetite Nanorods Stabilized by Polyaniline/Reduced Graphene Oxide as a Sensing Platform for Selective and Sensitive Non-enzymatic Hydrogen Peroxide Detection. *Electroanalysis* **2019**, *31*, 1507–1516. [CrossRef]
21. Subadra, S.T.U.I.; Sutiami, R.; Taufiq, A.; Diantoro, M.; Sunaryono; Arif; Hidayat; Mufti, N.; Hidayat, N.; Susanto, H.; et al. Preparation and Characterization of Magnetite Nanoparticles Combined with Polyaniline and Activated Carbon. *IOP Conf. Ser. Earth Environ. Sci.* **2019**, *276*, 012041. [CrossRef]
22. Xu, F.; Ma, L.; Huo, Q.; Gan, M.; Tang, J. Microwave absorbing properties and structural design of microwave absorbers based on polyaniline and polyaniline/magnetite nanocomposite. *J. Magn. Magn. Mater.* **2015**, *374*, 311–316. [CrossRef]
23. Batool, R.; Akhtar, M.A.; Hayat, A.; Han, D.; Niu, L.; Ahmad, M.A.; Nawaz, M.H. A nanocomposite prepared from magnetite nanoparticles, polyaniline and carboxy-modified graphene oxide for non-enzymatic sensing of glucose. *Microchim. Acta* **2019**, *186*, 267. [CrossRef]
24. Zanotto, C.; Ratuchne, F.; de Castro, E.G.; Marques, P.T. Structural Characterization of Magnetite and its Influence on the Formation of Composites with Polyaniline. *Orbital: Electron. J. Chem.* **2019**, *11*, 427–432. [CrossRef]
25. Farias-Mancilla, R.; Elizalde-Galindo, J.T.; Vigueras-Santiago, E.; Hernández-Escobar, C.A.; Vega-Rios, A.; Zaragoza-Contreras, E.A. Synthesis and Characterization of Polyaniline/Magnetite Nanocomposite. *Int. J. Theor. Appl. Nanotechnol.* **2016**, *4*, 1–8. [CrossRef]
26. Kim, J.H.; Fang, F.F.; Choi, H.J.; Seo, Y. Magnetic composites of conducting polyaniline/nano-sized magnetite and their magnetorheology. *Mater. Lett.* **2008**, *62*, 2897–2899. [CrossRef]
27. Benda, D.; Šňupárek, J.; Čermák, V. Inverse suspension polymerization of the hydrophilic acrylic monomers in the static continuous phase. *J. Dispers. Sci. Technol.* **1997**, *18*, 115–121. [CrossRef]
28. Wang, G.; Li, M.; Chen, X. Inverse suspension polymerization of sodium acrylate. *J. Appl. Polym. Sci.* **1997**, *65*, 789–794. [CrossRef]
29. Choudhary, M.S. Inverse Suspension Polymerization of Partially Neutralized and Lightly Cross-Linked Acrylic Acid: Effect of Reaction Parameters. *Macromol. Symp.* **2009**, *277*, 171–176. [CrossRef]
30. Meouche, W.; Branger, C.; Beurroies, I.; Denoyel, R.; Margaillan, A. Inverse Suspension Polymerization as a New Tool for the Synthesis of Ion-Imprinted Polymers. *Macromol. Rapid Commun.* **2012**, *33*, 928–932. [CrossRef]
31. Takail, Z.I.; Mustafa, M.K.; Asman, S.; Sekak, K.A. Preparation and Characterization of Magnetite (Fe_3O_4) nanoparticles By Sol-Gel Method. *Int. J. Nanoelectron. Mater.* **2019**, *12*, 37–46.
32. Ansari, F.; Sobhani, A.; Salavati-Niasari, M. Green synthesis of magnetic chitosan nanocomposites by a new sol–gel auto-combustion method. *J. Magn. Magn. Mater.* **2016**, *410*, 27–33. [CrossRef]
33. Corona-Rivera, M.A.; Ovando-Medina, V.M.; Martínez-Gutiérrez, H.; Silva-Aguilar, F.E.; Pérez, E.; Antonio-Carmona, I.D. Morphology and conductivity tuning of polyaniline using short-chain alcohols by heterophase polymerization. *Colloid Polym. Sci.* **2015**, *293*, 605–615. [CrossRef]
34. Huang, W.-Y.; Chang, M.-Y.; Wang, Y.-Z.; Huang, Y.-C.; Ho, K.-S.; Hsieh, T.-H.; Kuo, Y.-C. Polyaniline Based Pt-Electrocatalyst for a Proton Exchanged Membrane Fuel Cell. *Polymers* **2020**, *12*, 617. [CrossRef] [PubMed]

35. Wang, Y.Z.; Huang, W.Y.; Hsieh, T.H.; Jheng, L.C.; Ho, K.S.; Huang, S.W.; Chao, L. FeNxC Based Catalysts Prepared by the Calcination of Iron-Ethylenediamine@Polyaniline as the Cathode-Catalyst of Proton Exchange Membrane Fuel Cell. *Polymers* **2019**, *11*, 1368. [CrossRef]
36. Tseng, P.-H.; Wang, Y.-Z.; Hsieh, T.-H.; Ho, K.-S.; Huang, P.-C.; Lo, W.-T. Facile way to prepare one dimensional Ag@oligoaniline wires. *J. Taiwan Inst. Chem. Eng.* **2017**, *81*, 445–454. [CrossRef]
37. Hsieh, B.-Z.; Chuang, H.-Y.; Chao, L.; Li, Y.-J.; Huang, Y.-J.; Tseng, P.-H.; Hsieh, T.-H.; Ho, K.-S. Formation mechanism of a nanotubular polyanilines prepared by an emulsion polymerization without organic solvent. *Polymer* **2008**, *49*, 4218–4225. [CrossRef]
38. Sun, S.; Zeng, H. Size-controlled synthesis of magnetite nanoparticles. *J. Am. Chem. Soc.* **2002**, *124*, 8204–8205. [CrossRef]
39. Zhang, L.; He, R.; Gu, H.-C. Oleic acid coating on the monodisperse magnetite nanoparticles. *Appl. Surf. Sci.* **2006**, *253*, 2611–2617. [CrossRef]
40. Shen, S.-Y.; Wu, Y.-J.; Ho, K.-S.; Hsieh, T.-H.; Ho, T.-H.; Wang, Y.-Z.; Tseng, P.-H.; Hsu, Y.-C. Branched and curved nanotubular polyaniline synthesized by emulsion polymerization in presence of zinc salts of n-dodecylbenzenesulfonic acid. *Polymer* **2011**, *52*, 2609–2617. [CrossRef]
41. Grosvenor, A.P.; Kobe, B.A.; Biesinger, M.C.; McIntyre, N.S. Investigation of multiplet splitting of Fe 2p XPS spectra and bonding in iron compounds. *Surf. Interface Anal.* **2004**, *36*, 1564–1574. [CrossRef]

polymers

MDPI

Article

Fe, N-Doped Metal Organic Framework Prepared by the Calcination of Iron Chelated Polyimines as the Cathode-Catalyst of Proton Exchange Membrane Fuel Cells

Yu-Wei Cheng [1], Wen-Yao Huang [2], Ko-Shan Ho [3,*], Tar-Hwa Hsieh [3,*], Li-Cheng Jheng [3] and Yang-Ming Kuo [3]

1 Department of Chemical Engineering, Ming Chi University of Technology, New Taipei City 24301, Taiwan; louischengblue@gmail.com
2 Department of Photonics, National Sun Yat-Sen University, 70 Lienhai Rd., Kaohsiung 80424, Taiwan; wyhuang@faculty.nsysu.edu.tw
3 Department of Chemical and Materials Engineering, National Kaohsiung University of Science and Technology, 415, Chien-Kuo Road, Kaohsiung 80782, Taiwan; lcjheng@nkust.edu.tw (L.-C.J.); kuoyang1228@gmail.com (Y.-M.K.)
* Correspondence: hks@nkust.edu.tw (K.-S.H.); thh@nkust.edu.tw (T.-H.H.)

Citation: Cheng, Y.-W.; Huang, W.-Y.; Ho, K.-S.; Hsieh, T.-H.; Jheng, L.-C.; Kuo, Y.-M. Fe, N-Doped Metal Organic Framework Prepared by the Calcination of Iron Chelated Polyimines as the Cathode-Catalyst of Proton Exchange Membrane Fuel Cells. *Polymers* **2021**, *13*, 3850. https://doi.org/10.3390/polym13213850

Academic Editor: Arunas Ramanavicius

Received: 17 October 2021
Accepted: 4 November 2021
Published: 8 November 2021

Abstract: Aromatic polyimine (PIM) was prepared through condensation polymerization between p-phenylene diamine and terephthalaldehyde via Schiff reactions. PIM can be physically crosslinked with ferrous ions into gel. The gel-composites, calcined at two consecutive stages, with temperatures ranging from 600 to 1000 °C, became Fe- and N-doped carbonaceous organic frameworks (FeNC), which demonstrated both graphene- and carbon nanotube-like morphologies and behaved as an electron-conducting medium. After the two-stage calcination, one at 1000 °C in N_2 and the other at 900 °C in a mixture of N_2 and NH_3, an FeNC composite (FeNC-1000A900) was obtained, which demonstrated a significant O_2 reduction peak in its current–voltage curve in the O_2 atmosphere, and thus, qualified as a catalyst for the oxygen reduction reaction. It also produced a higher reduction current than that of commercial Pt/C in a linear scanning voltage test, and the calculated e-transferred number reached 3.85. The max. power density reached 400 mW·cm^{-2} for the single cell using FeNC-1000A900 as the cathode catalyst, which was superior to other FeNC catalysts that were calcined at lower temperatures. The FeNC demonstrated only 10% loss of the reduction current at 1600 rpm after 1000 redox cycles, as compared to be 25% loss for the commercial Pt/C catalyst in the durability test.

Keywords: FeNC catalyst; polyimine; two-stage calcination; oxygen reduction reaction

1. Introduction

The oxygen reduction reaction (ORR) is usually the bottleneck reaction for fuel cells, implying that catalysts, which are usually precious and expensive, are needed to lower the barrier of the reaction in order to improve the power and productivity of the fuel cells. To carry out ORR in a cheaper way without depressing the catalyzing capability of Fe, N-doped MOF (metal organic framework) composites are prepared, in which covalent-bonded iron nitrogen (Fe–N) can become an active center in the carbonaceous matrix after calcination.

The first MOF-based cathode catalyst was prepared using cobalt-coordinated with large cycled phthalocyanine [1], which was, over the following year, modified via high-temperature calcination to become Co-porphyrin. This did not increase the efficiency of the catalyst; however, the ORR in the cathode improved significantly [2]. Eventually, it was understood that calcination at a higher temperature than 800 °C is required to obtain an MOF-based cathode catalyst. Some iron- and nitrogen -containing carbonaceous materials [3] were calcined in the presence of N_2 or NH_3 to create the micro- or mesoporous areas of FeNC with defined numbers of active sites. FeNC was made available after calcination at temperatures that were higher than 950 in the argon and ammonia atmosphere in a study conducted by Mamtani [4]. Several closely related works for ORR and MOF can be found. Refs. [5–8] In a conventional iron- and nitrogen-doped MOF (FeNMOF), Fe ions are designed to be captured (complexed) with multi-nitrogen (usually 4-nitrogen) in a cyclic compound that contains 4-nitrogens [9–13] before being subjected to calcination. However, the preparation of FeNMOFs requires many tedious series of steps of organic reactions with limited yield. In other words, the expense of obtaining FeNMOF is close to that needed to purchase precious metals such as Pt or Pd. In this study, we attempted to avoid the tedious steps of organic synthesis and the associated expenses by directly polymerizing a nitrogen-containing polymer (polyaniline: PANI) in the presence of either Fe or Co ions to allow ions to complex with the nitrogen-containing monomers before the initiation of polymerization. New publications [14,15] on the application of PANI in the design of biosensors and biofuel cells are available. The polymerization of PANI on the carbonaceous surfaces demonstrated significant ORR in both acidic and alkaline media [16–18]. Additionally, the amino groups of the N-containing PANI on an XC72 (Vulcan) support could coordinate with Pt ions, resulting in the presence of a well-dispersed Pt-catalyst on the XC72 surfaces and the control of the pore sizes of the obtained PANI/XC72 composites [19,20]. However, PANI can easily be prepared by polymerizing aniline monomers (NH_4^+) in the acidic aqueous solution [21–24]. The repulsive force between NH_4^+ and Fe^{+2} or Co^{+2} hinders the coordination (complex) and decreases the amounts of metal ion captured by the resultant PANIs. Furthermore, due to the steric effect of the two huge benzene rings located on both sides of the amino group (-NH) in the backbone of PANI, most of the metal ions are not able to come close enough to induce the complexation with -NH, even when the PANI was synthesized following polymerization. Therefore, the degree of coordination with the metal ions by PANIs was too low to become an efficient cathode catalyst of fuel cell after calcination. Consequently, additional small N-containing molecules such as ethylene diamine (EDA) were added to capture most of the metal ions firmly in water before the addition of anilinium monomers to increase the degree of coordination before and after polymerization [25].

In this study, aromatic polyimine (PIM) replaced PANI to effectively remove the steric hindrance and allow the approach of metal ions to form robust coordination bonding before calcination, without the necessity of adding any small N-containing molecules before polymerization and calcination. Since PIM can usually be prepared quickly with a high yield, at temperatures slightly higher than room temperature, by means of Schiff condensation between the diamine and the dialdehyde, we used p-phenylene diamine (PDA) and terephthaldehyde (TPAl) as the monomers to obtain aromatic PIM. Compared to PANI, it was found that there was a large empty space around the imine groups (-N=CH-) without the presence of a huge benzene ring at one end, and that no hydrogen was connected to the nitrogen of the imine. Furthermore, no positive charge was present on PDA and TPAl monomers to repulse the positive metal ions before or after polymerization. Theoretically, the addition of metal ions could be carried out before polymerization with monomers or after polymerization with polymers, both of which could create a physically crosslinked gel if the degree of complexation is high enough. In other words, we were able to judge the degree of complexation according to whether or not gel was formed prior to being subjected to calcination.

2. Materials and Methods

2.1. Materials

p-Phenylene diamine (PDA) (Tokyo Kasei Kogyo Co., Ltd., Tokyo, Japan), terephthaldehyde (TPAl) (Tokyo Kasei Kogyo Co., Ltd., Tokyo, Japan), and iron(II) chloride hexahydrate ($FeCl_2 \cdot 6H_2O$, J.T. Baker, NJ, USA) were used in this study.

2.2. Preparation of FeNC Catalyst

Quantities of 1.34 g of PDA and 1.62 g of TPAl were placed in 80 and 50 mL of alcohol, respectively, before being mixed into a single solution. The mixture solution was stirred at room temperature for 12 h, during which the color changed to thick orange, indicating that the polymerization is complete. Then, 0.04 g of Iron(II) chloride hexahydrate was introduced into the solution and the viscosity of the mixture gradually increased before turning into a frozen gel. The gel-like composite was concentrated by centrifugation at 300 rpm for 10 min to obtain the precipitate in the bottom of the centrifugation tube. The precipitate was dried at 80 °C for 8 h before cooling to RT.

The obtained PIM, which was the precursor of the FeNC catalyst, was heated to 600 °C (700, 800, 900, 1000 °C) at 10 °C min^{-1} and maintained at 600 °C (700, 800, 900, 1000 °C) for 1 h in the argon atmosphere, then cooled to room temperature. The impurities and magnetic parts of the obtained materials were removed via washing in 9 M H_2SO_4 (aq.) at 80 °C for 36 h, followed by filtration, and the cake was washed with de-ionized water and alcohol before drying in a vacuum oven at 80 °C for 8 h. The acid-leached products were further calcined at 500 °C (600, 700, 800, 900 °C) in N_2 and NH_3 atmospheres, at 10 °C min^{-1} (named as FeNC-600A500), and washed again in 1 M H_2SO_4 (aq.) at 80 °C for 3 h, followed by drying in a vacuum oven at 60 °C. The sample was named FeNC-600A500 (-700A600, -800A700, -900A800, and -1000A900). The schematic diagram depicting the preparation of the FeNCs is shown in Scheme 1.

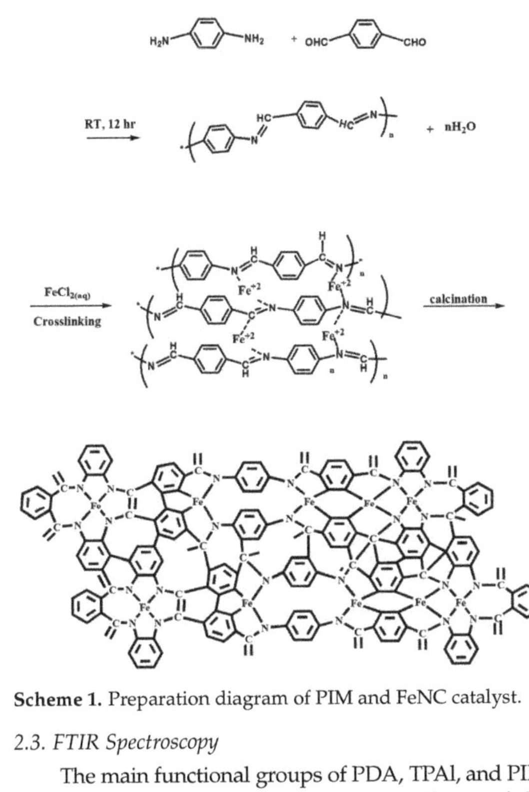

Scheme 1. Preparation diagram of PIM and FeNC catalyst.

2.3. FTIR Spectroscopy

The main functional groups of PDA, TPAl, and PIM were assigned in accordance with the FTIR spectra that were recorded on an IFS3000 v/s FTIR spectrometer (Bruker, Ettlingen, Germany) at room temperature with a resolution of 4 cm^{-1} and 16 scanning steps.

2.4. X-ray Photoelectron Spectroscopy (XPS)

The different binding energy spectra of N1s of various FeNCs were used to characterize the percentage of nitrogen in pyridine, pyrrole, graphenec, Fe-N., etc. after calcination with an XPS instrument produced by Fison (VG)-Escalab 210 (Fison, Glasgow, UK) using Al Ka X-ray source at 1486.6 eV. The pressure in the chamber was kept at 10^{-6} Pa or less during the measurement. The powered samples were shaped to become tablet samples using a stapler. The binding energies of the N1s around 400 eV were recorded.

2.5. Wide Angle X-ray Diffraction: Powder X-ray Diffraction (WXRD)

A copper target (Cu-Kα) Rigaku x-ray source (Rigaku, Tokyo, Japan), with a wavelength of 1.5402 Å, was the target for x-ray diffraction. The scanning angle (2θ) ranged from 10 to 90°, with a voltage of 40 kV and a current of 30 mA, and was operated at 1° min^{-1}.

2.6. Scanning Electronic Microscopy (SEM)

Using a SEM (field emission gun scanning electron microscope, AURIGAFE, Zeiss, Oberkochen, Germany), the sizes and morphologies of the FeNCs were obtained.

2.7. Transmission Electronic Microscopy (TEM)

Photos of the samples were taken using an HR-AEM field-emission transmission electron microscope (HITACHI FE-2000, Hitachi, Tokyo, Japan); the samples were dispersed in acetone, and were subsequently placed dropwise on carbonic-coated copper grids before being subjected to emissions.

2.8. Surface Area and Pore Size Measurement (BET Method)

Nitrogen adsorption–desorption isotherms (type IV) were obtained from an Autosorb IQ gas sorption analyzer (Micromeritics-ASAP2020, Norcross, GA, USA) at 25 °C. The samples were dried in a vacuum overnight at a temperature above 100 °C. The surface area was calculated according to the BET equation when a linear BET plot with a positive C value was in the relative pressure range. The pore size distribution was determined according to methods derived from the Quenched Solid Density Functional Theory (QSDFT), based on a model of slit/cylinder pores. The total pore volumes were determined at P/P_0 = 0.95.

2.9. Electrochemical Characterization

2.9.1. Current–Potential Polarization-Linear Scan Voltammetry (LSV)

The performance of the electrocatalyst support was implemented in a three-electrode system. The round working electrode, which had an area of 1.5 cm^2, was prepared as follows: Ag/AgCl, carbon graphite, and a Pt-strip were used as the reference, relative, and counter electrode, respectively. The electrochemical test was carried out in a potentiostat/galvanostat (Autolab-PGSTAT 30 Eco Chemie, KM Utrecht, The Netherlands) in 0.1 M $HClO_4$ solution, and C-V curves were obtained with scanning potentials from −0.2 to 1.0 V at a scanning rate of 50 mV·s^{-1}. The catalyst ink was prepared by mixing 3 mg support powder in isopropanol and stirring until it became uniform. Subsequently, 5% Nafion solution was added into the mixture as a binder, the mixture was ultra-sonicated for 1 h, and the obtained ink was uniformly spray-coated on the carbon paper for C-V testing.

The current-potential polarization curves obtained from LSV of the various FeNCs were measured using a rotating-disk electrode (RDE: Metrohm, FL, USA) operating at 900, 1200, 1600, 2500, and 3600 rpm in O_2-saturated 0.1 M $HClO_4$, respectively. The reduction current densities of various FeNCs, which were recorded at 1600 rpm within the measured voltage range (0.0~1.2 V), were chosen for comparison.

2.9.2. MEA Preparation

A Nafion® 212 sheet, purchased from Ion Power Inc., New Castle, DE, USA, was used as the proton exchange membrane. To remove the surface organic impurities and to convert the membranes into protonated (H^+) form, the Nafion-212 (4 × 4 cm), membrane was treated at 70 °C in 5 wt.% H_2O_2 aqueous solution for 1 h, and was then submerged in 1 M H_2SO_4 solution for 1 h. Subsequently, the treated membranes were dipped in distilled water for 15 min and were then stored in deionized water. The catalyst inks were prepared by mixing 20 mg of FeNC powders in isopropanol and were mechanically stirred until they became uniform, followed by the addition of 5% Nafion solution, before stirring again to reach uniformity. Eventually, the catalyst mixture was ultra-sonicated for 1h, followed by dropwise coating on both sides of the treated Nafion sheet, as the anode and cathode electrodes (2 × 2 cm), respectively, and hot-pressing at 140 °C with a pressure force of 70 kg cm^{-2} for 5 min to obtain the MEA.

2.9.3. Single-Cell Performance Testing

The MEA was installed in a fuel cell test station to measure the current and power densities of the assembled single cell using a single-cell testing device (model FCED-P50; Asia Pacific Fuel Cell Technologies, Ltd., Miaoli, Taiwan). The active cell area was 2 × 2 cm^2. The temperatures of the anode, cell, cathode and humidifying gas were maintained at around 70 °C. The flow rates of the anode input H_2 and the cathode input O_2 fuels were set at 200 and 100 mL·min^{-1}, respectively, based on stoichiometry. To test the electrochemical performance of FeNC cathode catalyst in the individual MEAs, both the C-V and output powers were measured.

3. Results

3.1. FTIR Spectra

The IR-spectra of the PDA, TPAl monomers, and PIM obtained from the Schiff condensation polymerization are demonstrated in Figure 1. The doublet peaks of the symmetric and asymmetric stretching modes of the primary amine, which belonged to PDA, can be clearly seen at around 3297 and 3201 cm^{-1}, respectively. The –C–N– bond is also visible at 1520 cm^{-1} and para-substituted benzene ring contributed to the peak at 835 cm^{-1}, which overlapped with the para-substituted ones of TPAl and PIM, indicating that the Schiff reaction was carried out at the para-positions for PDA and TPAl. The carbonyl group of the aldehyde of TPAl contributed the peak at 1700 cm^{-1}. The vanishing of the peaks of the carbonyl and primary amine in PIM revealed that the condensation reaction successfully occurred and that water was the by-product. The imine groups of the products of the Schiff reaction caused the sharp peak at 1620 cm^{-1}. The related polymerization reaction via Schiff condensation is included in the upper part of Scheme 1.

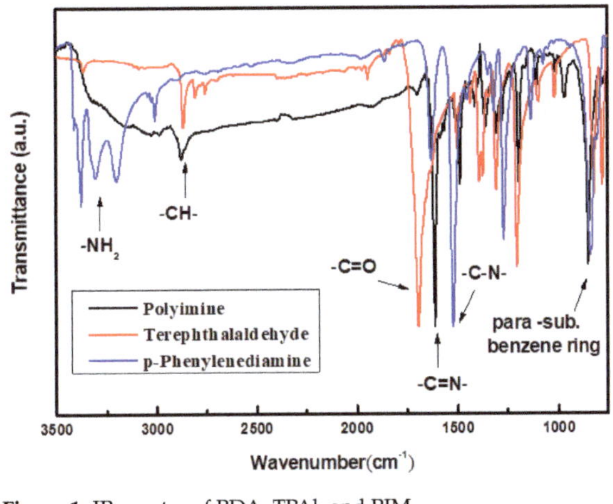

Figure 1. IR spectra of PDA, TPAl, and PIM.

After heating, the alcohol solution containing a mixture of PIM demonstrated clear swirls during stirring with a magnetic stirrer, as shown in Figure 2a. However, the liquid-like solution gradually started to freeze with the addition of $FeCl_2$ and eventually became a gel, as seen in Figure 2b. We concluded that the gel resulted from the formation of physically crosslinked PIM with Fe^{+2} ions, which could easily coordinate with the imine groups belonging to different PIM molecules to build up the crosslinking network of the gel, as depicted in Scheme 1. The gel was eventually calcined in the argon atmosphere to prepare the FeNC (Fe, N-doped MOF), as described in Scheme 1.

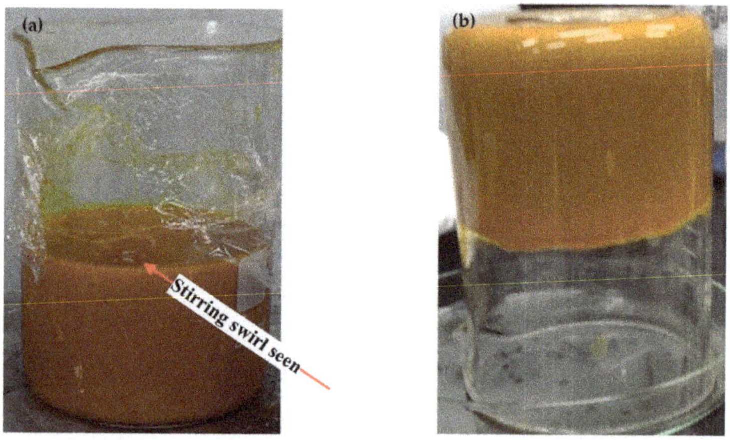

No $FeCl_{2(aq)}$ added in PIM solution stirred with a magnetic stirrer (dark yellow)

Mixed with $FeCl_{2(aq)}$ – physically crosslinking gel (light yellow) (beaker inverted)

Figure 2. Pictures of PIM alcohol solution (**a**) without $FeCl_{2(aq)}$ and (**b**) with $FeCl_{2(aq)}$.

3.2. XPS

The active sites of FeNC were able to absorb O_2 gas and form a peroxide that would dissociate in the presence of protons during reduction (Scheme 2). The formed O_2-captured Fe-N catalysts were reduced following two approaches [26,27], with one involving the Fe-N catalysts becoming diol and the other involving direct conversion into water following the 4-e route. The captured O_2 could proceed with another possible ORR with the involvement of two electrons, and H_2O_2, not H_2O, being the final product. The possible formation mechanism of H_2O_2, which is illustrated in Scheme 2, reveals that only two electrons were involved. Two possible mechanisms of the formation of H_2O_2, depending on the reduction reaction occurring before or after the proton doping, are also described in Scheme 2. The produced H_2O_2 could be further reduced to become H_2O, and an additional two electrons would have become involved if the reduction reaction continued. The catalytic mechanism followed the traditional six-coordinate catalytic reaction for Fe^{+2}.

Theoretically, the O_2 gas with two lone pairs could be attracted to the active sites of FeNC through the coordination, or could be trapped in the porous holes with various nitrogen-related bonds, in which case the increased polarity of the C-N bonding could improve the O_2 absorbing capability and cause the C-N bonds to behave as active sites, similarly to transitional metals (Fe). The formation of active nitrogen-containing compounds (–N), such as pyrrolic –N, graphitic –N, and pyridinic –N [28–33], is described in Scheme 3. At higher temperatures in the N_2 atmosphere, the first stage of calcination could create various –N-containing covalent bonds as active sites. In the mixed gases of N_2 and NH_3 at lower temperatures, the second calcination could create lots of micro- or mesopores on the FeNC surfaces, resulting in increased surface area and allowing more active –N and –Fe sites for the incoming O_2 gas.

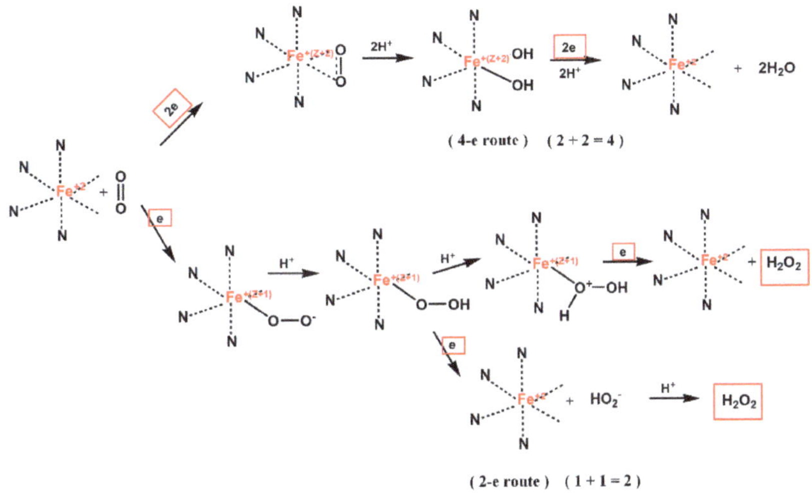

Scheme 2. Possible electrocatalytic mechanism by FeNC for ORR.

PIM could crosslink with each other into ladder-like polymers in the initial stage of thermal heating and higher temperature pyrolysis allowed the carbonization between the ladder-like polymers, which could create FeNMOF of graphitic –N, pyridinic –N, and pyrrolic –N (Scheme 3). Most of the pyridinic and pyrrolic –Ns were created on the edges of the calcinated PIM, while graphitic –Ns were mostly formed inside the network. The nitrogen-doped graphene (N-GF)-like structure of the calcined PIM also behaved as a conducting medium, transporting electrons from the anode. This made it possible to avoid the trouble of adding XC-72 during the preparation of the cathode ink. Depending on the sp3 or sp2 bonding of –N- in the aromatic matrix, there were two types of laddered PIMs formed, as illustrated in Scheme 3. For the ladder constructed mainly by sp3 –Ns, the strip of the crosslinked PIM wrinkled slightly, and a more planar strip of the crosslinked PIM formed for sp2 –Ns, as depicted in Scheme 3. The development of the laddered PIM with the increasing of the temperature was able to create an N-GF structure, which is depicted in the bottom of Scheme 3. These GF strips, which were either planar or wrinkled, could self-assemble into thicker slabs, as will be discussed in the SEM section.

The atomic concentration of FeNCs (Fe, N, C, and O) listed in Table 1 clearly demonstrates the increasing of the nitrogen and oxygen atom concentration upon higher temperature calcinations, as measured by XPS. This indicates that more nitrogen could dope into the carbonaceous matrix at higher temperatures, regardless of whether they came from the PIM or the influx of NH_3 gas, which also caused damage on the catalyst surface and led to an increased surface area, as will be discussed in the BET section. The N_{1s} XPS spectra of FeNCs calcined after acid leaching are presented in Figure S1 and the compositions of each type of nitrogen-doped (-N) group are shown in Figure 3 and Table 2. The covalent-bonded iron and nitrogen (Fe-N) were not found until calcination was higher than 700 °C, and graphitic and pyrrolic –Ns were predominant at temperatures below 600 °C in the second stage of calcination, according to Figure 3, Figure S1, and Table 1. Briefly, more pyridinic –N and active centers of Fe-N (bottom of Scheme 1) bonding were created at the second stage of calcination in the presence of mixed NH_3 and N_2 gases. Table 2 also illustrates two major –Ns (pyridinic-N and Fe-N) when the calcination was performed according to the 1000A900 procedure. The increasing temperature created more active sites, which led to a higher LSV current of the cathode and a higher power density of the single cell, which will be discussed in the electrochemical sections.

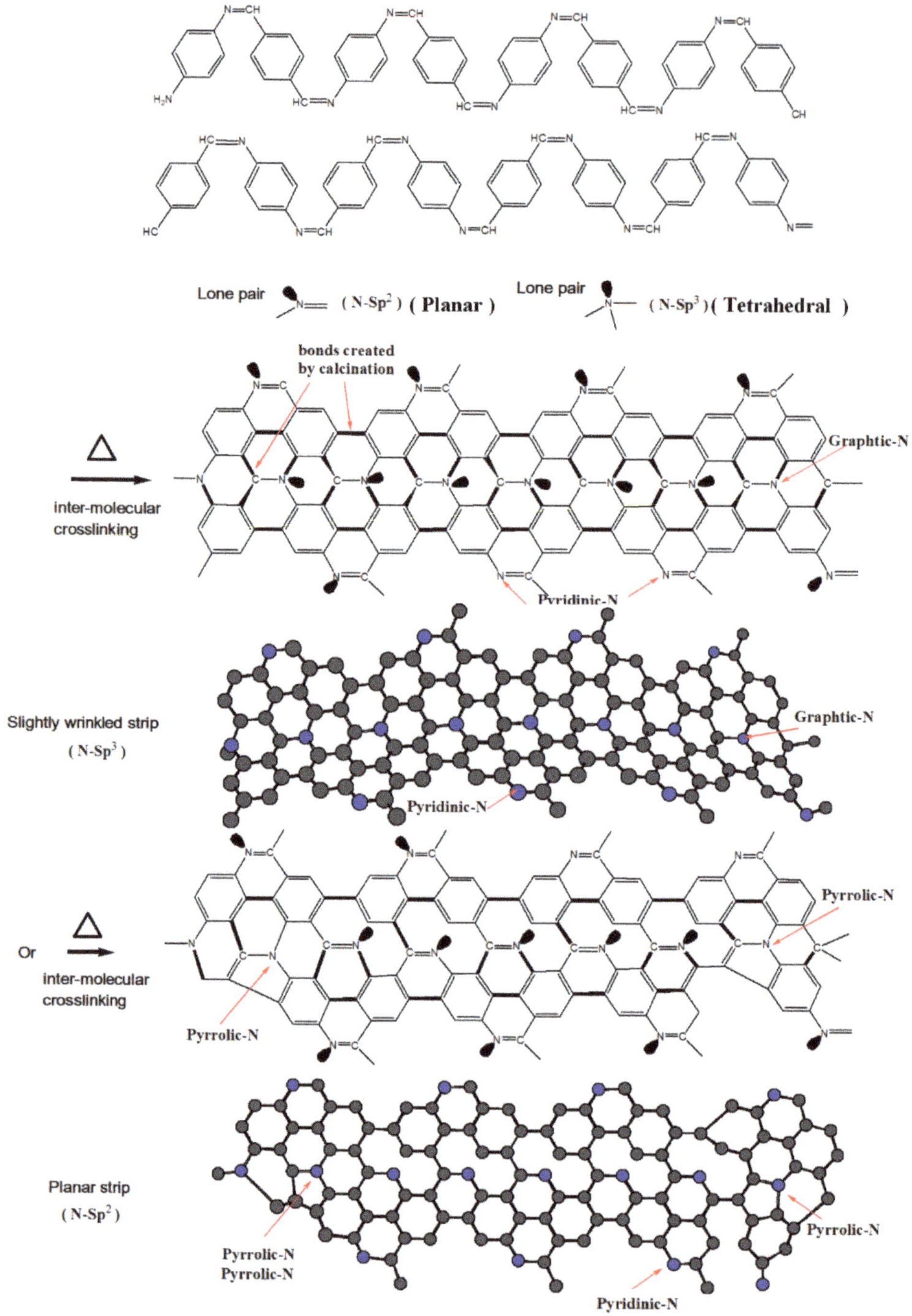

Scheme 3. Schematic diagram of the formation of nitrogen-containing active centers in the calcined PIM and N-GF networks.

Table 1. Atomic concentration obtained from XPS of FeNCs prepared with different calcination methods.

Catalysts	Atomic Concentration (at%)			
	C	O	N	Fe
FeNC-600A500	93.8	4.8	1.4	<0.1
FeNC-700A600	92.9	4.9	2.3	<0.1
FeNC-800A700	91.5	4.7	3.9	<0.1
FeNC-900A800	87.1	8.3	4.6	<0.1
FeNC-1000A900	84.2	10.2	5.6	<0.1

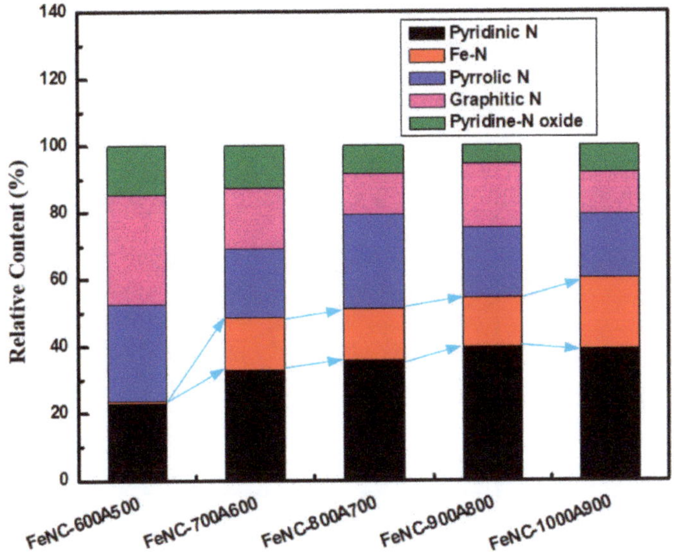

Figure 3. Relative compositions of each type of nitrogen-containing compound found in various FeNC matrices.

Table 2. Compositions of various –Ns of FeNCs determined by XPS

FeNC Catalysts	N 1s (at%)				
	Pyridinic N	Fe-N	Pyrrolic N	Graphitic N	Pyridine-N Oxide
FeNC-600A500	23.1	0.6	28.7	32.7	14.9
FeNC-700A600	33	15.7	20.3	18.3	12.7
FeNC-800A700	35.8	15.4	28.3	12.1	8.4
FeNC-900A800	39.6	14.8	21.1	19	5.5
FeNC-1000A900	39.1	21.1	19	12.8	8

3.3. XRD

The x-ray diffraction patterns, produced through the formation of GF after calcination after 700 °C during the first stage, in which a diffraction peak at 2θ = 26.5° gradually grew with the temperature, are seen in Figure 4. No significant peak is seen at 2θ = 26.5° for neat PIM in Figure 4 except for the characteristic diffraction peaks ((111), (110), (200), and

(210)) for pure, aromatic PIM before calcination. The PIM-related crystals were destroyed after 600 °C and only an amorphous pattern remained, demonstrating that the crosslinked PIM (ladder like) did not yet develop into GF or carbon nanotube (CNT) crystals. The Fe was covalently bonded in the amorphous carbon networks at this stage (600 °C), and both the carbonaceous and Fe domains started to create ordered domains after 700 °C, undergoing GF (or CNT)- and Fe-related crystallization (Fe_4N(111), Fe_3C(031), α-Fe(110)), respectively. For calcination temperatures over 700 °C, the solid crystallization resulted in the formation of the C(002) plane and more GF (or CNT) crystals started to build up. The characteristic diffraction peak (C(002)) of GF (or CNT) eventually became very sharp at 1000 °C, indicating that the ordered, conducting carbon matrix was entirely formed. Furthermore, the presence of Fe_3C and α-Fe seeds was able to induce the formation of CNT in the GF-dominating matrix with the increasing of the temperature [34], which will be discussed in the TEM section.

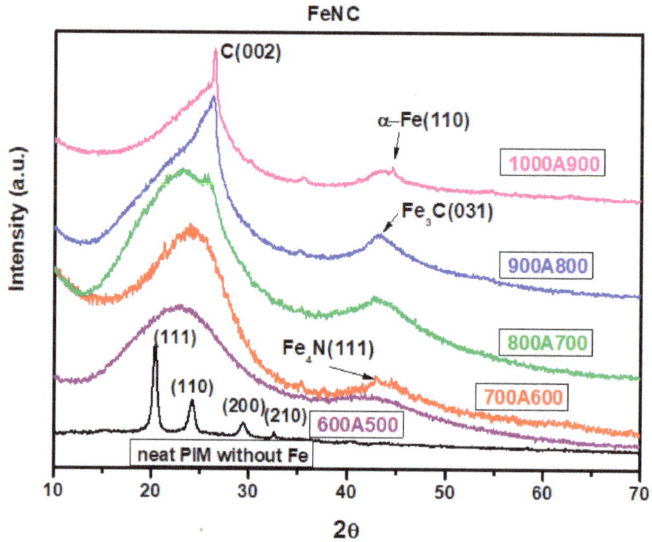

Figure 4. X-ray diffraction patterns of neat PIM and FeNCs prepared with different calcination methods.

3.4. Raman Spectroscopy

Although, as seen in Figure 4, the C(002) plane ($2\theta = 26.5°$), which was related to the formation of GF or CNT, became more and more significant with the increasing of the calcination temperature, the intensity of I_G (sp^2) decreased with the temperature, resulting in the increasing of the I_D/I_G ratio in the Raman spectra, as demonstrated in Figure 5. Carbons with sp^2 bonding outnumbered those with sp^3 bonding (smaller I_D/I_G ratio) for FeNC-600A500, indicating a more ordered form in their domain, as shown in Figure 5. However, these ordered domains did not contribute to the crystallization, and their x-ray diffraction spectra did not demonstrate significant crystallization peaks, as shown in Figure 4. With the increasing of the calcination temperature for FeNC-700A600, -800A700, the structures of the FeNCs were gradually destroyed by the active, large NH_3 molecules, which contributed to the increase in I_D/I_G when more sp^2 bonds were converted to sp^3 ones after the bombardment of NH_3 molecules, in accordance with the results shown in Figure 5. It seems that the damage on the structures of FeNCs did not occur on the crystalline region, which developed into GF or CNT at higher calcination temperatures according to the x-ray pattern shown in Figure 4. The robust crystalline structure of the GF (or CNT) formed at high calcination temperatures was able to withstand the attacking of NH_3 molecules, and to continuously grow into more ordered crystals, as a result of the higher energy provided at higher temperatures. In other words, at high temperatures, the active NH_3 molecules could only create more surface area for the FeNCs by destroying the amorphous part on the surface (see BET section); the conversion of sp^2 bonds to sp^3 but not cause any damage in the crystalline region, which could possibly have been located inside of the matrix.

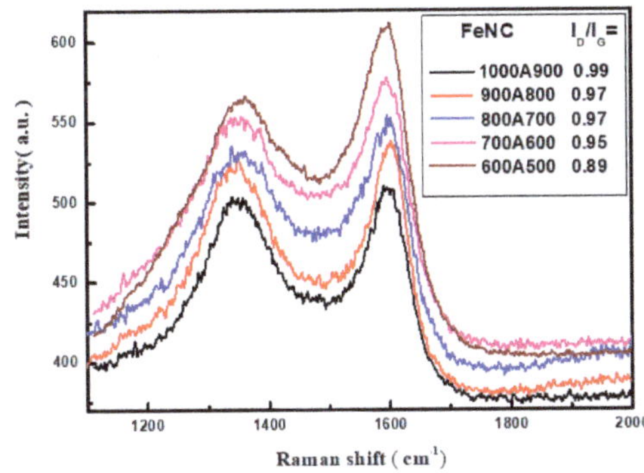

Figure 5. Raman spectra and I_D/I_G ratios of FeNCs prepared with different calcination methods.

3.5. SEM

Only particles with disordered surfaces and short rods are perceivable in the SEM micrographs of FeNC-600A500 and -700A600 demonstrated in Figure 6a,b. The short rods might have originated from the accumulation of a strip of crosslinked PIM, as described in Scheme 3. No significant flake-like self-assembled slabs of associated N-GF or CNT were found. With the increasing of the calcination temperature, these crosslinked strips were able to develop into N-GF planes that could have been associated with the thick slabs due to either the polarity provided by iron and nitrogen doping or the formation of covalent bonds between the planes (Figure 6c–e). The formation of Fe, N-doped GF slabs contributed to the 3D GF structure shown in Figure 6e and Figure S2.

Due to the attacking of the NH_3 molecules, more micro- and mesopores developed on the surfaces of FeNCs after calcination at temperatures above 800 °C.

Most of the Fe-related articles were actually on the surface of the GF slabs, as seen in the enlarged image in Figure 6f, where standing GF slabs are also perceivable and huge pores are present. These pores could accommodate more input O_2 molecules that were able to make contact with the active centers of Fe-N or various –N-doped carbon regions, catalyzing the ORR at the cathode. Furthermore, the highly conducting GF slabs that behaved as conducting carbon black (CB) in the Pt/C catalyst were capable of introducing more electrons that were transferred from the anode.

3.6. TEM

The TEM micrograph (Figure 7a) of the neat PIM calcined at 1000 °C demonstrates a thick layer morphology with no significant pores or broken sites found in the N-doped carbonaceous matrix. The introduction of iron doping could significantly break the thick layers and generate some short rod-like morphologies at calcination temperatures as low as 700 °C (Figure 7b). The iron doping created larger pores, and the iron atoms acted as the seeds of the formation of CNT from the carbonaceous matrix when the calcination temperature was over 800 °C, as shown in Figure 7c–e. A large number of generated winding carbon nanowires and tiny iron nanoparticles are visible in Figure 7c–e. The iron seed seen in Figure 7f was covered with carbon matrix, demonstrating the presence of the C(002) plane of either GF or CNT. Furthermore, the iron seed is characterized as α-Fe by its (110) plane according to its HR-TEM micrograph in Figure 7f. The full covering of α-Fe by the carbonaceous materials provides further evidence that these covering carbon domains (mainly C(002) planes) were actually growing from the α-Fe seed during calcination (>800 °C).

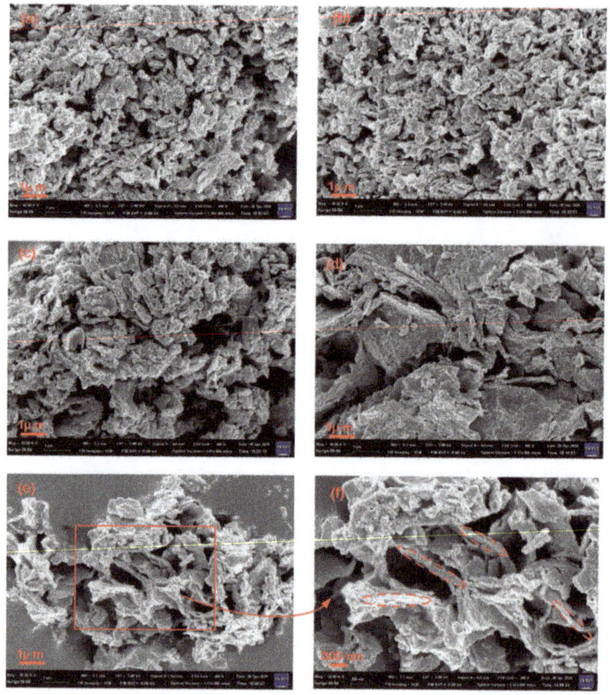

Figure 6. SEM micrographs of FeNCs: (**a**) 600A500, (**b**) 700600, (**c**) 800A700, (**d**) 900A800, (**e**) 1000A900, (**f**) enlarged image of 1000A900.

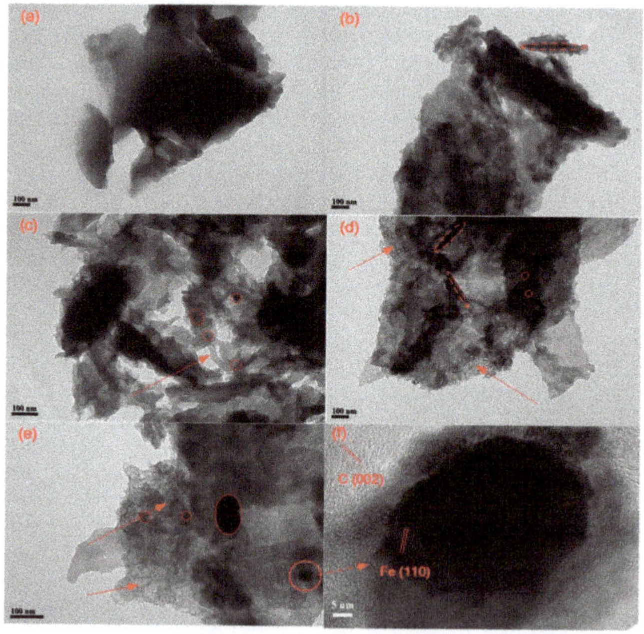

Figure 7. TEM micrographs of (**a**) neat PIM-1000, (**b**) FeNC-700A600, (**c**) FeNC-800A700, (**d**) FeNC-900A800, (**e**) FeNC-1000A900 and (**f**) HR-TEM of the tiny α-Fe particle in FeNC-1000A900.

3.7. BET Surface Area

The type-IV isotherm, which was related to the characteristic N_2 absorption and desorption curves of the mesopores, can be clearly seen in Figure 8. FeNC-1000A900 (Figure 8a) had a much higher specific volume than the other FeNCs at all relative pressures. Furthermore, in accordance with Figure 8a and Table 3, the surface area (specific volume) became higher and higher with the increasing of the calcination temperature after two-stage calcination in the NH_3 atmosphere. The collapsing effect caused by NH_3 at high calcination temperatures could have resulted in increased surface area and the exposure of more Fe-N active sites to O_2 gas in the cathode. The specific area was increased from 329.0 to 546.6 $m^2 \cdot g^{-1}$ when the FeNC was exposed to increasing calcination temperatures, as shown in Table 3 and Figure 8a.

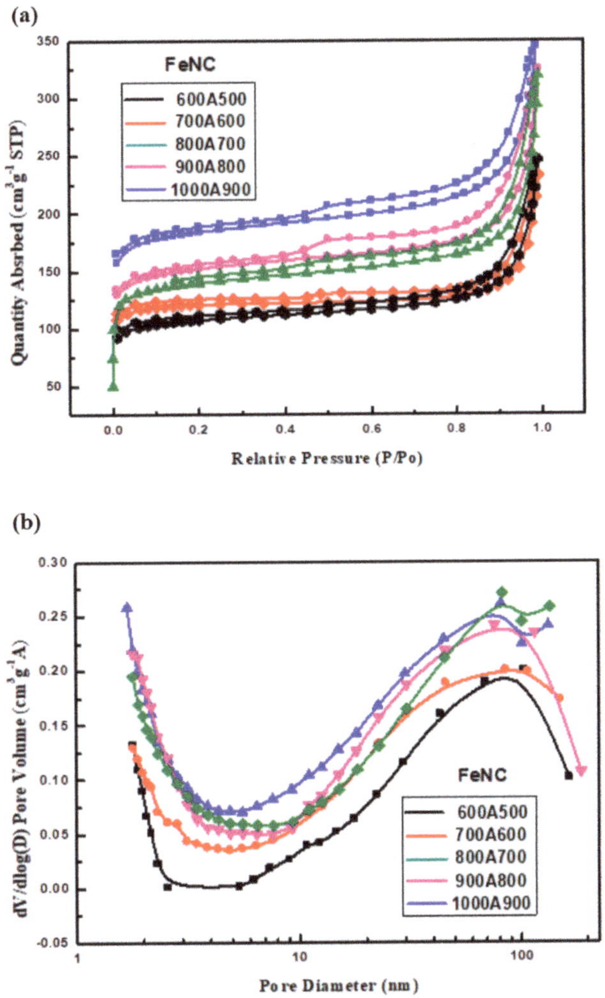

Figure 8. (**a**) N_2 adsorption–desorption isotherms and (**b**) pore size distribution of FeNCs prepared with different calcination methods.

Table 3. Specific area and average pore sizes of FeNCs.

Catalysts	BET Surface Area ($m^2 \cdot g^{-1}$)	Ave. Pore Size (nm)
FeNC-1000A900	564.6	4.95
FeNC-900A800	468.5	4.63
FeNC-800A700	406.6	4.12
FeNC-700A600	364.1	4.29
FeNC-600A500	329.0	3.77

The pore size distribution, measured via the Barrett–Joyner–Halenda (BJH) method, indicates the presence of both micro- and mesopores, as shown in Figure 8b. The increasing of the surface area with the temperature could have originated from the collapsing power of NH_3, which not only caused damage on the surfaces but also created more micro- and mesopores. The wide distribution of pore sizes indicates that the FeNCs were able to improve the ORR, since the micropores were able to unveil the active sites and confine the O_2 inside FeNC catalysts, significantly decreasing the diffusion path. [35]. The average pore sizes created on the FeNC surfaces ranged between 3.7 and 5.0 nm, as listed in Table 3; this allowed more O_2 molecules to stay inside.

3.8. CV and LSV Curve

The electrocatalytic activity of FeNCs, prepared at various temperatures, after acid-leaching was evaluated using the CV and LSV curves of the FeNC catalysts, as shown in Figures 9 and 10.

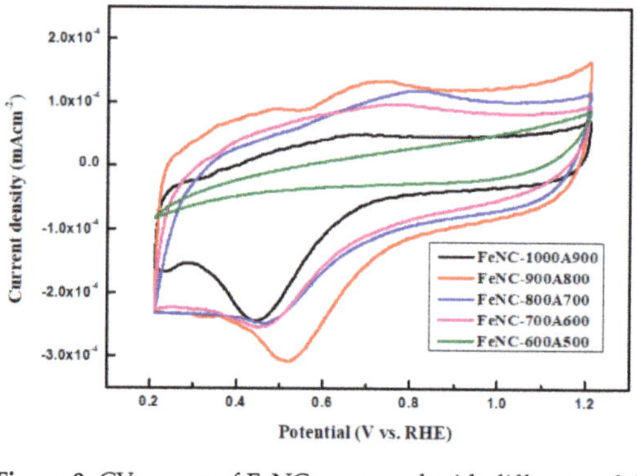

Figure 9. CV curves of FeNCs prepared with different calcination methods.

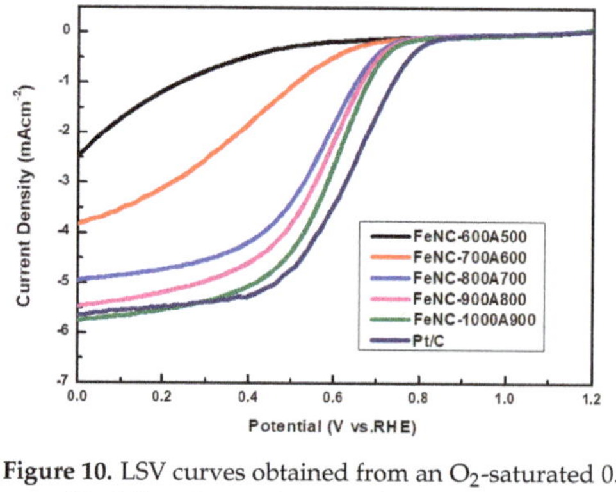

Figure 10. LSV curves obtained from an O_2-saturated 0.1 M $HClO_4$ aqueous solution at a scanning rate of 5 mV/s and a rotation rate of 1600 rpm for FeNCs prepared with different calcination methods.

Except FeNC-600A500, the CV curves of all catalysts demonstrated significant reduction peaks in the O_2 atmosphere at around 0.4~0.6 V, revealing their abilities, as the cathode catalysts of PEMFC, to cause ORR, as shown in Figure 9.

The LSV curves of all FeNC catalysts could be obtained in an O_2-saturated 0.1 M $HClO_4$ aqueous solution at a scanning rate of 5 mV/s and a rotation rate of 1600 rpm, as illustrated in Figure 10. The reduction current density at 0 voltage ranged from 2.5 to 5.8 $mAcm^{-2}$ when the FeNC was calcined from 500 to 900 after acid-leaching, as can be seen in Figure 10. In particular, the obtained reduction current density (5.8 $mAcm^{-2}$) of FeNC-100A900 was even higher than that of the commercial Pt/C catalyst (5.7 $mAcm^{-2}$), as illustrated in Figure 10. The high reduction current density could be attributed to the presence of more active sites and highly conducting GF (or CNT), as well as the high surface area, which were already discussed in the previous sections.

LSV curves for each FeNC could also be obtained from RDE at different rotating speeds. The potential was selected at the region where the current underwent mixed control by means of both kinetic and mass transfer (diffusion control) and the Koutecký–Levich (K–L) plot was linear, in accordance with Equation (1)

$$1/I = 1/I_K + 1/I_D \quad (1)$$

where:

I_K—the current contributed by kinetic control

I_D—the current contributed by diffusion control, which can be expressed in the form of Equation (2):

$$I_D = 0.62 \times AnFD^{2/3}\nu^{-1/6}C\sqrt{\omega} \quad (2)$$

A—the geometric area of the disk (cm^2);

F—Faraday's constant (C mol^{-1});

D—the diffusion coefficient of O_2 in the electrolyte ($cm^2\ s^{-1}$);

ν—the kinematic viscosity of the electrolyte ($cm^2\ s^{-1}$);

C—the concentration of O2 in the electrolyte (mol cm^{-3});

ω—the angular frequency of rotation (rad s^{-1});

n—the number of electrons involved in the reduction reaction.

The LSV curves of every type of FeNC are illustrated in Figure S3 and can be used to calculate I_D. After plotting I^{-1} vs. $\omega^{-1/2}$, the K–L lines of FeNC-1000A900 were established, and they are shown in Figure S4a. The slopes of these lines could be used to calculate the numbers of electrons involved in the reduction reaction (n). The electrons transferred for ORR differed from applied voltages and the average value was around 3.85 according to Figure S4b. If the rotating speed became faster than 2500 rpm, both FeNC-1000A900 and -900A800 demonstrated higher reduction currents than that of the commercial Pt/C catalyst, as seen in Figure S3, indicating that it is possible to prepare FeNC at lower temperatures to meet the requirement of gaining a comparable reduction in current density to that of Pt/C. Actually, the ORR phenomenon was already present for FeNC-700A600, as indicated by the CV curve shown in Figure 9. The low reduction current density for FeNCs prepared below 800 °C could be attributed to the morphologies that were significantly related to the performance of the catalysts in the ORR.

The numbers of e-transferred for each catalyst, at different potentials, were calculated, and are listed in Table 4, where it can be seen that the average e-transferred numbers increased significantly with the increasing of the calcination temperature. The average numbers ranged from 3.30 to 3.85, and less than one electron followed the 2-e route (Scheme 2); in other words, the ORR carried out via the 4-e route (Scheme 2) was between 65.0 and 92.5% according to Table 4.

Table 4. Numbers of transferred electrons of various FeNCs.

Potential \ FeNC	1000A900	900A800	800A700	700A600	600A500
0.4018	3.81	3.51	3.70	3.52	3.15
0.35053	3.82	3.93	3.72	3.58	3.24
0.30171	3.84	3.82	3.57	3.64	3.31
0.25044	3.87	3.65	3.54	3.66	3.37
0.20161	3.90	3.61	3.56	3.66	3.42
Ave.	**3.85**	**3.70**	**3.61**	**3.61**	**3.30**
% of 4-e route	**92.5**	**85.0**	**80.5**	**80.5**	**65.0**

3.9. Single Cell Testing

The limited max. power densities (Pmax less than 50 mWcm^{-2}) or current densities produced for the single cell with FeNC-700A600 as the cathode catalyst are shown in Figure 11. The Pmax-values grew with the increasing of the calcination temperature (700–1000 °C) from 40 to 400 mWcm^{-2}, and increased by 10 times due to the creation of more active centers and the increasing of the surface area. Even when calcined at a lower temperature of 900 °C, the single cell prepared with FeNC-900A800 as the cathode catalyst demonstrated a Pmax value equal to 310 mWcm^{-2}. The current density curve even extended to 1000 and 1300 mAcm^{-2} at 0.3 V for FeNC-900A800 and FeNC-1000A900, respectively. The high percentage of electrons (92.5%) that adopted the 4-e route of ORR for FeNC-1000A900 contributed to the higher power and current densities when they behaved as the cathode catalyst, which effectively promoted the ORR without large quantities of hydrogen peroxide being produced.

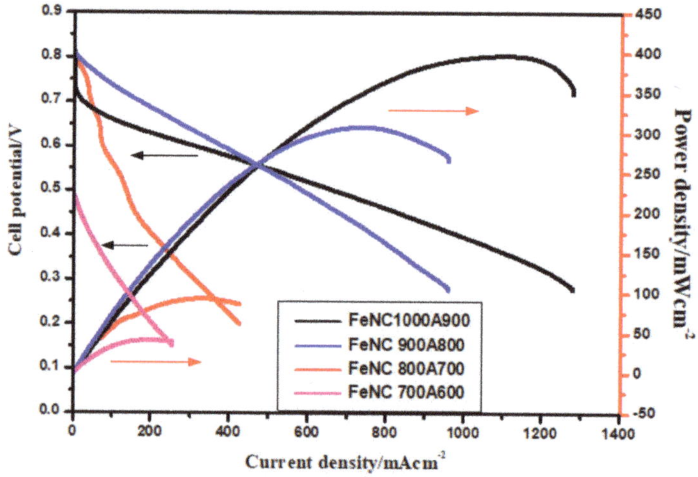

Figure 11. Power and current densities of single cells made of various FeNC cathode catalysts prepared with different calcination methods.

3.10. Durablity Test

A simple test of durability in strong acids was performed by measuring the LSV curves at various cycling times in O_2-saturated 0.1 M $HClO_4$ (aq.), which caused the FeNC catalyst to corrode, resulting in a decrease in the reduction current (Figure 12). The reduction current loss for the FeNC-1000A900 catalyst was only 10% compared to a loss of more than

20% for Pt/C after 1000 cycles, revealing that the non-precious FeNC catalyst was more acid-resistant as compared to Pt/C.

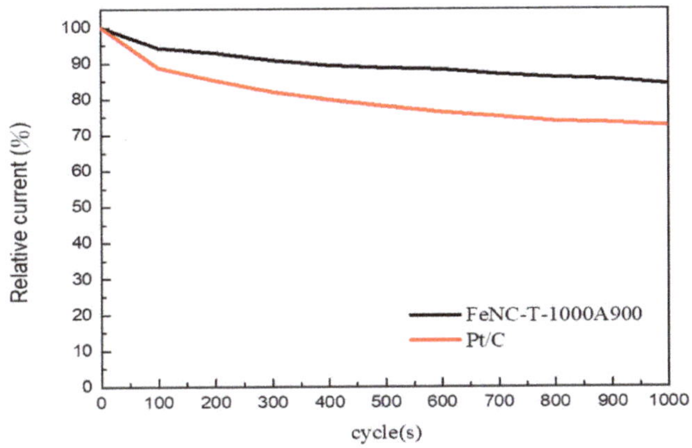

Figure 12. Relative current decay % of the LSV curves vs. cycling numbers recorded in O_2-saturated 0.1 M $HClO_4$ at 1600 rpm and 100 mV/s for the FeNC-1000A900 and Pt/C catalysts, respectively.

4. Conclusions

An aromatic PIM-based Fe- and N-doped organic carbonaceous framework (FeNMOF) was successfully synthesized by high temperature calcination. The non-precious FeNMOF proved to be a promising candidate to replace Pt/C as the most effective cathode catalyst in terms of improving the ORR in the PEMFC.

The complexation between the –Ns of PIM and Fe^{2+} led to the formation of FeNC networks after calcination. The second calcination in the NH_3 atmosphere, conducted after acid leaching, created high a surface area of 546.6 $m^2 \cdot g^{-1}$ that was composed of high concentrations of both micro- and mesopores, which exposed more Fe-active centers to the O_2, as characterized by SEM and TEM micrographs for FeNC-1000A900. The high concentration of active centers and the large surface area contributed to a higher reduction current of the cathode and a higher power density of the single cell as compared to the Pt/C catalyst. The non-precious FeNC catalyst even demonstrated higher stability than Pt/C in the durability test performed with 1000 cycles of redox reactions.

In the future, we intend to attempt the preparation of the FeNC catalyst at lower calcination temperatures while avoiding reductions in the current of the cathode and the power density of the prepared single cell.

Supplementary Materials: The following are available online at https://www.mdpi.com/article/10.3390/polym13213850/s1, Figure S1: XPS of FeNCs prepared with different calcination methods (a) 600A500 (b) 700A600 (c) 800A700 (d) 900A800 (e) 1000A900, Figure S2: Schematic diagram of the formation of 3D-GF, Figure S3: LSV curves of all FeNC and Pt/C catalysts measured at various rotating speeds, Figure S4: (a) Koutecký-Levich plots of FeNC-1000A900. (b) numbers of electrons transferred during ORR

Author Contributions: Conceptualization, K.-S.H.; Data curation, T.-H.H.; Formal analysis, Y.-M.K.; Funding acquisition, Y.-W.C.; Investigation, Y.-M.K. and L.-C.J.; Methodology, W.-Y.H. All authors have read and agreed to the published version of the manuscript.

Funding: This research received funding from MOST 108-2221-E-992-037 and MOST 109-2221-E-992-083, through the Minister of Science and Technology, Taiwan, ROC.

Institutional Review Board Statement: Not applicable.

Informed Consent Statement: Not applicable.

Data Availability Statement: The data presented in this study are available on request from the corresponding author.

Acknowledgments: Appreciation is expressed for the use of soft-matter TEM equipment belonging to the Instrument Center of National Cheng Kung University (NCKU), Ministry of Science and Technology, Taiwan, ROC.

Conflicts of Interest: The funders had no role in the design of the study; in the collection, analyses, or interpretation of data; in the writing of the manuscript, or in the decision to publish the results.

References

1. Jasinski, R. A new fuel cell cathode catalyst. *Nature* **1964**, *201*, 1212. [CrossRef]
2. Bagotzky, V.S.; Tarasevich, M.R.; Radyushkina, K.A.; Levina, O.A.; Andrusyova, S.I. Electrocatalysis of the oxygen reduction process on metal chelates in acid electrolyte. *J. Power Sources* **1978**, *2*, 233–240. [CrossRef]
3. Ratso, S.; Sahraie, N.R.; Sougrati, M.T.; Käärik, M.; Kook, M.; Saar, R.; Paiste, P.; Jia, Q.; Leis, J.; Mukerjee, S.; et al. Synthesis of highly-active Fe–N–C catalysts for PEMFC with carbide-derived carbons. *J. Mater. Chem. A* **2018**, *6*, 14663–14674. [CrossRef]
4. Mamtani, K.; Singh, D.; Tian, J.; Millet, J.M.; Miller, J.T.; Co, A.C.; Ozkan, U.S. Evolution of N-coordinated iron–carbon (FeNC) catalysts and their oxygen reduction (ORR) performance in acidic media at various stages of catalyst. *Catal. Lett.* **2016**, *146*, 1749–1770. [CrossRef]
5. Logeshwaran, N.; Ramakrishnan, S.; Chandrasekaran, S.S.; Vinothkannan, M.; Kim, A.R.; Sengodan, S.; Velusamy, D.B.; Varadhan, P.; He, J.-H.; Yoo, D.J. An efficient and durable trifunctional electrocatalyst for zinc–air batteries driven overall water splitting. *Appl. Catal. B. Environ.* **2021**, *297*, 120405. [CrossRef]
6. Elayappan, V.; Shanmugam, R.; Chinnusamy, S.; Yoo, D.J.; Mayakrishnan, G.; Kim, K.; Noh, H.S.; Kim, M.K.; Lee, H. Three-dimensional bimetal TMO supported carbon based electrocatalyst developed via dry synthesis for hydrogen and oxygen evolution. *Appl. Surf. Sci.* **2020**, *505*, 144642. [CrossRef]
7. Kannan, R.; Kim, A.R.; Yoo, D.J. Enhanced electrooxidation of methanol, ethylene glycol, glycerol, and xylitol over a polypyrrole/manganese oxyhydroxide/palladium nanocomposite electrode. *J. Appl Electrochem.* **2014**, *44*, 893–902. [CrossRef]
8. Vijayakumar, E.; Ramakrishnan, S.; Sathiskumar, C.; Yoo, D.J.; Balamurugan, J.; Noh, H.S.; Kwon, D.; Kim, Y.H.; Lee, H. MOF-derived CoP-nitrogen-doped carbon@ NiFeP nanoflakes as an efficient and durable electrocatalyst with multiple catalytically active sites for OER, HER, ORR and rechargeable zinc-air batteries. *Chem. Eng. J.* **2022**, *428*, 131115. [CrossRef]
9. Jaouen, F.; Proietti, E.; Lefèvre, M.; Chenitz, R.; Dodelet, J.P.; Wu, G.; Chung, H.T.; Johnston, C.M.; Zelenay, P. Recent advances in non-precious metal catalysis for oxygen reduction reaction in polymer electrolyte fuel cells. *Energy Environ. Sci.* **2011**, *4*, 114–130. [CrossRef]
10. Lefevre, M.; Proietti, E.; Jaouen, F.; Dodelet, J.P. Iron-based catalysts with improved oxygen reduction activity in polymer electrolyte fuel cells. *Science* **2009**, *324*, 71–74. [CrossRef] [PubMed]
11. Jaouen, F.; Lefevre, M.; Dodelet, J.-P.; Cai, M. Heat-treated Fe/N/C catalysts for O_2 electroreduction: Are active sites hosted in micropores? *J. Phys. Chem. B* **2006**, *110*, 5553–5558. [CrossRef]
12. Jaouen, F.; Herranz, J.; Lefèvre, M.; Dodelet, J.-P.; Kramm, U.I.; Herrmann, I.; Bogdanoff, P.; Maruyama, J.; Nagaoka, T.; Garsuch, A.; et al. Cross-laboratory experimental study of non-noble-metal electrocatalysts for the oxygen reduction reaction. *ACS Appl. Mater. Interfaces* **2009**, *1*, 1623–1639. [CrossRef]
13. Jaouen, F.; Marcotte, S.; Dodelet, J.-P.; Lindbergh, G. Oxygen reduction catalysts for polymer electrolyte fuel cells from the pyrolysis of iron acetate adsorbed on various carbon supports. *J. Phys. Chem. B* **2003**, *107*, 1376–1386. [CrossRef]
14. Ramanavicius, S.; Ramanavicius, A. conducting polymers in the design of biosensors and biofuel cells. *Polymers* **2021**, *13*, 49. [CrossRef] [PubMed]
15. Ramanavicius, S.; Ramanavicius, A. charge transfer and biocompatibility aspects in conducting polymer-based enzymatic biosensors and biofuel cells. *Nanomaterials* **2021**, *11*, 371. [CrossRef]
16. Gavrilov, N.; Dašić-Tomić, M.; Pašti, I.; Ćirić-Marjanović, G.; Mentus, S. Carbonized polyaniline nanotubes/nanosheets-supported Pt nanoparticles: Synthesis, characterization and electrocatalysis. *Mater. Lett.* **2011**, *65*, 962–965. [CrossRef]
17. Higgins, D.C.; Meza, D.; Chen, Z. Nitrogen-doped carbon nanotubes as platinum catalyst supports for oxygen reduction reaction in proton exchange membrane fuel cells. *J. Phys. Chem. C* **2010**, *114*, 21982–21988. [CrossRef]
18. Gong, K.; Du, F.; Xia, Z.; Durstock, M.; Dai, L. Nitrogen-doped carbon nanotube arrays with high electrocatalytic activity for oxygen reduction. *Science* **2009**, *323*, 760–764. [CrossRef]
19. Khomenko, V.G.; Barsukov, V.Z.; Katashinskii, A.S. The catalytic activity of conducting polymers toward oxygen reduction. *Electrochim. Acta* **2005**, *50*, 1675–1683. [CrossRef]
20. Chen, Z.; Higgins, D.; Chen, Z. Electrocatalystic activity of nitrogen doped carbon nanotubes with different morphologies for oxygen reduction reaction. *Electrochim. Acta* **2010**, *55*, 4799–4804. [CrossRef]
21. Chen, Z.; Higgins, D.; Tao, H.; Hsu, R.S.; Chen, Z. Highly active nitrogen-doped carbon nanotubes for oxygen reduction reaction in fuel cell applications. *J. Phys. Chem. C* **2009**, *113*, 21008–21013. [CrossRef]
22. Wang, Y.-Z.; Chang, K.-J.; Hung, L.-F.; Ho, K.-S.; Chen, J.-P.; Hsieh, T.-H.; Chao, L. Carboxylated carbonized polyaniline nanofibers as Pt-catalyst conducting support for proton exchange membrane fuel cell. *Synth. Met.* **2014**, *188*, 21–29. [CrossRef]

23. Wu, R.H.; Tsai, M.J.; Ho, K.S.; Wei, T.E.; Hsieh, T.H.; Han, Y.K.; Kuo, C.W.; Tseng, P.H.; Wang, W.Z. Sulfonated Polyaniline Nanofiber as Pt-Catalyst Conducting Support for Proton Exchange Membrane Fuel Cell. *Polymer* **2014**, *55*, 2035–2043. [CrossRef]
24. Wang, Y.Z.; Tsai, M.J.; Hsieh, T.H.; Tseng, P.H.; Ho, K.S. Studies on the 1D polyanilines prepared with n-dodecylbenzenesulfonic and camphorsulfonic acid. *Polym. Int.* **2015**, *64*, 1568–1577. [CrossRef]
25. Wang, Y.-Z.; Huang, W.-Y.; Hsieh, T.-H.; Jheng, L.-C.; Ho, K.-S.; Huang, S.-W.; Chao, L. FeNxC Based Catalysts Prepared by the Calcination of Iron-Ethylenediamine@Polyaniline as the Cathode-Catalyst of Proton Exchange Membrane Fuel Cell. *Polymers* **2019**, *11*, 1368. [CrossRef] [PubMed]
26. Yeager, E. Dioxygen electrocatalysis: Mechanisms in relation to catalyst structure. *J. Mol. Catal.* **1986**, *38*, 5–25. [CrossRef]
27. Kinoshita, K. *Electrochemical Oxygen Technology*; Interscience: New York, NY, USA, 1922.
28. Maldonado, S.; Stevenson, K.J. Influence of nitrogen doping on oxygen reduction electrocatalysis at carbon nanofiber electrodes. *J. Phys. Chem. B* **2005**, *109*, 4707–4716. [CrossRef]
29. Lai, L.; Potts, J.R.; Zhan, D.; Wang, L.; Poh, C.K.; Tang, C.; Gong, H.; Shen, Z.; Lin, J.; Ruo, R.S. Exploration of the active center structure of nitrogen-doped graphene-based catalysts for oxygen reduction reaction. *Energy Environ. Sci.* **2012**, *5*, 7936–7942. [CrossRef]
30. Lin, Z.; Song, M.K.; Ding, Y.; Liu, Y.; Liu, M.; Wong, C.P. Three-dimensional nitrogen-doped carbon nanotubes/graphene structure used as a metal-free electrocatalyst for the oxygen reduction reaction. *Chem. Phys.* **2012**, *14*, 3381–3387.
31. Zhao, Y.; Watanabe, K.; Hashimoto, K. self-supporting oxygen reduction electrocatalysts made from a nitrogen-rich network polymer. *J. Am. Chem. Soc.* **2012**, *134*, 9528–19531. [CrossRef]
32. Kramm, U.I.; Dodelet, J.P. Structure of the catalytic sites in Fe/N/C-catalysts for O_2-reduction in PEM fuel cells. *Chem. Phys.* **2012**, *14*, 11673–11688. [CrossRef]
33. Kattel, S.; Wang, G. Reaction pathway for oxygen reduction on FeN_4 embedded graphene. *J. Phys. Chem. Lett.* **2014**, *5*, 452–456. [CrossRef] [PubMed]
34. He, Z.; Maurice, J.-L.; Gohier, A.; Lee, C.S.; Pribat, D.; Cojocaru, C.S. Iron Catalysts for the Growth of Carbon Nanofibers: Fe, Fe_3C or Both? *Chem. Mater.* **2011**, *23*, 5379–5387. [CrossRef]
35. Liang, J.; Zheng, Y.; Chen, J.; Liu, J.; Hulicova-Jurcakova, D.; Jaroniec, M.; Qiao, S.Z. Facile Oxygen Reduction on a Three-Dimensionally Ordered Macroporous Graphitic C_3N_4/Carbon Composite Electrocatalyst. *Angew. Chem. Int. Ed.* **2012**, *51*, 3892–3896. [CrossRef] [PubMed]

polymers

MDPI

Review

Nanocomposite Polymer Electrolytes for Zinc and Magnesium Batteries: From Synthetic to Biopolymers

María Fernanda Bósquez-Cáceres, Sandra Hidalgo-Bonilla, Vivian Morera Córdova, Rose M. Michell and Juan P. Tafur *

School of Chemical Sciences & Engineering, Yachay Tech University, Urcuquí 100119, Ecuador; maria.bosquez@yachaytech.edu.ec (M.F.B.-C.); sahidalgo@yachaytech.edu.ec (S.H.-B.); vmorera@yachaytech.edu.ec (V.M.C.); rmichell@yachaytech.edu.ec (R.M.M.)
* Correspondence: jtafur@yachaytech.edu.ec

Abstract: The diversification of current forms of energy storage and the reduction of fossil fuel consumption are issues of high importance for reducing environmental pollution. Zinc and magnesium are multivalent ions suitable for the development of environmentally friendly rechargeable batteries. Nanocomposite polymer electrolytes (NCPEs) are currently being researched as part of electrochemical devices because of the advantages of dispersed fillers. This article aims to review and compile the trends of different types of the latest NCPEs. It briefly summarizes the desirable properties the electrolytes should possess to be considered for later uses. The first section is devoted to NCPEs composed of poly(vinylidene Fluoride-co-Hexafluoropropylene). The second section centers its attention on discussing the electrolytes composed of poly(ethylene oxide). The third section reviews the studies of NCPEs based on different synthetic polymers. The fourth section discusses the results of electrolytes based on biopolymers. The addition of nanofillers improves both the mechanical performance and the ionic conductivity; key points to be explored in the production of batteries. These results set an essential path for upcoming studies in the field. These attempts need to be further developed to get practical applications for industry in large-scale polymer-based electrolyte batteries.

Keywords: polymer electrolytes; composites; biopolymers; zinc batteries; magnesium batteries; properties

Citation: Bósquez-Cáceres, M.F.; Hidalgo-Bonilla, S.; Morera Córdova, V.; Michell, R.M.; Tafur, J.P. Nanocomposite Polymer Electrolytes for Zinc and Magnesium Batteries: From Synthetic to Biopolymers. *Polymers* **2021**, *13*, 4284. https://doi.org/10.3390/polym13244284

Academic Editors: Ting-Yu Liu and Yu-Wei Cheng

Received: 14 November 2021
Accepted: 28 November 2021
Published: 7 December 2021

1. Introduction

The development of new ways of obtaining energy from environmentally friendly materials is directly related to the need for developing devices capable of storing the power generated. Some devices are designed to store energy, such as rechargeable batteries, capacitors, sensors, and dye-sensitized solar cells (DSSC) [1]. Among these devices, batteries are the most used in everyday life around the world. The main disadvantage of the batteries currently in use, for example, lithium-ion batteries (LIBs), is that they can undergo thermal runaway, form protrusions, show low energy density, and low cycling efficiency [2–4]. Moreover, they are highly reactive, expensive, unsafe, and are a pollutant for the environment [5–10].

Some of the devices that could resolve the disadvantages identified for LIBs are redox flow batteries [11–13] and fuel cells [14–16], as they have no thermal runaway problem, and they are safe and less expensive. Among these energy storage options, zinc and magnesium are currently the multivalent ions in the sight of replacing lithium as the most reliable options to develop eco-friendly rechargeable batteries. Magnesium is the 8th most abundant metal in the Earth's crust [17], while zinc is the 24th most abundant [18], with an estimated 2800 million metric tons (Mt) of zinc contained in the Earth's crust [19]. Furthermore, magnesium and zinc can be recycled cheaply; in addition, they do not lose their physical properties [13,14] that contribute significantly to sustainability, aiming to reduce concentrate demand, energy consumption, and minimize waste disposal and pollutant emissions. On the other hand, it is known that lithium reserves present an amount of only 17 million Mt [20], and recycling lithium, which at present is heavily dependent on cobalt content, requires improvement due to environmental and economic concerns, besides the lower value of the recovered materials [21].

Zinc batteries present key features for battery' performance including high volume capacity [22] and little redox potential [23]. On the other hand, magnesium batteries possess a low electrode potential and a high volumetric capacity, almost double the Li-metal value [24]. Besides, zinc and magnesium have lower reactivity, lower cost, low toxicity, and intrinsic safety [25–29], critical characteristics for developing sustainable energy storage devices.

These batteries have a wide range of application fields in energy storage/release systems ranging from technological and military applications, to vehicles and wearable electronics [30–32]. To develop adequate energy storage devices for the end-users, one of the crucial features is whether the battery is only suitable for base station energy storage, or if it could also be employed for flexible devices. Hence, another point for the development of eco-friendly batteries is the physical state of the electrolyte. Currently, batteries use the electrolyte in a liquid state, which has safety, toxicity, flammability, and leakage drawbacks. In addition, other characteristics of current batteries, such as bulky design, electrode corrosion occurring at the interfaces, and dendrite growth on the metal electrode, reduce the capacity and life cycle of the device [33–35]. They can even lead to preferential nucleation and uneven currents during charging [36] and cause fires [37]. These issues are why current battery development strategies focus on using solid or gel-based electrolytes, improving the electrochemical properties.

Polymer electrolytes (PEs) have the most far-reaching advantages among all types of solid-state and gel-based electrolytes. They stand out for their high flexibility and good performance [38–40]. However, the main problem of these electrolytes is that they present low battery efficiency, insufficient ionic conductivities for practical applications, insufficient electrochemical stabilities, poor mechanical strength, and substantial interfacial resistance [10]. Therefore, recent research has focused on incorporating inorganic phases; these hybrid systems have higher ionic conductivity and mechanical stiffness and are non-flammable [29,41,42].

From this approach, nanocomposite polymer electrolytes (NCPEs) were born. The first report mentioning the addition of inorganic fillers in PEs was reported by Weston et al. in 1982 [43]. These authors showed how the combination of both phases reduced the drawbacks of electrolytes that did not combine inorganic/organic phases. Since then, several papers have been published to assemble PEs and NCPEs for Zn and Mg that could be considered for industrial applications [2,10,44–48].

Biopolymers are in the main scope of this review, focusing on developing electrolytes that can be considered environmentally friendly and biocompatible. However, it is worth mentioning that for a biopolymer to be considered environmentally friendly, the resource and the production method are vital characteristics to be taken into account [49]. On the other hand, biopolymers are characterized by their natural abundance, cost-effectiveness, high solvent compatibility, and film-forming ability. Some reviews focusing on biopolymer electrolytes have been published [50–52]. However, nanofillers are a new approach discussed here to get a precedent for further research and obtain better results for practical applications among this type of polymers.

NCPEs must meet specific requirements to be suitable rechargeable batteries. The polymer that acts as the host should possess an amorphous or low crystalline nature [53]. The cation–polymer interaction must be sufficiently strong to promote dissolution but not so strong as to inhibit ion exchange [54]. The designed electrolyte should be able to take up polar groups with a high molecular weight in its chain apart from sufficient electron-pair donors for coordination with cations [53], to achieve a good performance in cationic transport number, more significant than with anionic to reduce the concentration gradients for obtaining repeated charge-discharge steps and high-power density [55].

The electrolyte should undergo no net chemical changes of the battery. All Faradaic processes are expected to occur within the electrodes [56] since the electrolyte needs to be an inert battery part. Figure 1 presents the discharge scheme for a typical battery, whose circuit is closed so that electrons can get to the cathode. The performance level requires ionic conductivities values of at least 10^{-3} S·cm^{-1}. Moreover, it needs to show the lowest glass transition temperature (T_g) [54] possible, key for obtaining the highest conductivities, resulting in increased local segmental motion and, therefore, high diffusivity of the ions.

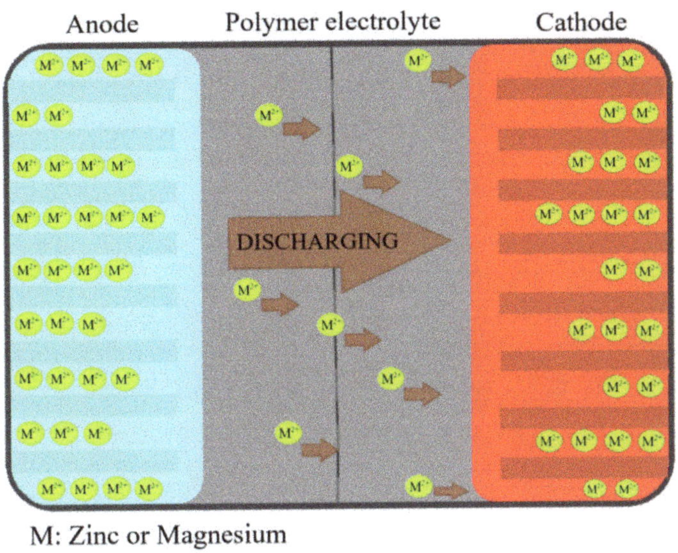

Figure 1. Schematic diagram of Zn-ion and Mg-ion battery discharge. Reproduced with permission from Renew. Sustain. Energy Rev., 65, Singh et al., Perspectives for solid biopolymer electrolytes in dye sensitized solar cell and battery application, 1098–1117, 2016 [50].

For good performance, it is also relevant to fulfill some electrochemical properties [57], such as high decomposition potential, low interfacial resistance, as well as some degree of stiffness, high chemical and thermal stability, to be durable for a long time under the conditions in which the device in which it is to be used operates [53]. Finally, rentability in the production process is indispensable since the main goal of the development is to take it to the industrial scale. As reviewed so far, these are the most important characteristics to consider when studying electrolytes used in zinc and magnesium batteries.

In this review, the recent advances of NCPEs for magnesium and zinc rechargeable batteries are overviewed, with a particular interest in the results regarding their ionic conductivities, electrochemical stabilities, and general performances in battery systems. This field continues thriving; still, new aspects of the nanoparticles' effects on the physical-chemical properties of the polymer electrolytes and their based power sources are ever discovered and need to be discussed to set an outline on future directions and challenges that come with the development of NCPEs for new batteries on worldwide demand.

2. Poly(vinylidene fluoride *co*-hexafluoropropylene)'s-Based Nanocomposite Polymer Electrolytes

Copolymerization is one of the most effective methods to improve the mechanical stability and electrical conductivity of materials [58–60]. Poly(vinylidene Fluoride-co-Hexafluoropropylene) polymer matrix (PVDF-co-HFP) has been extensively used for different purposes. It has an excellent performance in fuel cells, dye-synthetized solar cells, membrane distillation, and other electrochromic applications [61–64]. The block copolymer structure includes a crystallizable comonomer (-CH_2-CF_2-) and an amorphous HFP unit with a high dielectric constant (ε = 8.4), thanks to the presence of highly electronegative fluorine and the spontaneous alignment of C–F dipoles in the crystalline phases [65–67].

The copolymer presents a high solubility and lower crystallinity and glass transition temperature [68] than Poly(vinylidene Fluoride) (PVDF), making it a promising matrix for preparing nanocomposites, despite its non-biodegradability. Furthermore, the degree of crystallinity remaining in the system helps retain sufficient mechanical stability and structural rigidity to act as a separator between the battery's electrodes [69]. At the same time, the amorphous phase can serve as the conductive medium.

2.1. Magnesium-Ion Conduction

Within the field of rechargeable batteries, several studies have applied PVDF-co-HFP as a component of the electrolyte. The most recent ones were compiled in a review article, which focuses on lithium-sulfur batteries [70]. However, as reported, the balance between industrial development and environmental protection makes it essential to develop high energy density and non-pollutant rechargeable batteries, with magnesium ion and organic electrode batteries as the directions for post-lithium batteries [71]. For reference, Table 1 lists some properties of the nanocomposites employed along with PDVF-co-HFP as electrolytes for magnesium batteries.

Table 1. A summary of NCPEs composed of PVDF-co-HFP for magnesium batteries.

Nanocomposite	Ionic Salt	Conductivity ($S{\cdot}cm^{-1}$) 10^{-3}	Activation Energy (eV)	Electrochemical Stability Window (V)	State	Reference
No added	$Mg(Tf)_2$	0.15	-	5	Gel	[72]
	$Mg(ClO_4)_2$	0.293	0.33	4	Solid	[73]
SiO_2	$Mg(ClO_4)_2$	3.2	-	4.3	Gel	[74]
	$Mg(ClO_4)_2$	11	-	3.5	Gel	[75]
	$Mg(ClO_4)_2$	10	-	3.5	Gel	[76]
Al_2O_3	$Mg(Tf)_2$	3.3	-	3.3	Gel	[77]
$MgAl_2O_4$	$Mg(Tf)_2$	4.0	-	3.3	Gel	[77]
Al_2O_3 *	$Mg(NO_3)_2$	0.101	-	-	Solid	[78]
MgO	$Mg(ClO_4)_2$	8	0.235	3.5	Gel	[79]
	$Mg(ClO_4)_2$	6	0.032	3.5	Gel	[75]
MgO *	$Mg(NO_3)_2$	0.104	0.45	-	Solid	[80]
MgO and SiO_2	$Mg(ClO_4)_2$	10 and ~9	-	-	Gel	[81]
ZnO	$MgCl_2$	0.12	0.45	-	Solid	[82]
ZnO *	$Mg(NO_3)_2$	0.37	-	-	Solid	[83]
$BaTiO_3$	$Mg(Tf)_2$	0.411	-	-	Solid	[84]
TiO_2 *	$Mg(NO_3)_2$	0.010	0.30	-	Solid	[85]

* PVDF without copolymerization with HFP.

Maheshwaran et al. [72] studied the role that the salt added to the polymer had and how its amount affected the results. They developed a magnesium ion conducting gel polymer electrolytes (GPE) based on PVdF-co-HFP, magnesium triflate $Mg(Tf)_2$, in ethylene carbonate (EC), and diethyl carbonate (DEC). The analysis of this polymeric electrolyte by X-ray diffraction (XRD) showed a decrease in the crystallinity with the addition of salt. Moreover, FT-IR confirmed that magnesium triflate could suppress the nonpolar α crystalline phase of PVDF. Consequently, the electrolyte offered a predominant ionic character with a total ion transport number close to unity, making it considered to a certain extent for batteries because it was freestanding and stable. However, its considerably low ionic conductivity made it inappropriate, although it became a precedent for what can be achieved.

Solid polymer electrolytes (SPE) with PVDF-co-HFP as a polymer matrix were also studied. Ponmani et al. [73] blended this matrix with Poly(vinyl acetate) (PVAc) and added magnesium perchlorate $Mg(ClO_4)_2$ salt. The SPE film was found to be flexible, and the maximum ionic conductivity found was 0.293×10^{-3} S·cm^{-1}, obtained at 363 K. Cyclic voltammetry (CV) studies confirmed the Mg ion reversibility that demonstrated its conduction in the SPE.

One of the first electrolytes that employed PVDF-co-HFP was a magnesium-ion conducting GPE, composed of 15% of PVdF-co-HFP, 73% of $Mg(ClO_4)_2$ in EC/propylene carbonate (EC/PC), and 12% silicon dioxide (SiO_2) [74]. The cell in which this electrolyte was tested employed magnesium as anode and vanadium oxide (V_2O_5) as the cathode. The tests demonstrated low initial discharge capacity and poor cycling performances. These disadvantages could be attributed to high interfacial resistance at Mg anode [74]. The main problem identified from this research was the blocking of the charge transfer reaction, highlighting that further research on the interface should be conducted so that the cycling performance could be improved to a practical level.

Magnesium oxide (MgO) showed beneficial features in inducing consistent improvements in liquid electrolyte retention and the overall chemical, physical, and electrochemical properties in the work performed by Pandey et al. [79]. They presented novel research dispersing PVdF-co-HFP with nanosized MgO particles. It was analyzed by XRD patterns, obtaining a semi-crystalline structure with predominant peaks in 2θ = 14.6, 17, 20, 26.6, and 38°. These changes in the peaks showed the reduction in crystallinity of the PEs, caused by the entrapment of liquid electrolytes. FT-IR spectroscopic analysis was conducted to look over the ion-polymer interaction and the conformational changes, confirming the reduction of crystallinity. The T_g was observed at −65 °C for pure PVDF-co-HFP film, while with the addition of magnesium oxide, the value came down to −90 °C.

Electrodes play an important role when performing efficiency analysis. Pandey et al. [75] demonstrated the previous NCPE in a prototype cell of magnesium and multiwalled carbon nanotubes (MWCNT) composite as the negative electrode and the corresponding positive one with vanadium pentoxide (V_2O_5). The rechargeability of the cell was enhanced by substituting magnesium with Mg–MWCNT composite as the negative electrode. The discharge capacity faded away after ten cycles, attributed to the passivation of the negative electrode. Nevertheless, the electrolyte showed to be free-standing and flexible, with enough mechanical strength.

The role of fillers has been shown to be of great importance. These nanocomposites can form space-charge regions and induce a local electric field. This phenomenon was first approached by Kumar [86], who revealed the electric charge and area associated with the particle interact with the structure of the liquid electrolyte, provoking the space-charge region. It can be described as containing free electrons at the surface of the nanocomposite, and cations along with dipoles at the adjacent double-layer balance the surface electronic charge (Figure 2). Magnesium oxide is known to be slightly electronegative in nature. In the systems studied, a reversible reaction between the magnesium oxide and the magnesium (II) ion took place and formed the space-charge region, giving place to the $MgO:Mg^{2+}$ species [75].

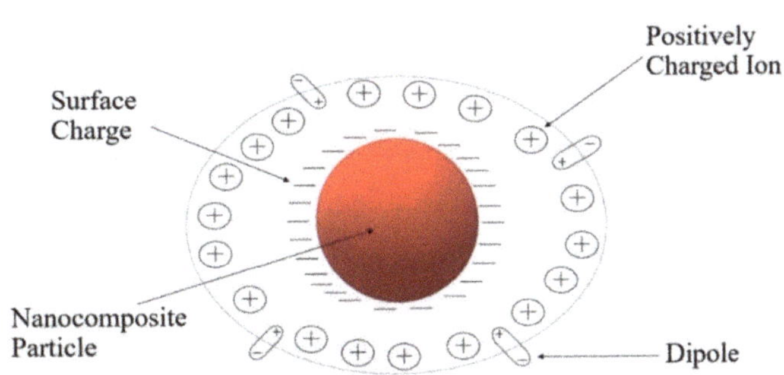

Figure 2. Schematic representation of the space charge and local electric field formation around a nanocomposite particle.

Magnesium oxide nanoparticles were combined with nano-sized silicon dioxide in a novel electrolyte by Pandey et al. [81]. When relating conductivity to filler content (Figure 3), the presence of two conductivity maxima was noticed, explained by the dissociation of ion aggregates/undissociated salt into free ions with the addition of filler particles (the first peak). The second maximum was described using the composite effect and based on a conducting interfacial space-charge double layer between the filler particles and the GPE. This local field was responsible for enhanced Mg^{2+} ion motion and enhanced transport number up to the addition of ~10 wt % of MgO. The cationic transport number measurements (t_+) also showed essential results, in which the best improvement was obtained by the presence of MgO particles (~0.44). On the other hand, with the addition of SiO_2 dispersion, the t_+ value did not increase substantially. Finally, for this polymer, it was pointed out that the nano-sized MgO supported the cationic motion. In contrast, the nano-sized SiO_2 supports the anion conduction in the filler/gel electrolyte interfacial regions.

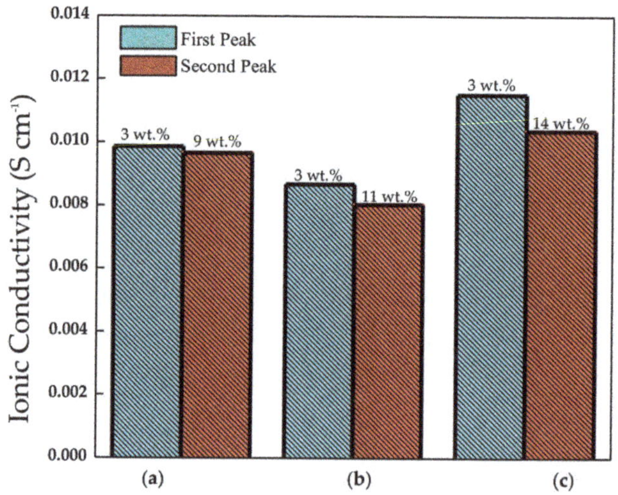

Figure 3. Room temperature conductivity peaks of composite gel polymer electrolyte films vs filler content: (**a**) nano-sized MgO, (**b**) micro-sized MgO, and (**c**) nano-sized SiO_2. Prepared from data in [81].

Nanosized silicon dioxide was ultimately tested with the addition of molybdenum trioxide (MoO_3) as the positive electrode in a posterior study [76]. This cell showed a discharge capacity of ~175 mAh·g^{-1} for an initial ten charge-discharge cycles. In addition, it presented the same conductivity value as the last cell. Finally, good thermal stability with a single-phase behavior was presented at a temperature range from −70 °C to 80 °C. Enhanced conductivity was attributed once again to the space-charge layers formed between the filler and GPE.

The effect of active and passive nanofillers, along with the copolymer, was studied for Mg NCPEs by Sharma et al. [77], incorporating Mg-triflate salt mixed with EC and PC, entrapped in PVDF-co-HFP. Aluminium oxide (Al_2O_3) and Mg aluminate ($MgAl_2O_4$) were used as passive and active fillers. The reduction of crystallinity was achieved as expected and confirmed by XRD, Field emission scanning electron microscopy (FESEM), and Differential Scanning Calorimetry (DSC) studies. By FESEM (Figure 4), it was observed that undispersed GPE showed larger grain sizes and possessed uniformly distributed pores. The incorporation of nanofillers in the undispersed GPE changes its morphology substantially. It was further proved that as the number of fillers increases its porosity leads to the entrapment of liquid electrolyte in the pores (demonstrated by the fillers not being seen in the NCPE (Figure 4c,f)), which further enhanced the ionic conductivity of the NCPE.

The addition of the passive filler conferred the cell to have a relatively good mechanical stability and thermally stability up to 100 °C. The active filler ensured an improvement in the ion transport number. The obtained electrochemical stability window (ESW) was key, showing their potential as electrolytes in ionic devices.

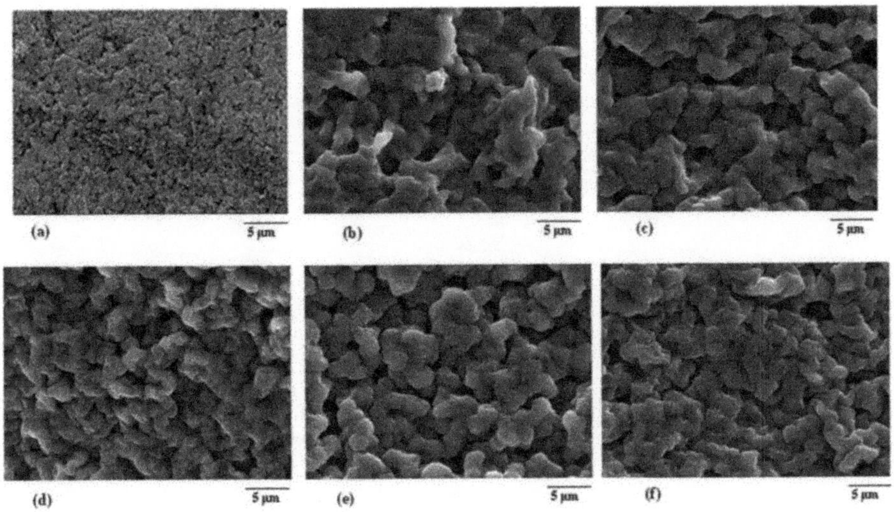

Figure 4. FESEM images of pure (**a**) PVdF-co-HFP film and nanocomposite GPE films containing (**b**) 0 wt %, (**c**) 6 wt % Al_2O_3, (**d**) 30 wt % Al_2O_3, (**e**) 6 wt % $MgAl_2O_4$, and (**f**) 20 wt % $MgAl_2O_4$. Reproduced with permission from Polym. Compos., 40, Sharma et al., Magnesium ion-conducting gel polymer electrolyte nanocomposites: Effect of active and passive nanofillers, 1295–1306, 2019 [77].

The latest report of PVDF-co-HFP electrolyte for magnesium batteries known so far implemented zinc oxide (ZnO) as nanofiller along with Magnesium chloride ($MgCl_2$) as ionic salt [82]. A transport number of 0.99 was achieved; the current change indicated that conductivity in the NCPE was predominantly ionic. PVDF was incorporated without copolymerizing it with HFP in a magnesium NCPE, along with MgO as a nanofiller [80]. The optimum nanofiller concentration (3wt %) was chosen to be the most suitable one with a conductivity of 1.04×10^{-4} S·cm^{-1}. On further increase in nanofiller concentrations, the ionic conductivity value decreased. Thermal stability and reduced melting point temperature were confirmed through thermogravimetric analysis (TGA) and XRD. Magnesium oxide nanoparticles enhanced the ionic conductivity and dielectric constant, confirmed by complex impedance spectroscopy [80]. The authors recently presented similar results with the addition of zinc oxide particles [83].

The results presented until now (Table 1) confirm the enhanced high ionic conductivity present in PVDF-co-HFP nanocomposite polymer electrolytes compared to the SPE systems and better thermal and mechanical stability compared to liquid systems. The enhancement in conductivity may be caused by the presence of the nanoparticles, facilitating the new kinetic path for ionic transport and polymer segmental motion. However, another conclusive characteristic is that when a specific percentage of nanofiller is added, a decrease in ionic conductivity is observed. Excessive fillers could provoke this in the NCPE that may trigger the formation of ion pairs and ion aggregation, such as the non-conducting phase presented as an electrically inert component blocking ion transport. So far, they are probably one of the best options to study and meet all the requirements for future use instead of lithium-ion conductive systems.

2.2. Zinc-Ion Conduction

Besides magnesium, zinc presents many advantages associated with zinc chemicals, as batteries of high specific/volumetric energy density can be fabricated. Ionic radii of Zn^{2+} (74 pm) and that of Li^+ (68 pm) are quite comparable, but Zn^{2+} has twice as much charge as Li^+ cation [87]. Furthermore, the natural resources of zinc are plentiful, and its stability makes it able to be handled safely in oxygen and humid atmosphere. The so-mentioned dielectric constant of PVDF-co-HFP is also known to generally assist in more significant ionization of zinc salts and then provide a high concentration of charge carriers. Consequently, Table 2 lists some properties of the nanocomposites employed along with PDVF-co-HFP as electrolyte for zinc batteries.

Table 2. A summary of NCPEs composed of PVDF-co-HFP for zinc batteries.

Nanocomposite	Ionic Salt	Conductivity ($S{\cdot}cm^{-1}$) 10^{-3}	Activation Energy (eV)	Electrochemical Stability Window (V)	State	Reference
No Added	$Zn(Tf)_2$	1.73	0.025	-	Gel	[88]
	$Zn(Tf)_2$	2.44×10^{-2}	0.380	3.45	Solid	[89]
	$Zn(Tf)_2$	0.144	-	4.14	Solid	[90]
TiO_2	$Zn(Tf)_2$	0.34	-	-	Solid	[91]
ZrO_2	$Zn(Tf)_2$	0.46	-	2.6	Solid	[92]
ZnO	$Zn(Tf)_2$	6.7	-	4.5	Gel	[93]
CeO_2SiO_2	$Zn(Tf)_2$	0.3	-	2.7	Solid	[57]
	$NH_4CF_3SO_3$	1.07	-	-	Solid	[94]

Tafur et al. [88] studied GPEs composed of PVdF-co-HFP with different ionic liquids, with and without zinc triflate salt. From attenuated total reflectance-Fourier transform infrared (ATR-FT-IR) and XRD spectroscopies, it was deduced that incorporating the ionic liquid and salt to the matrix produced more amorphous and polar membranes when comparing it to the original PVDF-co-HFP film. Besides, the electrical properties had shown to be dependent on the ionic liquid employed, aspect confirmed by measurements on ionic conductivity, impedance, and voltammetry. This report also studied the influence of the ionic liquid type on the performance of the GPE for Zn batteries. From ionic conductivity, impedance, and voltammetry measurements, changes in the results were observed when the salt was not added, or the added quantity was too low, indicating that the salt is the important charge carrier independent of the ionic liquid.

In another report, Tafur et al. [95] designed a battery employing manganese dioxide (MnO_2) as the cathode, zinc as the anode, and the GPE was assembled by the use of PVDF-co-HFP including 1-Butyl-3-methylimidazolium trifluoromethanesulfonate, 99% (BMIM Tf), and zinc triflate ($Zn(Tf)_2$). The electrolyte was then analyzed by X-ray photoelectron spectroscopy and Energy Dispersive X-Ray techniques. The remarkable results from the study showed the charge storage mechanism, which began with the reduction of Mn^{4+} to Mn^{2+} species, at the same time as Zn^{2+} cations, together with triflate ions, intercalate the cathode material during the discharge process. In the recharge process, it was evidenced that Mn^{2+} species returned to the positive electrode, and they were oxidized mainly to Mn^{4+}. Moreover, Mn^{2+} was not reduced to Mn^0 in the anode during the recharge process. Nevertheless, in every completed process, Zn^{2+} cations were not expulsed, remaining inside the electrode, probably stabilized by the triflate anions, which were not expulsed either. Besides, with the addition of an ionic liquid to the GPE, it was observed that the interaction between the zinc-ion and PVDF chains became weaker, enhancing ion mobility.

A zinc battery was designed with an SPE [89] based on PVDF-co-HFP with zinc triflate, obtaining a low crystallinity elucidated by XRD and scanning electron microscopy (SEM). CV of the SPE curve was flat without the presence of peaks. This fact indicated that the film presented excellent stability. No decomposition occurred in the operating voltage range and ESW of 3.45 V. These results were improved in a recent study performed by Liu et al. [90], where 1-ethyl-3-methylimidazolium trifluoromethanesulfonate (EMITf) was incorporated along with the PVDF-co-HFP membranes and zinc triflate salt. The ionic liquid has shown to reduce the crystallinity, enrich the nanopores' structure, and enhance the electrical and electrochemical properties of the electrolyte membranes. The ionic conductivity was enhanced by one order of magnitude. The electrolyte membrane was able to sustain a high thermal decomposition temperature of ~305 °C, and thus its mechanical performance was sufficient for considering it for practical applications.

Nanocomposites became first used with PVDF-co-HFP by Muda et al. [94], who designed a cell composed of Zn + $ZnSO_4{\cdot}7H_2O$ + polytetrafluoroethylene (PTFE) as the anode, MnO_2 + PTFE as the cathode, and the electrolyte composed by PVDF-co-HFP as host polymer, ammonium trifluoromethane sulfonate ($NH_4CF_3SO_3$) as ionic salt with silicon dioxide as filler. Herein was observed the so mentioned existence of two conductivity maxima at the concentration of 1 and 4 wt % of SiO_2, in this case, attributed to two percolation thresholds in the NCPE. The voltage was able to achieve a value of ~1.50 V to ~1.29 V. The assembled cell performed fairly well when discharged at the low current drain or with high load resistance, showing suitability for low current drain applications.

Titanium dioxide was incorporated in an NCPE designed by Johnsi et al. [91]. Differential scanning calorimetric results confirmed that with the addition of TiO_2, a reduction in the degree of crystallinity and the T_g value was obtained. The glass transition temperature is a fundamental parameter to grasp the structural changes of PEs occurring under various thermal conditions. The effect of filler content on the position of glass temperature could be evaluated from Figure 5. When more TiO_2 was added, these values increased slightly instead, which could be caused by the possible agglomeration of an excess amount of nanofillers. Furthermore, there was also evidence to ascertain that the decrease of the T_g value reflected an increase in flexibility of polymer chains, provoking enabling fast ion conduction within the NCPE system. The low value of conductivity limited the applications that could be obtained for this electrolyte.

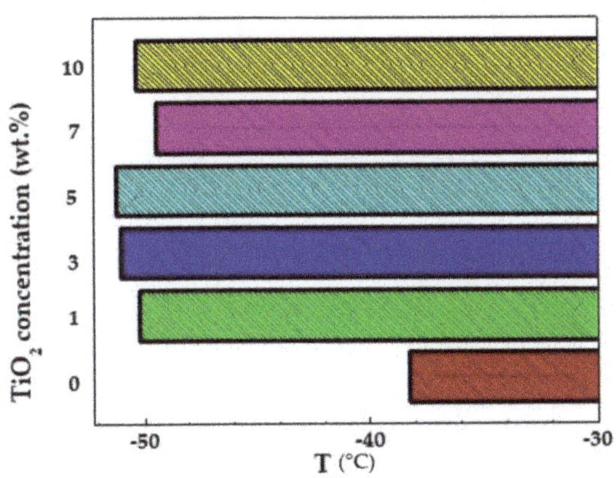

Figure 5. T_g (°C) of the NCPE systems vs the TiO_2 content. Prepared from data in [91].

Active fillers were studied by Hashmi [93] with an electrolyte composed by the addition of 1.0 M solution of zinc triflate in EC/PC immobilized in the host polymer and ZnO nanofiller. The morphological/structural changes for this gel electrolyte were monitored using SEM and XRD techniques. The micrographs obtained (Figure 6) showed that the texture and morphology of the gel polymer system had also been modified, revealing smaller crystallites and pores. For the highest amount (~25 wt %) sample, ZnO particles appeared to be disappeared in the SEM picture (Figure 6b), probably due to the polymer fully covering the particles, changing the texture of polymer network. On the other hand, with 10 wt % of filler, white spots could be seen, indicating the presence of nanoparticles in the gel network, forming a separate phase. Dark regions showed the micron-sized porosity from the micrographs corresponding to the undispersed GPE, where the liquid electrolyte could be retained. Thus, the polymer films established a semi-crystalline nature with enough porosity to maintain the ionic liquid in the electrolyte. For this gel, the increase in ionic conductivity with the dispersion of ZnO did not signify more than an order of magnitude, obtaining that for the undispersed sample, its value was $\sim 6.7 \times 10^{-3}$ S·cm^{-1}. Besides, with the increasing of temperature, for the sample of ~10 wt % ZnO particles, it offered ionic conductivities of 3.7×10^{-3} S·cm^{-1} at 30 °C and 1.4×10^{-2} S·cm^{-1} at 85 °C, which gave a precedent for potential application as an electrolyte in zinc batteries and other electrochemical applications over a wider temperature range. Furthermore, for this NCPE, a local electric field was detected due to the reversible reaction between ZnO and Zn^{2+}, homologous to the one previously presented in Figure 1.

Johnsi et al. [57] continued their work with passive fillers. They constructed a flexible, free-standing, transparent film composed of [75 wt % PVDF-co-HFP:25 wt % $ZnTf_2$]-x wt % cerium dioxide (CeO_2) where x = 1, 3, 5, 7, and 10, respectively. The film's detailed FT-IR spectral analysis indicated the feasibility of complexation between the host polymer matrix and the salt and nanofiller. The decomposition voltage that reached a range of 2.4 to 2.7 V was a key result for this cell. Johnsi's et al. [92] work continued with implementing zirconium dioxide (ZrO_2), with the same proportion of PVDF-co-HFP and $ZnTf_2$, with 7 wt % nanofiller. The obtained value of ionic conductivity represented an increase in an order of magnitude. XRD confirmed an amorphous phase present in the matrix. The cell achieved an ESW of 2.6 V with thermal stability up to 300 °C. The resulting cell exhibited many attractive and stable discharge characteristics for room temperature applications.

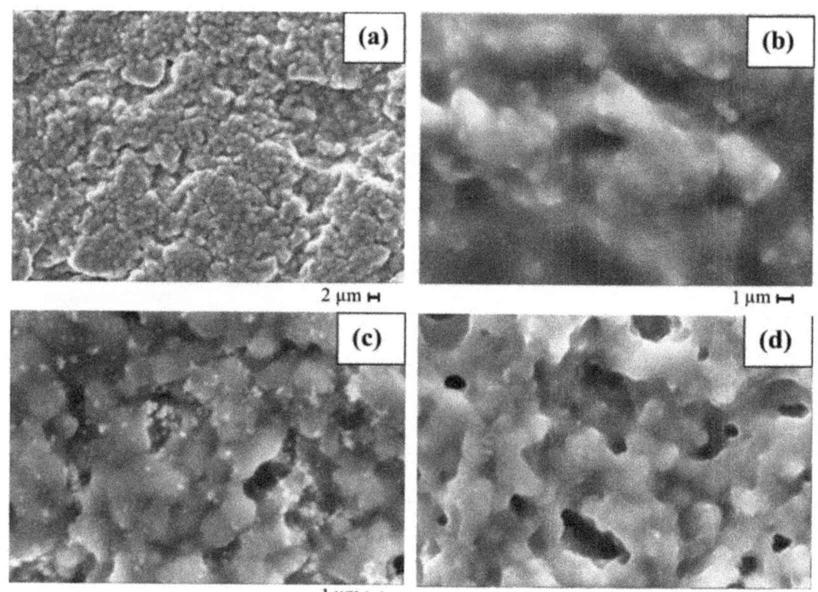

Figure 6. SEM micrographs of EC–PC–$Zn(Tf)_2$+PVdF-co-HFP (**a**) gel polymer electrolyte (magnification, ×1000); (**b**) gel electrolyte with magnification ×5000; and (**c**) its nanocomposites dispersed with ZnO particles of 10 wt % (magnification, ×5000); and (**d**) 25 wt % (magnification, ×5000) Reproduced with permission from J. Solid State Electrochem., 16, Sellam, Enhanced zinc ion transport in gel polymer electrolyte: Effect of nano-sized ZnO dispersion, 3105–3114, 2012 [93].

From the analysis presented in this section, it could be concluded that the addition of active fillers has shown more promising results for the applicability of NCPEs, than the ones with passive fillers. This assumption can be seen in Figure 7. The reports about

PVDF-co-HFP, with the addition of zinc triflate salt and various nanofillers, are presented versus their ionic conductivity obtained. Whereas the first bar, representing the NCPE without nanofillers, has a shallow and almost imperceptible value, the bar corresponding to the ZnO nanofiller has a clear advantage over the other nanofillers considered passive. This assumption is explained by the fact that zinc oxide nanoparticles have been shown to participate in the ionic conduction process. Hence, they deserve special attention for further studies on the field, even with other polymers and ionic salts.

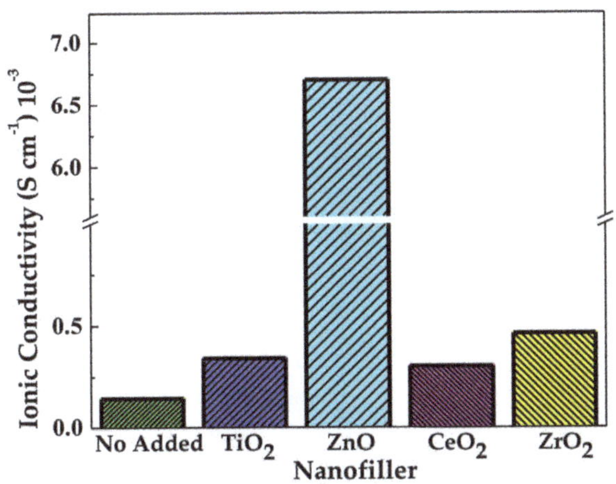

Figure 7. Plot of nanofiller versus ionic conductivity ($\times 10^{-3}$) for nanocomposite polymer electrolyte based on PVDF-co-HFP and zinc triflate salt. Prepared from data in [57,90–93].

3. The Poly(ethylene Oxide)'s Based Nanocomposite Polymer Electrolytes

Poly(ethylene oxide) (PEO) has been the focus of attention for many researchers among various polyethers because it is considered to be the best solvent medium for various ionic salts [96], and it is known to possess relatively high electrochemical stability [97]. It is water-soluble and in a semi-crystalline state at room temperature [98]. It presents a single helical structure, which also supports ionic conduction [99], favoring fast ion transport for the electrochemical processes in the batteries. The main chain of the polymer, known to be polar and flexible, owns vital electron-donating ether-oxygen groups, dissociating the salt, and generating carrier ions. These ions can migrate through the amorphous region of the polymer employing interchain/intrachain segmental motion. However, its high degree of crystallinity makes it necessary to incorporate metal salts that impede crystallization [100]. PEO shows a low ionic conductivity of 3.32×10^{-9} S·cm^{-1} at pure state [101], a fact that could be enhanced by the addition of nanofillers, along with other approaches discussed in this section. A review article was recently published by Feng et al. [102], where the interaction of ceramic fillers on the performance of PEO in lithium batteries is deeply studied. The authors concluded that composite SPEs are one of the most efficient ways to improve the electrolytes' ionic conductivity.

3.1. Magnesium-Ion Conduction

The primary purpose of the subsequent studies has been to elucidate the ion transfer mechanism in the nanocomposite polymer system and enhance the ion conductivity and mechanical strength. The present section discusses how the addition of nanocomposites decreases the degree of crystallinity in these electrolytes. Table 3 refers to the properties of the nanocomposites employed along with PEO as electrolytes for magnesium batteries.

Table 3. A summary of NCPEs composed of PEO for magnesium batteries in solid-state.

Nanocomposite	Ionic Salt	Conductivity ($S \cdot cm^{-1}$) 10^{-3}	Activation Energy (eV)	Electrochemical Stability Window (V)	Reference
No added	$Mg(Tf)_2$	56	0.49	4.6	[103]
	$Mg(Tf)_2$	0.277	0.40	-	[101]
	$Mg(ClO_4)_2$	0.277	0.30	-	[104]
MgO	$Mg(Tf)_2$	1.67	0.34	-	[101]
	$(CH_3COO)_2Mg \times 7H_2O$	363	0.013	7.6	[105]
	$Mg(ClO_4)_2$	1.04	0.29	-	[104]
TiO_2	$Mg(Tf)_2$	1.53	0.38	-	[101]
	$Mg(ClO_4)_2$	1.14	0.28	-	[104]
SiO_2	$Mg(Tf)_2$	0.586	0.36	-	[101]
	$Mg(ClO_4)_2$	0.87	0.28	-	[104]
B_2O_3	$MgCl_2$	0.716	-	-	[106]
Starch nanocrystals (SNCs)	$MgBr_2$	0.116	-	-	[107]

The studies involving PEO for magnesium-ion conduction began complexing magnesium triflate and incorporating EMITf ionic liquid, reported by Kumar et al. [103]. The ionic liquid happened to be vital in mediating the Mg^{2+} ion conduction and the gradual enhancement in the Mg^{2+} ion transport number. Raman studies evidenced the interaction of imidazolium cations with ether oxygen of PEO (Figure 8). The peak at 2871 cm^{-1} was found to be affected due to the complexation of PEO with $Mg(Tf)_2$ salt and the addition of ionic liquid. A peak appeared at 2848 cm^{-1} in curve b indicated the conformational changes of PEO chains after its complexation with Mg-salt. Besides, the PEO peak of 2871 cm^{-1} decreased and almost disappeared for the higher ionic liquid content. An additional peak (shoulder) appears at ~1025 cm^{-1} (Figure 8c–f), presumably because of the free triflate anions from EMITf. Such anions would be free only when EMI+ ions have the possibility of interaction with ether oxygen of PEO. Consequently, it has been shown to play an essential role in substantially enhancing the cation transfer value.

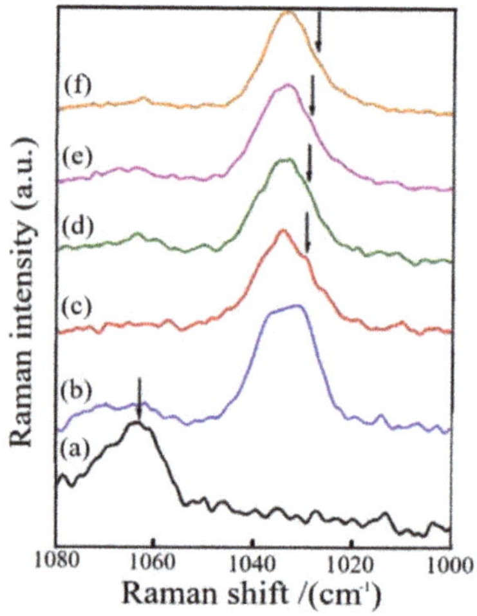

Figure 8. Raman spectra of (**a**) PEO pure film, (**b**) PEO25·$Mg(Tf)_2$ complex, and PEO25·$Mg(Tf)_2$ + x wt % EMITf system for (**c**) x = 5, (**d**) x = 10, (**e**) x = 20, and (**f**) x = 30 for spectral region of 1000–1080 cm^{-1}. Reproduced with permission from Electrochim. Acta, 56, Kumar et al. Ionic liquid mediated magnesium ion conduction in poly(ethylene oxide) based polymer electrolyte, 3864–3873, 2011 [103].

The approach of adding nanocomposites to PEO began with the report presented by Sundar et al. [106], who created an SPE of PEO with $MgCl_2$ as electrolytic salt and boron oxide (B_2O_3) as the filler. DSC and FT-IR characterized this cell. The best ionic conductivity was achieved with 2 wt % B_2O_3. The cell was assembled by adopting Mg as anode and MnO_2 as cathode, sandwiching the SPE between the electrodes. It got an open circuit voltage (OCV) of 1.9 V. The low ionic conductivity obtained made this cell unsuitable for practical applications. It indicated that this type of matrix could be considered for its cell performance, using fillers in the nanoscale.

In the work presented by Shao et al. [108] comprised a novel electrolyte based on PEO, magnesium borohydride ($Mg(BH_4)_2$), and MgO nanoparticles. A key feature presented by this work was a high coulombic efficiency of 98% for Mg plating/stripping and high cycling stability. The experiments and modeling performed established a correlation between improved solvation of the salt and solvent chain length, chelation, and oxygen denticity. A further development in experimentation with this polymer revealed that it could be used in NCPE for other multivalent chemistry to delineate the ionic association and solvation interactions within these electrolytes.

Another approach for a casting technique by dry/solution free hot press is presented in the following studies as an alternate way of synthesis since no organic solvent as the medium for mixing ingredients is needed, which is the most significant difference from solvent casting [102]. The main steps for solvent-casting and hot-pressing are shown in Figure 9a,b. Thermocompression avoids contact with air during the process, which results in more stable productions. On the other hand, the solvent-casting method can disperse the ceramic fillers more uniformly, resulting in more ductile films. Besides, the residual liquid in this procedure can act as a plasticizer for the further performance of the SPE. These features need to be taken into consideration depending on the application the electrolytes will have.

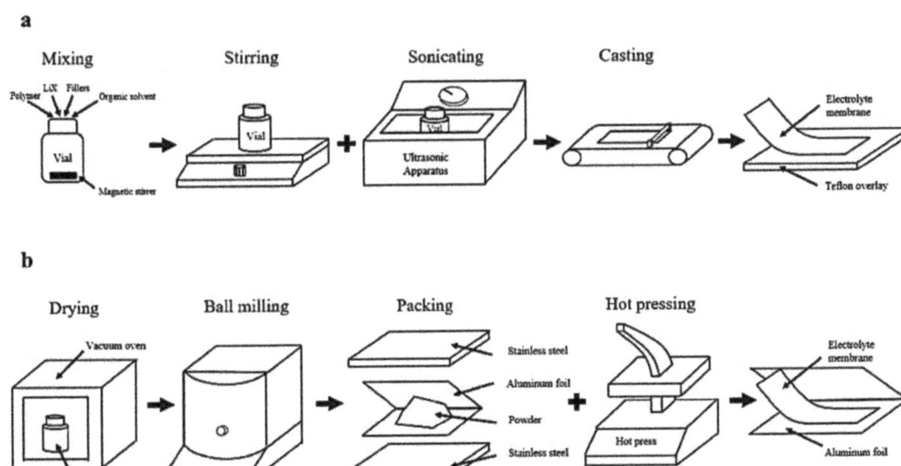

Figure 9. Schematic procedures for preparing solid polymer electrolytes by (**a**) solvent casting technique and (**b**) thermo-compression. Reproduced with permission from Nano Converg., 8, Feng et al.PEO based polymer-ceramic hybrid solid electrolytes: a review, 2, 2021 [102].

Thermocompression was implemented in the work done by Agrawal et al. [104]. The assembled electrolyte, composed of phases, has an SPE film of PEO and $Mg(ClO_4)_2$ salt, the first phase host matrix with a conductivity of $\sim 2.77 \times 10^{-6}$ S·cm^{-1}. The second phase had MgO nano/micro-sized particles as active fillers and TiO_2/SiO_2 nano-sized particles as passive fillers. The dispersion increased the room temperature conductivity of the SPE host by ~3–5-fold. The values of cation transport, however, remained in the range 0.21–0.30.

Casting by hot-press technique was performed as the synthesis process in the research developed by Agrawal et al. [101], with a primary phase host composed of PEO and magnesium triflate and micro/nano-sized materials TiO_2 (passive filler) and MgO (active filler). The employment of all of these substances together achieved an enhancement in the room temperature conductivity of the SPE host. Characterization was performed with XRD, FT-IR, DSC analysis. The total ionic transference number data (~0.98–0.99) showed the predominantly ionic character of the materials employed. Analyzing the concentration vs. log of conductivity (Figure 10) obtained two maximum peaks that suggested the presence of two conductivity mechanisms in the system, previously discussed in this report with reference [81]. The first peak could be related to dissociating undissociated salt and ion aggregates (if existed) into free ions. The second s-peak was then attributed to forming a high conducting interfacial space-charge double layer around insulating filler nanoparticles that could correspond to a filler particle percolation threshold. It is also concluded that nanofillers were more effective in increasing the cation transport number than micro fillers, increasing the cation transport number t_+.

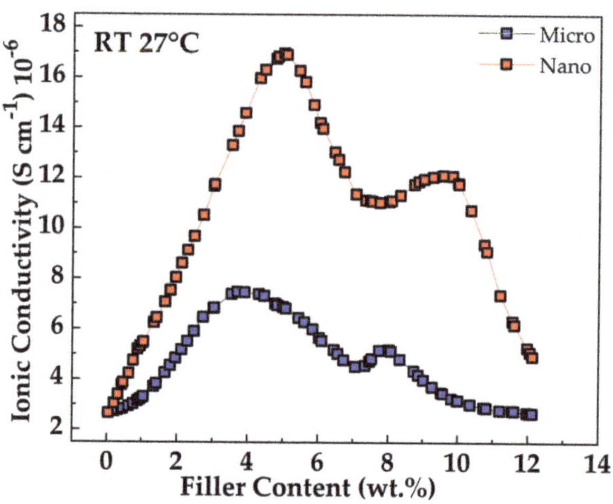

Figure 10. Active filler concentration-dependent conductivity variation for NCPE films: [80PEO: $20Mg(Tf)_2$] + xMgO micro/nano. Reproduced with permission from Mater. Chem. Phys., 139, Agrawal et al., Investigations on ion transport properties of hot-press cast magnesium ion conducting Nano-Composite Polymer Electrolyte (NCPE) films: Effect of filler particle dispersal on room temperature conductivity, 410–415, 2013 [101].

One of the latest reports on electrolytes based on PEO for Mg batteries was reported by Zaky et al. [105], with the incorporation of Mg salts treated with gamma irradiation to improve the PEO-Mg salt particle sizes. The electrical conductivity evaluated was more than three orders of magnitude than pure PEO, with a maximum value of 3.63×10^{-3} S·cm^{-1}. The optimum ionic conductivity of MgO in the irradiated sample was obtained with 20 mL, while 30 mL was the best for un-irradiated. The addition of MgO also improved the electrochemical potential window to about −3.2 to 4.4 V.

3.2. Zinc-Ion Conduction

In terms of zinc batteries, some works have been developed with the use of PEO. They are summarized in Table 4. For this metal, the first attempts were also designed without the addition of nanofillers. Therefore, an SPE with the addition of zinc chloride ($ZnCl_2$) was developed by Carrilho et al. [109] in a cell composed of zinc and niobium pentoxide

(Nb_2O_5) as electrodes. Their studies were performed at a temperature of 55 °C, obtaining a conductivity of 2.7×10^{-4} S·cm^{-1} and a cationic transference number value of 0.44 ± 0.05.

Table 4. A summary of NCPEs composed of PEO for zinc batteries in solid-state.

Nanocomposite	Ionic Salt	Conductivity (S·cm^{-1}) 10^{-3}	Activation Energy (eV)	Electrochemical Stability Window (V)	Reference
No added	$ZnCl_2$	2.7 *	-	2.60	[109]
Al_2O_3	$Zn(Tf)_2$	2.1	0.44	3.6	[110]
	$Zn(Tf)_2$	~0.101	0.19	-	[111]
TiO_2	$ZnCl_2$	~100	0.087	-	[112,113]
SiO_2	NH_4HSO_4	0.61	-	-	[114]
ZnO	$Zn(Tf)_2$	0.184	0.23	-	[115]
Zn Fe_2O_4	NH_4SCN	~10^{-3}	-	-	[116]

* Results obtained at 55 °C.

The cell testing showed a decrease in the cell voltage without attaining any constant value. This result suggested the discharge product was the result of a topochemical insertion. The capacity retention of the cell was observed to be very poor. After the second cycle, the cell was not able to retain its charge. Galvanostatic/potentiostat cycles were performed at a lower time (in discharges) and fixed potentials (in charges) to improve the latter result. After this, shallower discharges were obtained, resulting in a longer cycle life, with a less marked decrease in cell capacity at a constant voltage. Lifetime evaluation for the studied cell was 4.9 years if maintained at 55 °C, under non-operating conditions.

Agrawal et al. [114] designed two cells that employed PEO and NH_4HSO_4 and SiO_2 for the electrolyte that performed in two types of cells: MnO_2 + C and PbO_2 + V_2O_5 + C as cathodes, respectively. The researchers achieved an enhancement in the room temperature conductivity of polymer electrolyte approximately by an order of magnitude. Furthermore, it obtained a substantial increase in the mechanical strength of the films. The OCV was found to be in the range of 1.5–1.8 V for both batteries. The cell potential was stable through the discharges, but it discharged more quickly during higher current drain or low load resistance.

Gamma (γ) irradiation was presented as a novel technique to inhibit the crystalline phase in an electrolyte composed of PEO and $ZnCl_2$ as salt, with the addition of nanosized TiO_2 grains, by Turković et al. [112,113]. The polymer was subjected to γ-radiation from a Co-60 source. This approach was attempted since high-energy radiation could induce interchain linking of the polymer, inhibiting the crystalline phase in the polymer matrix. Small-angle X-ray scattering (SAXS) was recorded simultaneously with DSC, and wide-angle X-ray diffraction (WAXD) analyses were performed. Thanks to these techniques, it was obtained that the nanostructure of the γ-irradiated electrolyte changed during the crystalline-amorphous phase transition to a highly conductive superionic phase. Reduction in the T_g was observed, and ionic conductivity was enhanced, two desired changes in these processes. The conductivity of the nanocomposite prepared with irradiated powder ensured an improvement of two orders of magnitude compared to its homologous without irradiation.

An NCPE film was prepared using an SPE composed of PEO and zinc trifluoromethanesulfonate ($Zn(Tf)_2$) and then incorporating Al_2O_3 nano-filler particles by Karan et al. [111] using a completely dry hot-press cast technique. The complexation of the salt and the dispersal of filler in the host substantially increased the amorphous region, which supported the increase in ionic conductivity and cationic transport number. Nevertheless, the obtained values need to be improved for possible applications in high-energy batteries.

As another approach for employing nanocrystalline Al_2O_3, PEO was blended with polypropylene glycol (PPG) and $Zn(Tf)_2$ as dopant salt by Nancy et al. [110]. This matrix resulted in an enhanced ionic conductivity of one magnitude, compared to the previous work where this nanofiller was employed with alone PEO. This feature caused segmental flexibility and an increase in the amorphous phase. The SEM, XRD, and DSC measurements showed that conductivity was controlled by segmental motions of the polymer chain and ion hopping mechanism at Lewis acid-base sites and at elevated temperatures exhibited Arrhenius behavior which was satisfactorily explained by free volume theory.

Zinc ferrite nanoparticles were presented as a relatively new approach for NCPEs by Agrawal et al. [116] This nanofiller has been widely used in technological applications because of its high magnetic permeability in the radio frequency region and low core loss. In their research, they posited the changing of the bonding behavior of the system when compared to the original PEO. Herein, the decrease in crystallinity was confirmed by the DSC study of the system. The presumed hopping mechanism between coordinated sites, local structural relaxation, and segmental motion of the polymer was stated because of the increase of ionic conductivity with temperature. The rise in ~3–4 order of magnitude of ionic conductivity concerning the pure polymeric host confirmed its promising results for electrolyte applications.

The hot-press technique was attempted with ZnO active fillers by Karan et al. [115], who composed a two-layer electrolyte. The first layer obtained the highest conductivity of 1.09×10^{-6} S·cm^{-1}. A 5 wt % of ZnO revealed an optimum conduction composition with a conductivity of ~1.84×10^{-5} S·cm^{-1}, meaning that the filler's dispersal causes an enhancement of one order. The overall enhancement of four orders of magnitude from the pure PEO was obtained. The battery in which the films were assembled performed well under a low current drain state.

Ultimately, the outlook of incorporating branched aramid nanofibers (BANFs) to PEO was investigated by Wang et al. [117]. This combination enhanced the effective suppression of dendrites and fast cation transport due to the high stiffness of the BANF network combined with the high ionic conductivity of soft PEO, resulting in high tensile strength. The resulting battery showed the ability to withstand elastic deformation during bending and plastic deformation and remain functional (Figure 11). There have been different types of batteries that have shown to be capable of elastic deformations [118–120]; the ability to withstand plastic deformations while retaining the charge storage functions was a novel feature presented in this work. These features set it apart from other promising storage devices, improving the safety of the battery and its resistance to impact.

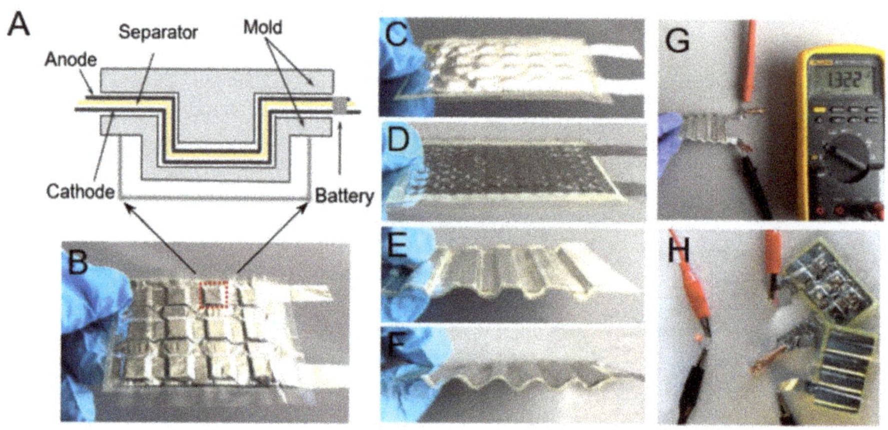

Figure 11. (**A**) Schematic of the mold used for plastic deformation studies. (**B**–**F**) Different plastically deformed shapes of Zn battery with solid-state biomimetic electrolyte. (**G**) The open-circuit voltage of Zn/PZB-931/γ-MnO_2 battery with square wave shape plastic deformation. (**H**) LED light powered by the two serial structural batteries. Reproduced with permission from ACS Nano, 13, Wang et al., Biomimetic Solid-State Zn^{2+} Electrolyte for Corrugated Structural Batteries, 1107–1115, 2019 [117].

The review in this section of the paper has shown that active fillers are supposed to be the first choice when choosing ceramic additives for PEO electrolytes. Moreover, the most optimized concentrations for these nanofillers are between 10–20 wt % to obtain the highest conductivity of each medium. Besides, the mechanical strength is shown to get better with the doping of ceramic particles. Moreover, the interfacial stability is assumed to be improved due to the water-scavenging effect of the nanofillers, previously reported in similar systems designed with PEO [121] in lithium batteries. Hence, the overviewed hybrid systems are in sight of being the most effective approach for improving the performance of solid-state electrolytes.

4. Nanocomposite Polymer Electrolytes Based on Other Synthetic Polymers

As discussed, the development of electrochemical devices that make use of polymer electrolytes has gotten considerable interest. It is currently developing PEs with sufficiently high room temperature conductivity. The choice of the polymer is then known to depend principally on the presence of polar groups with sufficient electron donor power to form coordination with cations and a low hindrance to bond rotation [122], besides biodegradability and recyclability. Some synthetic polymers have been successfully used as a host material to prepare PEs for specific applications (Table 5). Few studies in the field have developed electrolytes with these polymers. However, the results presented left a precedent that deserves to be discussed for later studies that imply their use (Table 6).

Table 5. Chemical structures of some synthetic polymers employed in nanocomposite polymer electrolytes NCPEs.

Name	Structure	Glass Transition Temperature (T_g) (°C) [123]
Poly(methyl methacrylate) (PMMA)	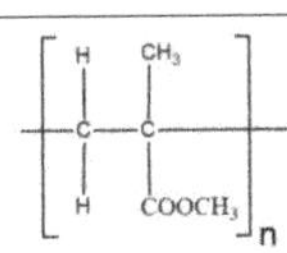	105
Poly(ethyl methacrylate) (PEMA)	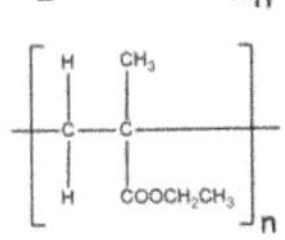	65
Poly(vinyl chloride) (PVC)	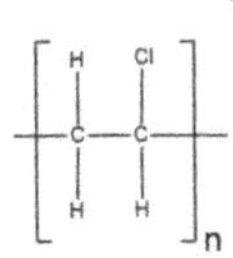	83
Poly(vinyl alcohol) (PVA)	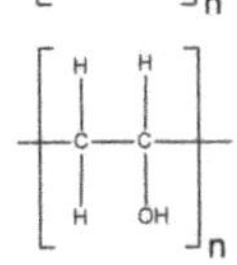	80
Poly(ε-caprolactone) (PCL)	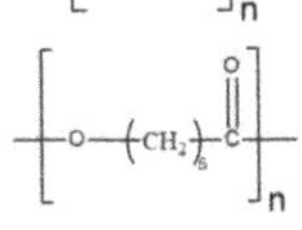	−66
Poly(vinylpyrrolidone) (PVP)	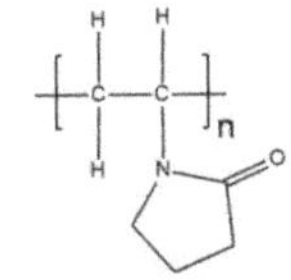	182

Table 6. NCPEs for magnesium and zinc batteries using other synthetic polymers.

Polymer	Nanocomposite	Ionic Salt	Conductivity ($S{\cdot}cm^{-1}$) 10^{-3}	Activation Energy (eV)	Electrochemical Stability Window (V)	State	Reference
			Magnesium				
PMMA + PVdF	MgO	$Mg(Tf)_2$	1.29×10^{-2}	-	-	Solid	[124]
PEMA	MgO	$Mg(Tf)_2$	0.12	0.46	3.4	Solid	[125]
PVP	Al_2O_3	$MgCl_2 \cdot 6H_2O$	1.22×10^{-2}	-	-	Solid	[126]
			Zinc				
PMMA/PVDF-co-HFP	SiO_2	NH_4SCN	43	0.196	3.2	Gel	[127]
	SiO_2	$Zn(Tf)_2$	6.71	-	5.07	Gel	[128]
PVC/PEMA	$Al_2O_3 + TiO_2$	$Zn(Tf)_2$	4.27	-	~4	Gel	[129]
	ZrO_2	$Zn(Tf)_2$	3.63	-	3.87	Gel	[130]
	SnO_2	$Zn(Tf)_2$	4.92	-	4.37	Gel	[131]
PCL	ODAMMT	$Zn(Tf)_2$	0.95	0.46	4.5	Gel	[132]
	Al_2O_3	$Zn(Tf)_2$	0.25	-	-	Gel	[133]
PVA	SiO_2	-	5.73×10^2	-	-	Gel	[134]
	ZnO	NH_4NO_3	4.71	0.92	-	Solid	[135]

Poly(methyl methacrylate) (PMMA) has been the focus of a few studies due to its beneficial effects on the stabilization of the electrode-electrolyte interface [136]. PMMA is non-biodegradable, and 100% recyclable [137]. Nevertheless, its recycling process is not environmentally viable due to the produced harmful products, limiting its use [138]. PMMA based GPEs happen to present very high transparency in the visible region. Furthermore, they present the ability to be diluted in various organic solvents [139]. However, they show poor dimensional stability. Although they appear solid-like, they exhibit flow properties. Poor mechanical properties offset a good conductivity achieved of such plasticized film at a high concentration of the plasticizer [140]. To overcome the drawbacks presented by PMMA film, it has been blended with other polymers to improve the segmental motion in polymer hybrid systems and hence a more flexible and elastic material.

Sarojini et al. [124] developed a blended polymer matrix of PMMA and PVdF for magnesium cells. It also included ethylene carbonate as a plasticizer, $Mg(Tf)_2$ as ionic salt and MgO as nanofiller. The best ionic conductivity increased the value by five orders of magnitude (~10^{-6} $S{\cdot}cm^{-1}$). This result was obtained thanks to the addition of the nanofiller, causing high conduction pathways. Nevertheless, the result was deficient to be considered for any application.

A blended polymer matrix composed of PMMA and PVdF-co-HFP was developed by Mishra et al. [127] for the design of an NCPE system for Zn cells. XRD and SEM studies confirmed the desirable amorphous and porous structure for the electrolyte. The best ionic conductivity, 4.3×10^{-3} $S{\cdot}cm^{-1}$, was obtained with 2 wt % of SiO_2. The conductivity variation for these films obeyed the behavior of having two maxima, previously reported in other works [81,101]. A proton battery was assembled with the electrolyte, employing $Zn/ZnSO_4{\cdot}7H_2O$ as anode and PbO_2/V_2O_5 as the cathode. The OCV for the battery was found at 1.55V. Besides, it showed rechargeability up to three cycles, and afterward, its discharge capacity faded away substantially.

Poly(ethyl methacrylate) (PEMA) is a very similar material to PMMA but with a lower T_g and has been shown to possess higher mechanical strength than PMMA [141]. Besides, PEMA shows excellent chemical and high surface resistance. In addition, it offers high optical transparency [142], a property that could be desired for devices where the electrolyte is located in a visible region of the device. PEMA was employed for an NCPE in work [125], along with magnesium triflate and 1-ethyl-3-methylimidazolium bis(trifluoromethylsulfonyl) imide (EMITFSI), dispersed with MgO for Mg cell electrolytes. SEM analysis confirmed the obtention of the amorphous nature of the films. In addition, TGA curves revealed that the more significant amount of MgO in NCPE slowed down the mass loss rate of decomposition products. However, the electrochemical potential window for the highest conducting sample assumed that magnesium ion was not predominantly the factor to the ionic conductivity enhancement of NCPE.

A blended polymer matrix was developed with PEMA and poly(vinyl chloride) (PVC), plasticized with zinc triflate, and EMIMTFSI ionic liquid was added for a novel NCPE by Candhadai et al. [128]. After that, it was doped with fumed SiO_2 as a nanofiller. This film exhibited the highest ionic conductivity value of 6.71×10^{-4} S·cm^{-1} for a 3 et al SiO_2. The increment of the amorphous phase was confirmed by XRD analysis. It resulted in slight progress in the zinc ion transport number and a wide ESW of ~5.07 V. This value ensured feasible zinc stripping/plating in the redox process involved. TG and DSC ascertained the improved thermal stability up to 180 °C and the reduction in T_g. The exact blend, PVC and PEMA, was filled with nano-sized fillers Al_2O_3, TiO_2 in the report by Prasanna et al. [129] for a zinc rechargeable battery. A high transport number value of 0.67 was obtained. From the studies analyzing glass transition temperature, the addition of fillers attenuated the values obtained, effect understood in terms of the obstruction of the polymer chains by the formation of cross-linking centers due to the interaction between the Lewis acid groups of the ceramic particles and the polar groups of the polymer chains.

Based on the previous work, Prasanna et al. [130] continued the research by changing the nanofiller employed, being zirconia (ZrO_2) the object of the study. The zinc ion transference number of 0.66 was almost the same obtained before with Al_2O_3 and TiO_2. DSC and TG analysis confirmed the improved thermal behavior of ZrO_2 added GPE compared to that of filler-free gel electrolytes. The interaction and complexation of the polymer components were probed by ATR-FT-IR analysis (Figure 12). The amplified coordination of Zn^{2+} cations and ceramic phase with C=O group was evidenced by the existence of a peak at 1721 cm^{-1} ascribed to the C=O group of PEMA in Figure 12b–e, upon the addition of 1 wt % nanofiller. The oxygen atoms of C=O group in PEMA generally acts as an electron donor resulting in the formation of a coordinate bond with zinc ions, and the addition of fillers enhances the intensity of this band, through hydrogen bonding between carbonyl oxygen (C=O) and the hydroxyl surface group (Zr-OH) of ZrO_2 thus forming -Zr-O . . . H . . . O=C- species. Ultimately, it was observed better thermal stability up to 180 °C, and ESW to 3.87 V.

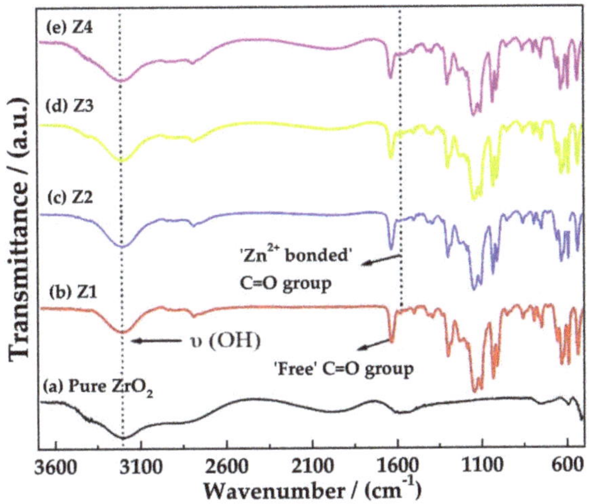

Figure 12. Room temperature ATR-FT-IR spectra of (**a**) pure ZrO_2 (**b**–**e**) NCPEs with varying concentrations of ZrO_2 in the wavenumber ranging from 4000 to 400 cm^{-1} at room temperature. Reproduced with permission from Polym. Compos., 40, Sai Prassana et al., PVC/PEMA-based blended nanocomposite gel polymer electrolytes plasticized with room temperature ionic liquid and dispersed with nano-ZrO_2 for zinc ion batteries, 3402–3411, 2019 [130].

These authors, in another work, incorporated the use of nano-sized tin oxide (SnO_2) [131]. XRD and SEM studies were performed, confirming the existence of porous morphologies.

Furthermore, the dispersion of SnO_2 improved the thermal behavior of the composite system to 185 °C, which was ascertained by TG analysis. The ESW was found to be 4.37 V. Together with a feasible zinc plating/stripping process of the gel composite sample, these features implied good potential applicability of such films as electrolytes.

Poly(vinylpyrrolidone) (PVP) is a biocompatible polymer. It is a virtually non-biodegradable polymeric lactam with an internal amide bond. The tertiary amide carbonyl groups of PVP present a Lewis base character such that PVP can form a variety of complexes with a wide range of inorganic salts [143]. It is also hygroscopic and easily soluble in water and organic solvents such as alcohol. It presents a high T_g of 170 °C, because of the rigid pyrrolidone group. However, water can be employed as a plasticizer lowering this value to below 40 °C [144]. Besides, it is inert, shows good environmental stability, easy processing, excellent transparency, and a strong tendency for complex formation with smaller molecules [145]. This polymer is studied because of its thermal stability and cross-linked composites having high mechanical strength. It also has good mechanical and electrical characteristics.

Basha et al. [126] developed an SPE composed of PVP and $MgCl_2x6H_2O$, with the addition of Al_2O_3 particles. Structural analysis showed orthorhombic lattice as evidence of a semi-crystalline nature present in the films. Optical analysis was used to identify the optical band gap of the material in the transmitting radiation. Graphs were plotted between absorption coefficient α, $(\alpha h\nu)^2$ and $(\alpha h\nu)^{1/2}$ as a function of $h\nu$ (Figure 13a–c) to calculate bandgap energy values. The optical properties revealed that for the composition of 15%, the bandgap energy was the lowest among all weight ratios. Hence, it was obtained that the films with the lowest activation energy had the highest conductivity. UV–Vis spectroscopy was performed in the 300–700 nm (Figure 13d). This tool showed to be helpful for the identification of intra molecular vibrations of inorganic complexes in solution. Two spectral peaks are observed at 350 nm, which is due to the π–π* transition. Besides, a small peak was at 425 nm, correlated with the benzene and quinone rings in the polymer chain.

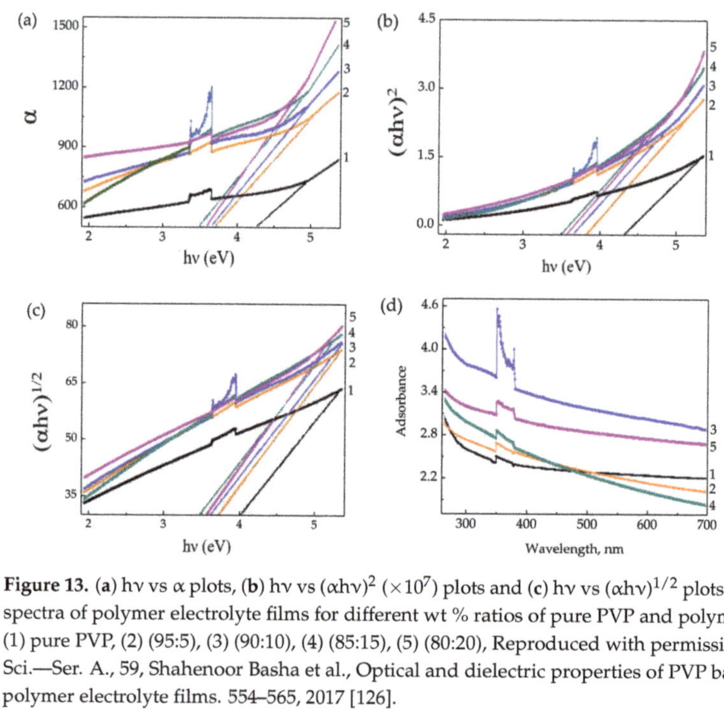

Figure 13. (**a**) hν vs α plots, (**b**) hν vs $(\alpha h\nu)^2$ ($\times 10^7$) plots and (**c**) hν vs $(\alpha h\nu)^{1/2}$ plots and (**d**) UV–Vis spectra of polymer electrolyte films for different wt % ratios of pure PVP and polymer electrolytes: (1) pure PVP, (2) (95:5), (3) (90:10), (4) (85:15), (5) (80:20), Reproduced with permission from Polym. Sci.—Ser. A., 59, Shahenoor Basha et al., Optical and dielectric properties of PVP based composite polymer electrolyte films. 554–565, 2017 [126].

Poly(vinyl alcohol) is a semi-crystalline synthetic biodegradable polymer from petroleum sources that presents various hydrophilic functional hydroxyl groups, which can favor water absorption. As a result, it shows a very high dielectric strength (>1000 kV mm^{-1}), good charge storage capacity, good mechanical properties, high tensile strength, abrasion resistance, and dopant-dependent electrical and optical properties [146]. In addition, PVA has several advantages, such as high hydrophilicity, high gel strength, nontoxicity, and low cost [147]. Fan et al. [134] designed a zinc-air battery (ZAB) assembled with a semi-solid/solid-state electrolyte constructed with PVA and the optimum addition of SiO_2. The ZAB presented excellent cycling stability over 48 h, stable discharge performance, and relatively high-power output. Flexibility was an outstanding feature obtained with no degradation through bending conditions. In the results, this cell was able to power a handheld electric fan, a light-emitting diode screen, or even a mobile phone (Figure 14), showing its promising potential for high-performance ZABs along with high safety, cost-effectiveness, excellent flexibility, electrolyte retention capability, as well as good thermal and mechanical properties.

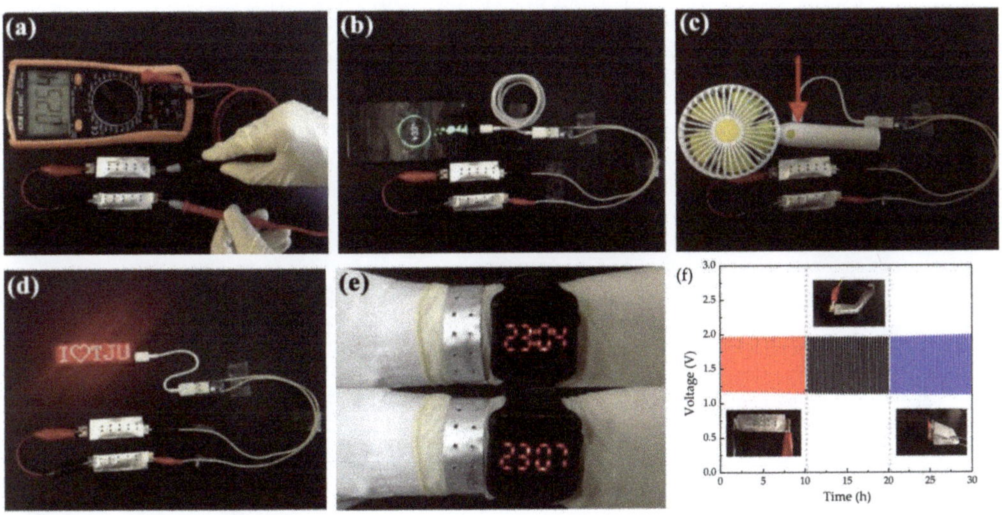

Figure 14. (**a**) Open circuit potential demonstration with two ZABs in series. A demonstration of (**b**) a mobile phone, (**c**) a handheld electric fan, and (**d**) an LED screen powered by two ZAB sets. (**e**) Photographs of an LED watch powered by a fabricated bracelet-type ZAB. (**f**) GCD curves of the ZAB under different bending conditions with corresponding photographs. Reproduced with permission from Nano Energy, 56, Fan et al., Porous nanocomposite gel polymer electrolyte with high ionic conductivity and superior electrolyte retention capability for long-cycle-life flexible zinc–air batteries, 454–462, 2019 [134].

The PVA electrolytes filled with nano ZnO transport parameters were conducted by Abdullah et al. [135] using the Rice–Roth model for proton-conducting batteries, explaining that the moderate addition of nanofiller enhances ionic conductivity by increasing mobility and number density of mobile proton ions. Another ZAB was designed to implement MWCNTs into the electrodes by Wang et al. [148], to improve the performance of the cell. It was found that MWCNTs were effective conductive additives in the anode as they bridged the zinc particles. The electrolyte was composed of poly(acrylic acid) (PAA) and PVA. A limitation for this NCPE was water evaporation because of the volatility character of the films.

Poly(ε-caprolactone) (PCL) is a biodegradable polymer that is nontoxic and widely used in biomedical applications because of its considerable degradation time in an aqueous medium and contact with microorganisms [149]. It is a synthetic thermoplastic polymer derived from crude oil and synthesized through the polymerization of ε-caprolactone monomer by a stannous octanoate catalyzed ring-opening mechanism [150]. It presents good mechanical properties [151]. Furthermore, it is a candidate polymer host for ionic conduction because it contains a Lewis base (ester oxygen) that can coordinate cations due to carbonyl functional groups in its backbone structure.

PCL was doped with zinc triflate and octadecylamine modified montmorillonite (ODAMMT) nano clay by Sownthari et al. [132]. The maximum electrical conductivity was 9.5×10^{-5} S·cm^{-1} for 15 wt % loadings of nano clay into the polymer-salt complex. XRD and DSC analysis confirmed the decrease in crystallinity. The electrolyte degradation happened in 90 days, making this electrolyte a promising candidate for battery applications.

An optimized NCPE composed of PCL, zinc triflate and the incorporation of Al_2O_3 was prepared by Sownthari et al. [133]. The complexation of polymer, salt, and filler was confirmed from FT-IR studies. The various relaxation processes associated with the conductivity mechanism were also analyzed during the investigation. From this, it was revealed that the chain length of polymer PCL was so long that the bond rotation was favorable only at low frequency, so the filler increased the amorphicity within the polymer network, and thus the rotation becomes feasible, making a shift toward higher frequency side which meant a shorter relaxation time. The increase of conductivity was mainly due to an apparent rise in the number density of charge carriers which was confirmed from FT-IR and dielectric studies. The increasing trend of dielectric constant matched well with the conductivity variation as a function of filler concentration.

5. Nanocomposite Polymer Electrolytes Based on Biopolymers

Most synthetic polymers are detrimental to the environment because of their non-biodegradability [152]. Consequently, the application of biodegradable polymers in energy storage devices is currently paramount in designing the next generation of batteries to reduce environmental impact. For a biobased polymer to be considered ecological, its origin and production technique are also of importance. Cellulose, starch, chitosan, agar, and carrageenan are some of the most common polymers used as hosts for batteries [51] (Table 7).

Table 7. Chemical structures of some biopolymers employed in nanocomposite polymer electrolytes NCPEs.

Name	Structure	Glass Transition Temperature (T_g) (°C) [123]
Agar	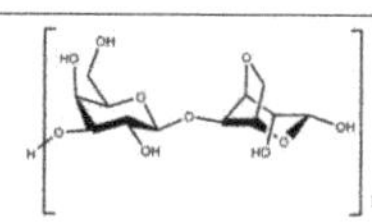	98
Carrageenan	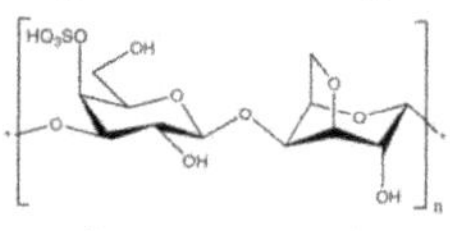	41
Cellulose	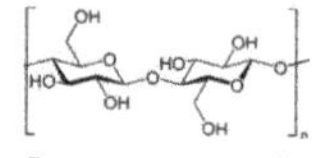	220
Chitosan	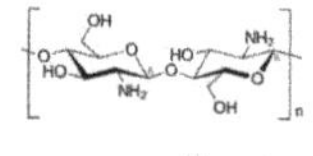	200
Starch	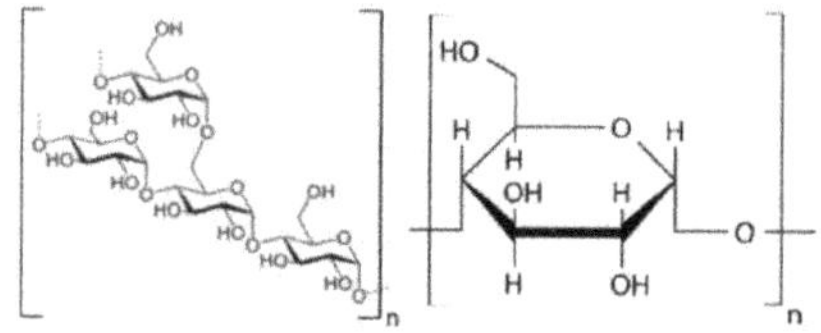 Amylopectin Amylose	227

Among the wide range of applications available for batteries, there is a need to design biocompatible batteries for implants that need a power source to perform their functions in the biomedical field. They go from sensing or stimulation to influencing critical biological processes like wound healing, tissue regeneration, or brain activity. Unfortunately, little work has been done so far to develop bioresorbable electronics or self-deployable power sources [153–155]. However, the present review pretends to show the promising results obtained so far with electrolytes that, with some modifications, could be employed in the biomedical field. Table 8 summarizes the features presented by these NCPEs.

Table 8. Polymer electrolytes based on biopolymers for electrochemical applications.

Polymer	Nanocomposite	Ionic Salt	Conductivity ($S{\cdot}cm^{-1}$) 10^{-3}	Activation Energy (eV)	Electrochemical Stability Window (V)	State	Reference
Cellulose (NFC)	No added	-	0.1	-	-	Hydrogel	[156]
Cellulose acetate	SiO_2	NH_4BF_4	7.9×10^{-3}	-	-	Gel	[157]
	TiO_2	NH_4BF_4	1.4×10^{-2}	0.12	-	Gel	[158]
Hexanoyl Chitosan	TiO_2	$LiClO_4$	3.06×10^{-4}	-	-	Solid	[159,160]
	TiO_2	$LiClO_4$	3.1×10^{-4}	0.08	-	Solid	[161]
	SiO_2	$LiClO_4$	1.96×10^{-4}	0.12	-	Solid	[161]
Chitosan	Al_2O_3	NH_4SCN	5.86×10^{-4}	-	-	Solid	[162]
	SiO_2	$Li(Tf)_2$	4.38×10^{-5}	0.26	-	Solid	[163]
	ZrO_2	$LiClO_4$	3.6×10^{-4}	-	-	Solid	[164]
Potato Starch	No added	$Mg(C_2H_3O_2)_2$	1.12×10^{-5}	-	-	Solid	[165]
Rice Starch	TiO_2	LiI	3.6×10^{-4}	0.22	-	Solid	[166]
Corn Starch	SiO_2	$LiClO_4$	1.23×10^{-4}	0.25	3.0	Solid	[167,168]
Corn Starch/Chitosan	No added	NH_4Cl	5.11×10^{-4}	-	-	Solid	[169]
Sago (starch)	No added	KOH	4.45×10^{-1}	-	-	Gel	[170]
κ-carrageenan	No added	-	3.32×10^{-2}	-	-	Solid	[171]
	No added	$MgCl_2$	4.76×10^{-3}	-	1.94	Solid	[172]
Agar	No added	NH_4SCN	1.03×10^{-3}	0.25	-	Solid	[173]
	No added	$Mg(Tf)_2$	1.0×10^{-3}	-	-	Solid	[174]
	TiO_2	LiI	5.12×10^{-4}	-	-	Solid	[175]

5.1. Cellulose

Cellulose is a biopolymer known to be the most abundant polymer in nature. It presents a molecular weight ranging from 300,000 to 500,000 Da. Its molecule offers three hydroxyls groups that can be modified to make the molecule water-soluble [176]. Many cellulose derivatives can be obtained from this modification, classified between cellulose ethers and cellulose esters. Cellulose has been applied in batteries for electrodes or separators as GPEs [176] and as binders/surface modifiers for graphite anodes for batteries. Moreover, synthetic polymers often need high-temperature processing stages to prepare GPEs, while with nanofiber cellulose (NFC) hydrogels, they can be cured at temperatures close to ambient temperature [177].

Cellulose was never intensively used as a polymer electrolyte in advanced batteries until Johari et al. [157] reported an NCPE based on cellulose acetate dispersed with SiO_2 for a battery with the configuration Zn/composite cellulose electrolytes/MnO_2. The results showed the expected increase in ionic conductivity and an OCV of 1.6V. The constancy of the assembled cell was tested for 24 hours. However, no further electrochemical studies were performed. In a posterior work [158], the same authors reported an NCPE based on cellulose acetate dispersed with nanosized TiO_2 particles for a battery with the same composition as the previous one. The OCV characteristic of the cell at room temperature showed that the initial voltage of the cell is 1.55 V, dropping to 1.40 V within the first two hours of assembly. The cell voltage was observed to have stabilized at this voltage, and the OCV remained constant at 1.40 V for a period of 24 h. The fabricated cell was reasonably stable in the open cell condition.

An NFC hydrogel was synthesized by Poosapati et al. [156] by adding gelatine, polyacrylic acid (PAA), and potassium hydroxide (KOH) as additives. The hydrogel with the most appropriate amounts of additives got an ionic conductivity of 0.1 $S{\cdot}cm^{-1}$, representing an increase of five orders of magnitude from the pristine hydrogel. This report concluded that the small amounts of additive present helped enhance the mechanical stability and ionic conductivity by changing the degree of crystallinity and ionic concentration in the hydrogel layers.

Another approach for cellulose applications has been made by employing it as a soaked separator electrolyte. Zhang et al. [178] studied a functionalized with quaternary ammonium (QA) laminate-structured nanocellulose/GO membrane, developed for a hydroxide-conducting electrolyte for zinc-air batteries. Herein, cellulose was utilized to interconnect the framework to integrate GO into a flexible membrane with higher water content. Achieving a laminated cross-linked structure eliminated the risk of pushing water out of the membrane when handling or bending, besides good adhesion to the electrodes. The membrane's enlarged d-spacing enhanced the mobility of hydroxide ions by vehicle mechanism, besides its lower activation energy. Water molecules also could have caused mobility by the Grotthus mechanism. At 70 °C, the ionic conductivity of 0.0588 $S{\cdot}cm^{-1}$ and an OCV of 1.4 V was achieved.

5.2. Chitosan

Chitosan is environmentally friendly and an excellent membrane-forming polymer material. It is known for being non-toxic, biodegradable, and biocompatible, making it a good solution for many electrochemical applications that can be modified to get electrolytes. Chitosan is produced from the deacetylation reaction of chitin. Chitin is a natural polysaccharide generally found in the exoskeleton of arthropods and various fungi [179]. Chitosan is widely applied in lots of fields, as in biotechnology, biomedicine [180]. Its molecule presents several polar groups, such as hydroxyl and amino groups, forming complexes with inorganic salts. However, pristine chitosan shows a very low ionic conductivity (10^{-9} $S{\cdot}cm^{-1}$) [181], the fact that it is tried to be enhanced by the addition of salts and fillers. Very little work has been done for chitosan NCPE applied in zinc or magnesium batteries. However, some authors have endeavored to implement nanofillers for lithium electrolyte applications [160,182–188], so the most noteworthy ones are now discussed.

Hexanoyl chitosan was employed as a polymer matrix with TiO_2 as filler and lithium perchlorate ($LiClO_4$) as doping salt to design NCPEs by Muhammad et al. [159]. The electrolyte system was characterized by impedance spectroscopy. It was shown that the increment in conductivity was caused by the increase in the mobility of free ions and the increase in the free ion concentration. The XRD results obtained for this electrolyte in a posterior work of the authors [189] confirmed the decrease in crystallinity, leading to the expected increase in conductivity that was modeled by the Rice and Roth model.

A system of the same composition based on hexanoyl chitosan + $LiClO_4$ + TiO_2 was reported by Winie et al. [190], who reported the complexation of the polymer and the salt as a result of the shift of $N(COR_2)$, O=C-NHR, and OCOR bands of hexanoyl chitosan to lower wavenumbers, and supported the use of chitosan as a polymer host in terms of the presence of lone pair electrons at the nitrogen and oxygen atoms where inorganic salts can be solvated. Results showed that both dielectric constant and dielectric loss decreased with increased frequency and increased with increased temperature.

Aziz et al. [162] reported novel chitosan-ammonium thiocyanate (NH_4SCN) complexes doped with nanosized Al_2O_3 filler. FT-IR and XRD confirmed the complexation between the cation of the salt and the donor atom in chitosan polymer. The high filler content increased the T_g value since it increased the crystallinity of the sample, as depicted by XRD. Alumina was employed in chitosan and lithium triflate ($Li(Tf)_2$) [191], where the AC conductivity studies showed the promising features already told for this kind of biopolymer system.

In a work by Navaratnam et al. [163] chitosan was used as the host polymer in a designed system consisting on $LiCF_3SO_3$ as the dopant salt, EC and PC as the plasticizers, and different concentrations of SiO_2 as the inorganic filler. The obtained ionic conductivity for this system was very low to consider for practical applications. However, in a following article by Rosli et al. [161], it was presented a study comparing the type of filler and their effect on the electrical properties in the polymer electrolyte, resulting in a higher conductivity enhancement brought about by TiO_2 compared to SiO_2 for the system hexanoyl chitosan-$LiClO_4$ polymer electrolyte. This finding can be understood by the more acidic nature of TiO_2, which promoted a greater degree of salt dissociation. Zirconia was employed as a nanofiller for the previous electrolyte composition of $LiClO_4$ as salt and chitosan by Sudaryanto et al. [164], and the films were characterized by XRD and EIS. The obtained results were quite comparable to the previously discussed work since the ionic conductivity was slightly higher. Besides, the obtained ion transference number of 0.55 was considered quite enough to apply it in an ion battery.

For magnesium batteries, a GPE based on chitosan, magnesium triflate, and EMITf was developed by Wang et al. [192]. The results showed that the Mg-ion mechanism was the complexation and decomplexation of Mg^{2+} with amine band (NH_2) from chitosan. The relaxation time of the electrolyte membrane was as low as 1.25×10^{-6} s, indicating that the mobility of ions was relatively high. The electrochemical properties of this GPE, presented in Table 7, were considered a precedent for future practical applications, and some latest reports were reported until the present date for chitosan polymer electrolytes for EDLC devices [193–195]. Still, they are out of the scope of this review. To the best of our knowledge, no studies are reporting NCPEs made of chitosan for magnesium batteries.

5.3. Starch

Starch is one of the most popular renewables and biodegradable polymers found as granules in plants. It is composed of a mixture of linear amylose (α(1,4) linked anhydroglucose) and branched amylopectin (α(1,6) linked anhydroglucose) polysaccharide chains. At the same time, it can undergo derivatization reactions. It is introduced some functional groups into the starch molecule, resulting in the alteration of its gelatinization, pasting, and retrogradation behavior [196]. Starch is known to be abundant in nature due to its wide variety of sources, consisting of several kinds of food plants from which it comes. Hence, besides its application in the food industry, it is applied industrially as binders and adhesives, absorbents and encapsulants, as coatings and sizes on paper, textiles, and carpets [197]. In addition, efforts have been made to use starch to create thermoplastic materials [198,199].

In terms of application for magnesium batteries, potato starch was doped with magnesium acetate ($Mg(C_2H_3O_2)_2$) [165], and the effect of glycerol and 1-butyl-3-methylimidazolium chloride (BmImCl) was studied in terms of conductivity and dielectric properties. It was concluded that too much plasticizer causes the salt to recrystallize, causing the cations to hardly coordinate at the polar atoms, decreasing the ionic conductivity. The effect of salt concentration in the biopolymer electrolyte matrix is demonstrated through dissociated ions model (Figure 15). Rice starch was employed for an NCPE composed of lithium iodide (LiI), 1-methyl-3-propylimidazolium iodide (MPII) as an ionic liquid, and TiO_2 nanopowder by Khanmirzaei et al. [166]. The resulting electrolyte was employed to build a DSSC, showing an efficiency of 0.17 at 1000 $W{\cdot}m^{-2}$ light intensity.

From the variety of sources available for starch, a biodegradable corn starch–lithium perchlorate ($LiClO_4$)-based SPE with the addition of nano-sized fumed silica [167] was prepared by solution casting technique. FT-IR results confirmed some complexation between corn starch, $LiClO_4$, and silica. Excessive SiO_2 content decreased the ionic conductivity through agglomeration of particles and cross-linking in the polymer, showed by DSC, TGA, and SEM studies. The same investigation group reported the NCPE applied in electric double-layer capacitors (EDLCs) [168]. The device was characterized by CV, galvanostatic charge-discharge, and AC impedance spectroscopy. The discharge characteristics were almost linear, which confirmed the capacitive behavior of the EDLC cell. The fabricated EDLC cells performed good cyclability up to 500 cycles with more than 90% coulombic efficiency.

A solid electrolyte designed by blending chitosan with corn starch for application in an electrochemical double-layer capacitor (EDLC) and proton batteries was reported [169]. From transference number measurements (TNM), the electrolytes' transference number of ion (t_{ion}) showed that ion is the dominant conducting species. The transference number of cation (t_+) for the highest conducting electrolyte was found to be 0.56. Linear sweep voltammetry (LSV) result confirmed the suitability of the highest conducting electrolyte to be used to fabricate EDLC and proton batteries. The open-circuit potential (OCP) of the primary proton batteries for 48 h was lasted at (1.54 ± 0.02) V, while that of secondary proton batteries lasted at (1.58 ± 0.01) V.

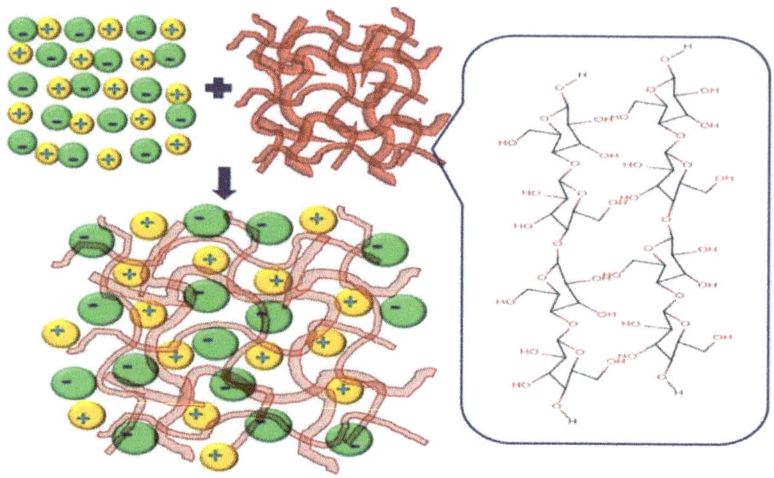

Figure 15. Schematic diagram showing ion dissociation in biopolymer salt matrix. Reproduced with permission from Renew. Sustain. Energy Rev., 65, Singh et al., Perspectives for solid biopolymer electrolytes in dye sensitized solar cell and battery application, 1098–1117, 2016 [50].

Masri et al. [170] used sago powder (starch from various tropical palms) to design a sago-KOH GPE. Then, it was employed in an experimental Zn-air battery using a porous Zn electrode as the anode. The battery showed outstanding discharge capacity and practical capacity obtained of 505 mAh·g^{-1}. In parallel, Zahid et al. [200] designed a GPE for zinc-air batteries based on cassava (Manihot esculenta), one of the most essential starch sources in tropical and subtropical areas. The highest ionic conductivity obtained was 4.34×10^{-3} S·cm^{-1}.

5.4. *Carrageenan*

Carrageenan is a linear sulfated polysaccharide polymer extracted from a marine red seaweed called Rhodophyceae and *Kappaphycus alvarezii*. It is consisted of repeating units of (1,3)-D-glucopyranose and (1,4)-3,6-anhydro-α-D-glucopyranose [201]. Besides, based on the number of sulfate groups, it is classified into three types: Kappa (κ)-carrageenan (one sulfate per disaccharide), iota (ι)-carrageenan (two sulfates per disaccharide, and lambda (λ)-carrageenan (three sulfates per disaccharide). Carrageenan is hydrophilic due to the presence of hydroxyl groups and the mentioned sulfate groups in it. In terms of electrochemical properties, this polymer is known for being in rich hydroxyl groups and oxygen atoms which are essential for interaction and coordination with cations [202]. Carrageenan has been used in various applications, including food, pharmaceutical, and cosmetic industries as viscosity builders, gelling agents, and stabilizers [203], even proving to have anti-tumor and anti-angiogenic activity [204], besides being applied in drug delivery systems and other biomedical applications [205]. In this field, Sabbagh et al. [206] have investigated the nanocomposite positive effects on structural, functional, morphological, and thermal properties of carrageenan hydrogels, obtaining promising results for drug-delivery systems. Moreover, it is known that carrageenan could be used as a prominent electrolyte in electrochemical devices with suitable modifications.

A rechargeable quasi-solid-state zinc ion battery using κ-carrageenan bio-polymer electrolyte was reported by Huang et al. [171]. The mechanical robustness of the electrolyte was reinforced by using a rice paper scaffold, which reduced the chances of short circuits as well. The κ-carrageenan electrolyte was found to be highly conductive. Furthermore, electrolyte production did not need water and oxygen-free environment or other protection measures, which is ideal for scaling up production. The zinc ion battery assembled with this biopolymer electrolyte also showed excellent cycling stability; 80% of its initial capacity still remained even the cyclic number extended to 450 cycles at 6.0 $A{\cdot}g^{-1}$ (Figure 16a). The morphology of cathode and anode materials remained after 450 cycles, almost unchanged compared to the morphology of the pristine MnO_2 and Zn (Figure 16b,c). Experimental results showed that the batteries maintained the discharge profile and AC impedance spectra after the test (Figure 16d). Besides, after 300 bending cycles, 95% capacity was retained (Figure 16e). This work brought new research opportunities in developing low-cost, flexible solid-state zinc ion batteries using green natural polymer, besides being capable of powering a timer under bending condition (Figure 16f), and also powered a timer when the entire device was immersed in water, demonstrating its good waterproofness (Figure 16g).

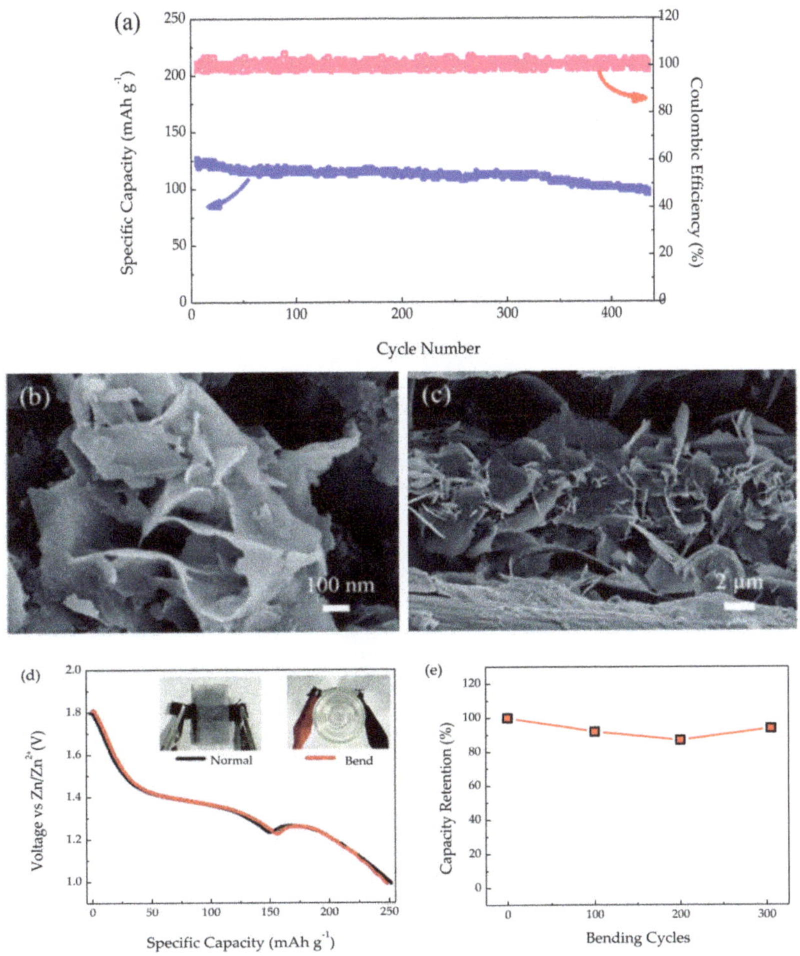

Figure 16. *Cont.*

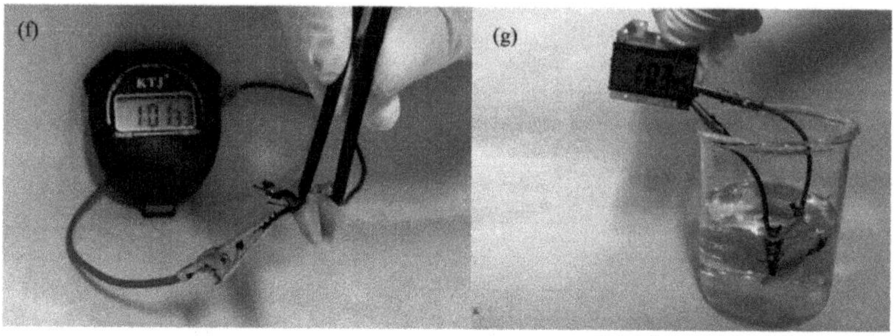

Figure 16. (**a**) Cycling stability of the solid-state ZIBs with KCR electrolyte cycled at 6.0 $A \cdot g^{-1}$ and corresponding coulombic efficiency. SEM images of (**b**) the MnO_2 cathode and (**c**) the electroplated Zn anode after 450 charge/discharge cycles. (**d**) Discharge curves under normal and bending conditions. (**e**) The bending test of solid-state ZIBs with KCR electrolyte for 300 cycles. (**f**) A solid-state ZIB with KCR electrolyte powers a timer under 180 degrees of bending conditions. (**g**) A solid-state ZIB with KCR electrolyte powers a timer when the battery is fully immersed in the water. Reproduced with permission from RSC Adv., 9, Flexible quasi-solid-state zinc ion batteries enabled by highly conductive carrageenan bio-polymer electrolyte, 16313–16319, 2019 [171].

For magnesium ion conduction, an SPE consisted of κ-carrageenan with $MgCl_2$ salt was designed by Sangeetha et al. [172], employing it in a primary magnesium battery. The resulting OCV for the battery was 2.17 V. In another report, κ-carrageenan containing tri-iodide/iodide redox couple was modified for dye-sensitized solar cell applications with nanofillers such as TiO_2, iron (III) oxide Fe_2O_3, and halloysite by Chan et al. [207]. For this system, the addition of various fillers to the PE system increased the dissociation of iodide ions and improved the ionic conductivity of the cells. DSSC characterization revealed a low efficiency due to relatively high charge transfer resistances at the TiO_2/dye/electrolyte interface.

5.5. Agar

Agar is defined as a strong gelling hydrocolloid from marine algae. It is constituted by repetitive units of D-galactose and 3,6-anhydro-L-galactose, with few variations and low content of sulfate esters [208]. Agar is widely known because of its gelling power, based exclusively on the hydrogen bonds formed among its linear galactan chains, providing excellent reversibility. Its unique properties make it suitable for many applications, especially in preparing microbiological culture media [209]. It is widely used in the food industry, in cosmetics, and in microbiology [210]. In terms of electrochemical properties, agar forms a slightly viscous solution on dissolving in hot water and then turns into a thermo-reversible gel when cooled down. Agar attracts attention because of its best film-forming capability, used in synthesizing agar hydrogel used as an electrode binder in fuel cells [211].

A novel PE based on agar and doped with NH_4SCN was prepared for zinc cells by Selvalakshmi et al. [173]. The obtained results ensured that it is a good candidate for a low-cost biopolymer electrolyte membrane for fuel cell applications and solid-state devices. Similarly, Alves et al. [174] prepared a PE-based on agar doped with magnesium triflate for magnesium ion conduction, but the obtained results were not suitable for practical applications.

In terms of the implementation of nanofillers, some studies of agar applied in electrolytes for dye-sensitized solar cells have been presented. A polysaccharide GPE composed of agar in 1-methyl-2-pyrrolidinone (NMP) as a polymer matrix, LiI)/iodine (I_2) as a redox couple, and TiO_2 nanoparticles as fillers were reported by Yang et al. [212]. Results showed that optimizing the electrolyte composition, such as agar and TiO_2 concentration, is necessary to improve the energy conversion efficiency of the DSSCs. Likewise, similar results were obtained for the same electrolyte system by Wang et al. [175], where the effects of LiI concentration were analyzed.

All the presented results so far show that NCPEs based on biopolymers solve the issues presented by all the kinds of PEs (increasing the low ionic conductivities and improving the mechanical properties). Furthermore, nanofillers have been shown to help prevent the dissolution of the salts provoked by the polar groups presented in the polysaccharide structure of most of the studied biopolymers, enhancing the ionic conductivity. Moreover, in contrast to the synthetic polymers mentioned in the previous sections, the presence of a wide variety of functional groups in biopolymers make them capable of showing various kinds of bonds and intermolecular forces. This feature is of vast importance in terms of enhancing the mechanical stability of the matrix. Nevertheless, to confirm the benefits of these kinds of forces, more attention to mechanical testing is recommended to determine the best application for these matrixes after paying attention to the alternates of material processing available so far.

6. Conclusions

According to the literature available so far, the key aspects related to nanocomposite polymer electrolytes for batteries composed of zinc or magnesium have been presented and discussed along with the review to know their suitability for application in rechargeable cells.

The copolymer matrix composed of PVDF-co-HFP has been shown to hold on to the ionic liquid and retain it in the membranes. Moreover, according to the results by the number of publications so far, the improved results in conductivity obtained for this matrix along with the addition of nanofillers, provoke a space-charge region, understood by the evistence of free electrons at the surface of the nanocomposite, facilitating the new kinetic path for ionic transport and polymer segmental motion. This mechanism ensures the electrolyte to be capable of ion transference. Improved high ionic conductivity and better thermal and mechanical stability compared to liquid electrolyte systems have been confirmed. Moreover, when a specific percentage of nanofiller is added, a decrease in ionic conductivity is observed. Excessive fillers could provoke this in the NCPE that may trigger the formation of ion pairs and ion aggregation, such as the non-conducting phase presented as an electrically inert component blocking ion transport.

By adding nanoparticles, it has been possible to reduce the degree of crystallinity of the polymers, an aspect proven in the investigations on PEO electrolytes. This feature is a crucial issue because the membrane's amorphous degree is responsible for conduction along with the electrolyte. Furthermore, conductivity is improved because the nanoparticles can act as a solid plasticizer, ensuring electrochemical properties and mechanical strength. Besides, it is discussed the NCPEs assembled with other synthetic polymers. Despite some minor variations, the results presented replicate the argument of improving the electrolyte properties by adding nanofillers, setting a solid precedent regarding the applicability of these polymers, where not much research has been found with the discussed approach.

Biopolymer electrolytes based on natural molecules have shown comparable ion conduction and electrochemical properties with traditional synthetic polymer electrolytes as the ones discussed in previous sections. Besides, when discussing stability, the natural polymer-based electrolytes are comparable, if not better, than the synthetic ones. The processes for enhancing their properties are reachable by the same methods. The conduction mechanism in both electrolytes is the same and is explained in terms of the exchange of ions between complexed sites. Moreover, their abundance, low cost, and easier processing ability make biopolymer electrolytes expected to bring a better future of green technologies than non-biodegradable, toxic, and harmful materials used in commercial batteries today.

When choosing which state is better for the electrolyte, between making it a solid or a gel, it is paramount to consider the device's application. Gel polymer electrolytes have been shown to be capable of being employed in conditions where flexibility is well appreciated. However, GPEs currently need mobile liquids to perform the conduction process, and the current ones present concerns in terms of stability, safety, and general sustainability. Hence, the search for more benign and environmentally-friendly mobile liquids is a current issue to increase the expectation on developing batteries based on sustainable components.

In general, for all the polymer electrolytes, the addition of nanoparticles (ZnO, MgO, TiO_2, Al_2O_3, SnO_2) has been proved to enhance the electrical conductivity by, in the least of the cases, one order of magnitude. Moreover, conductivity is improved, but cationic species' electrochemical properties, mechanical strength, and transport properties are. Ultimately, the awareness of the addition of nanofillers improving the mechanical stability and ionic conductivity is a crucial point to be explored in the production of batteries. However, the state-of-the-art is still lacking in terms of the development of NCPEs based on biopolymers. The few investigations overviewed so far set a precedent for the demand for further research with this specific approach. These attempts need to be further developed to get practical applications for the industry in large scale of polymer-based electrolyte batteries, as well as other electrochemical devices, such as bioresorbable electronic devices that include biobatteries, offering an innovative solution to the problems currently faced by biomedical applications, generating positive impacts to the wellness of human beings and the environment.

The research in this field needs to continue developing. Still, zinc and magnesium are absolutely the future of batteries that present electrolytes in solid-state. These metals are likely to replace lithium, thanks to their high energy potential, inherent safety, cost-effectiveness, and environmental-friendliness, along with the employment of the biodegradable biopolymers discussed in this article for the electrolyte. These features set the path for developing novel environment-friendly battery systems in the present world that urges for more sustainable options.

Author Contributions: Conceptualization, formal analysis, data curation, investigation, writing—original draft preparation, M.F.B.-C.; writing—review and editing, S.H.-B., V.M.C. and R.M.M.; Conceptualization, methodology, supervision, project administration, funding acquisition, writing—review and editing, J.P.T. All authors have read and agreed to the published version of the manuscript.

Funding: This research was funded by Corporación Ecuatoriana para el Desarrollo de la Investigación y Academia, CEDIA, through its CEPRA program, grant number CEPRA 2021–012 Hacia la economía circular: desarrollo de eco-empaques a partir de desechos agroindustriales and The APC was funded by Corporación Ecuatoriana para el Desarrollo de la Investigación y Academia, CEDIA.

Institutional Review Board Statement: Not applicable.

Informed Consent Statement: Not applicable.

Data Availability Statement: Not applicable.

Conflicts of Interest: The authors declare no conflict of interest. The funders had no role in the design of the study; in the collection, analyses, or interpretation of data; in the writing of the manuscript, or in the decision to publish the results.

References

1. Lorca, S.; Santos, F.; Fernández Romero, A.J. A review of the use of GPEs in zinc-based batteries. A step closer to wearable electronic gadgets and smart textiles. *Polymers* **2020**, *12*, 1–79. [CrossRef] [PubMed]
2. Jaschin, P.W.; Gao, Y.; Li, Y.; Bo, S.H. A materials perspective on magnesium-ion-based solid-state electrolytes. *J. Mater. Chem. A* **2020**, *8*, 2875–2897. [CrossRef]
3. Qiu, L.; Xiang, W.; Tian, W.; Xu, C.L.; Li, Y.C.; Wu, Z.G.; Chen, T.R.; Jia, K.; Wang, D.; He, F.R.; et al. Polyanion and cation co-doping stabilized Ni-rich Ni–Co–Al material as cathode with enhanced electrochemical performance for Li-ion battery. *Nano Energy* **2019**, *63*, 103818. [CrossRef]
4. Xu, Y.D.; Xiang, W.; Wu, Z.G.; Xu, C.L.; Li, Y.C.; Guo, X.D.; Lv, G.P.; Peng, X.; Zhong, B.H. Improving cycling performance and rate capability of Ni-rich $LiNi_{0.8}Co_{0.1}Mn_{0.1}O_2$ cathode materials by $Li_4Ti_5O_{12}$ coating. *Electrochim. Acta* **2018**, *268*, 358–365. [CrossRef]
5. Etacheri, V.; Marom, R.; Elazari, R.; Salitra, G.; Aurbach, D. Challenges in the development of advanced Li-ion batteries: A review. *Energy Environ. Sci.* **2011**, *4*, 3243–3262. [CrossRef]
6. Wang, Y.; Yi, J.; Xia, Y. Recent progress in aqueous lithium-ion batteries. *Adv. Energy Mater.* **2012**, *2*, 830–840. [CrossRef]
7. Yi, J.; Wang, C.; Xia, Y. Comparison of thermal stability between micro- and nano-sized materials for lithium-ion batteries. *Electrochem. Commun.* **2013**, *33*, 115–118. [CrossRef]
8. Yi, J.; Hou, M.Y.; Bao, H.L.; Wang, C.X.; Wang, J.Q.; Xia, Y.Y. In-situ generation of Li_2FeSiO_4/C nanocomposite as cathode material for lithium ion battery. *Electrochim. Acta* **2014**, *133*, 564–569. [CrossRef]

9. Kim, H.; Hong, J.; Park, K.Y.; Kim, H.; Kim, S.W.; Kang, K. Aqueous rechargeable Li and Na ion batteries. *Chem. Rev.* **2014**, *114*, 11788–11827. [CrossRef]
10. Wu, K.; Huang, J.; Yi, J.; Liu, X.; Liu, Y.; Wang, Y.; Zhang, J.; Xia, Y. Recent Advances in Polymer Electrolytes for Zinc Ion Batteries: Mechanisms, Properties, and Perspectives. *Adv. Energy Mater.* **2020**, *10*, 1–32. [CrossRef]
11. Xiong, P.; Zhang, L.; Chen, Y.; Peng, S.; Yu, G. A Chemistry and Microstructure Perspective on Ion-Conducting Membranes for Redox Flow Batteries. *Angew. Chemie Int. Ed.* **2021**, *60*, 24770–24798. [CrossRef] [PubMed]
12. Arévalo-Cid, P.; Dias, P.; Mendes, A.; Azevedo, J. Redox flow batteries: A new frontier on energy storage. *Sustain. Energy Fuels* **2021**. [CrossRef]
13. Zhang, H.; Sun, C. Cost-effective iron-based aqueous redox flow batteries for large-scale energy storage application: A review. *J. Power Sources* **2021**, *493*. [CrossRef]
14. Pankratova, G.; Bollella, P.; Pankratov, D.; Gorton, L. Supercapacitive biofuel cells. *Curr. Opin. Biotechnol.* **2022**, *73*, 179–187. [CrossRef]
15. Wang, Y.; Ruiz Diaz, D.F.; Chen, K.S.; Wang, Z.; Adroher, X.C. Materials, technological status, and fundamentals of PEM fuel cells—A review. *Mater. Today* **2020**, *32*, 178–203. [CrossRef]
16. Wang, Y.; Seo, B.; Wang, B.; Zamel, N.; Jiao, K.; Adroher, X.C. Fundamentals, materials, and machine learning of polymer electrolyte membrane fuel cell technology. *Energy AI* **2020**, *1*, 100014. [CrossRef]
17. Pollock, T.M. Weight loss with magnesium alloys. *Science* **2010**, *328*, 986–987. [CrossRef] [PubMed]
18. Wu, X.F.; Neumann, H. Zinc-catalyzed organic synthesis: C-C, C-N, C-O bond formation reactions. *Adv. Synth. Catal.* **2012**, *354*, 3141–3160. [CrossRef]
19. Anyadike, N. *Lead and Zinc*; Woodhead Publishing Ltd.: Sawston, UK, 2015; pp. 2010–2011.
20. USGS. Lithium statistics and information. *U.S. Geol. Surv.* **2020**, *53*, 98–99.
21. Chen, M.; Ma, X.; Chen, B.; Arsenault, R.; Karlson, P.; Simon, N.; Wang, Y. Recycling End-of-Life Electric Vehicle Lithium-Ion Batteries. *Joule* **2019**, *3*, 2622–2646. [CrossRef]
22. Pan, H.; Shao, Y.; Yan, P.; Cheng, Y.; Han, K.S.; Nie, Z.; Wang, C.; Yang, J.; Li, X.; Bhattacharya, P.; et al. Reversible aqueous zinc/manganese oxide energy storage from conversion reactions. *Nat. Energy* **2016**, *1*, 16039. [CrossRef]
23. Kundu, D.; Adams, B.D.; Duffort, V.; Vajargah, S.H.; Nazar, L.F. A high-capacity and long-life aqueous rechargeable zinc battery using a metal oxide intercalation cathode. *Nat. Energy* **2016**, *1*, 16119. [CrossRef]
24. Saha, P.; Datta, M.K.; Velikokhatnyi, O.I.; Manivannan, A.; Alman, D.; Kumta, P.N. Rechargeable magnesium battery: Current status and key challenges for the future. *Prog. Mater. Sci.* **2014**, *66*, 1–86. [CrossRef]
25. Wang, F.; Borodin, O.; Gao, T.; Fan, X.; Sun, W.; Han, F.; Faraone, A.; Dura, J.A.; Xu, K.; Wang, C. Highly reversible zinc metal anode for aqueous batteries. *Nat. Mater.* **2018**, *17*, 543–549. [CrossRef]
26. Deivanayagam, R.; Ingram, B.J.; Shahbazian-Yassar, R. Progress in development of electrolytes for magnesium batteries. *Energy Storage Mater.* **2019**, *21*, 136–153. [CrossRef]
27. Liu, F.; Chen, Z.; Fang, G.; Wang, Z.; Cai, Y.; Tang, B.; Zhou, J.; Liang, S. V_2O_5 Nanospheres with Mixed Vanadium Valences as High Electrochemically Active Aqueous Zinc-Ion Battery Cathode. *Nano-Micro Lett.* **2019**, *11*, 1–11. [CrossRef]
28. Fang, G.; Zhu, C.; Chen, M.; Zhou, J.; Tang, B.; Cao, X.; Zheng, X.; Pan, A.; Liang, S. Suppressing Manganese Dissolution in Potassium Manganate with Rich Oxygen Defects Engaged High-Energy-Density and Durable Aqueous Zinc-Ion Battery. *Adv. Funct. Mater.* **2019**, *29*. [CrossRef]
29. Boaretto, N.; Meabe, L.; Martinez-Ibañez, M.; Armand, M.; Zhang, H. Review—Polymer Electrolytes for Rechargeable Batteries: From Nanocomposite to Nanohybrid. *J. Electrochem. Soc.* **2020**, *167*, 070524. [CrossRef]
30. Zhao, J.; Zha, J.; Zeng, Z.; Tan, C. Recent advances in wearable self-powered energy systems based on flexible energy storage devices integrated with flexible solar cells. *J. Mater. Chem. A* **2021**, *9*, 18887–18905. [CrossRef]
31. Volontsevich, D.; Strimovskyi, S.; Veretennikov, I.; Sivykh, D.; Karpov, V. The Choice of the Electric Energy Storage Device Type for the Hybrid Power Drive of Military Wheeled Vehicles. In *International Conference Innovation in Engineering*; Springer: Cham, Switzerland, 2022; pp. 201–212.
32. Luo, Y.; Wu, Y.; Li, B.; Qu, J.; Feng, S.P.; Chu, P.K. Optimization and cutting-edge design of fuel-cell hybrid electric vehicles. *Int. J. Energy Res.* **2021**, *45*, 18392–18423. [CrossRef]
33. He, W.; Zuo, S.; Xu, X.; Zeng, L.; Liu, L.; Zhao, W.; Liu, J. Challenges and strategies of zinc anode for aqueous zinc-ion batteries. *Mater. Chem. Front.* **2021**, *5*, 2201–2217. [CrossRef]
34. Ye, T.; Li, L.; Zhang, Y. Recent Progress in Solid Electrolytes for Energy Storage Devices. *Adv. Funct. Mater.* **2020**, *30*, 1–20. [CrossRef]
35. Zhao, C.; Liu, L.; Qi, X.; Lu, Y.; Wu, F.; Zhao, J.; Yu, Y.; Hu, Y.S.; Chen, L. Solid-State Sodium Batteries. *Adv. Energy Mater.* **2018**, *8*, 1703012. [CrossRef]
36. Wood, K.N.; Kazyak, E.; Chadwick, A.F.; Chen, K.-H.; Zhang, J.-G.; Thornton, K.; Dasgupta, N.P. Dendrites and Pits: Untangling the Complex Behavior of Li Metal Anodes through Operando Video Microscopy. *ECS Meet. Abstr.* **2017**, *MA2017-01*, 518. [CrossRef] [PubMed]
37. Wu, F.; Yuan, Y.X.; Cheng, X.B.; Bai, Y.; Li, Y.; Wu, C.; Zhang, Q. Perspectives for restraining harsh lithium dendrite growth: Towards robust lithium metal anodes. *Energy Storage Mater.* **2018**, *15*, 148–170. [CrossRef]

38. Zhang, H.; Li, C.; Piszcz, M.; Coya, E.; Rojo, T.; Rodriguez-Martinez, L.M.; Armand, M.; Zhou, Z. Single lithium-ion conducting solid polymer electrolytes: Advances and perspectives. *Chem. Soc. Rev.* **2017**, *46*, 797–815. [CrossRef]
39. Hallinan, D.T.; Villaluenga, I.; Balsara, N.P. Polymer and composite electrolytes. *MRS Bull.* **2018**, *43*, 775–781. [CrossRef]
40. Mindemark, J.; Lacey, M.J.; Bowden, T.; Brandell, D. Beyond PEO—Alternative host materials for Li+-conducting solid polymer electrolytes. *Prog. Polym. Sci.* **2018**, *81*, 114–143. [CrossRef]
41. Wan, J.; Xie, J.; Mackanic, D.G.; Burke, W.; Bao, Z.; Cui, Y. Status, promises, and challenges of nanocomposite solid-state electrolytes for safe and high performance lithium batteries. *Mater. Today Nano* **2018**, *4*, 1–16. [CrossRef]
42. Yu, J.; Lyu, Y.Q.; Liu, J.; Effat, M.B.; Kwok, S.C.T.; Wu, J.; Ciucci, F. Enabling non-flammable Li-metal batteries via electrolyte functionalization and interface engineering. *J. Mater. Chem. A* **2019**, *7*, 17995–18002. [CrossRef]
43. Weston, J.E.; Steele, B.C.H. Effects of inert fillers on the mechanical and electrochemical properties of lithium salt-poly(ethylene oxide) polymer electrolytes. *Solid State Ionics* **1982**, *7*, 75–79. [CrossRef]
44. Zhao, Q.; Stalin, S.; Zhao, C.Z.; Archer, L.A. Designing solid-state electrolytes for safe, energy-dense batteries. *Nat. Rev. Mater.* **2020**, *5*, 229–252. [CrossRef]
45. Arya, A.; Sharma, A.L. Electrolyte for energy storage/conversion (Li^+, Na^+, Mg^{2+}) devices based on PVC and their associated polymer: A comprehensive review. *J. Solid State Electrochem.* **2019**, *23*, 997–1059. [CrossRef]
46. Park, B.; Schaefer, J.L. Review—Polymer Electrolytes for Magnesium Batteries: Forging Away from Analogs of Lithium Polymer Electrolytes and Towards the Rechargeable Magnesium Metal Polymer Battery. *J. Electrochem. Soc.* **2020**, *167*, 070545. [CrossRef]
47. Huy, V.P.H.; So, S.; Hur, J. Inorganic fillers in composite gel polymer electrolytes for high-performance lithium and non-lithium polymer batteries. *Nanomaterials* **2021**, *11*, 1–40. [CrossRef]
48. Lu, K.; Jiang, T.; Hu, H.; Wu, M. Hydrogel Electrolytes for Quasi-Solid Zinc-Based Batteries. *Front. Chem.* **2020**, *8*, 1–12. [CrossRef] [PubMed]
49. Nakajima, H.; Dijkstra, P.; Loos, K. The recent developments in biobased polymers toward general and engineering applications: Polymers that are upgraded from biodegradable polymers, analogous to petroleum-derived polymers, and newly developed. *Polymers* **2017**, *9*, 523. [CrossRef] [PubMed]
50. Singh, R.; Polu, A.R.; Bhattacharya, B.; Rhee, H.W.; Varlikli, C.; Singh, P.K. Perspectives for solid biopolymer electrolytes in dye sensitized solar cell and battery application. *Renew. Sustain. Energy Rev.* **2016**, *65*, 1098–1117. [CrossRef]
51. Rayung, M.; Aung, M.M.; Azhar, S.C.; Abdullah, L.C.; Su'ait, M.S.; Ahmad, A.; Jamil, S.N.A.M. Bio-based polymer electrolytes for electrochemical devices: Insight into the ionic conductivity performance. *Materials* **2020**, *13*, 838. [CrossRef]
52. Lizundia, E.; Kundu, D. Advances in Natural Biopolymer-Based Electrolytes and Separators for Battery Applications. *Adv. Funct. Mater.* **2021**, *31*, 2005646. [CrossRef]
53. Austin Suthanthiraraj, S.; Johnsi, M. Nanocomposite polymer electrolytes. *Ionics* **2017**, *23*, 2531–2542. [CrossRef]
54. Armand, M.B.; Bruce, P.G.; Forsyth, M.; Scrosati, B.; Wieczorek, W. Polymer Electrolytes. In *Energy Materials*; Bruce, D.W., O'Hare, D., Walton, R.I., Eds.; Wiley: West Sussex, UK, 2011; pp. 1–31. ISBN 9780470997529.
55. Abdullah, M.; Lenggoro, W.; Okuyama, K. Polymer Electrolyte Nanocomposites. In *Encyclopedia of Nanoscience and Nanotechnology*; Nalwa, H.S., Ed.; American Scientific Publishers: Stevenson Ranch, CA, USA, 2004; pp. 731–762. ISBN 1-58883-064-0.
56. Li, Q.; Chen, J.; Fan, L.; Kong, X.; Lu, Y. Progress in electrolytes for rechargeable Li-based batteries and beyond. *Green Energy Environ.* **2016**, *1*, 18–42. [CrossRef]
57. Johnsi, M.; Austin Suthanthiraraj, S. Electrochemical and structural properties of a polymer electrolyte system based on the effect of CeO_2 nanofiller with PVDF-co-HFP for energy storage devices. *Ionics* **2016**, *22*, 1075–1083. [CrossRef]
58. Abbrent, S.; Plestil, J.; Hlavata, D.; Lindgren, J.; Tegenfeldt, J.; Wendsjö, Å. Crystallinity and morphology of PVdF-HFP-based gel electrolytes. *Polymer* **2001**, *42*, 1407–1416. [CrossRef]
59. Brigandi, P.J.; Cogen, J.M.; Pearson, R.A. Electrically conductive multiphase polymer blend carbon-based composites. *Polym. Eng. Sci.* **2014**, *54*, 1–16. [CrossRef]
60. Kaur, G.; Adhikari, R.; Cass, P.; Bown, M.; Gunatillake, P. Electrically conductive polymers and composites for biomedical applications. *RSC Adv.* **2015**, *5*, 37553–37567. [CrossRef]
61. Puguan, J.M.C.; Chung, W.J.; Kim, H. Ion-conductive and transparent PVdF-HFP/silane-functionalized ZrO_2 nanocomposite electrolyte for electrochromic applications. *Electrochim. Acta* **2016**, *196*, 236–244. [CrossRef]
62. Lalia, B.S.; Guillen, E.; Arafat, H.A.; Hashaikeh, R. Nanocrystalline cellulose reinforced PVDF-HFP membranes for membrane distillation application. *Desalination* **2014**, *332*, 134–141. [CrossRef]
63. Prabakaran, K.; Mohanty, S.; Nayak, S.K. PEO/PVdF–HFP electrolytes for natural dye sensitized solar cell applications: Effect of modified nano-TiO_2 on electrochemical and photovoltaic performance. *J. Mater. Sci. Mater. Electron.* **2015**, *26*, 3887–3897. [CrossRef]
64. Shin, J.; Nho, Y.C.; seon Hwang, I.; Fei, G.; Kim, A.R.; Nahm, K.S. Irradiated PVdF-HFP-tin oxide composite membranes for the applications of direct methanol fuel cells. *J. Memb. Sci.* **2010**, *350*, 92–100. [CrossRef]
65. Xie, L.; Huang, X.; Yang, K.; Li, S.; Jiang, P. "Grafting to" route to PVDF-HFP-GMA/$BaTiO_3$ nanocomposites with high dielectric constant and high thermal conductivity for energy storage and thermal management applications. *J. Mater. Chem. A* **2014**, *2*, 5244–5251. [CrossRef]
66. Zhu, L.; Wang, Q. Novel ferroelectric polymers for high energy density and low loss dielectrics. *Macromolecules* **2012**, *45*, 2937–2954. [CrossRef]

67. Xiong, M.; Tang, H.; Wang, Y.; Lin, Y.; Sun, M.; Yin, Z.; Pan, M. Expanded polytetrafluoroethylene reinforced polyvinylidenefluoride-hexafluoropropylene separator with high thermal stability for lithium-ion batteries. *J. Power Sources* **2013**, *241*, 203–211. [CrossRef]
68. Ataollahi, N.; Ahmad, A.; Hamzah, H.; Rahman, M.Y.A.; Mohamed, N.S. Preparation and Characterization of PVDF-HFP/MG49 Based Polymer Blend Electrolyte. *Int. J. Electrochem. Sci* **2012**, *7*, 6693–6703.
69. Ma, T.; Cui, Z.; Wu, Y.; Qin, S.; Wang, H.; Yan, F.; Han, N.; Li, J. Preparation of PVDF based blend microporous membranes for lithium ion batteries by thermally induced phase separation: I. Effect of PMMA on the membrane formation process and the properties. *J. Memb. Sci.* **2013**, *444*, 213–222. [CrossRef]
70. Yang, Q.; Deng, N.; Chen, J.; Cheng, B.; Kang, W. The recent research progress and prospect of gel polymer electrolytes in lithium-sulfur batteries. *Chem. Eng. J.* **2020**, 127427. [CrossRef]
71. Zhou, H. New energy storage devices for post lithium-ion batteries. *Energy Environ. Sci.* **2013**, *6*, 2256. [CrossRef]
72. Maheshwaran, C.; Mishra, K.; Kanchan, D.K.; Kumar, D. Mg^{2+} conducting polymer gel electrolytes: Physical and electrochemical investigations. *Ionics* **2020**, *26*, 2969–2980. [CrossRef]
73. Ponmani, S.; Prabhu, M.R. Development and study of solid polymer electrolytes based on PVdF-HFP/PVAc: Mg $(ClO_4)_2$ for Mg ion batteries. *J. Mater. Sci. Mater. Electron.* **2018**, *29*, 15086–15096. [CrossRef]
74. Oh, J.S.; Ko, J.M.; Kim, D.W. Preparation and characterization of gel polymer electrolytes for solid state magnesium batteries. *Electrochim. Acta* **2004**, *50*, 903–906. [CrossRef]
75. Pandey, G.P.; Agrawal, R.C.; Hashmi, S.A. Performance studies on composite gel polymer electrolytes for rechargeable magnesium battery application. *J. Phys. Chem. Solids* **2011**, *72*, 1408–1413. [CrossRef]
76. Pandey, G.P.; Agrawal, R.C.; Hashmi, S.A. Magnesium ion-conducting gel polymer electrolytes dispersed with fumed silica for rechargeable magnesium battery application. *J. Solid State Electrochem.* **2011**, *15*, 2253–2264. [CrossRef]
77. Sharma, J.; Hashmi, S. Magnesium ion-conducting gel polymer electrolyte nanocomposites: Effect of active and passive nanofillers. *Polym. Compos.* **2019**, *40*, 1295–1306. [CrossRef]
78. Patel, S.; Kumar, R. Effect of Al_2O_3 on electrical properties of polymer electrolyte for electrochemical device application. *Mater. Today Proc.* **2021**, *46*, 2175–2178. [CrossRef]
79. Pandey, G.P.; Agrawal, R.C.; Hashmi, S.A. Magnesium ion-conducting gel polymer electrolytes dispersed with nanosized magnesium oxide. *J. Power Sources* **2009**, *190*, 563–572. [CrossRef]
80. Patel, S.; Kumar, R. Synthesis and characterization of magnesium ion conductivity in PVDF based nanocomposite polymer electrolytes disperse with MgO. *J. Alloys Compd.* **2019**, *789*, 6–14. [CrossRef]
81. Pandey, G.P.; Agrawal, R.C.; Hashmi, S.A. Electrical and electrochemical properties of magnesium ion conducting composite gel polymer electrolytes. *J. Phys. D. Appl. Phys.* **2010**, *43*, 255501. [CrossRef]
82. Nidhi Sandhya, P.; Kumar, R. PVDF-HFP based nanocomposite polymer electrolytes for energy storage devices dispersed with various nano-fillers. *AIP Conf. Proc.* **2020**, *2220*, 080044. [CrossRef]
83. Patel, S.; Kumar, R. Effect of nanoparticles on electrical properties of PVDF-based Mg^{2+} ion conducting polymer electrolytes. *Bull. Mater. Sci.* **2021**, *44*, 1–9. [CrossRef]
84. Jayanthi, S.; Kalapriya, K. Structural, transport, morphological, and thermal studies of nano barium titanate–incorporated magnesium ion conducting solid polymer electrolytes. *Polym. Polym. Compos.* **2021**. [CrossRef]
85. Patel, S.; Kumar, R. Effect of dispersion of ceramic filler on thermal, structural and transport properties of polymer electrolyte for electrochemical applications. *AIP Conf. Proc.* **2021**, *2352*, 020004. [CrossRef]
86. Kumar, B. From colloidal to composite electrolytes: Properties, peculiarities, and possibilities. *J. Power Sources* **2004**, *135*, 215–231. [CrossRef]
87. Polu, A.R.; Kumar, R.; Joshi, G.M. Effect of zinc salt on transport, structural, and thermal properties of PEG-based polymer electrolytes for battery application. *Ionics* **2014**, *20*, 675–679. [CrossRef]
88. Tafur, J.P.; Fernández Romero, A.J. Electrical and spectroscopic characterization of PVdF-HFP and TFSI-ionic liquids-based gel polymer electrolyte membranes. Influence of $ZnTf_2$ salt. *J. Memb. Sci.* **2014**, *469*, 499–506. [CrossRef]
89. Liu, J.; Khanam, Z.; Muchakayala, R.; Song, S. Fabrication and characterization of Zn-ion-conducting solid polymer electrolyte films based on PVdF-HFP/$Zn(Tf)_2$ complex system. *J. Mater. Sci. Mater. Electron.* **2020**, *31*, 6160–6173. [CrossRef]
90. Liu, J.; Ahmed, S.; Khanam, Z.; Wang, T.; Song, S. Ionic liquid-incorporated zn-ion conducting polymer electrolyte membranes. *Polymers* **2020**, *12*, 1755. [CrossRef]
91. Johnsi, M.; Suthanthiraraj, S.A. Preparation, zinc ion transport properties, and battery application based on poly(vinilydene fluoride-*co*-hexa fluoro propylene) polymer electrolyte system containing titanium dioxide nanofiller. *High Perform. Polym.* **2015**, *27*, 877–885. [CrossRef]
92. Johnsi, M.; Suthanthiraraj, S.A. Compositional effect of ZrO_2 nanofillers on a PVDF-*co*-HFP based polymer electrolyte system for solid state zinc batteries. *Chin. J. Polym. Sci.* **2016**, *34*, 332–343. [CrossRef]
93. Hashmi, S.A. Enhanced zinc ion transport in gel polymer electrolyte: Effect of nano-sized ZnO dispersion. *J. Solid State Electrochem.* **2012**, *16*, 3105–3114. [CrossRef]
94. Muda, N.; Ibrahim, S.; Kamarulzaman, N.; Mohamed, N.S. PVDF-HFP-$NH_4CF_3SO_3$-SiO_2 nanocomposite polymer electrolytes for protonic electrochemical cell. *Key Eng. Mater.* **2011**, *471–472*, 373–378. [CrossRef]
95. Tafur, J.P.; Abad, J.; Román, E.; Fernández Romero, A.J. Charge storage mechanism of MnO_2 cathodes in Zn/MnO_2 batteries using ionic liquid-based gel polymer electrolytes. *Electrochem. Commun.* **2015**, *60*, 190–194. [CrossRef]

96. Jaipal Reddy, M.; Chu, P.P. Ion pair formation and its effect in PEO:Mg solid polymer electrolyte system. *J. Power Sources* **2002**, *109*, 340–346. [CrossRef]
97. Agrawal, R.C.; Pandey, G.P. Solid polymer electrolytes: Materials designing and all-solid-state battery applications: An overview. *J. Phys. D. Appl. Phys.* **2008**, *41*, 223001. [CrossRef]
98. Prodduturi, S.; Manek, R.V.; Kolling, W.M.; Stodghill, S.P.; Repka, M.A. Solid-State Stability and Characterization of Hot-Melt Extruded Poly(ethylene oxide) Films. *J. Pharm. Sci.* **2005**, *94*, 2232–2245. [CrossRef]
99. Chawla, P.; Trivedi, S.; Pandey, K.; Tripathi, M. Dielectric Studies of [PEO: CH3COOLi]: Graphite System Synthesized by Hot Press and Solution Cast Technique. *Proc. Natl. Acad. Sci. India Sect. A Phys. Sci.* **2018**, *88*, 187–193. [CrossRef]
100. Lascaud, S.; Perrier, M.; Vallee, A.; Besner, S.; Prud'homme, J.; Armand, M. Phase Diagrams and Conductivity Behavior of Poly(ethylene oxide)-Molten Salt Rubbery Electrolytes. *Macromolecules* **1994**, *27*, 7469–7477. [CrossRef]
101. Agrawal, R.C.; Sahu, D.K.; Mahipal, Y.K.; Ashrafi, R. Investigations on ion transport properties of hot-press cast magnesium ion conducting Nano-Composite Polymer Electrolyte (NCPE) films: Effect of filler particle dispersal on room temperature conductivity. *Mater. Chem. Phys.* **2013**, *139*, 410–415. [CrossRef]
102. Feng, J.; Wang, L.; Chen, Y.; Wang, P.; Zhang, H.; He, X. PEO based polymer-ceramic hybrid solid electrolytes: A review. *Nano Converg.* **2021**, *8*, 1–12. [CrossRef] [PubMed]
103. Kumar, Y.; Hashmi, S.A.; Pandey, G.P. Ionic liquid mediated magnesium ion conduction in poly(ethylene oxide) based polymer electrolyte. *Electrochim. Acta* **2011**, *56*, 3864–3873. [CrossRef]
104. Agrawal, R.C.; Sahu, D.K.; Mahipal, Y.K.; Ashrafi, R. Ion transport property of hot-press cast Mg^{2+}-ion conducting nano-composite polymer electrolyte membranes: Study of effect of active/passive filler particle dispersal on conductivity. *Indian J. Pure Appl. Phys.* **2013**, *51*, 320–323.
105. Zaky, M.M.; Eyssa, H.M.; Sadek, R.F. Improvement of the magnesium battery electrolyte properties through gamma irradiation of nano polymer electrolytes doped with magnesium oxide nanoparticles. *J. Vinyl Addit. Technol.* **2019**, *25*, 243–254. [CrossRef]
106. Sundar, M.; Selladurai, S. Effect of fillers on magnesium-poly(ethylene oxide) solid polymer electrolyte. *Ionics* **2006**, *12*, 281–286. [CrossRef]
107. Koduru, H.K.; Marinov, Y.G.; Kaleemulla, S.; Rafailov, P.M.; Hadjichristov, G.B.; Scaramuzza, N. Fabrication and characterization of magnesium—ion-conducting flexible polymer electrolyte membranes based on a nanocomposite of poly(ethylene oxide) and potato starch nanocrystals. *J. Solid State Electrochem.* **2021**, *25*, 2409–2428. [CrossRef]
108. Shao, Y.; Rajput, N.N.; Hu, J.; Hu, M.; Liu, T.; Wei, Z.; Gu, M.; Deng, X.; Xu, S.; Han, K.S.; et al. Nanocomposite polymer electrolyte for rechargeable magnesium batteries. *Nano Energy* **2015**, *12*, 750–759. [CrossRef]
109. Carrilho-Plancha, M.J.; Rangel, C.M.; Correia De Sequeira, C.A. Electrochemical characterisation of a $Zn/(PEO)_4ZnCl_2/Nb_2O_5$ solid-state cell. *J. Solid State Electrochem.* **2012**, *16*, 665–671. [CrossRef]
110. Nancy, A.C.; Suthanthiraraj, S.A. Effect of Al2O3 nanofiller on the electrical, thermal and structural properties of PEO:PPG based nanocomposite polymer electrolyte. *Ionics* **2017**, *23*, 1439–1449. [CrossRef]
111. Karan, S.; Agrawal, R.C. Ion Transport and Materials Characterization Studies on Hot-Press Cast Zn^{2+} Conducting Nano-Composite Polymer Electrolyte (NCPE) Films: [90 PEO: 10Zn $(CF_3SO_3)_2$] + xAl_2O_3. *J. Ravishankar Univ.* **2019**, *32*, 76–83. [CrossRef]
112. Turković, A.; Pavlović, M.; Dubček, P.; Lučić-Lavčević, M.; Etlinger, B.; Bernstorff, S. SAXS/DSC Study of Polymer Electrolyte for Zn Rechargeable Nanostructured Galvanic Cells. *J. Electrochem. Soc.* **2007**, *154*, A554. [CrossRef]
113. Turković, A.; Dubček, P.; Pavlović, M.; Bernstorff, S. SAXS/DSC/WAXD Study of γ-irradiated Polymer Electrolyte for Zn Rechargeable Nanostructured Galvanic Cells. *ECS Trans.* **2009**, *16*, 437–443. [CrossRef]
114. Agrawal, R.C.; Hashmi, S.A.; Pandey, G.P. Electrochemical cell performance studies on all-solid-state battery using nano-composite polymer electrolyte membrane. *Ionics (Kiel)* **2007**, *13*, 295–298. [CrossRef]
115. Karan, S.; Sahu, M.; Sahu, T.B.; Mahipal, Y.K.; Sahu, D.K.; Agrawal, R.C. Investigations on materials and ion transport properties of Zn^{2+} conducting nano-composite polymer electrolytes (NCPEs): [(90 PEO: $10Zn(CF_3SO_3)_2$)+ xZnO]. *Mater. Today Commun.* **2017**, *13*, 269–274. [CrossRef]
116. Agrawal, S.L.; Singh, M.; Dwivedi, M.M.; Pandey, K. Investigation on ion conduction behaviour in Zn-ferrite based polymer nanocomposite electrolyte. *Fibers Polym.* **2011**, *12*, 864–874. [CrossRef]
117. Wang, M.; Emre, A.; Tung, S.; Gerber, A.; Wang, D.; Huang, Y.; Cecen, V.; Kotov, N.A. Biomimetic Solid-State Zn2+ Electrolyte for Corrugated Structural Batteries. *ACS Nano* **2019**, *13*, 1107–1115. [CrossRef]
118. Liu, J.; Guan, C.; Zhou, C.; Fan, Z.; Ke, Q.; Zhang, G.; Liu, C.; Wang, J. A Flexible Quasi-Solid-State Nickel–Zinc Battery with High Energy and Power Densities Based on 3D Electrode Design. *Adv. Mater.* **2016**, *28*, 8732–8739. [CrossRef] [PubMed]
119. Li, H.; Han, C.; Huang, Y.; Huang, Y.; Zhu, M.; Pei, Z.; Xue, Q.; Wang, Z.; Liu, Z.; Tang, Z.; et al. An extremely safe and wearable solid-state zinc ion battery based on a hierarchical structured polymer electrolyte. *Energy Environ. Sci.* **2018**, *11*, 941–951. [CrossRef]
120. Fu, J.; Zhang, J.; Song, X.; Zarrin, H.; Tian, X.; Qiao, J.; Rasen, L.; Li, K.; Chen, Z. A flexible solid-state electrolyte for wide-scale integration of rechargeable zinc–air batteries. *Energy Environ. Sci.* **2016**, *9*, 663–670. [CrossRef]
121. Li, Q.; Sun, H.Y.; Takeda, Y.; Imanishi, N.; Yang, J.; Yamamoto, O. Interface properties between a lithium metal electrode and a poly(ethylene oxide) based composite polymer electrolyte. *J. Power Sources* **2001**, *92*, 201–205. [CrossRef]

122. Rao, C.V.S.; Ravi, M.; Raja, V.; Bhargav, P.B.; Sharma, A.K.; Rao, V.V.R.N. Preparation and characterization of PVP-based polymer electrolytes for solid-state battery applications. *Iran. Polym. J.* **2012**, *21*, 531–536. [CrossRef]
123. CROW Polymer Database. Available online: http://polymerdatabase.com/index.html (accessed on 11 April 2021).
124. Sarojini, S.; Anjalai, C. AC Impedance Studies on Magnesium Ion Conducting Polymer Electrolyte System with Ethylene Carbonate as Plasticizer and MgO as Nanofiller. *Chem. Sci. Trans.* **2016**, *5*, 56–60. [CrossRef]
125. Zain, N.F.; Zainal, N.; Mohamed, N.S. The effects of MgO nanofiller to the physicochemical and ionic liquid retention properties of PEMA-MgTf2-EMITFSI nanocomposite polymer electrolytes. *Polym. Compos.* **2018**, *39*, 1500–1506. [CrossRef]
126. Shahenoor Basha, S.K.; Sunita Sundari, G.; Vijay Kumar, K.; Rao, M.C. Optical and dielectric properties of PVP based composite polymer electrolyte films. *Polym. Sci. Ser. A* **2017**, *59*, 554–565. [CrossRef]
127. Mishra, K.; Hashmi, S.A.; Rai, D.K. Nanocomposite blend gel polymer electrolyte for proton battery application. *J. Solid State Electrochem.* **2013**, *17*, 785–793. [CrossRef]
128. Candhadai Murali, S.P.; Samuel, A.S. Zinc ion conducting blended polymer electrolytes based on room temperature ionic liquid and ceramic filler. *J. Appl. Polym. Sci.* **2019**, *136*, 1–14. [CrossRef]
129. Sai Prasanna, C.M.; Austin Suthanthiraraj, S. Investigations of Zinc Ion Dissociation in Gel Polymer Electrolytes Based on Poly(vinyl chloride) and Poly(ethyl methacrylate) Blend on the Addition of Two Different Ceramic Nanofillers. *J. Inorg. Organomet. Polym. Mater.* **2019**, *29*, 483–501. [CrossRef]
130. Sai Prasanna, C.M.; Austin Suthanthiraraj, S. PVC/PEMA-based blended nanocomposite gel polymer electrolytes plasticized with room temperature ionic liquid and dispersed with nano-ZrO_2 for zinc ion batteries. *Polym. Compos.* **2019**, *40*, 3402–3411. [CrossRef]
131. Sai Prasanna, C.M.; Austin Suthanthiraraj, S. Improved zinc ion transportation in gel polymer electrolyte upon the addition of nano-sized SnO_2. *Polym. Polym. Compos.* **2020**, *28*, 54–65. [CrossRef]
132. Sownthari, K.; Suthanthiraraj, S.A. Preparation and properties of biodegradable polymer-layered silicate nanocomposite electrolytes for zinc based batteries. *Electrochim. Acta* **2015**, *174*, 885–892. [CrossRef]
133. Sownthari, K.; Suthanthiraraj, S.A. Structural and AC impedance studies on nanocomposite polymer electrolytes based on poly(ε-caprolactone). *J. Appl. Polym. Sci.* **2014**, *131*. [CrossRef]
134. Fan, X.; Liu, J.; Song, Z.; Han, X.; Deng, Y.; Zhong, C.; Hu, W. Porous nanocomposite gel polymer electrolyte with high ionic conductivity and superior electrolyte retention capability for long-cycle-life flexible zinc–air batteries. *Nano Energy* **2019**, *56*, 454–462. [CrossRef]
135. Abdullah, O.G.; Salman, Y.A.K.; Tahir, D.A.; Jamal, G.M.; Ahmed, H.T.; Mohamad, A.H.; Azawy, A.K.; Abdullah, C.; Salman, O.G.; Tahir, Y.A.K.; et al. Effect of ZnO nanoparticle content on the structural and ionic transport parameters of polyvinyl alcohol based proton-conducting polymer electrolyte membranes. *Membranes* **2021**, *11*, 163. [CrossRef] [PubMed]
136. Chen, H.W.; Lin, T.P.; Chang, F.C. Ionic conductivity enhancement of the plasticized PMMA/LiClO4 polymer nanocomposite electrolyte containing clay. *Polymer (Guildf)* **2002**, *43*, 5281–5288. [CrossRef]
137. Jäger, M.; Wilke, A. Comprehensive biocompatibility testing of a new PMMA-HA bone cement versus conventional PMMA cement in vitro. *J. Biomater. Sci. Polym. Ed.* **2003**, *14*, 1283–1298. [CrossRef]
138. Kikuchi, Y.; Hirao, M.; Ookubo, T.; Sasaki, A. Design of recycling system for poly(methyl methacrylate) (PMMA). Part 1: Recycling scenario analysis. *Int. J. Life Cycle Assess.* **2014**, *19*, 120–129. [CrossRef]
139. Su'Ait, M.S.; Ahmad, A.; Hamzah, H.; Rahman, M.Y.A. Preparation and characterization of PMMA-MG49-$LiClO_4$ solid polymeric electrolyte. *J. Phys. D. Appl. Phys.* **2009**, *42*. [CrossRef]
140. Ahmad, S.; Ahmad, S.; Agnihotry, S.A. Nanocomposite electrolytes with fumed silica in poly(methyl methacrylate): Thermal, rheological and conductivity studies. *J. Power Sources* **2005**, *140*, 151–156. [CrossRef]
141. Han, H.S.; Kang, H.R.; Kim, S.W.; Kim, H.T. Phase-separated polymer electrolyte based on poly(vinyl chloride)/poly(ethyl methacrylate) blend. *J. Power Sources* **2002**, *112*, 461–468. [CrossRef]
142. Reiter, J.; Krejza, O.; Sedlaříková, M. Electrochromic devices employing methacrylate-based polymer electrolytes. *Sol. Energy Mater. Sol. Cells* **2009**, *93*, 249–255. [CrossRef]
143. Turner, D.T.; Schwartz, A. The glass transition temperature of poly(N-vinyl pyrrolidone) by differential scanning calorimetry. *Polymer* **1985**, *26*, 757–762. [CrossRef]
144. Ramaswamy, M.; Malayandi, T.; Subramanian, S.; Srinivasalu, J.; Rangaswamy, M. Magnesium ion conducting polyvinyl alcohol–polyvinyl pyrrolidone-based blend polymer electrolyte. *Ionics* **2017**, *23*, 1771–1781. [CrossRef]
145. Majhi, P.R.; Moulik, S.P.; Burke, S.E.; Rodgers, M.; Palepu, R. Physicochemical investigations on the interaction of surfactants and salts with polyvinylpyrrolidone in aqueous medium. *J. Colloid Interface Sci.* **2001**, *235*, 227–234. [CrossRef] [PubMed]
146. Rajendran, S.; Sivakumar, M.; Subadevi, R. Investigations on the effect of various plasticizers in PVA-PMMA solid polymer blend electrolytes. *Mater. Lett.* **2004**, *58*, 641–649. [CrossRef]
147. Dubal, D.P.; Chodankar, N.R.; Kim, D.H.; Gomez-Romero, P. Towards flexible solid-state supercapacitors for smart and wearable electronics. *Chem. Soc. Rev.* **2018**, *47*, 2065–2129. [CrossRef] [PubMed]
148. Wang, Z.; Meng, X.; Wu, Z.; Mitra, S. Development of flexible zinc–air battery with nanocomposite electrodes and a novel separator. *J. Energy Chem.* **2017**, *26*, 129–138. [CrossRef]
149. Aziz, S.B. Li+ ion conduction mechanism in poly (ε-caprolactone)-based polymer electrolyte. *Iran. Polym. J.* **2013**, *22*, 877–883. [CrossRef]

150. Flieger, M.; Kantorová, M.; Prell, A.; Řezanka, T.; Votruba, J. Biodegradable plastics from renewable sources. *Folia Microbiol.* **2003**, *48*, 27–44. [CrossRef]
151. Ray, S.S.; Bousmina, M. Biodegradable polymers and their layered silicate nanocomposites: In greening the 21st century materials world. *Prog. Mater. Sci.* **2005**, *50*, 962–1079. [CrossRef]
152. Salleh, N.S.; Aziz, S.B.; Aspanut, Z.; Kadir, M.F.Z. Electrical impedance and conduction mechanism analysis of biopolymer electrolytes based on methyl cellulose doped with ammonium iodide. *Ionics* **2016**, *22*, 2157–2167. [CrossRef]
153. Huang, X.; Wang, D.; Yuan, Z.; Xie, W.; Wu, Y.; Li, R.; Zhao, Y.; Luo, D.; Cen, L.; Chen, B.; et al. A Fully Biodegradable Battery for Self-Powered Transient Implants. *Small* **2018**, *14*, 1800994. [CrossRef] [PubMed]
154. Majdecka, D.; Drami ska, S.; Stolarczyk, K.; Kizling, M.; Krysiski, P.; Golimowski, J.; Bilewicz, R. Sandwich Biobattery with Enzymatic Cathode and Zinc Anode for Powering Sensors. *ECS Trans.* **2014**, *61*, 1–7. [CrossRef]
155. Huang, X. Materials and applications of bioresorbable electronics. *J. Semicond.* **2018**, *39*, 011003. [CrossRef]
156. Poosapati, A.; Jang, E.; Madan, D.; Jang, N.; Hu, L.; Lan, Y. Cellulose hydrogel as a flexible gel electrolyte layer. *MRS Commun.* **2019**, *9*, 122–128. [CrossRef]
157. Johari, N.A.; Kudin, T.I.T.; Ali, A.M.M.; Winie, T.; Yahya, M.Z.A. Studies on cellulose acetate-based gel polymer electrolytes for proton batteries. *Mater. Res. Innov.* **2009**, *13*, 232–234. [CrossRef]
158. Johari, N.A.; Kudin, T.I.T.; Ali, A.M.M.; Yahya, M.Z.A. Effects of TiO_2 on conductivity performance of cellulose acetate based polymer gel electrolytes for proton batteries. *Mater. Res. Innov.* **2011**, *15*, s229–s231. [CrossRef]
159. Muhammad, F.H.; Subban, R.H.Y.; Winie, T. Electrical studies on hexanoyl chitosan-based nanocomposite polymer electrolytes. *AIP Conf. Proc.* **2009**, *1136*, 61–65. [CrossRef]
160. Muhammad, F.H.; Subban, R.H.Y.; Winie, T. Structural and electrical characterization of hexanoyl chitosan- $LiClO_4$-TiO_2-DMC polymer electrolytes. *Key Eng. Mater.* **2014**, *594–595*, 608–612. [CrossRef]
161. Rosli, N.H.A.; Muhammad, F.H.; Chan, C.H.; Winie, T. Effect of filler type on the electrical properties of hexanoyl chitosan- based polymer electrolytes. *Adv. Mater. Res.* **2014**, *832*, 224–227. [CrossRef]
162. Aziz, N.A.; Majid, S.R.; Yahya, R.; Arof, A.K. Conductivity, structure, and thermal properties of chitosan-based polymer electrolytes with nanofillers. *Polym. Adv. Technol.* **2011**, *22*, 1345–1348. [CrossRef]
163. Navaratnam, S.; Ramesh, K.; Basirun, W.J. Investigation of ion conducting behaviour of composite chitosan based polymer electrolytes. *Mater. Res. Innov.* **2011**, *15*, s184–s186. [CrossRef]
164. Sudaryanto; Yulianti, E.; Patimatuzzohrah. Structure and properties of solid polymer electrolyte based on chitosan and ZrO_2 nanoparticle for lithium ion battery. *AIP Conf. Proc.* **2016**, *1710*, 020003. [CrossRef]
165. Shukur, M.F.; Ithnin, R.; Kadir, M.F.Z. Ionic conductivity and dielectric properties of potato starch-magnesium acetate biopolymer electrolytes: The effect of glycerol and 1-butyl-3-methylimidazolium chloride. *Ionics* **2016**, *22*, 1113–1123. [CrossRef]
166. Khanmirzaei, M.H.; Ramesh, S. Nanocomposite polymer electrolyte based on rice starch/ionic liquid/TiO_2 nanoparticles for solar cell application. *Meas. J. Int. Meas. Confed.* **2014**, *58*, 68–72. [CrossRef]
167. Teoh, K.H.; Ramesh, S.; Arof, A.K. Investigation on the effect of nanosilica towards corn starch-lithium perchlorate-based polymer electrolytes. *J. Solid State Electrochem.* **2012**, *16*, 3165–3170. [CrossRef]
168. Teoh, K.H.; Lim, C.S.; Liew, C.W.; Ramesh, S.; Ramesh, S. Electric double-layer capacitors with corn starch-based biopolymer electrolytes incorporating silica as filler. *Ionics* **2015**, *21*, 2061–2068. [CrossRef]
169. Shukur, M.F.; Kadir, M.F.Z. Hydrogen ion conducting starch-chitosan blend based electrolyte for application in electrochemical devices. *Electrochim. Acta* **2015**, *158*, 152–165. [CrossRef]
170. Masri, M.N.; Nazeri, M.F.M.; Mohamad, A.A. Sago Gel Polymer Electrolyte for Zinc-Air Battery. *Adv. Sci. Technol.* **2010**, *72*, 305–308. [CrossRef]
171. Huang, Y.; Liu, J.; Zhang, J.; Jin, S.; Jiang, Y.; Zhang, S.; Li, Z.; Zhi, C.; Du, G.; Zhou, H. Flexible quasi-solid-state zinc ion batteries enabled by highly conductive carrageenan bio-polymer electrolyte. *RSC Adv.* **2019**, *9*, 16313–16319. [CrossRef]
172. Sangeetha, P.; Selvakumari, T.M.; Selvasekarapandian, S.; Srikumar, S.R.; Manjuladevi, R.; Mahalakshmi, M. Preparation and characterization of biopolymer K-carrageenan with $MgCl_2$ and its application to electrochemical devices. *Ionics* **2020**, *26*, 233–244. [CrossRef]
173. Selvalakshmi, S.; Vijaya, N.; Selvasekarapandian, S.; Premalatha, M. Biopolymer agar-agar doped with NH4SCN as solid polymer electrolyte for electrochemical cell application. *J. Appl. Polym. Sci.* **2017**, *134*. [CrossRef]
174. Alves, R.D.; Rodrigues, L.C.; Andrade, J.R.; Pawlicka, A.; Pereira, L.; Martins, R.; Fortunato, E.; Silva, M.M. Study and characterization of a novel polymer electrolyte based on agar doped with magnesium triflate. *Mol. Cryst. Liq. Cryst.* **2013**, *570*, 1–11. [CrossRef]
175. Wang, W.; Guo, X.; Yang, Y. Lithium iodide effect on the electrochemical behavior of agarose based polymer electrolyte for dye-sensitized solar cell. *Electrochim. Acta* **2011**, *56*, 7347–7351. [CrossRef]
176. Sudhakar, Y.N.; Selvakumar, M.D.; Krishna, B. *Biopolymer Electrolytes Fundamentals and Applications in Energy Storage*, 1st ed.; Elsevier: Amsterdam, The Netherlands; Oxford, UK, 2018; ISBN 9780128136119.
177. Wang, W.; Zhang, X.; Teng, A.; Liu, A. Mechanical reinforcement of gelatin hydrogel with nanofiber cellulose as a function of percolation concentration. *Int. J. Biol. Macromol.* **2017**, *103*, 226–233. [CrossRef]
178. Zhang, J.; Fu, J.; Song, X.; Jiang, G.; Zarrin, H.; Xu, P.; Li, K.; Yu, A.; Chen, Z. Laminated Cross-Linked Nanocellulose/Graphene Oxide Electrolyte for Flexible Rechargeable Zinc–Air Batteries. *Adv. Energy Mater.* **2016**, *6*, 1600476. [CrossRef]

179. Yahya, M.Z.A.; Arof, A.K. Effect of oleic acid plasticizer on chitosan-lithium acetate solid polymer electrolytes. *Eur. Polym. J.* **2003**, *39*, 897–902. [CrossRef]
180. Yang, R.; Li, H.; Huang, M.; Yang, H.; Li, A. A review on chitosan-based flocculants and their applications in water treatment. *Water Res.* **2016**, *95*, 59–89. [CrossRef]
181. Mohamed, N.S.; Subban, R.H.Y.; Arof, A.K. Polymer batteries fabricated from lithium complexed acetylated chitosan. *J. Power Sources* **1995**, *56*, 153–156. [CrossRef]
182. Winie, T.; Jamal, A.; Hanif, N.S.M.; Shahril, N.S.M. Hexanoyl chitosan-polystyrene blend based composite polymer electrolyte with surface treated TiO2 fillers. *Key Eng. Mater.* **2014**, *594–595*, 656–660. [CrossRef]
183. Winie, T.; Hanif, N.S.M.; Chan, C.H.; Arof, A.K. Effect of the surface treatment of the TiO_2 fillers on the properties of hexanoyl chitosan/polystyrene blend-based composite polymer electrolytes. *Ionics* **2014**, *20*, 347–352. [CrossRef]
184. Winie, T.; Mohd Shahril, N.S. Conductivity enhancement by controlled percolation of inorganic salt in multiphase hexanoyl chitosan/polystyrene polymer blends. *Front. Mater. Sci.* **2015**, *9*, 132–140. [CrossRef]
185. Aziz, S.B.; Rasheed, M.A.; Abidin, Z.H.Z. Optical and Electrical Characteristics of Silver Ion Conducting Nanocomposite Solid Polymer Electrolytes Based on Chitosan. *J. Electron. Mater.* **2017**, *46*, 6119–6130. [CrossRef]
186. Aziz, S.B. Role of dielectric constant on ion transport: Reformulated Arrhenius equation. *Adv. Mater. Sci. Eng.* **2016**, *2016*, 2527013. [CrossRef]
187. Aziz, S.B.; Abidin, Z.H.Z. Ion-transport study in nanocomposite solid polymer electrolytes based on chitosan: Electrical and dielectric analysis. *J. Appl. Polym. Sci.* **2015**, *132*, 41774. [CrossRef]
188. Muhammad, F.H.; Azmar, A.; Winie, T. Transport properties of hexanoyl chitosan-$LiClO_4$-TiO_2 composite polymer electrolyte. *AIP Conf. Proc.* **2015**, *1674*, 020029. [CrossRef]
189. Rosli, N.H.A.; Muhammad, F.H.; Subban, R.H.Y.; Winie, T. Structural and electrical studies of hexanoyl chitosan based electrolyte system. *Mater. Res. Innov.* **2011**, *15*, s94–s96. [CrossRef]
190. Winie, T.; Han, C.C.; Subban, R.H.Y. Ac conductivity and dielectric properties of hexanoyl chitosan-$LiClO_4$-TiO_2 composite polymer electrolytes. *Adv. Mater. Res.* **2011**, *335–336*, 873–880. [CrossRef]
191. Winie, T.; Hanif, N.S.M.; Rosli, N.H.A.; Subban, R.H.Y. Ac Conductivity Study of Hexanoyl Chitosan-$LiCF_3SO_3$-EC-Al_2O_3 Nanocomposite Polymer Electrolytes. *Adv. Mater. Res.* **2013**, *667*, 93–98. [CrossRef]
192. Wang, J.; Song, S.; Gao, S.; Muchakayala, R.; Liu, R.; Ma, Q. Mg-ion conducting gel polymer electrolyte membranes containing biodegradable chitosan: Preparation, structural, electrical and electrochemical properties. *Polym. Test.* **2017**, *62*, 278–286. [CrossRef]
193. Dannoun, E.M.A.; Aziz, S.B.; Brza, M.A.; Nofal, M.M.; Asnawi, A.S.F.M.; Yusof, Y.M.; Al-Zangana, S.; Hamsan, M.H.; Kadir, M.F.Z.; Woo, H.J. The study of plasticized solid polymer blend electrolytes based on natural polymers and their application for energy storage EDLC devices. *Polymers* **2020**, *12*, 1–19. [CrossRef] [PubMed]
194. Aziz, S.B.; Dannoun, E.M.A.; Hamsan, M.H.; Abdulwahid, R.T.; Mishra, K.; Nofal, M.M.; Kadir, M.F.Z.; Appetecchi, B.; Kim, D. Improving EDLC Device Performance Constructed from Plasticized Magnesium Ion Conducting Chitosan Based Polymer Electrolytes via Metal Complex Dispersion. *Membranes* **2021**, *11*, 289. [CrossRef] [PubMed]
195. Hamsan, M.H.; Aziz, S.B.; Nofal, M.M.; Brza, M.A.; Abdulwahid, R.T.; Hadi, J.M.; Karim, W.O.; Kadir, M.F.Z. Characteristics of EDLC device fabricated from plasticized chitosan:$MgCl_2$ based polymer electrolyte. *J. Mater. Res. Technol.* **2020**, *9*, 10635–10646. [CrossRef]
196. Pawlicka, A.; Sabadini, A.C.; Raphael, E.; Dragunski, D.C. Ionic conductivity thermogravimetry measurements of starch-based polymeric electrolytes. *Mol. Cryst. Liq. Cryst.* **2008**, *485*, 804–816. [CrossRef]
197. Khiar, A.S.A.; Arof, A.K. Conductivity studies of starch-based polymer electrolytes. *Ionics* **2010**, *16*, 123–129. [CrossRef]
198. Sen, A.; Bhattacharya, M. Residual stresses and density gradient in injection molded starch/synthetic polymer blends. *Polymer* **2000**, *41*, 9177–9190. [CrossRef]
199. Wang, J.; Liang, Y.; Zhang, Z.; Ye, C.; Chen, Y.; Wei, P.; Wang, Y.; Xia, Y. Thermoplastic starch plasticized by polymeric ionic liquid. *Eur. Polym. J.* **2021**, *148*, 110367. [CrossRef]
200. Zahid, A.R.M.; Masri, M.N.; Hussin, M.H.; Bakar, M.B.A. The preliminary study on cassava (*Manihot esculenta*) as gel polymer electrolyte for zinc-air battery. *AIP Conf. Proc.* **2018**, *2030*, 020278. [CrossRef]
201. Mobarak, N.N.; Jumaah, F.N.; Ghani, M.A.; Abdullah, M.P.; Ahmad, A. Carboxymethyl Carrageenan Based Biopolymer Electrolytes. *Electrochim. Acta* **2015**, *175*, 224–231. [CrossRef]
202. Moniha, V.; Alagar, M.; Selvasekarapandian, S.; Sundaresan, B.; Boopathi, G. Conductive bio-polymer electrolyte iota-carrageenan with ammonium nitrate for application in electrochemical devices. *J. Non. Cryst. Solids* **2018**, *481*, 424–434. [CrossRef]
203. De Ruiter, G.A.; Rudolph, B. Carrageenan biotechnology. *Trends Food Sci. Technol.* **1997**, *8*, 389–395. [CrossRef]
204. Yao, Z.; Wu, H.; Zhang, S.; Du, Y. Enzymatic preparation of κ-carrageenan oligosaccharides and their anti-angiogenic activity. *Carbohydr. Polym.* **2014**, *101*, 359–367. [CrossRef] [PubMed]
205. Pacheco-Quito, E.M.; Ruiz-Caro, R.; Veiga, M.D. Carrageenan: Drug Delivery Systems and Other Biomedical Applications. *Mar. Drugs* **2020**, *18*, 583. [CrossRef]
206. Sabbagh, F.; Kiarostami, K.; Khatir, N.M.; Rezania, S.; Muhamad, I.I.; Hosseini, F. Effect of zinc content on structural, functional, morphological, and thermal properties of kappa-carrageenan/NaCMC nanocomposites. *Polym. Test.* **2021**, *93*, 106922. [CrossRef]

207. Chan, S.; Paolo, J.; Bantang, O.; Bantang, J.P.; Camacho, D. Influence of Nanomaterial Fillers in Biopolymer Electrolyte System for Squaraine-based Dye-Sensitized Solar Cells. *Int. J. Electrochem. Sci* **2015**, *10*, 7696–7706.
208. Armisén, R.; Galatas, F. Agar. In *Handbook of Hydrocolloids: Second Edition*; Elsevier Inc.: Amsterdam, The Netherlands, 2009; pp. 82–107, ISBN 9781845695873.
209. Kato, S.; Yamagishi, A.; Daimon, S.; Kawasaki, K.; Tamaki, H.; Kitagawa, W.; Abe, A.; Tanaka, M.; Sone, T.; Asano, K.; et al. Isolation of previously uncultured slowgrowing bacteria by using a simple modification in the preparation of agar media. *Appl. Environ. Microbiol.* **2018**, *84*, e00807-18. [CrossRef] [PubMed]
210. Selvalakshmi, S.; Mathavan, T.; Selvasekarapandian, S.; Premalatha, M. Effect of ethylene carbonate plasticizer on agar-agar: NH4Br-based solid polymer electrolytes. *Ionics* **2018**, *24*, 2209–2217. [CrossRef]
211. An, L.; Zhao, T.S.; Zeng, L. Agar chemical hydrogel electrode binder for fuel-electrolyte-fed fuel cells. *Appl. Energy* **2013**, *109*, 67–71. [CrossRef]
212. Yang, Y.; Hu, H.; Zhou, C.H.; Xu, S.; Sebo, B.; Zhao, X.Z. Novel agarose polymer electrolyte for quasi-solid state dye-sensitized solar cell. *J. Power Sources* **2011**, *196*, 2410–2415. [CrossRef]

polymers

MDPI

Article

Replica of Bionic Nepenthes Peristome-like and Anti-Fouling Structures for Self-Driving Water and Raman-Enhancing Detection

Yen-Ting Lin [1], Chun-Hao Wu [1], Wei-Lin Syu [1], Po-Cheng Ho [1], Zi-Ling Tseng [2], Ming-Chien Yang [2], Chin-Ching Lin [3], Cheng-Chen Chen [4,*], Cheng-Cheung Chen [5,6,*] and Ting-Yu Liu [1,*]

1 Department of Materials Engineering, Ming Chi University of Technology, New Taipei City 243303, Taiwan; ken851028@gmail.com (Y.-T.L.); m10188005@mail2.mcut.edu.tw (C.-H.W.); shewinnie43@gmail.com (W.-L.S.); gf183183@gmail.com (P.-C.H.)

2 Department of Materials Science and Engineering, National Taiwan University of Science and Technology, Taipei 106335, Taiwan; grace5032731@gmail.com (Z.-L.T.); myang@mail.ntust.edu.tw (M.-C.Y.)

3 Senguan Tech Co., Ltd., Tainan City 71076, Taiwan; chinchinglin@hotmail.com

4 Department of Architecture, National Taipei University of Technology, Taipei 106344, Taiwan

5 Institute of Preventive Medicine, National Defense Medical Center, New Taipei City 23742, Taiwan

6 Graduate Institute of Medical Science, National Defense Medical Center, Taipei 11490, Taiwan

* Correspondence: coolhas@ntut.edu.tw (C.-C.C.); chengcheung_chen@mail.ndmctsgh.edu.tw (C.-C.C.); tyliu0322@gmail.com (T.-Y.L.)

Abstract: The flexible, anti-fouling, and bionic surface-enhanced Raman scattering (SERS) biochip, which has a Nepenthes peristome-like structure, was fabricated by photolithography, replicated technology, and thermal evaporation. The pattern of the bionic Nepenthes peristome-like structure was fabricated by two layers of photolithography with SU-8 photoresist. The bionic structure was then replicated by polydimethylsiloxane (PDMS) and grafting the zwitterion polymers (2-methacryloyloxyethyl phosphorylcholine, MPC) by atmospheric plasma polymerization (PDMS-PMPC). The phospholipid monomer of MPC immobilization plays an important role; it can not only improve hydrophilicity, anti-fouling and anti-bacterial properties, and biocompatibility, but it also allows for self-driving and unidirectional water delivery. Ag nanofilms (5 nm) were deposited on a PDMS (PDMS-Ag) substrate by thermal evaporation for SERS detection. Characterizations of the bionic SERS chips were measured by a scanning electron microscope (SEM), optical microscope (OM), X-ray photoelectron spectrometer (XPS), Fourier-transform infrared spectroscopy (FTIR), and contact angle (CA) testing. The results show that the superior anti-fouling capability of proteins and bacteria (*E. coli*) was found on the PDMS-PMPC substrate. Furthermore, the one-way liquid transfer capability of the bionic SERS chip was successfully demonstrated, which provides for the ability to separate samples during the flow channel, and which was detected by Raman spectroscopy. The SERS intensity (adenine, 10^{-4} M) of PDMS-Ag with a bionic structure is ~4 times higher than PDMS-Ag without a bionic structure, due to the multi-reflection of the 3D bionic structure. The high-sensitivity bionic SERS substrate, with its self-driving water capability, has potential for biomolecule separation and detection.

Keywords: Nepenthes structure; bionic replica; zwitterion polymers; self-driving water; unidirectional water delivery; surface-enhanced Raman scattering (SERS) detection

Citation: Lin, Y.-T.; Wu, C.-H.; Syu, W.-L.; Ho, P.-C.; Tseng, Z.-L.; Yang, M.-C.; Lin, C.-C.; Chen, C.-C.; Chen, C.-C.; Liu, T.-Y. Replica of Bionic Nepenthes Peristome-like and Anti-Fouling Structures for Self-Driving Water and Raman-Enhancing Detection. *Polymers* **2022**, *14*, 2465. https://doi.org/10.3390/polym14122465

Academic Editor: Seong J. Cho

Received: 9 May 2022
Accepted: 15 June 2022
Published: 17 June 2022

1. Introduction

There are many micro- or nano-scale structures with different functions and special characteristics in nature, which have not yet been discovered because of the limitations of existing detection technologies. A benefit of the progress of manufacturing and the advancement in detection techniques in recent years has been that the micro-scale structure can now easily be observed, encouraging more researchers to invest in micro biomaterial structures. For example, lotus leaves possess a super-hydrophobic and self-cleaning ability due to the unique nanostructure of their surface [1,2]. This nanostructure contains structural wax on the surface layer, which not only increases surface roughness, but also traps air between the solid and liquid interface [3]. Biomimetic technology [4–12] is a technology that imitates the 3.8 billion years of evolutionary experience of living things. Every lifestyle, growth process, and ecosystem is a source of inspiration for simulation. It has the two major contradictions of "nature but primitive" and "technology but pollution", which are currently significant [13]. Biomimetic technology has three advantages: low costs, high efficiency, and low pollution. With the proper utilization of this technology, we can achieve low resource and energy consumption while obtaining optimal production capacity and benefits. For example, aircraft coatings based on shark skin resistance [6,14–16], radar sonar based on bats [17], solar applications inspired by photosynthesis, bullet train heads based on kingfisher beaks [18,19], antifouling coatings based on frog skins [20], etc. Biomaterials have also been widely applied in SERS substrates due to their 3D periodic microstructures, such as butterfly wings, cicada wings, and rose petals [21–23]. The best case for the continuation, this study focuses on the surface structure of nepenthes.

Nepenthes lives in a barren environment. Uniquely, they can capture insects with their peristome to meet their fundamental nutrient needs [2]. Most species obtain their nutrients from trapped and food animals. The inner surface of the insect trap is covered with slippery and fragile wax [24], forming an effective trap. The peristome on the top aims to attract and capture prey by forming a smooth liquid film. Previous studies indicated that the peristome structure of Nepenthes consists of radially arranged ribs. Rain, dew, and honey form a layer of liquid film on the surface, transforming the peristome in a super-hydrophilic surface. The surface of the epidermal cells of Nepenthes peristome is smooth and wax-free. The addition of wax-free crystals and hygroscopic honey increases the capillary force and promotes the formation of the liquid film, further enhancing the liquid transport speed on the surface of the Nepenthes peristome [25], which even resists gravity. Nevertheless, the continuous radial arrangement of the peristome grooved structure exhibits liquid transmission characteristics without other external forces. Fabricating a surface with the same structural features as the replica molding method is ideal to achieve the one-way liquid transfer (self-driving water) capability, which has potential to apply in microfluidic devices without pumping. Directly using the natural peristome as a template, the peristome structure can be replicated by artificial poly (dimethylsiloxane) (PDMS); however, the surface is curved and difficult to reprocess. Therefore, we propose a high-precision photolithography process to imitate the Nepenthes peristome.

The plane structure is reproduced on PDMS through transfer-printing technology [13,26,27]. Most studies choose PDMS as the transfer material due to its excellent pattern reproduction accuracy, easy preparation, and ease of observation, etc. A microfluidic chip is a very promising analytical platform for the sample pre-treatment, separation, and detection [28]; for example, a microfluidic based surface-enhanced Raman scattering (SERS) chip was applied to detect creatinine in blood for 2 min [28,29]. The microfluidic SERS chip was proven to effectively detect bio-samples. However, the replica PDMS is an extremely hydrophobic material, whereas Nepenthes peristome has hydrophilic structure; to address this, we propose a hydrophilic modified material with 2-methacryloyloxyethyl phosphorylcholine (MPC), which contains a phospholipid structure that was polymerized and grafted on the PDMS surface by atmospheric plasma. MPC is a zwitterionic monomer with good antithrombotic and blood compatibility [30–32]. This polymer is highly hydrophilic due to the phospholipid polar tail in the MPC molecular structure. Moreover, the inhibitory effect on the adhesion of cells, platelets, enzymes, and proteins in the blood can also reduce the chances of the organism recognizing the material as a foreign object, and increases the biocompatibility of the material [33,34]. Therefore, the hydrophilic bionic structure was imitated to exhibit the same liquid transport capability as the Nepenthes peristome in this study.

2. Materials and Methods

2.1. *Photolithography*

The photolithography process of a Nepenthes peristome-like structure is shown in Figure 1. The silicon wafer substrate was consecutively cleaned with acetone, ethanol, and deionized water, and cleaned by vacuum plasma for 10 min after being dried by N_2 gas. After the plasma treatment, the surface achieved temporary hydrophilicity to increase the adhesion of the substrate and photoresist. SU-8 negative photoresist (KAYAKU SU8-2025) was spin-coated at a gradual acceleration speed of 500 rpm for 10 s, 1500 rpm for 10 s, and 3000 rpm for 30 s, which could make the homogenous coating for the SU-8 photoresist. Then, incubation occurred for 30 min before soft baking. Soft baking following gradual heating (65–90 °C) can avoid bubbles caused by an excessive heating rate. SU-8 negative photoresist was cross-linked with a mercury lamp (365 nm) for 25 s, with an exposure process of two layers of photomask patterns. After exposure, adhesion of the exposed pattern and the substrate were improved by post-exposure baking at 90 °C, and then immersed in the developer to remove the un-crosslinked SU-8 photoresist. Following developments, we rinsed with deionized water and hard baked for 3 min to remove the last remaining solvent. The SU-8 mold of the Nepenthes peristome-like structure was then used in the further replica of the PDMS membranes. The easy-cleaning coating solution, Naegix E720 (Senguan Tech Co., Ltd., Tainan City, Taiwan), was used as the control to evaluate the surface energy of a replica of a bionic Nepenthes peristome-like structure.

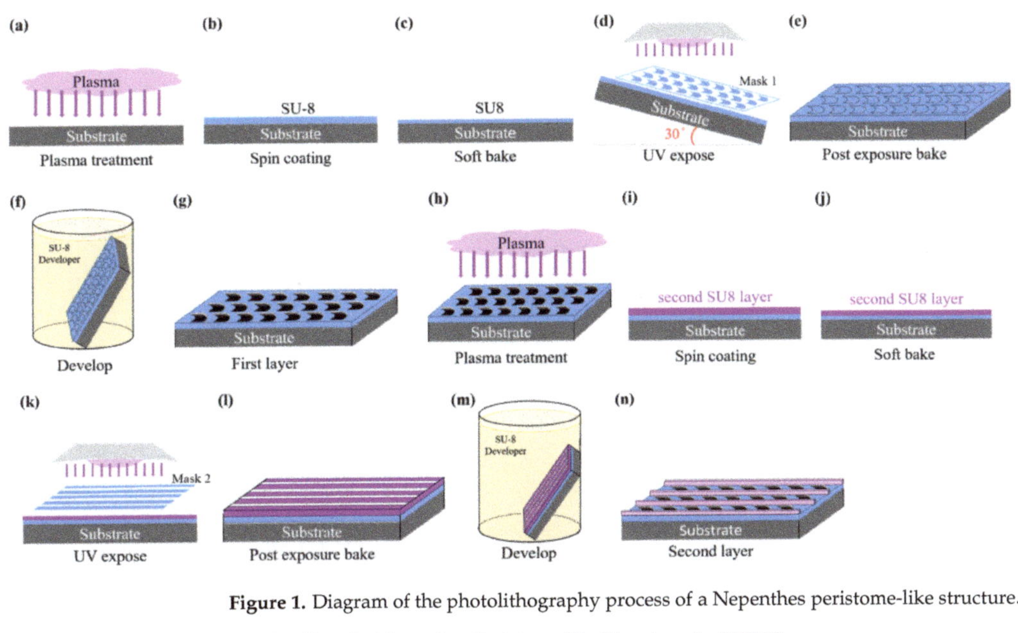

Figure 1. Diagram of the photolithography process of a Nepenthes peristome-like structure.

2.2. *Replica of a Nepenthes Peristome-like Structure by PDMS*

Polydimethylsiloxane (PDMS, Dow Corning® Sylgard 184, part A) and a crosslinking agent (Sylgard 184, part B) were mixed with a 10:1 PDMS base to a curing agent ratio. The stiffness of the PDMS can be manipulated by the addition of a curing agent. To avoid bubbles, the mixed PDMS must be placed in a vacuum desiccator until the bubbles disappear. The PDMS solution was then poured into a 35 mm dish and placed in the oven at 60 °C for 4 h until the structure stabilized. SU-8 based bionic (Nepenthes peristome-like) structure was cleaned with deionized water before plasma cleaning for 1 min. Moreover, the platinum was coated on the surface of the SU-8 based bionic structure. After all these steps were completed, the mixed PDMS solution was poured into the SU-8 mold and vacuumed to remove extra bubbles. Then, it stood in the oven at 65 °C for 4 h. The bionic replica of the PDMS membrane was slit to generate the Nepenthes peristome-like structure (Figure 2).

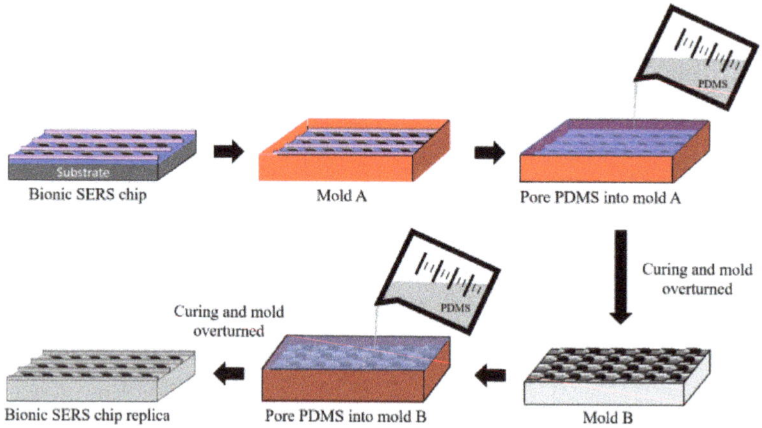

Figure 2. Diagram of the replicated Nepenthes peristome-like structure process by polydimethylsiloxane (PDMS).

2.3. MPC Immobilization by Atmospheric Plasma

The zwitterion polymer (2-methacryloyloxyethyl phosphorylcholine, MPC) modification flowchart is shown in Figure 3. The bionic replica of the PDMS membrane was pre-treated by plasma cleaning for 1 min (Figure 3a) and then immersed in MPC solution (Figure 3b) [27,28]. In addition to activating the surface functional groups, the hydrophilic treatment can effectively improve the uniformity of the modification. Subsequently, the pre-treated PDMS was treated by oxygen atmospheric plasma (AP Plasma Jet, Feng Tien Electronic Co., Ltd., Taipei, Taiwan) (Figure 3c) to polymerize the MPC monomers to MPC polymer brushes (Figure 3d) at an operating power of 1.2 kW and oxygen flux rate of 10 slm (L/min) for 10 s, and a 1.2 cm distance was maintained from the plasma torch.

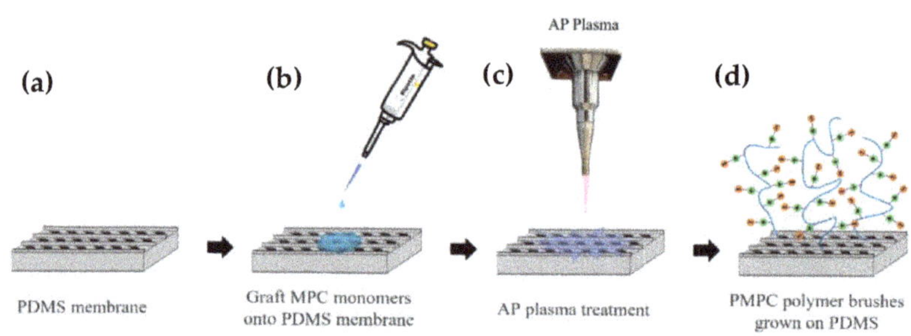

Figure 3. Schematic diagram of the replica of the PDMS-based bionic Nepenthes peristome-like structure with 2-methacryloyloxyethl phosphorylcholine (MPC) immobilization by atmospheric plasma. (**a**) The replica PDMS membrane was pre-treated by plasma cleaning for 1 min; (**b**) PDMS immersed in MPC solution; (**c**) MPC monomers were polymerized by oxygen atmospheric plasma and formed (**d**) MPC polymer brushes.

2.4. Anti-Bacterial Adhesion Capability

The antibacterial effect of PDMS and PDMS-PMPC against bacteria (*E. coli*) was inspected with the bacterial adhesion method. The samples were shaken with the lysate of *E. coli* at 37 °C. After 24 h, the samples were stained with SYTO 9 (green fluorescence nucleic acid stain) and stood for 5 min. The results were obtained with fluorescence microscope.

2.5. Anti-Protein Adhesion Capability

The PDMS and PDMS-PMPC were incubated in a 10 mL phosphate-buffered solution (PBS) solution of albumin from human serum (HSA) in the 24-well tissue culture plate at 37 °C for 1 h. Afterward, the samples were gently washed 3 times using PBS in the 24-well tissue culture plate. Then, the samples were incubated with 1 wt% aqueous solution of sodium dodecyl sulfate (SDS). The BCA kit was used to determine the concentration of the proteins in the SDS solution, and the concentration was detected by UV-vis spectroscopy.

2.6. Cell Attachment Tests

The cell attachment was detected with cells adhesion. The PDMS and PDMS-PMPC were placed in the 24-well plate, injected with NIH 3T3 mouse embryonic fibroblast cells (3T3 cells) to the substrate surface, and incubated at 37 °C for 24 h. The cells' attachment behavior would be evaluated by fluorescent staining (nucleus staining and cell membrane staining) to observe the amount of cell adhesion by fluorescent microscopy.

2.7. Biocompatibility

Biocompatibility is defined as the ability of a material to perform with an appropriate host response in a specific application [28]. The biocompatibility of pristine PDMS and PDMS-PMPC was evaluated with the proliferation of 3T3 fibroblast. The substrates (1 cm × 1 cm) and 1 mL solution of 3T3 cells (10^5 cells/mL) were placed in the 24-well plate under a 5% CO_2 atmosphere The cell viability was determined at 37 °C for 24, 48, and 72 h by thiazolyl blue tetrazolium bromide (MTT) assay and absorbance at 570 nm.

2.8. One-Way Liquid Transfer Capability

One-way liquid transfer capability of PDMS and PDMS-PMPC substrates was measured by the stained deionized water dropped on the PDMS and PDMS-PMPC substrates. The distance of stained deionized water flowing was recorded after 4 min. The flowing distance would be evaluated between pristine PDMS and PDMS-PMPC substrates. The longer flowing distance shows the greater one-way liquid transfer capability.

2.9. Characterizations

The morphology of the bionic SERS substrate was observed by scanning electron microscopy (SEM) (JEOL JSM-6701F, Tokyo, Japan). FT-IR spectroscopy (FT-IR, Perkin-Elmer Spectrum-One, Shelton, CT, USA) was used to differentiate the composition and chemical structure of the bionic SERS substrate. The chemical-binding energy of the bionic SERS substrate was carried out by a K-Alpha X-ray photoelectron spectrometer (XPS) (Thermo Fisher Scientific, Waltham, MA, USA). The hydrophilicity of the bionic SERS substrates was recorded by a contact angle goniometer (DSA 100, Krüss GmbH, Hamburg, Germany). Raman spectroscopy (HORIBA, LabRAM HR Evolution) was used to evaluate the SERS spectra of the bionic SERS substrate, with a 632.8 nm He-Ne laser, operated under a 10× objective lens with a detection range of 400–2000 cm^{-1}.

3. Results and Discussion

3.1. Optical Microscope (OM) Analysis

The photomask patterns of the Nepenthes peristome-like structure were shown in Figure 4a–c. The arrow arrays (Figure 4a) were fabricated in the first (bottom) layer. Then, the straight pattern arrays (Figure 4b) were covered as the second layer to develop the bionic Nepenthes peristome-like structure (Figure 4c). Figure 4d–f show the OM images of the first layer (Figure 4d), the second layer (Figure 4e), and the overlaid layers pattern (Figure 4f) of SU-8 photoresist during the photolithographic process. The resulting three patterns are similar to the three photomask patterns. The line point A to point B (Figures 4d and 5) passes through the plane and the pattern represents the segmented plane of the cross-section from the first layer of SU-8 photoresist. Although the arrow array (Figure 4d) becomes broader during the photolithographic process, compared to the pristine photomask patterns (Figure 4a), the gap channels between the arrow arrays do not change too much. Furthermore, the cross-section of the first layer replicated by PDMS was observed in Figure 5. The depth of the gap channel could be clearly observed in the cross-section image, showing the pattern integrity after the photolithographic process.

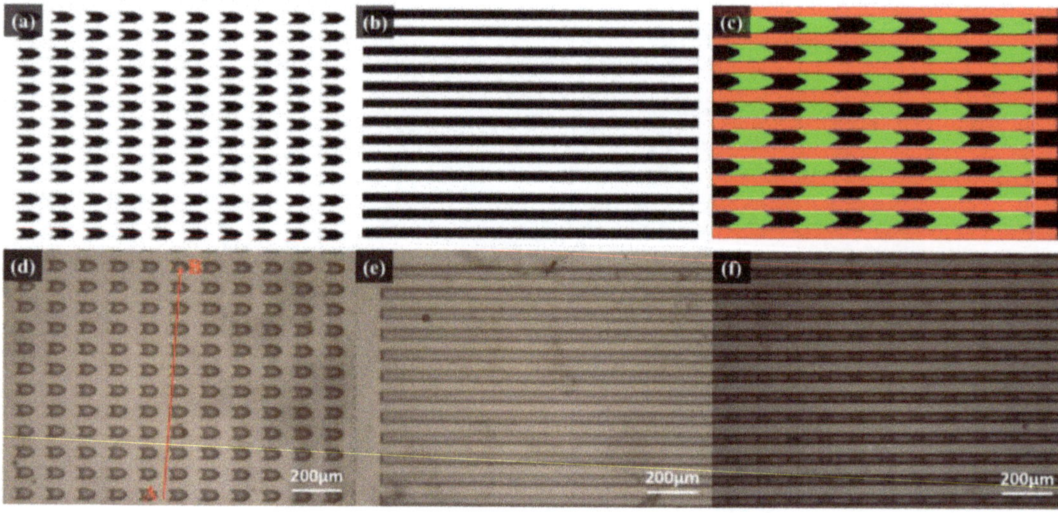

Figure 4. Diagram of (**a**) the first and (**b**) the second layers of photomask patterns; (**c**) overlaid photomask patterns. Optical microscope (OM) image of (**d**) first layer pattern, (**e**) second layer pattern, and (**f**) overlaid layers pattern of SU-8 photoresist. The line point A to point B passes through the plane and the pattern indicates the segmented plane of the cross-section from the first layer of SU-8 photoresist.

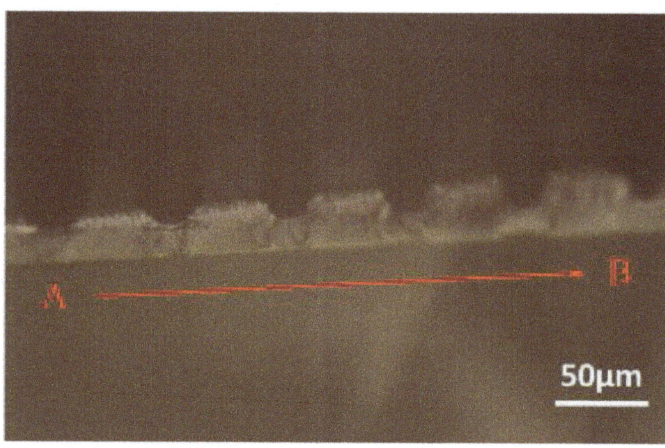

Figure 5. The OM image of the cross-section of the first layer of SU-8 photoresist.

3.2. Scanning Electron Microscope (SEM) Observation

From Figure 6a–c, the continuous radially arranged furrows were observed with grooves at the real (pristine) Nepenthes peristome structure. The width was 30–40 μm and the length was 130–160 μm. In comparison, the replica of the PDMS-based bionic Nepenthes peristome-like structure is shown in Figure 6d–f. The double layer structure was successfully fabricated by the replicated process, and the detail structure is very similar to the pristine Nepenthes peristome.

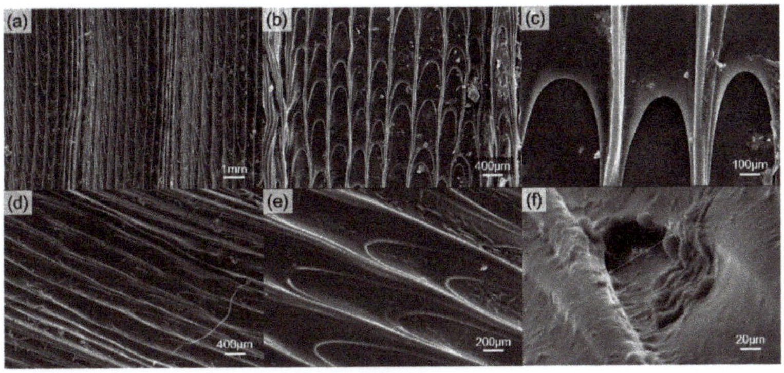

Figure 6. SEM images of (**a–c**) the real (pristine) Nepenthes peristome and (**d–f**) the replica of the Nepenthes peristome-like structure of PDMS.

3.3. Hydrophilicity by Contact Angle Measurements

The contact angle of the pristine Nepenthes peristome exhibiting a super-hydrophilic surface leads to an unapparent water contact angle (almost 0°), as shown in Figure 7a,b. However, the replica of the bionic PDMS surface (without modified MPC) shows the super-hydrophobicity (CA: ~142°) in Figure 7c,d, which is higher than the coating of commercial self-cleaning products (Naegix E720) (~90°). The structure of the replica of bionic PDMS is similar to the pristine Nepenthes peristome, but the surface energy is completely different. The reason for that is because the hydrophilic functional group was found on the pristine Nepenthes peristome, whereas the replica of the bionic PDMS surface is more hydrophobic. Therefore, MPC polymer brushes (PMPC) were grafted on the PDMS substrate by atmospheric plasma to improve hydrophilicity. The contact angle displayed was ~28° after PMPC was immobilized on the PDMS-replicated Nepenthes peristome, which is closer to the pristine Nepenthes peristome (Figure 7e,f).

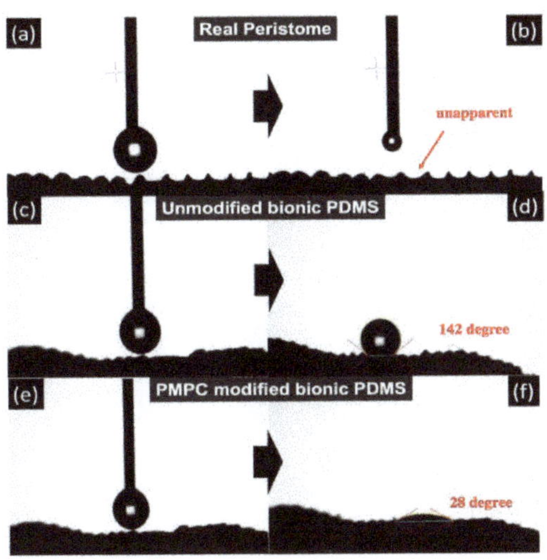

Figure 7. The contact angle of (**a**,**b**) the pristine Nepenthes peristome (almost 0°), (**c**,**d**) Nepenthes peristome-like structure replicated by PDMS without MPC modification, and (**e**,**f**) the Nepenthes peristome-like structure replicated by PDMS with MPC modification.

3.4. FTIR Spectrum Analysis

Figure 8 shows the FTIR spectra of pristine PDMS and PDMS coated with PMPC. The FTIR spectrum of PDMS shows the characteristic peaks at a stretching frequency of 789 cm^{-1} bending vibration modes of Si-CH_3 and Si-$(CH_2)_n$-Si, the characteristic peaks at the stretching frequency of 1020 cm^{-1} due to the bending vibration modes of the asymmetric Si–O–Si stretching vibrations, and the characteristic peaks at 1259 cm^{-1} due to the bending vibration modes of Si-CH_3 symmetric bending [35,36]. PDMS coated with PMPC has the characteristic peaks of 971 cm^{-1}, 1090 cm^{-1}, and 1247 cm^{-1} due to P-O, R-$N^+(CH_3)_3$, and the P=O stretching vibrational band [37,38]. The results confirmed that PMPC polymer brushes were successfully immobilized on the PDMS substrate.

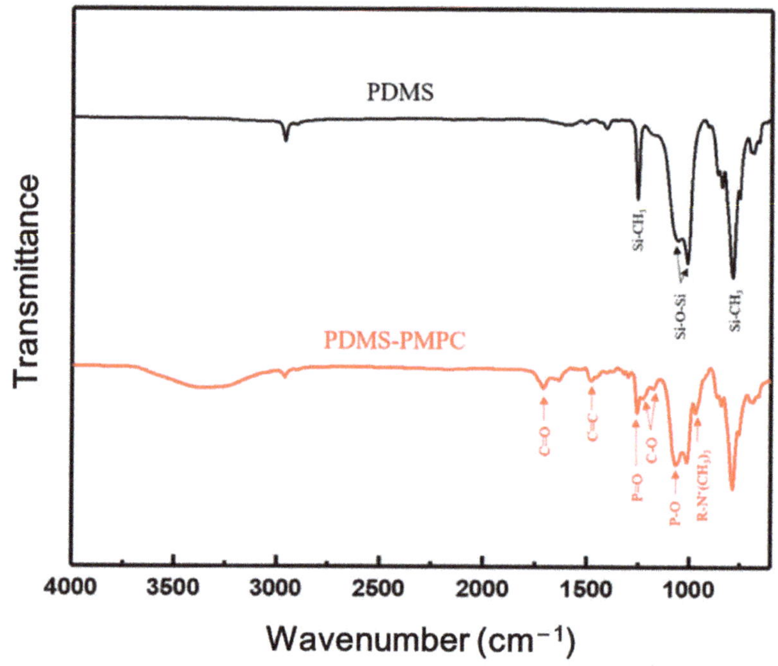

Figure 8. FTIR spectrum of PDMS and MPC polymer brushes (PMPC) were grafted on the PDMS substrate and PDMA-PMPC substrate.

3.5. XPS Spectrum Analysis

The binding energy change between the PDMS and PDMS-PMPC substrates was measured by an XPS analysis (Figure 9). The full spectra (Figure 9a) show that the surface primarily contained O, Si, C, P, and N. Compared with the PDMS surface, the characteristic peaks of N-1s and P-2s orbitals were observed at the PDMS-PMPC substrates, which contribute to the phosphorus and nitrogen bonds of MPC polymer brushes. The element ratios of N and P element contents of PDMS-PMPC were 4.43% and 4.66%, whereas the characteristic peaks of N-1s and P-2s were barely observed on the pristine PDMS. From the XPS spectra of C-1s (Figure 9b), new characteristic peaks were observed at C-O (286 eV) and C=O (288.5 eV) for the PDMS-PMPC substrates. In addition, an additional peak of N-1s spectra (Figure 9c) was found at 402.4 eV, which is the functional group of $N^+(CH_3)_3$ on the PDMS-PMPC substrate. These results are similar to the FTIR analysis (Figure 8), indicating that the PMPC polymer brushes can grow on the replica of the bionic PDMS surface after atmospheric plasma polymerization.

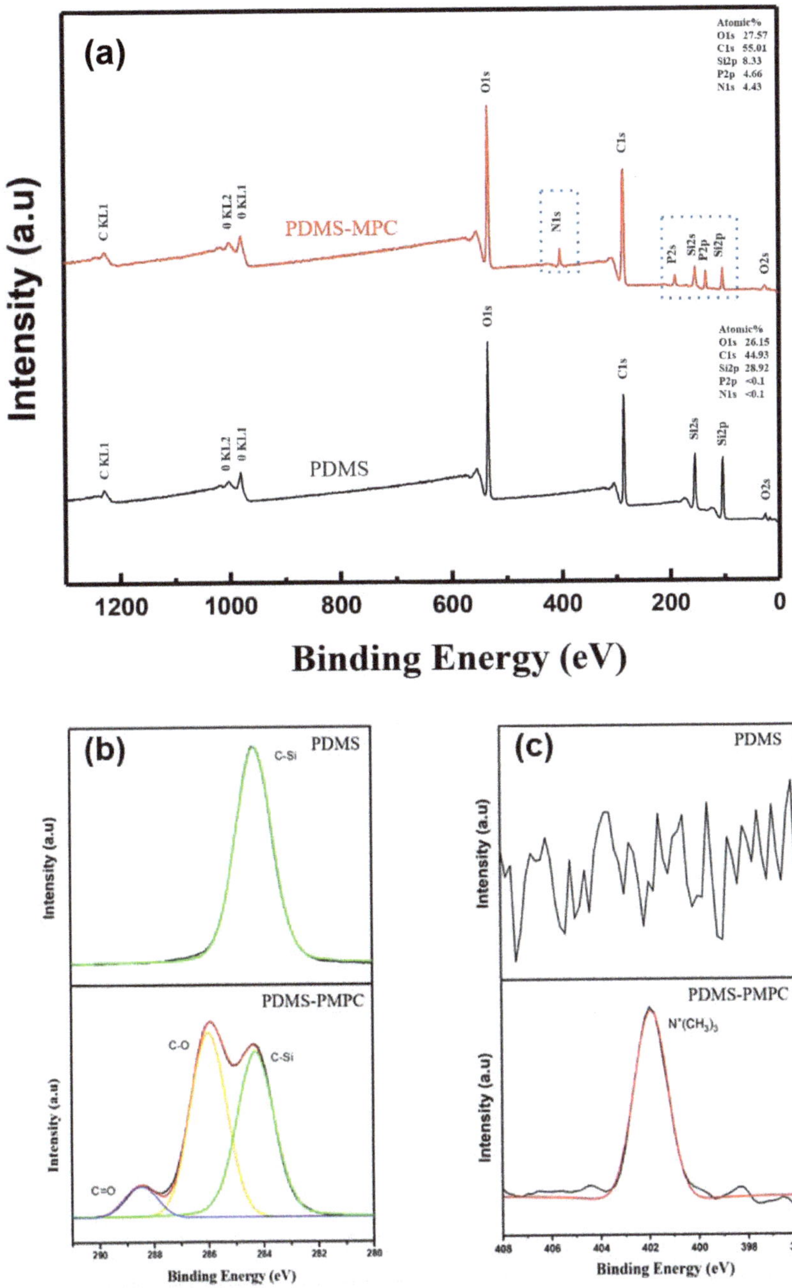

Figure 9. (**a**) XPS full spectra, (**b**) C-1s spectra, and (**c**) N-1s spectra of the PDMS and PDMS-PMPC substrates.

3.6. Antibacterial Adhesion Test

The antibacterial effect of PDMS and PDMS-PMPC against bacteria (*E. coli*) was assessed using the bacterial adhesion method. The fluorescence images of bacterial morphology (Figure 10) demonstrate that the bacteria are attached to the sample surface. We noticed a higher number of bacteria on the pristine PDMS substrate (Figure 10a) than on the PMPC-coated PDMS substrate (Figure 10b). This is because the PMPC structure forms the steric hindrance, and the steric hindrance enables the PDMS surface to display antibacterial properties [28].

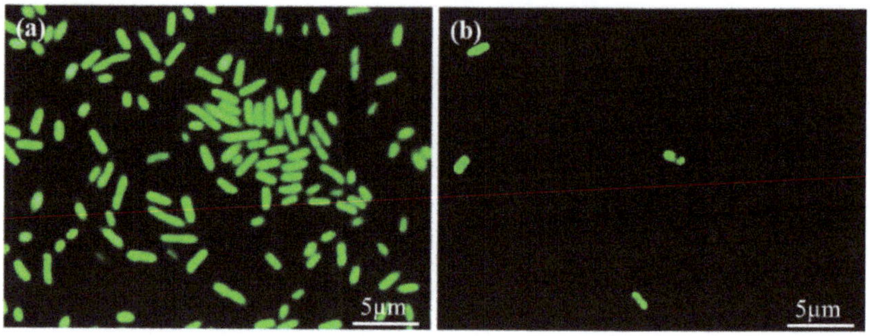

Figure 10. The anti-fouling capability of bacteria (*E. coli*) on the (**a**) pristine PDMS and (**b**) PDMS-PMPC substrate.

3.7. Anti-Protein Adhesion Test

The anti-protein adsorption performance of the PDMS and PDSA-PMPC surfaces were quantitatively evaluated by a BCA protein assay (Thermo Fisher, USA). Figure 11 shows the amounts of human serum albumin (HSA), which adhered to the pristine and PMPC-coated PDMS substrates. The adsorbed amount of HSA on the pristine PDMS was 60% higher than the PDMS-PMPC substrate. The PMPC was very effective in preventing the protein from adsorption. This result corresponded with the contact angle. The smaller the water contact angle after PMPC grafting, the greater the anti-fouling capability of proteins [34].

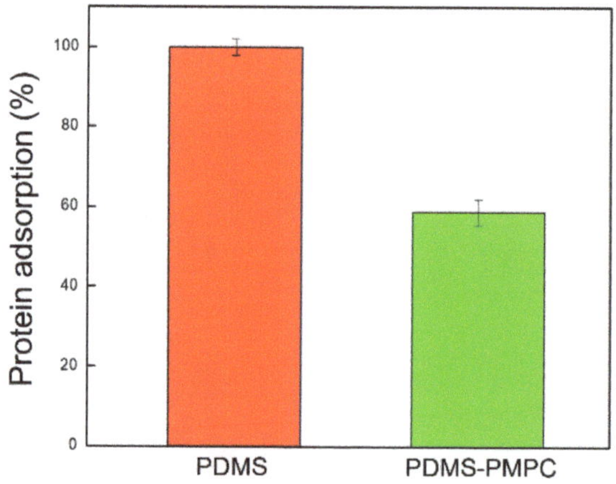

Figure 11. The anti-fouling capability of proteins (human serum albumin, HSA) on the pristine PDMS and PDMS-PMPC substrates ($n = 3$).

3.8. Cells Attachment Test

The results of 3T3 cells attachment on PDMS and PDMS-PMPC substrate are shown in Figure 12. Although the PDMS surface was hydrophobic, the amount of surface cell adhesion was concentrated and dense (Figure 12a). The PDMS grafted on PMPC (Figure 12b) made it difficult for the cells to attach to the substrate due to the change in surface charge, reduced the number of cells attached, and kept the cell adhesion survival rate.

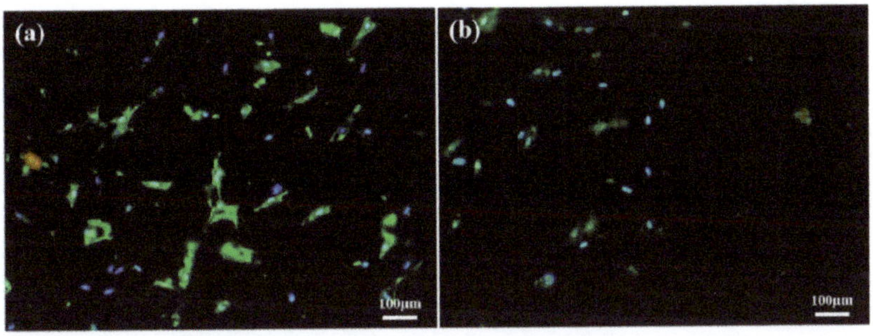

Figure 12. 3T3 cells attachment test on the (**a**) pristine PDMS and (**b**) PDMS-PMPC substrates.

3.9. Biocompatibility Test

Biocompatibility is defined as the ability of a material to perform with an appropriate host response in a specific application [31]. That is, the material will respond appropriately when it comes into contact with the host. In this study, the biocompatibility of pristine PDMS and PDMS-PMPC was evaluated with the proliferation of NIH 3T3 cells, which are mouse embryonic fibroblasts and have been widely used in biocompatibility tests. Figure 13 shows that 3T3 cells proliferated on the control, PDMS, and PDMS-PMPC substrates at 1–3 days. The results show that cell viability (%) at day 1 was almost the same in the three samples, but pristine PDMS and PDMS-MPC dropped to about 70% at day 2. At day 3, the cell viability (%) of 3T3 cells on the PDMS continued to decline, whereas that on the PDMS-PMPC increased to 85%, similarly with the control group. This demonstrates that PMPC immobilization on the surface can effectively improve biocompatibility at day 3, compared to the pristine PDMS.

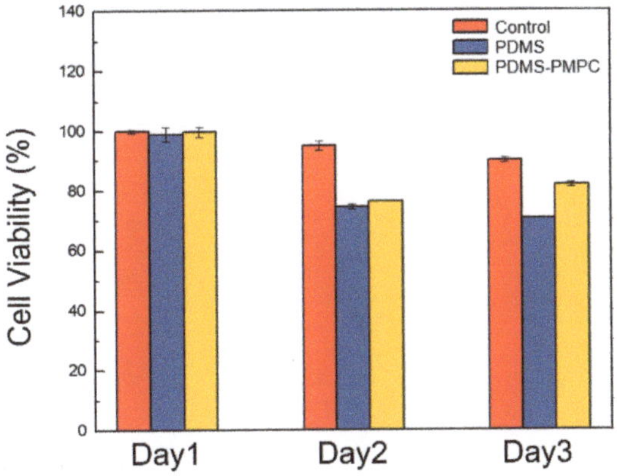

Figure 13. Biocompatibility tests of control, pristine PDMS, and PDMS-PMPC substrates (n = 3).

3.10. One-Way Liquid Transfer Capability

The liquid channels (20 mm) were fabricated by imitating channels of the real Nepenthes peristome, as shown in Figure 14a. The stained deionized water (10 μL) was utilized for its one-way liquid transfer capability. The results indicated that the ability of self-driving and unidirectional water delivery was found in the PDMS-PMPC substrate (red-stained water) after 4 min incubation (running distance: ~10 mm), compared with the sticky blue stained water in the PDMS substrate (Figure 14b). The delivery rate of PDMS-PMPC was about 2.5 mm/min, which provides the chromatographic capability of the complex samples. The complex samples could be separated in the chromatographic process (depending on the polarity or molecular weight of the samples), and then detected using the SERS technique.

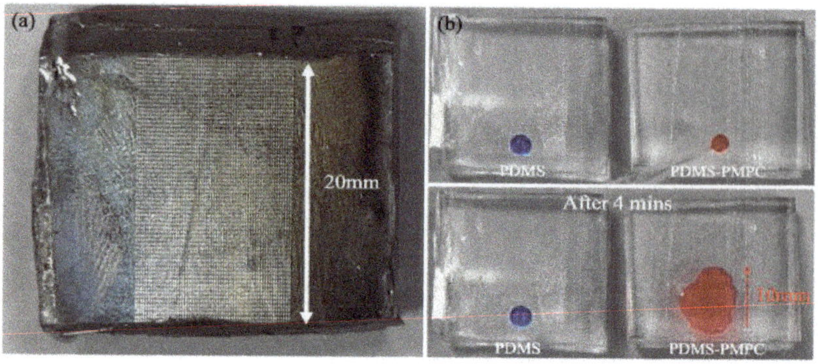

Figure 14. (**a**) The liquid channels (20 mm) of the PDMS-based substrate were used for the channels of the real Nepenthes peristome; (**b**) the stained deionized water (10 μL) was used for the one-way liquid transfer capability (blue water for PDMS and red water for PDMS-PMPC).

3.11. SERS Detection by Raman Spectroscopy

To enable the SERS detection (Figure 15), the replica PDMS-PMPC with a bionic structure was deposited on 5 nm of Ag nanofilm (PDMS-PMPC-Ag (W/)) by thermal evaporation, and pristine PDMS-PMPC with Ag (PDMS-PMPC-Ag (W/O)) deposition was used as the control. The 633 nm laser was used for Raman spectroscopy, and adenine (10^{-4} M) was used as the model biomolecules analytes. The results indicated a characteristic peak of adenine at 733 cm^{-1}, and the SERS intensity of PDMS-PMPC-Ag (W/) was ~4 times stronger than that of PDMS-PMPC-Ag (W/O), which indicated that multi-reflection by the 3D bionic structure enhanced SERS intensity. The pristine PDMS without Ag deposition and bionic structure (PDMS (W/O)) is displayed as a black line in the background.

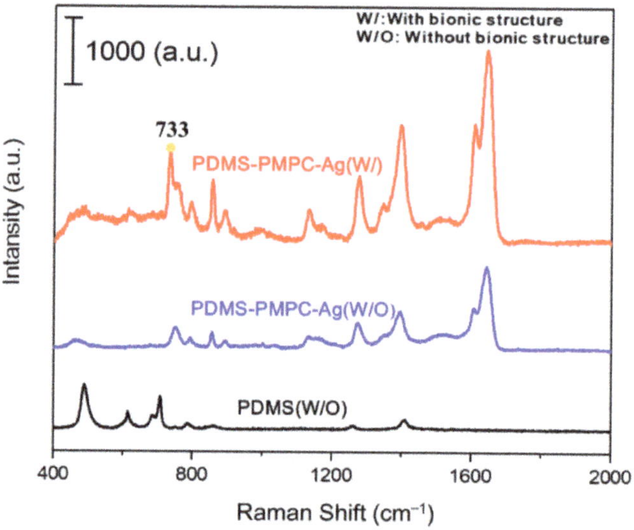

Figure 15. Raman spectrum of PDMS without (W/O) bionic structure, PDMS-PMPC-Ag without (W/O) bionic structure, and PDMS-PMPC-Ag with (W/) bionic structure.

4. Conclusions

In this study, we successfully fabricated a bionic Nepenthes peristome-like structure by photolithography with SU-8 photoresist. Furthermore, the bionic structure was replicated by a flexible PDMS substrate and then immobilized with MPC polymer brushes by an atmospheric plasma treatment to improve its hydrophilicity, antibacterial attachment, anti-protein adsorption, and biocompatibility. The self-driving capability of SERS detection and the one-way liquid transfer were demonstrated on the replica of the bionic PDMS-PMPC substrate. A benefit of the Nepenthes peristome-like structure with PMPC modification was that the delivery rate was about 2.5 mm/min, unlike the sticky stained water on the unmodified PMPC surface. This can provide the chromatographic capability to separate complex samples, and further detect them by Raman spectroscopy. The flexible and multi-functional SERS chips could potentially be applied in a wearable device for biomedical and environmental detection.

Author Contributions: Conceptualization, M.-C.Y., C.-C.L. and C.-C.C. (Cheng-Chen Chen); methodology, Y.-T.L., C.-H.W., W.-L.S., P.-C.H. and Z.-L.T.; software, W.-L.S. and P.-C.H.; validation, C.-C.C. (Cheng-Cheung Chen), C.-C.L. and T.-Y.L.; formal analysis, W.-L.S., P.-C.H. and Z.-L.T.; investigation, Y.-T.L. and C.-H.W.; resources, M.-C.Y. and C.-C.L.; data curation, Y.-T.L., C.-H.W. and Z.-L.T.; writing—original draft preparation, Y.-T.L., C.-H.W., W.-L.S., P.-C.H., C.-C.C. (Cheng-Chen Chen), C.-C.C. (Cheng-Cheung Chen) and T.-Y.L.; writing—review and editing, M.-C.Y., C.-C.C. (Cheng-Chen Chen), C.-C.C. (Cheng-Cheung Chen) and T.-Y.L.; visualization, W.-L.S. and P.-C.H.; supervision, C.-C.C. (Cheng-Chen Chen), C.-C.C. (Cheng-Cheung Chen) and T.-Y.L.; project administration, M.-C.Y. and T.-Y.L.; funding acquisition, C.-C.L. and C.-C.C. (Cheng-Cheung Chen) All authors have read and agreed to the published version of the manuscript.

Funding: This work was financially supported by the Ministry of Science and Technology in Taiwan (MOST 109-2622-E-131-009, MOST 110-2221-E-131-009, MOST 111-2622-E-131-003, MOST 108-2623-E-016-003-D).

Informed Consent Statement: Not applicable.

Conflicts of Interest: The authors declare no conflict of interest.

References

1. Yan, Y.Y.; Gao, N.; Barthlott, W. Mimicking natural superhydrophobic surfaces and grasping the wetting process: A review on recent progress in preparing superhydrophobic surfaces. *Adv. Colloid Interface Sci.* **2011**, *169*, 80–105. [CrossRef] [PubMed]
2. Zhang, P.; Zhang, L.; Chen, H.; Dong, Z.; Zhang, D. Surfaces inspired by the Nepenthes peristome for unidirectional liquid transport. *Adv. Mater.* **2017**, *29*, 1702995. [CrossRef] [PubMed]
3. Bico, J.; Marzolin, C.; Quéré, D. Pearl drops. *EPL* **1999**, *47*, 220. [CrossRef]
4. Liu, G.; Yuan, Z.; Qiu, Z.; Feng, S.; Xie, Y.; Leng, D.; Tian, X. A brief review of bio-inspired surface technology and application toward underwater drag reduction. *Ocean Eng.* **2020**, *199*, 106962. [CrossRef]
5. Russell, T. Surface-responsive materials. *Science* **2002**, *297*, 964–967. [CrossRef]
6. Willermet, C. Biological Anthropology in 2015: Open access, biocultural interactions, and social change. *Am. Anthropol.* **2016**, *118*, 317–329. [CrossRef]
7. Lan, J.; Han, J. Research on the radiation characteristics of aerodynamic noises of a simplified bogie of the high-speed train. *J. Vibroeng.* **2017**, *19*, 2280–2293.
8. Jaggessar, A.; Shahali, H.; Mathew, A.; Yarlagadda, P.K. Bio-mimicking nano and micro-structured surface fabrication for antibacterial properties in medical implants. *J. Nanobiotechnol.* **2017**, *15*, 64. [CrossRef]
9. Currey, J.D. *Bones: Structure and Mechanics*; Princeton University Press: Princeton, NJ, USA, 2006.
10. Vincent, J. Structural biomaterials. In *Structural Biomaterials*; Princeton University Press: Princeton, NJ, USA, 2012.
11. Meyers, M.A.; Chen, P.-Y.; Lin, A.Y.-M.; Seki, Y. Biological materials: Structure and mechanical properties. *Prog. Mater. Sci.* **2008**, *53*, 1–206. [CrossRef]
12. Fayemi, P.-E.; Maranzana, N.; Aoussat, A.; Bersano, G. Bio-inspired design characterisation and its links with problem solving tools. In Proceedings of the DS 77: Proceedings of the DESIGN 2014 13th International Design Conference, Dubrovnik, Croatia, 19–22 May 2014; pp. 173–182.
13. Zhang, P.; Chen, H.; Li, L.; Liu, H.; Liu, G.; Zhang, L.; Zhang, D.; Jiang, L. Bioinspired smart peristome surface for temperature-controlled unidirectional water spreading. *ACS Appl. Mater. Interfaces* **2017**, *9*, 5645–5652. [CrossRef]
14. Gestrelius, S.; Lyngstadaas, S.; Hammarström, L. Emdogain–periodontal regeneration based on biomimicry. *Clin. Oral Investig.* **2000**, *4*, 120–125. [CrossRef] [PubMed]
15. Guo, Z.; Liu, W.; Su, B.-L. Superhydrophobic surfaces: From natural to biomimetic to functional. *J. Colloid Interface Sci.* **2011**, *353*, 335–355. [CrossRef] [PubMed]
16. Herminghaus, S. Roughness-induced non-wetting. *EPL* **2007**, *79*, 59901. [CrossRef]

17. Feng, L.; Li, S.; Li, Y.; Li, H.; Zhang, L.; Zhai, J.; Song, Y.; Liu, B.; Jiang, L.; Zhu, D. Super-hydrophobic surfaces: From natural to artificial. *Adv. Mater.* **2002**, *14*, 1857–1860. [CrossRef]
18. Cassie, A.; Baxter, S. Superhydrophobic Surfaces and Emerging Applications: Non-Adhesion. *Energy Green Eng. Trans. Faraday Soc.* **1944**, *40*, 546–551. [CrossRef]
19. Liang, Y.; Yang, K.; Bai, Q.; Chen, J.; Wang, B. Modeling and experimental analysis of microburr formation considering tool edge radius and tool-tip breakage in microend milling. *J. Vac. Sci. Technol. B Microelectron. Nanometer Struct. Processing Meas. Phenom.* **2009**, *27*, 1531–1535. [CrossRef]
20. Chen, T.L.; Lin, Y.P.; Chien, C.H.; Chen, Y.C.; Yang, Y.J.; Wang, W.L.; Chien, L.F.; Hsueh, H.Y. Fabrication of Frog-Skin-Inspired Slippery Antibiofouling Coatings Through Degradable Block Copolymer Wrinkling. *Adv. Funct. Mater.* **2021**, *31*, 2104173. [CrossRef]
21. Zhang, D.; Zhang, W.; Gu, J.; Fan, T.; Liu, Q.; Su, H.; Zhu, S. Inspiration from butterfly and moth wing scales: Characterization, modeling, and fabrication. *Prog. Mater. Sci.* **2015**, *68*, 67–96. [CrossRef]
22. Cheng, Y.-W.; Hsiao, C.-W.; Lin, C.-H.; Huang, L.-Y.; Chen, J.-S.; Yang, M.-C.; Liu, T.-Y. Bionic 3D periodic nanostructures by Ag nano-islands deposited on cicada wings for rapid SERS detection. *Surf. Coat. Technol.* **2022**, *436*, 128323. [CrossRef]
23. Zhang, C.; Chen, S.; Wang, J.; Shi, Z.; Du, L. Reproducible Flexible SERS Substrates Inspired by Bionic Micro-Nano Hierarchical Structures of Rose Petals. *Adv. Mater. Interfaces* **2022**, *9*, 2102468. [CrossRef]
24. Li, C.; Wu, L.; Yu, C.; Dong, Z.; Jiang, L. Peristome-Mimetic Curved Surface for Spontaneous and Directional Separation of Micro Water-in-Oil Drops. *Angew. Chem.* **2017**, *129*, 13811–13816. [CrossRef]
25. Yu, C.; Zhang, L.; Ru, Y.; Li, N.; Li, C.; Gao, C.; Dong, Z.; Jiang, L. Drop cargo transfer via unidirectional lubricant spreading on peristome-mimetic surface. *ACS Nano* **2018**, *12*, 11307–11315. [CrossRef] [PubMed]
26. Li, Z.; Zhang, D.; Wang, D.; Zhang, L.; Feng, L.; Zhang, X. A bioinspired flexible film fabricated by surface-tension-assisted replica molding for dynamic control of unidirectional liquid spreading. *ACS Appl. Mater. Interfaces* **2019**, *11*, 48505–48511. [CrossRef] [PubMed]
27. Zhou, S.; Yu, C.; Li, C.; Jiang, L.; Dong, Z. Droplets Crawling on Peristome-Mimetic Surfaces. *Adv. Funct. Mater.* **2020**, *30*, 1908066. [CrossRef]
28. He, X.; Ge, C.; Zheng, X.; Tang, B.; Chen, L.; Li, S.; Wang, L.; Zhang, L.; Xu, Y. Rapid identification of alpha-fetoprotein in serum by a microfluidic SERS chip integrated with Ag/Au Nanocomposites. *Sens. Actuators B Chem.* **2020**, *317*, 128196. [CrossRef]
29. Su, X.; Xu, Y.; Zhao, H.; Li, S.; Chen, L. Design and preparation of centrifugal microfluidic chip integrated with SERS detection for rapid diagnostics. *Talanta* **2019**, *194*, 903–909. [CrossRef]
30. Feng, W.; Brash, J.L.; Zhu, S. Non-biofouling materials prepared by atom transfer radical polymerization grafting of 2-methacryloloxyethyl phosphorylcholine: Separate effects of graft density and chain length on protein repulsion. *Biomaterials* **2006**, *27*, 847–855. [CrossRef]
31. Yang, W.; Chen, S.; Cheng, G.; Vaisocherova, H.; Xue, H.; Li, W.; Zhang, J.; Jiang, S. Film thickness dependence of protein adsorption from blood serum and plasma onto poly (sulfobetaine)-grafted surfaces. *Langmuir* **2008**, *24*, 9211–9214. [CrossRef]
32. Chen, J.-S.; Liu, T.-Y.; Tsou, H.-M.; Ting, Y.-S.; Tseng, Y.-Q.; Wang, C.-H. Biopolymer brushes grown on PDMS contact lenses by in situ atmospheric plasma-induced polymerization. *J. Polym. Res.* **2017**, *24*, 69. [CrossRef]
33. Chen, J.-S.; Ting, Y.-S.; Tsou, H.-M.; Liu, T.-Y. Highly hydrophilic and antibiofouling surface of zwitterionic polymer immobilized on polydimethylsiloxane by initiator-free atmospheric plasma-induced polymerization. *Surf. Coat. Technol.* **2018**, *344*, 621–625. [CrossRef]
34. Lin, Y.-T.; Ting, Y.-S.; Chen, B.-Y.; Cheng, Y.-W.; Liu, T.-Y. Bionic shark skin replica and zwitterionic polymer brushes functionalized PDMS membrane for anti-fouling and wound dressing applications. *Surf. Coat. Technol.* **2020**, *391*, 125663. [CrossRef]
35. Ito, H.; Onitsuka, S.; Gappa, R.; Saitoh, H.; Roacho, R.; Pannell, K.; Suzuki, T.; Niibe, M.; Kanda, K. Fabrication of amorphous silicon carbide films from decomposition of tetramethylsilane using ECR plasma of Ar. *Proc. J. Phys. Conf. Ser.* **2013**, *441*, 012039. [CrossRef]
36. Hoek, I.; Tho, F.; Arnold, W.M. Sodium hydroxide treatment of PDMS based microfluidic devices. *Lab Chip* **2010**, *10*, 2283–2285. [CrossRef]
37. Zhao, Y.; Wen, J.; Ge, Y.; Zhang, X.; Shi, H.; Yang, K.; Gao, X.; Shi, S.; Gong, Y. Fabrication of stable biomimetic coating on PDMS surface: Cooperativity of multivalent interactions. *Appl. Surf. Sci.* **2019**, *469*, 720–730. [CrossRef]
38. D'Angelo, A.J.; Panzer, M.J. Decoupling the ionic conductivity and elastic modulus of gel electrolytes: Fully zwitterionic copolymer scaffolds in lithium salt/ionic liquid solutions. *Adv. Energy Mater.* **2018**, *8*, 1801646. [CrossRef]

polymers

MDPI

Article

Nonlinear Conductivity and Space Charge Characteristics of SiC/Silicone Rubber Nanocomposites

Ming-Ze Gao [1], Zhong-Yuan Li [2,*] and Wei-Feng Sun [3]

1 Department of Electrical Engineering and Electronics, School of Medical Imaging, Mudanjiang Medical University, Mudanjiang 157011, China; gaomingze00@163.com
2 Electric Power Research Institute, State Grid Heilongjiang Electric Power Co., Ltd., Harbin 150080, China
3 School of Electrical and Electronic Engineering, Nanyang Technological University, Singapore 639798, Singapore; weifeng.sun@ntu.edu.sg
* Correspondence: li8zhongyuan@126.com; Tel.: +86-0451-56394438

Abstract: To achieve a preferable compatibility between liquid silicone rubber (LSR) and cable main insulation in a cable accessory, we developed SiC/LSR nanocomposites with a significantly higher conductivity nonlinearity than pure LSR, whilst representing a notable improvement in space charge characteristics. Space charge distributions in polarization/depolarization processes and surface potentials of SiC/LSR composites are analyzed to elucidate the percolation conductance and charge trapping mechanisms accounting for nonlinear conductivity and space charge suppression. It is verified that SiC/LSR composites with SiC content higher than 10 wt% represent an evident nonlinearity of electric conductivity as a function of the electric field strength. Space charge accumulations can be inhibited by filling SiC nanoparticles into LSR, as illustrated in both dielectric polarization and depolarization processes. Energy level and density of shallow traps increase significantly with SiC content, which accounts for expediting carrier hopping transport and surface charge decay. Finite-element multiphysics simulations demonstrate that nonlinear conductivity acquired by 20 wt% SiC/LSR nanocomposite could efficiently homogenize an electric field distributed in high-voltage direct current (HVDC) cable joints. Nonlinear conductivities and space charge characteristics of SiC/LSR composites discussed in this paper suggest a feasible modification strategy to improve insulation performances of direct current (DC) cable accessories.

Keywords: cable joint; liquid silicone rubber; nonlinear conductivity; space charge

Citation: Gao, M.-Z.; Li, Z.-Y.; Sun, W.-F. Nonlinear Conductivity and Space Charge Characteristics of SiC/Silicone Rubber Nanocomposites. *Polymers* **2022**, *14*, 2726. https://doi.org/10.3390/polym14132726

Academic Editors: Ting-Yu Liu and Yu-Wei Cheng

Received: 5 June 2022
Accepted: 29 June 2022
Published: 3 July 2022

1. Introduction

Recent electrical energy supply raises the requirements for delivering electricity over long distances in which traditional high-voltage alternating current (AC) transmission systems are facing inevitable challenges. Compared with conventional AC transmission, the major advantages of high-voltage direct current (HVDC) transmissions, such as long transmission distances, low losses, and large transmission capacities, need to competently provide long-distance transmissions [1–4].

HVDC cable accessories are essential connecting devices in power transmission and transformation systems. However, the multi-layer structures of composite insulation in cable accessories lead to great discrepancies and incompatibilities in the electrical conductivities of internal insulation materials. In particular, the conductivity of accessory insulation is far lower than that of the main insulation cross-linked polyethylene (XLPE), where the electric field is distributed inversely proportional to the material conductivity [5,6]. As a result, the poor accessory insulation results in local electric field distortions and space charge accumulations under long-term polarization electric fields, which degrades the insulation and breaks down the resistance of cable accessories. Therefore, cable accessories are the weakest insulation points in direct current (DC) cable lines.

Recently, homogenizing the electric field by optimizing the geometry of the insulation structure and using nonlinear-conductivity composite materials has been found capable of alleviating local electric field concentration, improving the insulation performances of cable accessories [7,8]. With the development of large-capacity and high-voltage power equipment, nonlinear materials have shown significant advantages for improving practical efficiency and ensuring electrical compatibility in insulation equipment. The SiC/silicone rubber (SiR) composites with SiC content greater than 20 vol% exhibit evident conductivity nonlinearity which is attributed to carrier hopping conductance introduced by SiC fillers [9]. The dielectric properties of ZnO/SIR composites show significant conductivity nonlinearity when the volume fraction of spherical ZnO fillers is higher than 30 vol%, with relative permittivity being increased by five times [10]. It has been reported that calcium copper titanate ($CaCu_3Ti_4O_1$) nanofibers/liquid silicone rubber composites present notable nonlinear conductivity and an improvement of space charge characteristics due to the introduced charge traps by $CaCu_3Ti_4O_1$ nanofillers [11]. SiC nanofillers with nonlinear conductivity can effectively suppress interface charge accumulations in SiC/polymer nanocomposites by introducing shallower traps [12]. In addition, coatings prepared by filling inorganic nonlinear nanoparticles (e.g., SiC and ZnO) into polymers exhibit self-adaptive conductivity in response to the applied electric field. The coating of SiC/Epoxy resin on insulator surfaces could promote the dissipation of surface charge and effectively improve the flashover voltage [13].

The previous studies concerning electric field homogenization using nonlinear additive fillers mainly focused on the influence of the type and content of fillers on the nonlinear conductivity of internal insulation, without elucidating clear pictures of nonlinear conductance in combination with space charge characteristics. Therefore, in this study, SiC nanoparticles filled into liquid silicone rubber (LSR) were used to render nanocomposites with appreciable nonlinear conductivities. The abnormally varying carrier transport and space charge suppression are analyzed to reveal the underlying mechanism of the comprehensive improvements in both nonlinear conductance and insulation strength, which provides a basic reference for the engineering application of nonlinear composites.

2. Materials and Methods

2.1. Material Preparation

Two-component (A and B) liquid silicone rubber (LSR, Wacker, Germany) was selected as the raw material for the composite matrix. SiC material (Deke Daojin, Beijing, China) with a particle size of 40 nm was adopted as the filling additive. SiC/LSR nanocomposites with SiC contents of 0, 5.0, 10.0, and 20.0 wt% were prepared by the mechanical stirring method. The preparation process was as follows: (1) same masses of A and B, and appropriate mass of SiC nanoparticles were accurately weighed according to filling content to be poured into an beaker and stirred continuously at room temperature; (2) the SiC/LSR mixture was poured into a 10 mm × 10 mm × 0.2 mold in a vacuum environment at a constant room temperature for 2 h; and (3) the mold containing the SiC/LSR mixture was placed into a flat vulcanizing instrument and vulcanized for 15 min under 15 MPa at 393 K, and then cured at 473 K for 4 h to finally obtain the material samples.

2.2. Microstructure Characterization

The insulation properties of nanodielectrics are closely related to the dispersion of the doped nano-phases in a polymer matrix. Thus, the dispersion of the SiC nanofillers in an EPDM polymer matrix was characterized by a scanning electron microscope (SEM) (SU8020, Hitachi High-Technologies Corporation, Tokyo, Japan). The specimens were cold-brittle ruptured in a vessel containing liquid nitrogen to acquire the cross section to be observed by SEM after being sprayed with a gold film.

2.3. Mechanical Tensile Test

According to the standard of ISO 37:2005, the stress–strain characteristics were measured with an elongation speed of 5 mm/min. The tested sample was made into dumbbell type sample 1 mm thick, with a total length of 75 mm and an effective test width of 4 ± 0.2 mm. The surface was smooth. without visible defects.

2.4. Electric Conductivity

Electric conductivities of the SiC/LSR nanocomposites in samples of 50 mm diameter and 300 μm thickness were measured with a three-electrode system consisting of a high-voltage DC power supply (HB-Z103-2AC, Tianjin Hengbo High Voltage Power Supply Co., Ltd., Tianjin, China), a picoammeter (Est122), three electrodes, and an oven at temperatures from 30 to 70 °C. The current values were predicted approach a quasi-steady sate after applying voltage for 30 min. Three aluminum electrodes were evaporated in a vacuum in which the measuring disc-shaped electrode with a 50 mm diameter was encircled by a protective annular-shaped electrode with a 75 mm diameter on one side of the film samples; a high-voltage circular electrode with a diameter of 78 mm was placed on the other side. Each group of samples was measured several times, and the average value was taken to ensure the accuracy and reliability of the tested results.

2.5. Space Charge Distribution

The pulsed electro-acoustic (PEA) method was used to measure the space charge characteristics of the SiC/LSR composites in samples with a dimension of 50 mm × 50 mm × 0.3 mm at room temperature in which the polarization electric field, the pulse voltage and width, and the input impedance were specified as 30 kV/mm, 400 V and 8 ns, and 1.368 Ω, respectively. After voltage was applied for 30 min to test the space charge accumulation in the polarization process, the tested sample was short-circuited to obtain depolarization decaying of the space charge distribution. The total space charge $Q(t)$ was calculated as follows [14]:

$$Q(t) = \int_0^h |\rho(x,t)| S \mathrm{d}x \tag{1}$$

where $Q(t)$ denotes the total charge quantity internal material at time t, h signifies the sample thickness, $\rho(x,t)$ symbolizes the space charge density at the position of x in the sample at t, and S is the electrode area.

Considering the small space charge injection under a low electric field, the waveform at a low electric field intensity of 3 kV/mm was adopted as the waveform reference for the space charge measurement system to restore the measured signal of a space charge distribution under a high electric field.

2.6. Isothermal Surface Potential Decay

The isothermal surface potential decay (ISPD) method can characterize the energy level distribution of charge traps and the charge transport in a detrapping process in which the corona-discharge method is used to charge the film sample, as shown by the schematic measurement system in Figure 1.

The distance between the needle tip and the grid was 5 mm, and the distance between the grid and the sample surface was 5 mm. The needle and grid electrodes were connected to a high-voltage source with an aluminum back electrode being grounded during measurement. The electrostatic probe was fixed on the sliding guide bracket about 3 mm from the sample surface. Under negative corona charging, the needle and grid electrodes were charged by −8 kV and −5 kV voltage, respectively, for 5 min. Then, the sample was transferred below an electrostatic probe to be measured for 25 min.

Double exponential function was used to numerically fit the decaying surface potential over time very well, as expressed by:

$$U = Ae^{\frac{-t}{\tau_1}} + Be^{\frac{-t}{\tau_2}} \quad (2)$$

where A, τ_1, B, and τ_2 are fitting parameters by which the trap energy level distribution can be calculated by the surface potential decaying curve [14,15] which could be considered as the volume trap characteristics. The trap energy level and trap density are calculated by the following formulas:

$$E_t = k_b T \ln(\nu t),\ N(E) = \frac{4\varepsilon_0\varepsilon_1}{qk_b T d^2}\left| t\frac{dU}{dt} \right| \quad (3)$$

where E_t denotes trap energy level, $N(E)$ indicate trap density, d represents sample thickness, ε_0 and ε_1 are vacuum permittivity and relative permittivity, respectively, v symbolizes the electron escape frequency, and q and k_b signify the electron charge and Boltzmann constant, respectively.

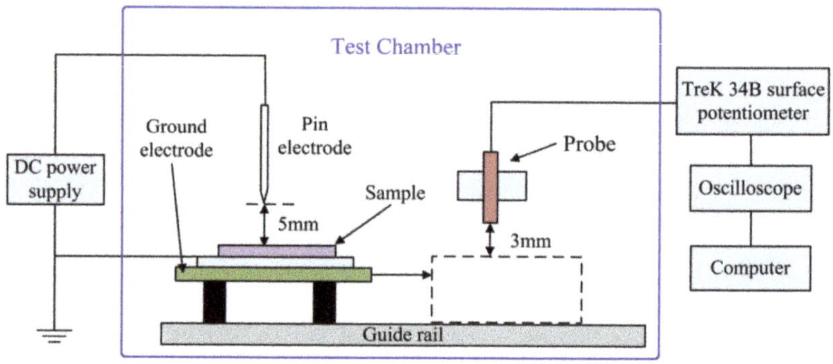

Figure 1. Schematic test system of isothermal surface potential decay.

2.7. Finite-Element Electric Field Simulation

We investigates the electric field distributions in the 200 kV DC cable joint with the SiC/LSR composites used as reinforced insulation, as schematically shown by the geometrically modeled structure in Figure 2. The diameter of the cable core was 30 mm, and the thicknesses of the main insulation (XLPE material), inner shield, outer shield, and reinforced insulation were specified as 16, 2, 1, and 68 mm, respectively. The length of the simulation model was 2600 mm. To make the simulation more identical to the actual situation, an air environment domain was added to the simulation model. The convective heat transfer parameter at the interface of the cable joint and the air was set to 10 W/(m^2·K). DC high voltage was loaded to the edge of the cable core. The inner edge of the outer semi-conductive shield layer of insulation and the boundary of the stress–cone were set as the electric ground boundary. For thermal simulations, which were coupled to the electrical field based on the electrical properties of each conduction or insulation component, the core and ambient (outside outer protective layer) temperatures were set as 70 °C and 30 °C, respectively. The thermal–electrical coupling simulations were implemented by COMSOL Multiphysics according to reference [16], employing material parameters of electrical and thermal parameters for each constituent as listed in Table 1.

Numerical simulations with the finite-element method were performed to calculate the electric field distributed in the cable accessory using the modified electrical properties of reinforced insulation materials characterized by the prepared SiC/LSR nanocomposites, such as the nonlinear conductivity and the higher relative permittivity than cable main insulation materials as listed in Table 1. A free triangular cell was utilized for finite-element meshing by the Delaunay triangulation algorithm, which was refined locally at the positions where the electric field strength varies significantly in the cable terminal. The maximum and minimum numbers of elements were adjusted until no obtuse angles in the triangulation meshing process appeared. The element growth rate was specified as 1.5, which means that the element size increased by about 50% from one element to another. The slack in narrow regions was set as 1.0 to prevent the triangular meshing from generating different sizes of elements.

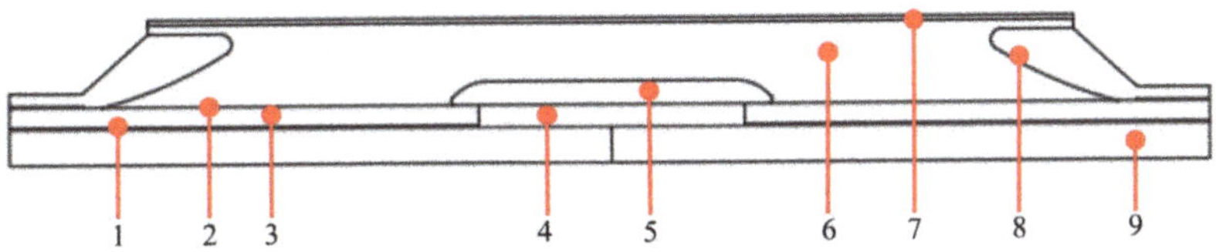

Figure 2. Geometry model of cable joint: 1, inner shield; 2, outer shield; 3, XLPE; 4, connection fittings; 5-high-voltage shielding tube; 6, reinforced insulation; 7, outer protective layer; 8, stress cone; 9, conductor core.

Table 1. Electrical and thermal parameters of materials specified in the electric field simulations.

Materials	Density/(g·cm^{-3})	Relative Permittivity	Coefficient of Thermal Conductivity/(W·m^{-1}·K^{-1})
XLPE	910	2.27	1640
Inner Shield	950	100	2500

3. Results and Discussion

3.1. Microstructure and Mechanical Property

The microscopic morphologies of the SiC/LSR nanocomposites observed by SEM are shown in Figure 3 in which the highlighted areas indicate SiC nanofillers. For 5.0 wt% filling content, SiC nanofillers are evenly dispersed without any obvious agglomeration. With the increase of SiC content up to 20.0 wt% content, the highlighted parts inside the nanocomposites increase slightly in size, while retaining a high dispersivity of homogeneous spatial distributions, as shown in Figure 3. The SEM images verify that the SiC/LSR nanocomposites were successfully prepared as expected to maintain a high dispersion of SiC nanofillers even for the filling content approaching 20.0 wt%.

The tensile modulus and broken-elongation of the SiC/LSR nanocomposites, as illustrated by the stress–strain characteristics in Figure 4, increase and decrease, respectively, with the increase of SiC content. When SiC content approached 20.0 wt%, the tensile strength and broken-elongation of SiC/LSR nanocomposites were maintained at 6 MPa and 350%, respectively, which meet the mechanical requirement for cable accessories. However, due to the fact that the mechanical performances of cable accessories are greatly influenced by operating conditions and cable manufacturing technology, the applications of the SiC/LSR nanocomposites need to be further investigated for improving mechanical properties.

3.2. Electrical Conductance

Electrical conductivity was normalized as the dependence of the current density on the electric field (*J*-*E* variation curves) to investigate the charge transport characteristics of the SiC/LSR nanocomposites, as shown by the *J*-*E* curves of the SiC/LSR composites in Figure 5. The critical points arise around electric field strength of 10^7 V/m, as a threshold E_{th} at which point the charge transport mechanism changes from Ohmic conductance to space charge limited conductance (SCLC). Nonlinear coefficients β_1 and β_2 distinguished by E_{th} are obtained by the linear fitting for the logarithm *J*-*E* curves, as listed in Table 2.

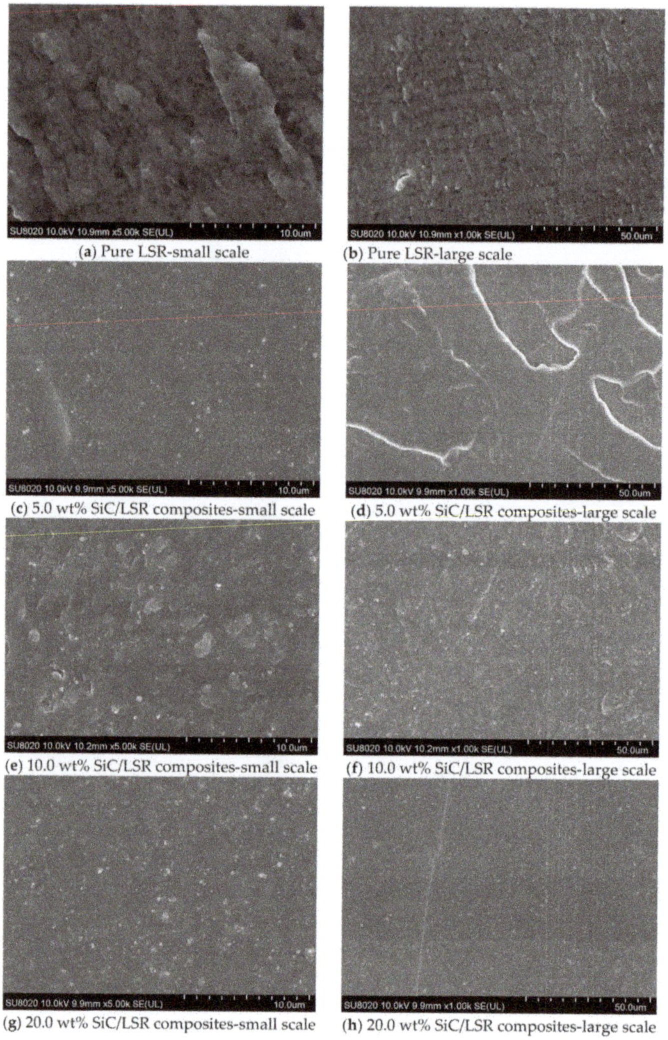

(**a**) Pure LSR-small scale (**b**) Pure LSR-large scale
(**c**) 5.0 wt% SiC/LSR composites-small scale (**d**) 5.0 wt% SiC/LSR composites-large scale
(**e**) 10.0 wt% SiC/LSR composites-small scale (**f**) 10.0 wt% SiC/LSR composites-large scale
(**g**) 20.0 wt% SiC/LSR composites-small scale (**h**) 20.0 wt% SiC/LSR composites-large scale

Figure 3. SEM images of SiC/LSR nanocomposites.

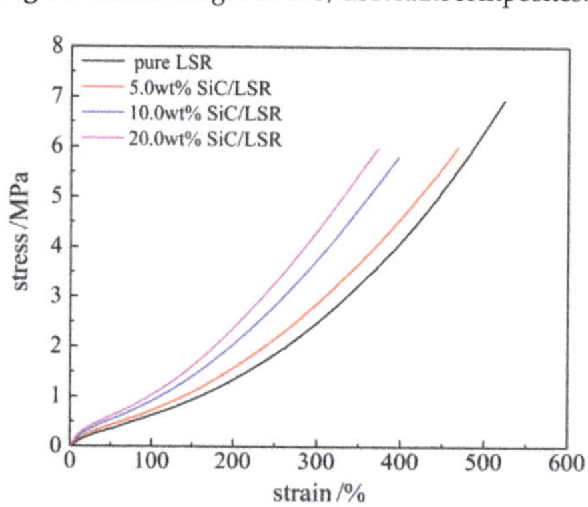

Figure 4. Stress–strain characteristics of SiC/LSR nanocomposites.

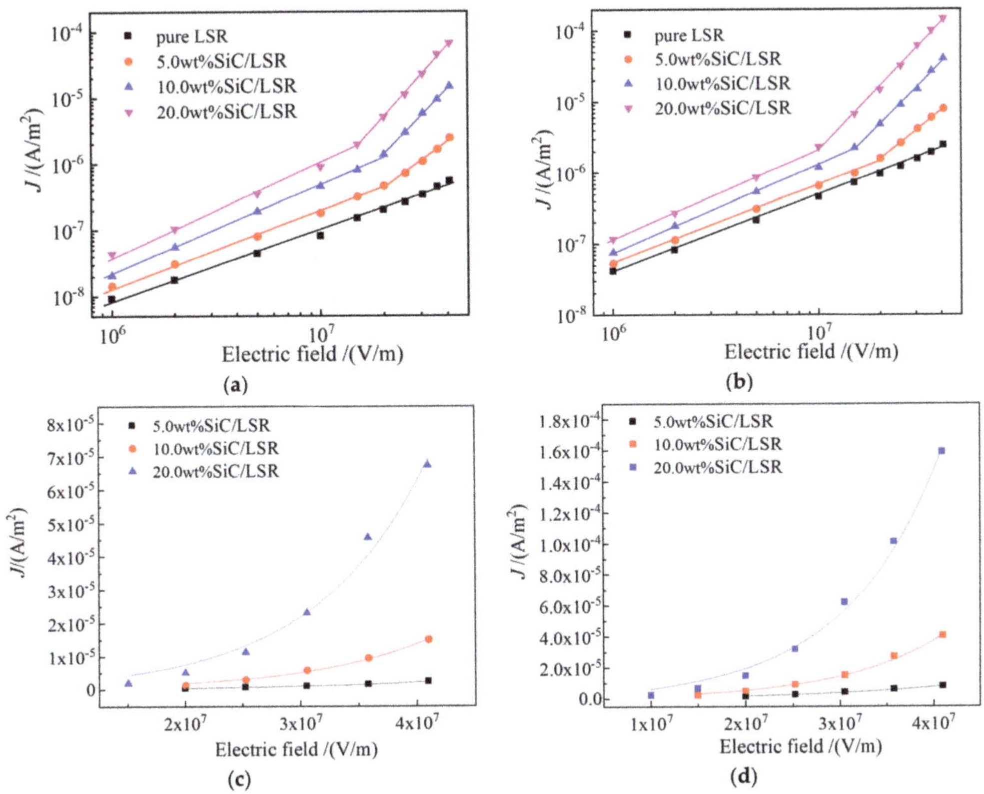

Figure 5. Electric conductance characteristics in the logarithm *J-E* curves of pure LSR and SiC/LSR nanocomposites at diverse temperatures of: (**a**) 30 °C; and (**b**) 70 °C; hopping conductance characteristics of fitting conductivities for the SiC/LSR nanocomposites at (**c**) 30 °C; and (**d**) 70 °C.

Table 2. Nonlinear coefficients of the SiC/LSR nanocomposites.

Materials	30 °C		70 °C	
	E_{th} (kV/mm)	β_1, β_2	E_{th} /(kV/mm)	β_1, β_2
pure LSR	-	-	-	-
5.0 wt%	21.1	1.14, 2.31	10.1	1.10, 2.26
10.0 wt%	20.4	1.37, 3.31	14.9	1.23, 2.85
20.0 wt%	14.8	1.37, 3.56	10.5	1.29, 2.99

When $E < E_{th}$, the nonlinear coefficient $\beta_1 < 2$ complies with the Ohmic conductance mechanism that the charge transport is produced by impurity ionization in the SiC/LSR composites. In contrast, when $E > E_{th}$, the nonlinear coefficient $\beta_2 > 2$ which means the carrier concentration and carrier mobility increase drastically. The process of carrier increment in polymers is a thermal excitation process. According to hopping conductance theory, most of the charges involved in conducting current are in local states, and the charge transfer between localized states is the main process. The process of charge passing from one localized state to another is graphically described as hopping conductance, the current density formula of the jump conductance model can be described as follows:

$$J_n = 2ndve \exp\left(-\frac{\chi}{k_b T}\right) \sinh\left(\frac{elE}{2k_b T}\right) \tag{4}$$

where n represents carrier concentration, χ denotes activation energy, l symbolizes hopping distance of the carrier, and T is thermodynamic temperature. To judge whether the conduction mechanism under a high electric field is dominated by hopping transport, the J-E curves of Figure 5a,b are fitted by Equation (4) according to the hopping conductance model, as shown in Figure 5c,d. The hopping distance of the SiC/LSR composites can be seen in Table 3. The fitting curves are highly consistent with the tested conductivities of the SiC/LSR nanocomposites. Based on the fitting curves in the hopping conductance model, the hopping distance of 5.0 wt%, 10.0 wt%, and 20.0 wt% for the SiC/LSR nanocomposites are calculated by Equation (4), as shown in Table 3. The hopping distance becomes evidently larger with the increase in SiC content, implying that SiC nanofillers facilitate the charge detrapping and thus expedite hopping transports to acquire or increase conductivity nonlinearity. Therefore, the higher the temperature, the higher the density of trap that can be disentangled, and the shorter the hopping distance. Moreover, according to the classical percolation theory [17], when the content of SiC nanoparticles in the LSR matrix exceeds the percolation threshold, the SiC nanofillers are close enough to form a random conductive network, resulting in the macroscopic performance of nonlinear conductivity of the SiC/LSR nanocomposites.

Table 3. Hopping distance of percolation conductance in the SiC/LSR composites.

Materials	30 °C	70 °C
5.0 wt% SiC/LSR	4.0	3.5
10.0 wt% SiC/LSR	5.0	4.9
20.0 wt% SiC/LSR	5.4	5.2

3.3. *Space Charge Characteristics*

In high-voltage DC cable systems, the XLPE cable and its cable accessories bear a high electric field for long periods in which the insulation layer suffers space charge accumulation under electric field distortion at the interface of the multilayered composite insulation structure, accelerating dielectric aging and finally causing insulation failures. For cable accessories, the maximum electric field intensity resides at the stress-cone root of reinforced insulation. This is the part that requires high resistance to space charge accumulations. Space charge characteristics of pure LSR and SiC/LSR nanocomposites in polarization and depolarization processes are shown in Figure 6. Space charge accumulation appears inside the pure LSR sample in the polarization process, which can be significantly inhibited by filling SiC nanoparticles and becomes more obvious for higher SiC content, as comparatively illustrated by the left panels of Figure 6a. Due to the lower mass of negative carriers compared to positive carriers, it is more likely that the negative charge can approach the anode without neutralization in the electrical migration process, accounting for the evident heterocharge accumulations arising near the anode which can be inhibited by filling SiC particles and alleviated with the increase of SiC content.

Homocharge and heterocharge as two classes of space charges are accumulated by trapping the electrode-injected carriers and the impurity ionized carriers that form the applied electrode and the dielectric material interior, respectively [18–20]. Therefore, in the depolarization process, heterocharges decay more quickly than homocharges under a short-circuit as shown in the right panels of Figure 6. There, the space charge decaying near the anode is rather faster than that of the homocharges near the cathode. Meanwhile, the charge density of pure LSR at the initial stage of short-circuit depolarization is much higher than that of the SiC/LSR nanocomposites in which charge density has been further decreased by the increasing SiC content.

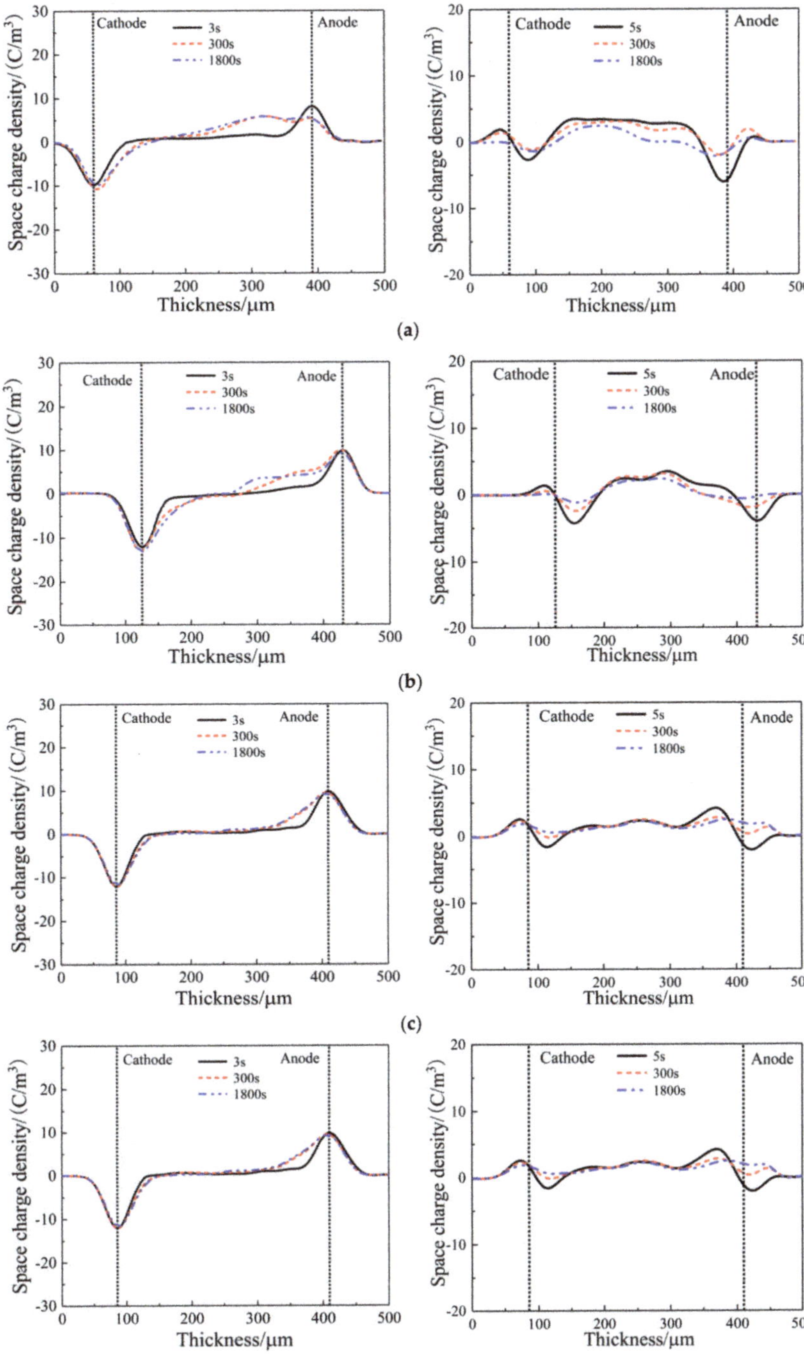

Figure 6. Space charge distribution in: (**a**) pure LSR; (**b**) 5 wt%; (**c**) 10 wt%; and (**d**) 20 wt% SiC contents in the polarization process of applying voltage (left panels) and in the depolarization process under a short-circuit (right panels).

Mean space charge densities in pure LSR and SiC/LSR nanocomposites during a depolarization process are calculated by Equation (1), with the results being shown in Figure 7. The space charges decaying with an almost constant rate in the initial depolarization stage of 0~300 s derives mainly from the detrapping of the charges captured in shallow traps, while the significantly lower decaying rate of the space charge density after depolarization for 300 s implies that the space charge decaying is derived from detrapping the charges captured in deeper traps. In particular, it is evident that the space charges density of the SiC/LSR nanocomposites decreases with the increase in SiC content. In comparison with pure LSR, SiC/LSR nanocomposites acquire a higher resistance to space charge accumulation, while persisting at a higher depolarization rate due to the higher electrical conductance derived from carrier hopping transport through the conductive channels formed between SiC nanofillers.

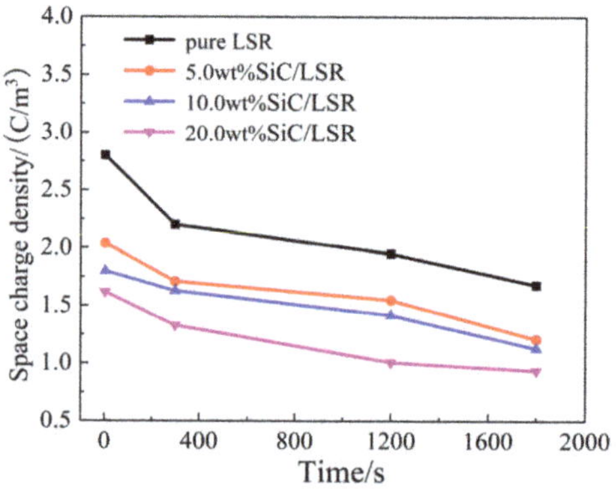

Figure 7. Space charge density of pure LSR and SiC/LSR nanocomposites during depolarization.

3.4. Charge Trap Characteristics

Surface charge dissipation is generally fulfilled by transporting along the material surface or from interior material toward electrode neutralization with charged particles in gas material. As indicated by Figure 8a, the SiC/LSR nanocomposites have achieved a notably higher rate of surface charge dissipation than pure LSR. For the electrode structure and the sample size used for our study, the strength of the normal electric field between the sample surface and the ground electrode is much higher than that of tangential electric fields along the sample and electrode surfaces in the process of surface charge dissipation. This process is dominated by the surface charge transports through the trapping and detrapping processes in the sample interior (volume conduction) to the ground electrode. The potential decay rate of the material surface increases significantly with the increase in SiC content due to the hopping transport of the charges percolated from the traps introduced by the SiC nanofillers. This is manifested by the higher threshold electric field of the SiC/LSR nanocomposites compared with pure LSR at which the electric conduction alters into the highly nonlinear region. Space charge characteristics or surface charge dissipation of the SiC/LSR nanocomposites is dominated by carrier trapping and detrapping processes. Meanwhile, the energy level and spatial density of trap distributions determine the activation energy (carrier concentration) and hopping distance (carrier mobility) of percolation conductance in the nonlinear region. According to ISPD curves, we calculate trap level distributions of pure LSR and the SiC/LSR nanocomposites by Equations (2) and (3), as shown by the results in Figure 8b.

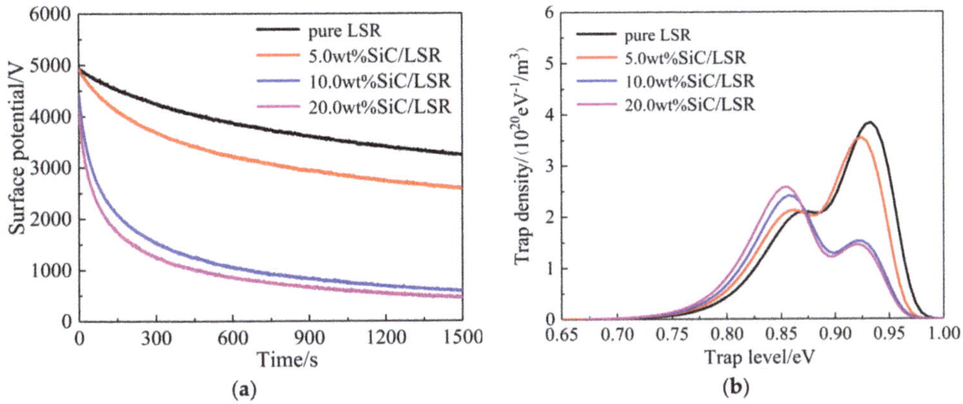

Figure 8. (**a**) Isothermal surface potential decay (ISPD) curves; and (**b**) trap level distributions of SiC/LSR nanocomposites.

At about 0.86 and 0.93 eV of trapping depth, two peaks of trap density appear in the energetic spectra for the SiC/LSR nanocomposites, which could be referenced as the shallow and deep trap levels, respectively. It is noted that the SiC/LSR nanocomposites present distinctly lower and considerably higher densities of deep and shallow traps, respectively, compared with pure LSR, which is more obvious for higher SiC content. The multi-core model [21,22], which is competent to characterize the interface region between the nanofillers and the polymer matrix in nanodielectrics, is composed of a bonding layer, a binding layer, and a loose layer. In the bonding and binding layers, deep traps are derived from the imperfect chemical bonding in the molecular structure, while the amorphous structure in the loose layer mainly contributes to shallow traps. When the filling content is raised to a sufficient level, generally higher than 5.0 wt%, the interface areas as described by the loose layer between the SiC nanofillers overlap to form conductive channels for hopping transports of the charge carriers detrapped from shallow traps, as described by the percolation effect. The significant increase of shallow trap density by filling SiC nanoparticles results in the substantial percolation conductance at room temperature, which will be aggravated by increasing temperature or SiC content, accounting for the extraordinary conductivity nonlinearity given by the SiC/LSR nanocomposites.

3.5. Electric Field Simulation

For DC cable in operation, the temperature in the metallic cable core would be raised by Joule heating, leading to a specific temperature gradient from the cable core to the external insulation. The conductivity of polymer insulation materials, which determines the electric field distribution of the DC cable joint, depends greatly on operation temperature and heat production. For pure LSR or SiC/LSR composites as the reinforced insulation material, the electric field distributions in a 200 kV DC cable joint are simulated under a thermal–electrical fully coupling condition, as shown in Figure 9.

Electric field distribution is inversely proportional to conductivity under DC voltage, and the conductivity of pure LSR is much lower than that of XLPE. Thus, high electric field is distributed in LSR reinforced insulation, as shown in Figure 9a. Maximum electric field strength appearing at the stress-cone root approach 43.40 kV/mm when pure LSR is used for reinforced insulation, which will be effectively reduced by using SiC/LSR nanocomposites as reinforced insulation material, as shown in Figure 9b,c. The position of maximum electric field strength can be ameliorated by locating it in the main XLPE insulation when using the SiC/LSR nanocomposite with a high SiC content approaching 20 wt%, as shown in Figure 9d.

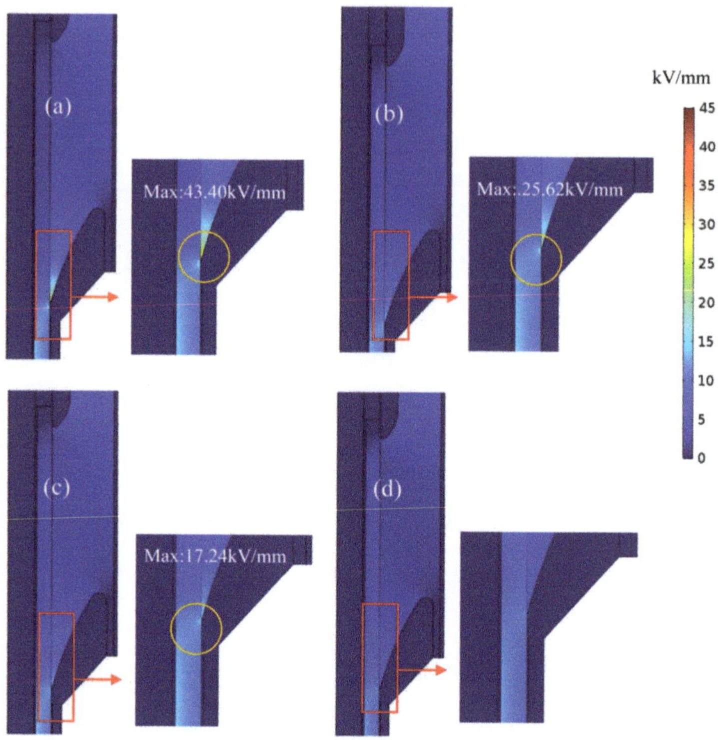

Figure 9. Steady-state electric field distributions in a cable joint with reinforce insulation of: (**a**) pure LSR and SiC/LSR nanocomposites of (**b**) 5 wt%; (**c**) 10 wt%; and (**d**) 20 wt% SiC contents.

4. Conclusions

The effect of SiC nano fillers on the conductivity and the charge trap characteristics of LSR composites was investigated in this study. Conductivity nonlinearity and threshold electric field strength of SiC/LSR nanocomposites increases and decreases, respectively, with the increase of SiC content, which will be intensified at a higher temperature. The highest density peak of trap level distribution shifts from deeper traps to shallow traps by introducing SiC nanofillers into the LSR matrix, which accounts for hopping conductance and impeding space charge injection. Based on the established cable joint model of finite-element multiphysics simulations, it is concluded that the highest electric field strength at the stress-cone root can be effectively reduced by using SiC/LSR nanocomposites as reinforced insulation, compared with using pure LSR material, which can even be intensified by raising the SiC content. As a consequence, the approximate SiC particles doped LSR can suppress the accumulation of space charge and improve conductivity nonlinearity, which provides a possible method for the improvement of cable accessory performance.

Author Contributions: Conceptualization, W.-F.S.; data curation and formal analysis, M.-Z.G.; writing—original draft preparation, M.-Z.G.; writing—review and editing, Z.-Y.L. All authors have read and agreed to the published version of the manuscript.

Funding: This work was supported by Research Start-up Fund of Mudanjiang Medical University (Grant No.: 2021-MYBSKY-057).

Institutional Review Board Statement: Not applicable.

Informed Consent Statement: Not applicable.

Data Availability Statement: Theoretical methods and results are available from the authors.

Conflicts of Interest: The authors declare no conflict of interest.

References

1. Mazzanti, G. Life and Reliability Models for High Voltage DC Extruded Cables. *IEEE Electr. Insul. Mag.* **2017**, *33*, 42–52. [CrossRef]
2. Li, J.; Gao, Y.; Song, Z. On the Crosslinking Temperature Dependence of Charge Transport Behavior in XLPE Film Deduced from Surface Potential Decay Measurements. *J. Electrost.* **2021**, *109*, 103539. [CrossRef]
3. Li, Z.; Su, J. Inhibition Effect of Graphene on Space Charge Injection and Accumulation in Low-Density Polyethylene. *Nanomaterials* **2018**, *8*, 956. [CrossRef] [PubMed]
4. Morshuis, P.; Cavallini, A.; Fabiani, D. Stress Conditions in HVDC Equipmentand Routes to in Service Failure. *IEEE Trans. Dielectr. Electr. Insul.* **2015**, *22*, 81–91. [CrossRef]
5. Khalil, M.S. International Research and Development Trends and Problems of HVDC Cables with Polymeric Insulation. *IEEE Electr. Insul. Mag.* **2002**, *13*, 35–47. [CrossRef]
6. Lan, L.; Wu, J.D.; Wang, Y.N.; Yin, Y. Space Charge Property at The Interface in Low Density Polyethylene/ethylene Propylene Rubber double-layered Insulation. *Proc. Chin. Soc. Electr. Eng.* **2015**, *35*, 1266–1272.
7. Tavernier, K.; Varlow, B.R.; Auckland, D.W. Improvement in Electrical Insulators by Non-Linear Fillers. *IEE Proc.-Sci. Meas. Technol.* **1999**, *146*, 88–94. [CrossRef]
8. Donzel, L.; Greuter, F.; Christen, T. Nonlinear Resistive Electric Field Grading Part 2: Materials and Applications. *IEEE Electr. Insul. Mag.* **2011**, *27*, 18–29. [CrossRef]
9. Wang, X.; Nelson, J.K.; Schadler, L.S. Mechanisms Leading to Nonlinear Electrical Response of a Nano p-SiC/Silicone Rubber Composite. *IEEE Trans. Dielectr. Electr. Insul.* **2010**, *17*, 1687–1696. [CrossRef]
10. Hu, J.; Zhao, X.L.; Yang, X. Improving the Electric Field Strength Distribution of Cable Terminals by Stress Cone of Nonlinear Conductivity Material. *High Volt. Eng.* **2017**, *43*, 398–404.
11. Chi, Q.G.; Li, Z.; Zhang, T.D. Nonlinear conductivity of copper calcium titanate nanofibers/liquid silicone rubber composite. *Acta Mater. Compos. Sin.* **2019**, *36*, 2248–2258.
12. Li, J.; Du, B.; Kong, X.X. Nonlinear Conductivity and Interface Charge Behaviors between LDPE and EPDM/SiC Composite for HVDC Cable Accessory. *IEEE Trans. Dielectr. Electr. Insul.* **2017**, *24*, 1566–1573. [CrossRef]
13. Xue, J.Y.; Chen, J.H.; Dong, J.H. The regulation mechanism of SiC/epoxy coatings on surface charge behavior and flashover performance of epoxy/alumina spacers. *J. Phys. D Appl. Phys.* **2019**, *52*, 405502. [CrossRef]
14. Du, B.; Li, X.; Jiang, J. Surface Charge Accumulation and Decay on Direct-fluorinated Oil-impregnated Paper. *IEEE Trans. Dielectr. Electr. Insul.* **2016**, *23*, 3094–3101. [CrossRef]
15. Chen, G.; Xu, Z.Q. Charge Trapping and Detrapping in Polymeric Materials. *J. Appl. Phys.* **2009**, *106*, 123707. [CrossRef]
16. Li, C.M.; Wu, G.F.; Li, C.Y. Effect of the defects inside XLPE insulated HVDC cable termination on the electric field distribution. *Electr. Mach. Control* **2018**, *12*, 63–67.
17. Strümpler, R.; Glatz-Reichenbach, J. Conducting polymer composites. *J. Electroceram.* **1999**, *3*, 329–346. [CrossRef]
18. Zhao, H.; Xi, C.; Zhao, X.; Sun, W.F. Elevated Temperature Space Charge Characteristics and Trapping Mechanism of Cross-Linked Polyethylene Modified by UV-Initiated Grafting MAH. *Molecules* **2020**, *25*, 3973. [CrossRef]
19. Xh, A.; Ir, B.; Ps, C. PP/PP-HI/Silica Nanocomposites for HVDC Cable Insulation: Are Silica Clusters Beneficial for Space Charge Accumulation. *Polym. Test.* **2021**, *98*, 107186.
20. Zhou, Y.; Wang, W.; Guo, T. Space Charge Accumulation Characteristics in HVDC Cable under Temperature Gradient. *Energies* **2020**, *13*, 5571. [CrossRef]
21. Tanaka, T.; Kozako, M.; Fuse, N. Proposal Multi-Core Model for Polymer Nanocomposite Dielectrics. *IEEE Trans. Dielectr. Electr. Insul.* **2005**, *12*, 669–681. [CrossRef]
22. Tanaka, T.; Montanari, G.C.; Mulhaupt, R. Polymer Nanocomposites as Dielectrics and Electrical Insulation-perspectives for Processing Technologies, Material Characterization and Future Applications. *IEEE Trans. Dielectr. Electr. Insul.* **2004**, *11*, 763–784. [CrossRef]

polymers

MDPI

Article

Enhanced Thermal Stability of Mesoporous Carbon Microbeads-Based Lithium-Ion Batteries by Propargyl Methacrylate as Electrolyte Additive

Yu-Ruei Kung [1,*], Jing-Tang Su [1], Chiung-Cheng Huang [1], Yaoming Xiao [2] and Jeng-Yu Lin [3,*]

1 Department of Chemical Engineering and Biotechnology, Tatung University, Taipei 104327, Taiwan
2 College of Chemical Engineering and Materials Science, Quanzhou Normal University, Quanzhou 362046, China
3 Department of Chemical and Materials Engineering, Tunghai University, Taichung City 407224, Taiwan
* Correspondence: yrkung@gm.ttu.edu.tw (Y.-R.K.); jylin@thu.edu.tw (J.-Y.L.); Tel.: +886-2-77364660 (Y.-R.K.); +886-4-23590121 (ext. 33208) (J.-Y.L.)

Abstract: In this current work, propargyl methacrylate (PMA) was successfully adopted to be an efficient electrolyte additive to stabilize the formation of a solid electrolyte interface (SEI) layer on mesoporous carbon microbeads (MCMB) in Li-ion batteries, especially at elevated temperatures. According to a series of material and electrochemical characterizations, the optimized concentration of PMA additive in the electrolyte was found to be 0.5 wt.%. The MCMB electrode cycled with the optimized 0.5 wt.% PMA-containing electrolyte exhibited impressive capacity retention of 90.3% after 50 cycles at 0.1C at elevated temperature, which was remarkably higher than that using the PMA-free electrolyte (83.5%). The improved electrochemical stability at elevated temperature could be ascribed to the rapid formation of stable and thin SEI layer on MCMB surface, which were investigated and suggested to be formed via PMA copolymerization reactions.

Keywords: propargyl methacrylate (PMA); mesoporous carbon microbeads (MCMB); solid electrolyte interface (SEI)

Citation: Kung, Y.-R.; Su, J.-T.; Huang, C.-C.; Xiao, Y.; Lin, J.-Y. Enhanced Thermal Stability of Mesoporous Carbon Microbeads-Based Lithium-Ion Batteries by Propargyl Methacrylate as Electrolyte Additive. *Polymers* **2022**, *14*, 4491. https://doi.org/10.3390/polym14214491

Academic Editors: Ting-Yu Liu and Yu-Wei Cheng

Received: 29 September 2022
Accepted: 20 October 2022
Published: 24 October 2022

1. Introduction

Lithium-ion battery (LIB) has been turned into one of the promising energy storage devices due to its excellent performance related to energy density, lifetime, ease of use, and low cost of maintenance over other traditional batteries [1–3]. Nevertheless, conventional graphite-based anodes still suffer from severe issues in the significant concerns of safety, cyclic performance, and high-rate performance although they commonly possess high energy density, low operating potential, and long cycle life during the charge and discharge process. The low working potential of graphite generally results in the formation of dendrites due to the continuous lithium deposition, which would therefore cause the safety issue during charging and discharge. During the initial charging/discharging process, an electrically insulating and ionically conductive surface passive film is formed by electrolyte decomposition, which is named a solid electrolyte interface (SEI) [4]. SEI layer formation process would suppress the efficiency of LIBs since it consumes a considerable part of the anode material and electrolyte. However, it also effectively prevents the electrolyte from further reduction reaction on the surface of the anode, thereby protecting the structure of the graphite anode and improving the cycling life and safety of the graphite-based LIBs [5]. Since the properties of SEI film would be necessary for achieving better battery performance, it is necessary to form a suitable SEI passive layer. Consequently, great deals of electrolyte additives have been used in the electrolyte to improve the quality of the SEI passive film and efficiently suppress the electrolyte decomposition [6,7].

Several electrolyte additive systems have been widely reported for LIB application for improving their cycliability and/or coulombic efficiency as collected and illustrated in

Table S1, such as vinylene carbonate (VC) [8–12], borate/boronic acid [13,14], vinyl acetate (VA) [15], propane sultone (PS) [16], ethylene sulfite (ES) [17,18], propargyl methanesulfonate (PMS) [18–20] and other sulfur-containing compounds [21–25]. Most of the electrolyte additive compounds are with sulfonyl (-SO_2-), carbonyl (-C=O), and vinyl (CH_2=CH_2) functional groups for tuning the polarity. In general, electron-withdrawing oxygen and sulfur units make the vinyl group as electrophilic, which facilitates the reduction process to stabilize the formation of the SEI film [26,27]. Among the various electrolyte additive materials, VC is found to produce a durable, stable, cohesive network SEI formation, which can efficiently prevent from cracking and continuously re-exposing of the active materials to the electrolyte during charging and discharging processes [28]. As forming the SEI layer on graphite electrode by introducing PMS electrolyte additive agent, the layer consists of organic components produced by the decomposition of the alkyne moiety, indicating that the related low contents of inorganic components, such as LiF, Li_2CO_3 inside the SEI layer [15]. Owing to the dense structure and low impedance characterization of the SEI layer, the freshly formed SEI by alkyne additive could provide efficient electrochemical behavior as a passivated layer in Li-ion batteries. They also observed that the SEI layer grown in the presence of VC can efficiently facilitate suppressing the salt anion (PF_6) decomposition. Normally unsaturated terminal functionality like vinyl, allyl, and propargyl groups can efficiently promote the polymerization of electrolyte additives on the electrode surface and thereby, increase the resistance. This process generally forms a passive film or a network structure on the electrode surface to improve the long-term performance of LIBs. The formation of SEI film can efficiently suppress the electrolyte-electrode reaction, which results in reduced gaseous ethylene generation during elevated temperature storage and cycling operations.

By taking the advantage of electrolyte additives, we systemically investigated the role of propargyl methacrylate (PMA) with a redox-active functional group as an efficient electrolyte additive (Scheme 1) to address the formation of SEI films on MCMB anodes and improve their long-term electrochemical stability, especially at elevated temperature. Moreover, the polymerization mechanism of the PMA additive during cycling operation was investigated in this research.

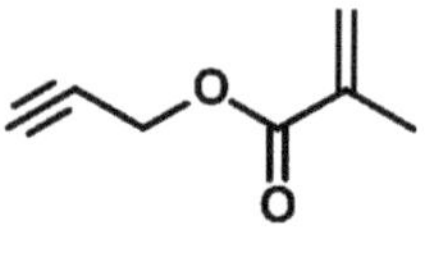

PMA

Scheme 1. Chemical structure of propargyl methacrylate (PMA).

2. Materials and Methods

2.1. Materials

MCMB and PMA were purchased from China Steel Chemical Corporation, Taiwan. Conductive carbon black, Super P, polyvinylidene fluoride (PVDF) Kynar 7200 were purchased from Arkema Technical Polymers. Anhydrous *N*-methyl-2-pyrrollidine (NMP) was obtained from Merck. Ethylene carbonate (EC) and diethyl carbonate (DEC) were mixed with 1.0 M $LiPF_6$ obtained from Formosa Plastics Group, Taiwan. All reagents, as mentioned above, were in the analytical grade and used directly without any further purification.

2.2. Fabrication of Coin Cells

The MCMB electrodes were fabricated by stirring continuously with 85 wt.% MCMB, 10 wt.% super P and 5 wt.% PVDF in NMP for 2 h, and subsequently the well-mixed slurry was evenly coated on a thin Cu foil by using a blade-coating method with an automatic coater, and then dried at 120 °C for 12 h in a vacuum to remove the most residual NMP solvent and water. Prior to the assembly of the coin cells, the coated MCMB foils were punched into circular electrodes of 10 mm in diameter. Afterward, the 2032-type coin cells were assembled using the MCMB electrodes as the anodes and the lithium metal foils as cathodes in addition to Celgard 2500 monolayer polypropylene (PP) was used as the membrane separator. The assembly of the coin cells was performed in a glove box filled with argon. The as-fabricated coin cells were kept at room temperature for 12 h to ensure the complete impregnation of the electrodes and separators with electrolyte. The base electrolyte solution without PMA additive was composed of 1M $LiPF_6$ in ethylene EC/DEC with a volume ratio of 1:1. Additionally, the base electrolytes containing different weight ratios of PMA from 0.5 to 2 wt.% were prepared and employed.

2.3. Methods

The morphology and microstructure of the samples were analysed by field-emission scanning electron microscope (FE-SEM, JEOL JSM-7000F, JEOL Ltd., Tokyo, Japan). The chemical bonding of the samples was examined via an attenuated total reflection–Fourier transform infrared (ATR-FTIR) spectroscopy (Jasco 6700, Jasco Inc., Tokyo, Japan). The surface chemical state was characterized using X-ray photoelectron spectroscopy (XPS, VG ESCA Scientific Theta Probe). PHI Quantera SXM X-ray photoelectron spectrometer (ULVAC-PHI Inc., Kanagawa, Japan) with an Al Ka X-ray radiation. The negative ion depth profiling method from the secondary ion mass spectrometry (SIMS) analyzed the pristine and cycled electrode materials. A pulsed 30 keV Bi^+ primary ion source was used on 100 × 100 μm area for testing with the current density of 1 pA for all electrodes. Depth profiling was measured by sputtering with 1 keV Cs^+ ion gun, which produces the 40 nA target current over a 260 × 260 μm area for all electrodes. The galvanostatic charge/discharge tests of the assembled coin cells were performed with a battery autotest system (BAT-750B, AcuTech Systems Co., Ltd.) between 0.005 V and 2 V at room temperature (25 °C) and elevated temperature (55 °C). The cyclic voltammetry (CV) was conducted using a CHI 614D electrochemical workstation at a sweep rate of 0.01 mV s^{-1} between 0.005 and 2 V. Electrochemical impedance spectroscopy (EIS) was performed using a Zahner IM6 potentiostat in the frequency range of 100 kHz to 10 MHz with an AC amplitude of 10 mV.

3. Results and Discussion

To investigate the SEI formation of the PMA-modified MCMB electrodes with different concentrations, CV experiments were conducted and depicted in Figure 1. The PMA-free system in Figure 1a shows an obviously irreversible cathodic peak generated around 0.2 V after the first cycle scan, which represents the reasonable processes of lithium intercalation into the MCMB electrode [29]. Figure 1b–d shows all the CV curves of MCMB electrodes measured in the base electrolyte containing different concentrations of PMA additive. As represented in Figure 1a, the well-defined peaks appeared at 0.66 and 0.73 V vs. Li^+/Li in the first cycle and then disappeared in the subsequent cycles due to the rapid SEI formation from EC and DEC electrolytes [11]. More importantly, the cathodic peak below 0.2 V and anodic peak between 0.2 V to 0.5 V are representative peaks corresponding to the lithium intercalation into and de-intercalation from the MCMB electrode, respectively [30]. All PMA-modified MCMB electrodes represent the same peak caused by SEI formation at around 0.66 V in addition to the peak at around 1.20 V. The intensities of peaks at 1.20 V increase with the amount of PMA, confirming that it should be related to the electrochemical reduction behavior of PMA additives. The anodic peak between 0.6 and 0.7 V can be assigned to the two-electron reduction of EC converted into Li_2CO_3 and ethylene as following equation 1 and the peak at 0.73 V is associated with EC one electron reduced into alkyl lithium carbonate ($ROCO_2Li$) as shown in Equations (2) and (3), respectively [31].

$$\text{EC} + 2\,e^- + 2\,\text{Li}^+ \rightarrow \text{Li}_2\text{CO}_3 + \text{CH}_2{=}\text{CH}_2 \quad (1)$$

$$\text{EC} + e^- + \text{Li}^+ \rightarrow [\text{Li}^+\ \text{EC}^-] \quad (2)$$

$$2\,[Li^+\,EC^-] \rightarrow LiOCO_2CH_2CH_2CO_2OLi + CH_2{=}CH_2 \quad (3)$$

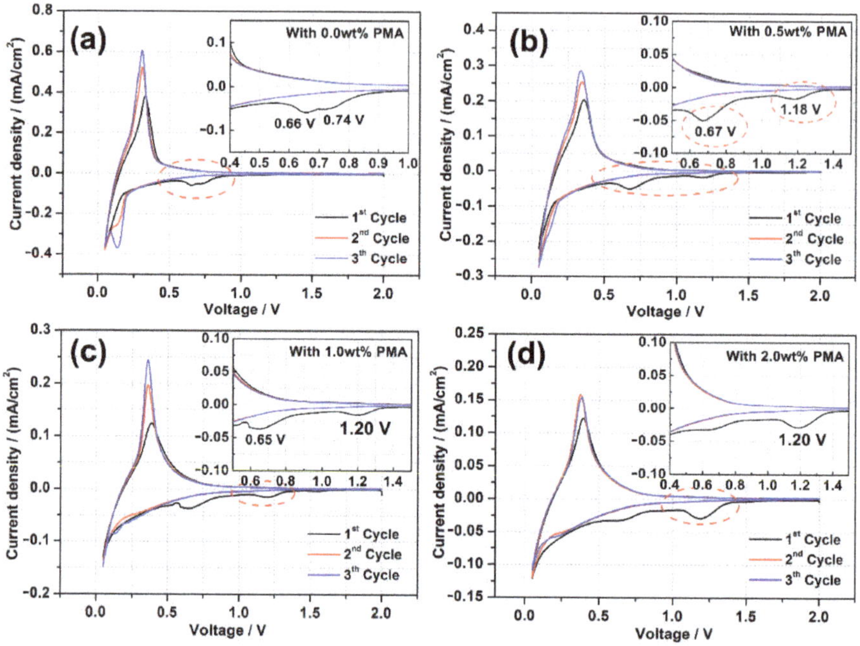

Figure 1. CV curves and their enlarged regions of the MCMB electrodes recorded in the base electrolyte containing (**a**) 0.0 wt.%, (**b**) 0.5 wt.%, (**c**) 1.0 wt.% and (**d**) 2.0 wt.% of PMA additive at a scan rate of 0.1 mV/s.

On the other hand, an additional reduction peak at the higher potential of around 1.20 V vs. Li^+/Li is observed for the MCMB anodes in the PMA-containing electrolytes, which can be ascribed to the initiation of SEI formation on the MCMB electrode surface due to the decomposition of PMA. However, the reduction peaks are found to be vanishhed in the consequent CV cycles. Since the reduction potential of PMA (1.20 V vs. Li^+/Li) is higher than that of the EC/DEC based electrolyte (0.7 V vs. Li^+/Li), indicating that PMA is preferentially reduced on the MCMB surface before the reduction of the EC/DEC based electrolyte. Moreover, the current density of the reduction peak due to the PMA decomposition at the first cycle is found to be increased with increasing the PMA concentration in the electrolyte. This signifies that more decomposition of PMA on MCMB surface could occur with increasing PMA concentration. On the other hand, the current density of the reduction peak derived from the decomposition of the EC/DEC based electrolyte considerably diminishes upon the presence of the PMA additive, which can dynamically retard the further decomposition of the EC/DEC electrolyte. Furthermore, it is observed that the cathodic peak below 0.2 V and an anodic peak between 0.2 V to 0.5 V become decreased with increasing the amount of PMA additive. This reveals that the thick SEI derived from the decomposition of high PMA concentration would be a barrier for reducing lithium intercalation/de-intercalation in the MCMB electrode.

Figure 2a presents the ex-situ ATR FT-IR spectra of the pure PMA, EC/DEC electrolyte, and their mixture. As depicted in Figure 2a, the pure PMA reveals a strong peak at 3298 cm^{-1} and a medium-intensive peak near 2130 cm^{-1}, which can be attributed to the alkyne H-C≡C and C≡C triple bond stretching vibrations, respectively. Moreover, the C=C stretching and =C-H out-of-plane bending vibration peaks at 1637 cm^{-1} and 943 cm^{-1}, respectively, are associated with the alkene group of PMA. Furthermore, the vibration peak located at 1723 cm^{-1} is responsible for the acrylate ester functional group (O-C=O) of

PMA. As for the EC/DEC electrolyte, the strong transmittance peaks at 1772 and 1805 cm^{-1} related to the C=O stretching vibrations from alkyl (for DEC) and cyclic (for EC) carbonate nature, can be obviously found, in addition to a vibration peak at around 1150 cm^{-1} correlated to the C-O bending vibrations. Figure 2b presents the ex-situ ATR FT-IR spectra of the fresh MCMB electrode and the cycled MCMB electrode in the electrolyte with various concentrations of PMA additive. As shown in Figure 2b, the three peaks located around 1700 to 1800 cm^{-1} for different carbonyl (C=O) groups, 1635 cm^{-1} (C=C), 1401 cm^{-1} ($-CH_2$), 1034 cm^{-1} (C–O) are apparently found for all samples except the pristine MCMB electrode. These peaks mainly correspond to the formation of lithium alkyl carbonate ($ROCO_2Li$) species [31]. Due to its relatively less content of PMA additive on the MCMB samples, the signals of alkyne group (H-C≡C and C≡C) are not easy to be observed in the FT-IR spectra, as shown in Figure 2b. Furthermore, the carbonyl groups (C=O, C-O) of EC and DEC are found to be a little shifted to the lower frequency for the cycled MCMB electrode with PMA additive. This phenomenon indicates that the introduction of PMA additive could enhance the solvation interactions of Li^+ ions with the oxygen elements of carbonyl group (C=O, C-O) from EC/DEC-based electrolyte instead of the hydrogen bonding behaviors [32]. According to Aurbach et al. [11,33] Li_2CO_3 can be formed over the surface of Li electrode from decomposed open ring carbonates of electrolyte. They also demonstrated that EC and DEC could undergo two-electron redox reactions to form the nucleophilic carbonate intermediates and alkylene. Furthermore, unstable nucleophilic carbonates can readily react with the remained EC and DEC electrolytes to form alkyl carbonates, as described in Equations (4) and (5), respectively.

$$EC + DEC + 2e^- \rightarrow CO_2^- + CH_2{=}CH_2 \text{ (gas)} \quad (4)$$

$$CO_2^- + DEC \rightarrow CH_3CH_2(OCO_2^-)_2 \quad (5)$$

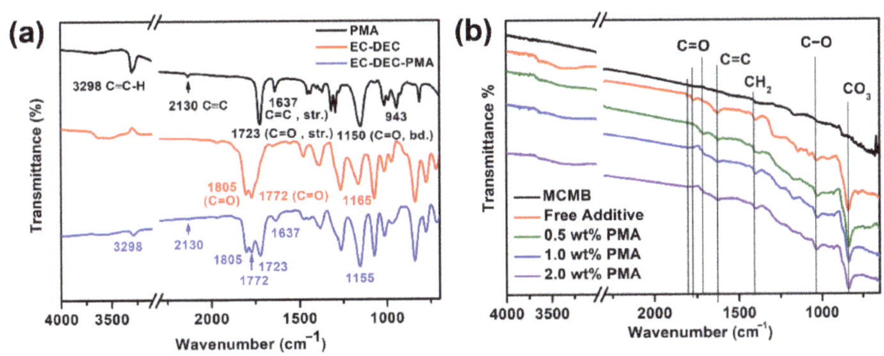

Figure 2. (**a**) ATR FT-IR spectra of PMA, EC/DEC electrolyte and EC/DEC containing PMA, and (**b**) ATR FT-IR spectra of fresh MCMB electrode and cycled MCMB electrode in the base electrolyte containing different PMA concentrations.

These dissociated carbonates combined with the Li^+ ions would precipitate as well-known SEI thin films. Due to the low concentration of electrolyte near to counter electrode, the nucleophilic reaction can effectively be suppressed. Hence, the carbonate ion easily reacts with the Li^+ ions to form a major surface species Li_2CO_3 and become as an important passive agent to stabilize the Li counter electrode. The ATR FT-IR spectra of the MCMB electrode cycled in the presence of PMA electrolyte additive show that the peaks associated with the $ROCO_2Li$ become much smaller than that in the absence of PMA additive, confirming that the formation of $ROCO_2Li$ corresponds to the two-electron electrochemical reduction process of EC, which exhibits a good agreement with the CV studies.

Figure 3a presents the cycling performance of MCMB electrodes charged/discharged in the base electrolyte containing various PMA concentrations at 0.1 C for 50 cycles at room temperature. It is clearly seen that the MCMB anodes in PMA-free and 0.5 wt.% PMA-containing electrolytes displayed superior electrochemical stability compared to those tested in 1.0 wt.% and 2.0 wt.% PMA-containing electrolytes. To further investigate their long-term cycling performance, the galvanostatic charge-discharge tests were performed at 0.1 C for 100 cycles. As can be seen in Figure 3b, the discharge capacity of the MCMB electrode in the absence of PMA was decreased along with the significant capacity decay after 80 cycles. The capacity retention of the MCMB electrode in the PMA-free electrolyte was estimated as ca. 80.4%. However, with the introduction of 0.5 wt.% PMA in the base electrolyte, the capacity retention of the MCMB electrode was then improved to ca. 86.9%. This signifies that the introduction of 0.5 wt.% PMA can slightly improve the electrochemical stability of MCMB anodes when operated at room temperature.

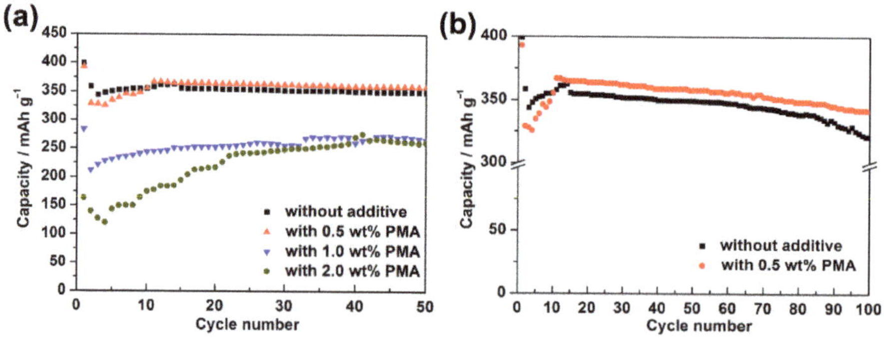

Figure 3. (**a**) Cycling performance of the MCMB electrodes charged and discharged in the base electrolyte containing various PMA concentrations at 0.1C for 50 cycles at room temperature. (**b**) The comparison of long-term cycling performance between the MCMB anodes tested in PMA-free and 0.5 wt.% PMA-containing electrolytes at 0.1 C for 100 cycles at room temperature.

Figure 4a further shows the initial charge/discharge curves of the MCMB electrodes in the base electrolyte containing various PMA concentrations at elevated temperature. During the discharging process, the MCMB electrode in the additive-free electrolyte showed an obvious voltage plateau at around 0.75 V, which is highly associated with the reduction of the ECDEC-based electrolyte. However, that typical behavior almost disappeared in the charge/discharge curves of the cells in the presence of PMA additive. This confirms the findings from the CV curve that the PMA decomposition film could efficiently suppress the formation of SEI film due to the reduction of EC/DEC base electrolyte. The MCMB electrode with the PMA-free base electrolyte and 1.0 wt.% PMA-containing electrolyte exhibited a discharge capacity of 393.35 and 392.16 mAh g^{-1}, respectively. When the PMA concentration was increased to 1.0 and 2.0 wt.%, the discharge capacity was further increased to 400.44 and 411.12 mAh g^{-1}, respectively. The increased initial discharge capacity of the MCMB in the presence of high PMA concentration can be ascribed to the decomposition of PMA in the first discharge process. As depicted in Figure 4b, the cycling stability of the MCMB electrodes in the base electrolyte containing various PMA concentrations at elevated temperature was verified. It is worth noting that the capacity retention of the MCMB electrode was achieved up to 90.3% in the based electrolyte containing 0.5 wt.% PMA, which was significantly superior to that of using the additive-free electrolyte (83.5%). Nevertheless, the capacity retention of the MCMB electrode became decreased with further increasing the PMA concentration to 1.0 wt.% (86.3%) and 2.0 wt.% (81.3%). After 50 cycles, the discharge capacity of the MCMB electrode was found around 354.04 mAh g^{-1} in the base electrolyte containing 0.5 wt.% PMA, which was even significantly higher than that in the additive-free electrolyte (328.48 mAh g^{-1}). As a result, the cycling performance of the MCMB electrode at elevated temperature could be remarkably improved when the base electrolyte contained 0.5 wt.% PMA.

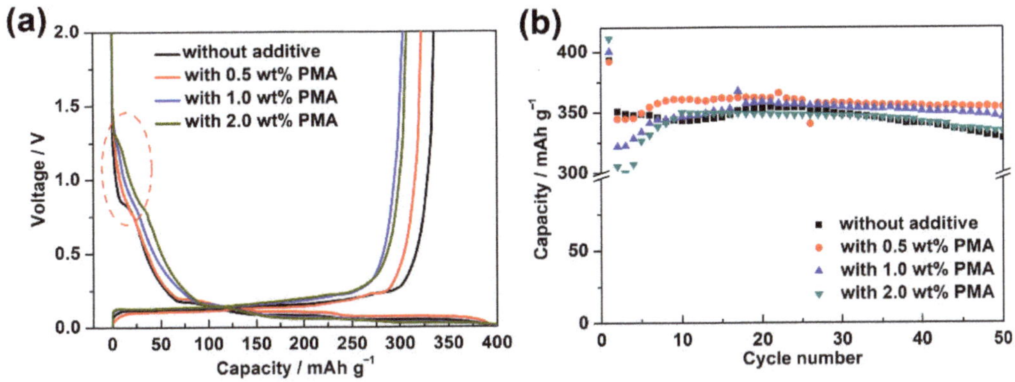

Figure 4. (**a**) Initial galvanostatic charge and discharge curves and (**b**) cycling performance of MCMB electrodes in the base electrolyte containing various PMA concentrations at 0.1C for 50 cycles at elevated temperature.

To investigate the morphology, change of the MCMB electrode in the presence of PMA electrolyte additive, the FE-SEM images were taken for the MCMB electrodes charged/discharged for 3 cycles at a current density of 0.1 C at elevated temperature. As can be seen in Figure 5a,b, the surface morphology of the MCMB electrodes before and after charged/discharged in the PMA-free electrolyte did not have an obvious difference, suggesting that the growth of the SEI film from the decomposition of EC/DEC electrolyte is relatively slow. Nevertheless, the surface evolution of the MCMB electrodes was obviously observed when the PMA was introduced. It was found that the surface of the MCMB electrodes became smoother with increasing the PMA concentration, as depicted in Figure 5b–e. This confirms the decomposition of PMA additive, as shown in the CV studies. Figure S1 further presents the FE-SEM images of the MCMB electrodes after 100-cycle charged/discharged tests in the PMA-free and 0.5 wt.% PMA-containing electrolytes. As depicted in Figure S1a, the MCMB electrode in the absence of PMA additive reveals uneven and rough surface morphology, which could be due to the continued/uncontrolled growth during the prolonged cycles. In contrast, the smooth and compact SEI layer is found to be formed on the MCMB surface (Figure S1b) when operated in the 0.5 wt.% PMA-containing electrolyte. This suggests that the small amount of PMA additive could efficiently stabilize the SEI layer growth.

Lots of previous studies reported that the SEI layer would be rapidly formed at elevated temperature [11,31,33]. Therefore, the surface morphology of the MCMB electrodes after being charged/discharged at 0.1C for 50 cycles at elevated temperature in the base electrolyte containing various PMA concentrations was investigated. Figure 6a presents that the surface of the MCMB electrode tested in the PMA-free electrolyte at elevated temperature was covered with a relatively thicker SEI layer compared with that tested at room temperature (Figure S2a). Additionally, the relatively smooth and even MCMB surface was observed with increasing PMA concentration in the base electrolyte, as depicted in Figure 6b–d. This signifies that the decomposition rate of the PMA additive was also significantly accelerated at elevated temperature.

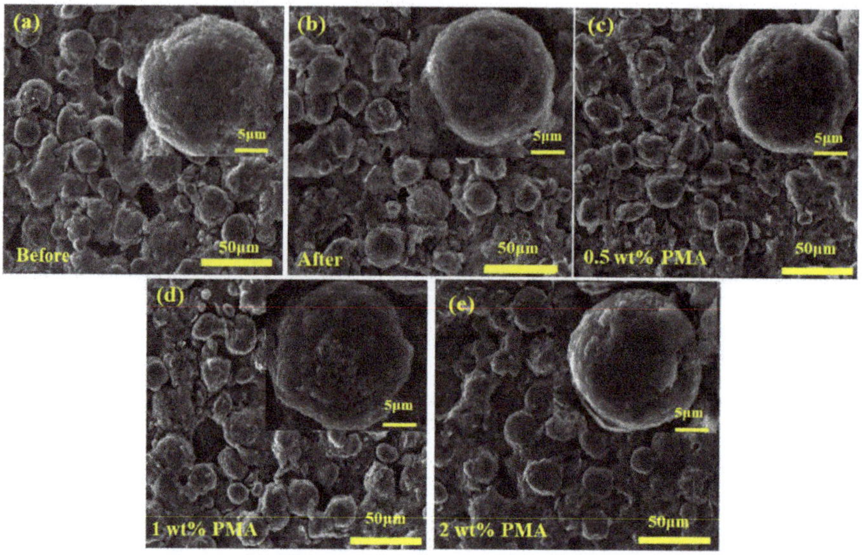

Figure 5. FE-SEM image of MCMB (**a**) before or after 3 cycles with (**b**) free additive, (**c**) 0.5 wt.%, (**d**) 1.0 wt.% and (**e**) 2.0 wt.% of PMA electrolyte additives.

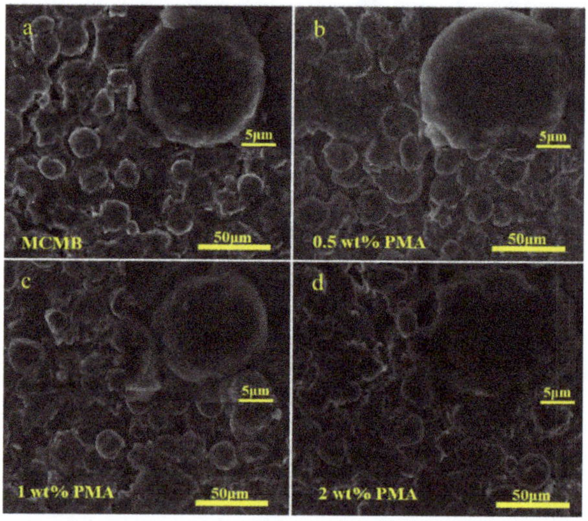

Figure 6. FE-SEM images of MCMB electrodes tested in the base electrolyte containing (**a**) 0.0 wt.%, (**b**) 0.5 wt.% (**c**), 1.0 wt.% and (**d**) 2.0 wt.% PMA at elevated temperature.

To further investigate the effect of PMA additive on the chemical composition of the SEI films on the MCMB electrode surface, the 3rd and 50th cycled MCMB electrodes at elevated temperature were examined by using XPS analyses, as shown in Figure 7. The selected and detailed XPS results are also summarized in Figure 7 and Figure S3, respectively. The obtained C 1s spectra reveal that the decomposition products of the SEI film consist of C-O and C-C containing groups. As shown in C 1s spectra, 284.5 eV belongs to the sp^2 carbon-carbon bonding, and it confirms the presence of inherent as graphite nature. The broad peak around 285 to 288 eV can be assigned to the presence of hydrocarbon (C-H) (285.8 eV), which could be possibly derived from alkyl carbonate

($R\text{-}CH_2OCO_2\text{-}Li$) (286.8 eV) and PVDF (286.0 eV) binder, respectively. Moreover, the peak observed at 286.8 eV is associated with C=O bond of lithium alkyl carbonate when cycled in the base electrolyte with 0.5 wt.% PMA. When compared to the MCMB cycled in the PMA-free electrolyte, the MCMB electrode tested in the base electrolyte containing 0.5 wt.% PMA displays noticeable weaker signals of C-C, $R\text{-}CH_2OCO_2\text{-}Li$, C=O bonds. Moreover, C=O located at binding energies of 289.0 and 290.5 eV mainly corresponds to Li_2CO_3 and $ROCO_2\text{-}Li$, which are generally regarded as a pair of the essential decomposition products of PMA [34]. Interestingly, it can be found that the intensity of the Li_2CO_3 for the MCMB electrode cycled in the base electrolyte in the presence of PMA is relatively weaker when compared to that of the MCMB electrode test in the PMA-free electrolyte. This clearly reveals that the PMA can efficiently retard the formation of organic and inorganic carbonate compounds.

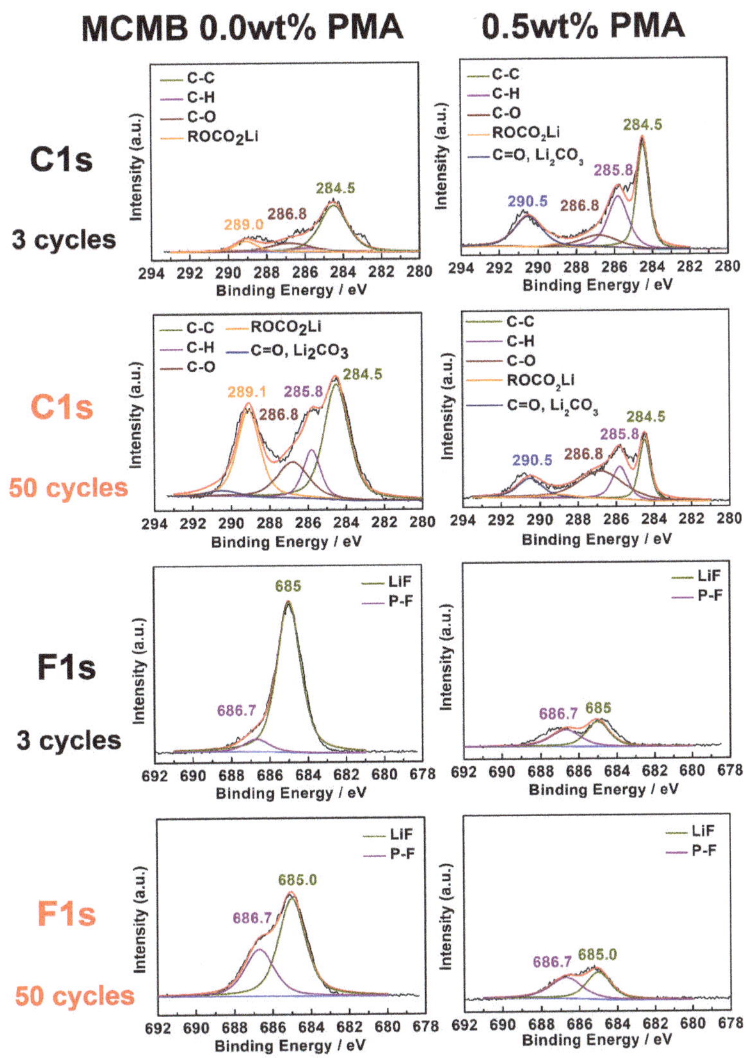

Figure 7. XPS spectra of selected C 1s and F 1s for MCMB electrodes cycled in the base electrolyte with and without 0.5 wt.% PMA additive.

Figure S3 presents the O 1s spectra of MCMB electrode cycled in the base electrolyte with and without 0.5 wt.% PMA additive. It reveals the main C-O peak at 531.3 eV originating from R-OCO_2-Li and a relatively weak peak at 532.5 eV due to the formation of Li_2CO_3 compound in the SEI film. As for the F 1s spectra depicted in Figure 7, a clear and strong peak appears at 685.0 eV for the MCMB electrode cycled in the PMA-free electrolyte, which would be inorganic LiF resulting from electrolyte salt decomposition. It is found that the intensity of LiF for the MCMB electrode in the PMA-free electrode is relatively higher than that in the 0.5% PMA-containing electrolyte. This signifies that LiF easily grows on the surface of MCMB electrode in the absence of the PMA additive. The large amount of deposited LiF on the MCMB electrode generally leads to high interfacial impedance [17]. In the case of P 2p spectra shown in Figure S3, the peak at 137.2 eV and 133.9 corresponds to that of the LiP_xOF_y and Li_xPF_y, respectively. The MCMB electrode cycled in the PMA-free electrolyte shows a much higher amount of lithium alkyl carbonate compounds compared to that in the PMA-containing electrolyte, suggesting that a stable interfacial structure of SEI to prevent the electrolyte from further decomposition could be well established in the presence of PMA additive. This indicates that the addition of 0.5 wt.% PMA can efficiently promote the rapid growth of stable SEI film on MCMB surface and therefore improve its cyclability, especially when operated at elevated temperature.

To investigate the deep profiles of composition for the SEI layers formed on MCMB electrodes, SIMS analyses were conducted. Figure 8 depicts the content of C, F, O, Li, P elements of the MCMB electrodes charged/discharged in the base electrolyte containing various concentrations of PMA additive for consecutive 50 cycles. According to the variation in depth of the C, F, O, Li, P signals, the SEI thickness on the MCMB surface could be approximately evaluated. The thickness of the SEI layer is estimated to be 110, 36, 45 and 60 nm for the MCMB electrode cycled in the base electrolyte containing 0, 0.5, 1 and 2 wt.% PMA, respectively. It is interesting to note that MCMB electrodes cycled in the base electrolyte containing 0.5 and 1 wt.% of PMA exhibit the SEI layer in the thickness of around 16~18 nm. Hence, it is noteworthy to mention that the addition of 0.5 and 1 wt.% of PMA can help to grow the thin SEI layers on the MCMB surface when compared to that with PMA-free and 2 wt.% PMA-containing electrolytes.

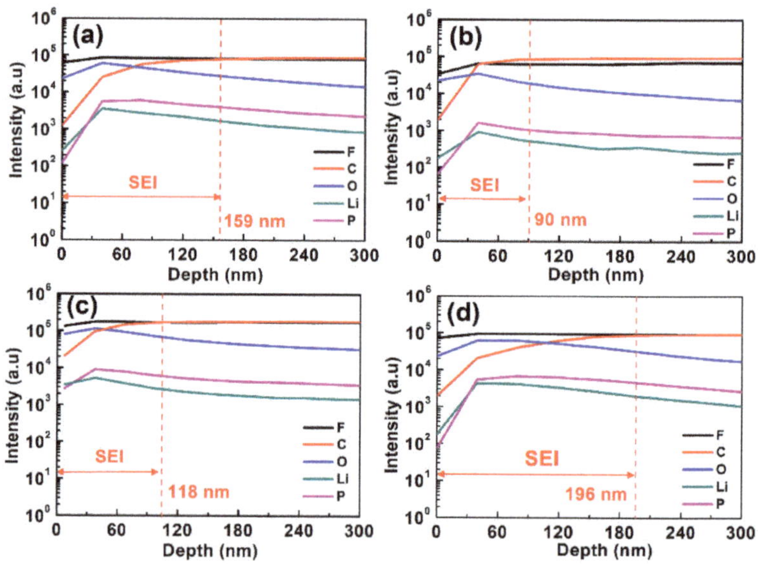

Figure 8. SIMS depth profile of the MCMB electrode with additive-free electrolyte and different concentrations of PMA added electrolyte at 55 °C. (**a**) additive-free electrolyte, (**b**) 0.5 wt.% (**c**) 1.0 wt.% and (**d**) 2.0 wt.% of PMA additive.

Additionally, EIS measurements for the MCMB electrodes cycled in the base electrolyte containing different PMA concentrations at elevated temperature were carried out and the resultant Nyquist plots are depicted in Figure 9. The inset of Figure S4a is the equivalent circuit model used to fit the obtained EIS spectra [31,34,35]. The equivalent circuit parameters obtained from the Nyquist plots are given in Table 1. R_s represents the electrolyte resistance or solution resistance, the resistance at the first depressed high-frequency semicircle was related to the SEI film formed on the electrode surface (R_{sei}), and the resistance at the second depressed mid-frequency difference is related to the charge transfer resistance of both cathode and anode (R_{ct}), and the inclined line in the lower frequency represents the Warburg impedance (W). After cycling for 50 cycles, it can be found that the R_{sei} value of the MCMB electrode is remarkably decreased from 10.34 Ω to 4.56 Ωwhen only introducing 0.5 wt.% PMA into the electrolyte. Nevertheless, the R_{sei} value of MCMB electrode is observed to be significantly increased with further increasing the PMA concentration from 0.5 wt.% to 2 wt.%. When introducing a large amount of PMA, the increased R_{sei} value at a high concentration of PMA could be ascribed to the severe decomposition of PMA. This EIS result also confirms that the optimized addition of 0.5 wt.% PMA in the base electrolyte can rapidly stabilize the SEI growth and form a compact and thin SEI layer on the MCMB surface. Therefore, it would efficiently promote the electron transfer between the electrolyte and MCMB electrode and facilitate the rapid Li-ion diffusion process, thus resulting in the low R_{ct} and Z_W values.

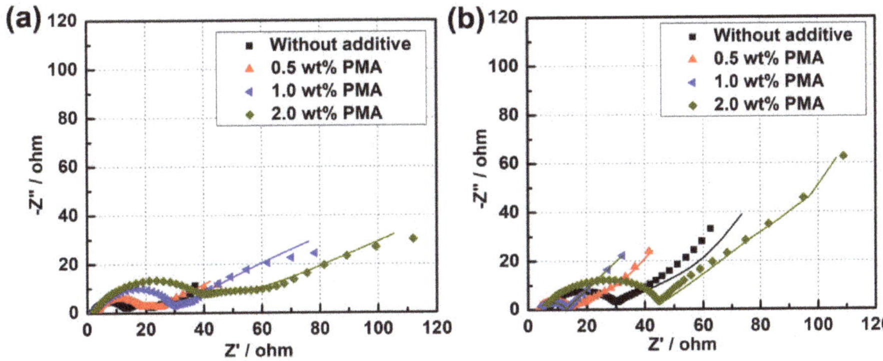

Figure 9. Nyquist plots of MCMB electrodes after cycled in the base electrolyte containing different concentrations of PMA additive for (**a**) 3 cycles and (**b**) 50 cycles at 55 °C.

Table 1. EIS fitting results obtained from Nyquist plots of MCMB.

PMA Content (wt.%)	After 3 Cycles				After 50 Cycles			
	R_s (Ω)	R_{sei} (Ω)	R_{ct} (Ω)	Z_W (Ω)	R_s (Ω)	R_{sei} (Ω)	R_{ct} (Ω)	Z_W (Ω)
0.0	2.167	9.837	7.101	49.54	6.68	10.34	10.41	84.19
0.5	3.046	14.04	4.432	96.1	3.72	4.60	4.39	43.40
1.0	2.106	26.43	4.993	107.1	3.45	7.13	2.16	122.00
2.0	2.487	34.65	10.49	305.6	5.25	22.91	15.30	188.90

According to the aforementioned results, we postulated a possible mechanism of SEI formation in the base electrolyte in the presence of PMA, as illustrated in Scheme 2. As depicted in Scheme 2, the dominant SEI components are suggested, which include the Li_2CO_3, RCO_2Li, $ROCO_2Li$, etc. Since PMA additives consist of both triple bond and double bond, which are highly prone to decompose and produce both the radical and anion species, and it leads to the uniform formation of the SEI film via route A, B and C, respectively. Simultaneously, the decomposition of the carbonate-based solvent and electrolytes would be effectively suppressed, especially operated at elevated temperature.

The triple and double-bonded (CH_2=CH_2, CH=CH-CH_2-COO-CH=CH(CH_3)) compounds and the inorganic (Li_2CO_3) SEI products could possibly originate from the PMA additives. As illustrated in Scheme 2, the triple-bond in the PMA compound decomposed by the reduction process would produce the propynyl and methyl methacrylate radicals on the negative electrode. At the initial charge/discharge process, as also shown in Scheme 2, copolymerization reactions of double and triple bonds took place on the negative electrode surface, and it forms a stable SEI film before the electrolyte decomposition. The humidity sensitive $LiPF_6$ electrolytes exhibit inferior thermal stability at elevated temperatures, and it can produce the PF_5, HF and LiOH combined with the trace amount of H_2O as shown in Equation (6) [36].

$$LiPF_6 + H_2O \rightarrow PF_5 + HF + LiOH \tag{6}$$

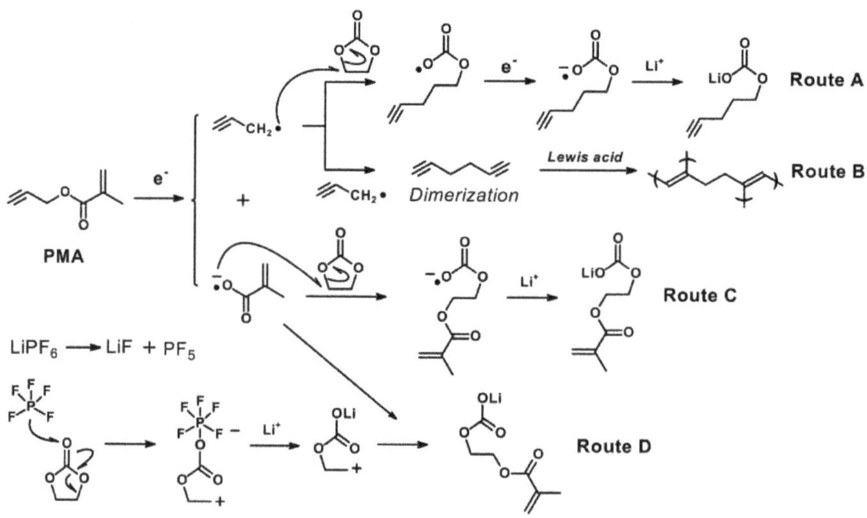

Scheme 2. The possible mechanism pathways for the decomposition of PMA on the MCMB electrode.

Moreover, $LiPF_6$, $LiBF_4$, and $LiAsF_6$ have been shown to generate the Lewis acid PF_5, BF_3, and AsF_5 which are known to catalyse the ring-opening polymerization of poly (ethylene carbonate) (PEC) (as shown in route D) or poly (ethylene oxide) (PEO)-like products with accompanying production of CO_2 gas [30]. On the other hand, DEC-based electrolyte attacked by PF_5 is decomposed, resulting in producing HF by side reactions [37].

Due to the copolymerization reactions, the reduction potential of PMA is higher than that of the base electrolyte, therefore the stable SEI film formation by PMA could also protect MCMB electrode from HF attack and result in improved cyclability. On the other hand, part of the propynyl and methyl methacrylate radicals react with EC and Li cations to form lithium alkyl carbonates. Therefore, an excessive amount of PMA added to the electrolyte system causes additional side reactions, which would result in the reduced cycling performance of LIBs [36,38]. The XPS results in Figure 7 also further support and confirm the SEI products formed by the proposed reaction mechanism.

4. Conclusions

In summary, the PMA was successfully used as an electrolyte additive to stabilize the SEI formation on MCMB electrode for LIB applications. It is found that the compact and thin SEI layer was formed on MCMB electrode surface when introducing optimized 0.5 wt.% PMA in the base electrolyte. The MCMB electrode cycled with 0.5 wt.% PMA-containing electrolyte exhibited an impressive discharge capacity retention of 86.9% (after 100 cycles, 0.1C rate), and 90.3% (after 50 cycles, 0.1C rate) at room temperature and 55 °C, respectively. The improved cyclability of MCMB electrode in the presence of 0.5 wt. % PMA was ascribed to the formation of a stable and thin SEI layer. The formation of the stable SEI layer was then suggested via a rapid copolymerization reaction of PMA compounds. Therefore, in view of these results, PMA can be regarded as an efficient electrolyte additive for stabilizing the SEI layer on MCMB anodes in LIBs.

Supplementary Materials: The following supporting information can be downloaded at: https://www.mdpi.com/article/10.3390/polym14214491/s1, Figure S1. FE-SEM images of (a) Bare MCMB, MCMB material (b) after charge/discharge tests of 100 cycles using additive free electrolyte (c), 0.5 wt.% (d), 1 wt.% (e) and 2 wt.%; Figure S2. FE-SEM images of MCMB anodes after 100 cycles (a) without additive and (b) 0.5 wt.% PMA. Figure S3. XPS spectra of C 1s, F 1s, O 1s and P 2p for MCMB cells containing 0.0 and 0.5 wt.% PMA additive in electrolyte at 3 cycles and 50 cycles; Figure S4. The electrical equivalent circuit (EEC) was employed for simulating the obtained EIS spectra. Table S1. Comparative results of commonly used electrolyte additives to stabilize the graphite anode

Author Contributions: Investigation, J.-T.S. and C.-C.H.; writing—original draft preparation, Y.-R.K. and J.-Y.L.; writing—review and editing, Y.-R.K., Y.X. and J.-Y.L. All authors have read and agreed to the published version of the manuscript.

Funding: This research was funded by National Science and Technology Council, Taiwan, Republic of China (R.O.C.), grant number 107-2218-E-036-003-MY2 and Tatung University, Taipei, Taiwan, Republic of China (R.O.C.), grant number B110-C03-009.

Institutional Review Board Statement: Not applicable.

Informed Consent Statement: Not applicable.

Data Availability Statement: Not applicable.

Acknowledgments: The authors are grateful for the financial support from the Ministry of Science and Technology, Taiwan, Republic of China (R.O.C.) and Tatung University, Taipei, Taiwan, Republic of China (R.O.C.).

Conflicts of Interest: The authors declare no conflict of interest.

References

1. Beck, F.; Rüetschi, P. Rechargeable batteries with aqueous electrolytes. *Electrochim. Acta* **2000**, *45*, 2467–2482. [CrossRef]
2. Palacin, M.R. Recent advances in rechargeable battery materials: A chemist's perspective. *Chem. Soc. Rev.* **2009**, *38*, 2565–2575. [CrossRef] [PubMed]
3. Nitta, N.; Wu, F.; Lee, J.T.; Yushin, G. Li-ion battery materials: Present and future. *Mater. Today* **2015**, *18*, 252–264. [CrossRef]
4. Banerjee, A.; Wang, X.; Fang, C.; Wu, E.A.; Meng, Y.S. Interfaces and interphases in all-solid-state batteries with inorganic solid electrolytes. *Chem. Rev.* **2020**, *120*, 6878–6933. [CrossRef] [PubMed]
5. Zhang, S.S.; Xu, K.; Jow, T.R. EIS study on the formation of solid electrolyte interface in Li-ion battery. *Electrochim. Acta* **2006**, *51*, 1636–1640. [CrossRef]
6. Zhang, S.S. A review on electrolyte additives for lithium-ion batteries. *J. Power Sources* **2006**, *162*, 1379–1394. [CrossRef]
7. Haregewoin, A.M.; Wotango, A.S.; Hwang, B.-J. Electrolyte additives for lithium ion battery electrodes: Progress and perspectives. *Energy Environ. Sci.* **2016**, *9*, 1955–1988. [CrossRef]
8. Xia, J.; Aiken, C.; Ma, L.; Kim, G.-Y.; Burns, J.; Chen, C.L.; Dahn, J. Combinations of ethylene sulfite (ES) and vinylene carbonate (VC) as electrolyte additives in Li $(Ni_{1/3}Mn_{1/3}Co_{1/3})O_2$/graphite pouch cells. *J. Electrochem. Soc.* **2014**, *161*, A1149–A1157. [CrossRef]
9. Park, S.; Jeong, S.Y.; Lee, T.K.; Park, M.W.; Lim, H.Y.; Sung, J.; Cho, J.; Kwak, S.K.; Hong, S.Y.; Choi, N.-S. Replacing conventional battery electrolyte additives with dioxolone derivatives for high-energy-density lithium-ion batteries. *Nat. Commun.* **2021**, *12*, 838. [CrossRef]
10. McMillan, R.; Slegr, H.; Shu, Z.X.; Wang, W.D. Fluoroethylene carbonate electrolyte and its use in Lithium ion batteries with graphite anodes. *J. Power Sources* **1999**, *81*, 20–26. [CrossRef]
11. Aurbach, D.; Gamolsky, K.; Markovsky, B.; Gofer, Y.; Schmidt, M.; Heider, U. On the use of vinylene carbonate (VC) electrolyte solutions for Li-ion as an additive to batteries. *Electrochim. Acta* **2002**, *47*, 1423–1439. [CrossRef]
12. Lee, Y.; Jung, C. Quantitative and qualitative study on the solid electrolyte interface formed by 2-(5h) Furanone: A novel additive for propylene carbonate-based Lithium-ion battery electrolytes. *Electrochim. Acta* **2018**, *265*, 662–669. [CrossRef]
13. Zhang, S.S.; Xu, K.; Jow, T.R. Enhanced performance of Li-ion cell with $LiBF_4$−PC based electrolyte by addition of small amount of LiBOB. *J. Power Sources* **2006**, *156*, 629–633. [CrossRef]

14. Wang, B.; Qu, Q.T.; Xia, Q.; Wu, Y.P.; Li, X.; Gan, C.L.; Van Ree, T. Effects of 3,5-bis(trifluoromethyl)benzeneboronic acid as an additive on electrochemical performance of propylene carbonate-based electrolytes for Lithium ion batteries. *Electrochim. Acta* **2008**, *54*, 816–820. [CrossRef]
15. Abe, K.; Yoshitake, H.; Kitakura, T.; Hattori, T.; Wang, H.; Yoshio, M. Additives-containing functional electrolytes for suppressing electrolyte decomposition in lithium-ion batteries. *Electrochim. Acta* **2004**, *49*, 4613–4622. [CrossRef]
16. Lin, F.-W.; Tran, N.T.T.; Hsu, W.-D. Effect of 1,3-propane sultone on the formation of solid electrolyte interphase at Li-ion battery anode surface: A first-principles study. *ACS Omega* **2022**, *5*, 13541–13547. [CrossRef] [PubMed]
17. Zheng, X.; Wang, W.; Huang, T.; Fang, G.; Pan, Y.; Wu, M. Evaluation of di(2,2,2-trifluoroethyl) sulfite as a film-forming additive on the MCMB anode of lithium-ion batteries. *J. Power Sources* **2016**, *329*, 450–455. [CrossRef]
18. Tong, B.; Song, Z.; Wan, H.; Feng, W.; Armand, M.; Liu, J.; Zhang, H.; Zhou, Z. Sulfur-containing compounds as electrolyte additives for lithium-ion batteries. *InfoMat* **2021**, *3*, 1364–1392. [CrossRef]
19. Abe, K.; Hattori, T.; Kawabe, K.; Ushigoe, Y.; Yoshitake, H. Functional electrolytes triple-bonded compound as an additive for negative electrode. *J. Electrochem. Soc.* **2007**, *154*, A810–A815. [CrossRef]
20. Högström, K.C.; Malmgren, S.; Hahlin, M.; Rensmo, H.; Thébault, F.; Johansson, P.; Edström, K. The influence of PMS-additive on the electrode/electrolyte interfaces in $LiFePO_4$/graphite Li-ion batteries. *J. Phys. Chem. C* **2013**, *117*, 23476–23486. [CrossRef]
21. Li, B.; Xu, M.; Li, B.; Liu, Y.; Yang, L.; Li, W.; Hu, S. Properties of solid electrolyte interphase formed by prop-1-ene-1,3-sultone on graphite anode of Li-ion batteries. *Electrochim. Acta* **2013**, *105*, 1–6. [CrossRef]
22. Xu, M.Q.; Li, W.S.; Zuo, X.X.; Liu, J.S.; Xu, X. Performance improvement of Lithium ion battery using PC as a solvent component and BS as an SEI forming additive. *J. Power Sources* **2007**, *174*, 705–710. [CrossRef]
23. Yao, W.H.; Zhang, Z.R.; Gao, J.; Li, J.; Xu, J.; Wang, Z.C.; Yang, Y. Vinyl ethylene sulfite as a new additive in propylene carbonate-based electrolyte for Lithium ion batteries. *Energy Environ. Sci.* **2009**, *2*, 1102–1108. [CrossRef]
24. Sano, A.; Maruyama, S. Decreasing the initial irreversible capacity loss by addition of cyclic sulfate as electrolyte additives. *J. Power Sources* **2009**, *192*, 714–718. [CrossRef]
25. Ming, J.; Cao, Z.; Wu, Y.; Wahyudi, W.; Wang, W.; Guo, X.; Cavallo, L.; Hwang, J.-Y.; Shamim, A.; Li, L.-J.; et al. New insight on the role of electrolyte additives in rechargeable Lithium ion batteries. *ACS Energy Lett.* **2019**, *4*, 2613–2622. [CrossRef]
26. Chang, C.-C.; Hsu, S.-H.; Jung, Y.-F.; Yang, C.-H. Vinylene carbonate and vinylene trithiocarbonate as electrolyte additives for lithium ion battery. *J. Power Sources* **2011**, *196*, 9605–9611. [CrossRef]
27. Jo, S.-C.; Hong, J.-W.; Choi, I.-H.; Kim, M.-J.; Kim, B.G.; Lee, Y.-J.; Choi, H.Y.; Kim, D.; Kim, T.Y.; Baeg, K.-J.; et al. Multimodal capturing of polysulfides by phosphorus-doped carbon composites for flexible high-energy-density lithium–sulfur batteries. *Small* **2022**, *18*, 2200326. [CrossRef]
28. Kennedy, T.; Mullane, E.; Geaney, H.; Osiak, M.; O'Dwyer, C.; Ryan, K.M. High-performance germanium nanowire-based lithium-ion battery anodes extending over 1000 cycles through in situ formation of a continuous porous network. *Nano Lett.* **2014**, *14*, 716–723. [CrossRef]
29. Pan, Q.; Guo, K.; Wang, L.; Fang, S. Novel modified graphite as anode material for lithium-ion batteries. *J. Electrochem. Soc.* **2002**, *149*, A1218–A1223. [CrossRef]
30. Vogdanis, L.; Martens, B.; Uchtmann, H.; Hensel, F.; Heitz, W. Synthetic and thermodynamic investigations in the polymerization of ethylene carbonate. *Makromol. Chem.* **1990**, *191*, 465–472. [CrossRef]
31. An, S.J.; Li, J.; Daniel, C.; Mohanty, D.; Nagpure, S.; Wood III, D.L. The state of understanding of the lithium-ion-battery graphite solid electrolyte interphase (SEI) and its relationship to formation cycling. *Carbon* **2016**, *105*, 52–76. [CrossRef]
32. Akita, Y.; Segawa, M.; Munakata, H.; Kanamura, K. In-situ Fourier transform infrared spectroscopic analysis on dynamic behavior of electrolyte solution on $LiFePO_4$ cathode. *J. Power Sources* **2013**, *239*, 175–180. [CrossRef]
33. Aurbach, D.; Ein-Eli, Y.; Chusid, O.; Carmeli, Y.; Babai, M.; Yamin, H. The correlation between the surface chemistry and the performance of Li-carbon intercalation anodes for rechargeable 'Rocking-Chair' type batteries. *J. Electrochem. Soc.* **1994**, *141*, 603–611. [CrossRef]
34. Pradanawati, S.A.; Wang, F.-M.; Rick, J. In situ formation of pentafluorophosphate benzimidazole anion stabilizes high-temperature performance of lithium-ion batteries. *Electrochim. Acta* **2014**, *135*, 388–395. [CrossRef]
35. Shim, E.-G.; Nam, T.-H.; Kim, J.-G.; Kim, H.-S.; Moon, S.-I. Diphenyloctyl phosphate as a flame-retardant additive in electrolyte for Li-ion batteries. *J. Power Sources* **2008**, *175*, 533–539. [CrossRef]
36. Wang, R.; Wang, Z.; Li, X.; Zhang, H. Electrochemical analysis the influence of propargyl methanesulfonate as electrolyte additive for spinel LTO interface layer. *Electrochim. Acta* **2017**, *241*, 208–219. [CrossRef]
37. Okamoto, Y. Ab Initio calculations of thermal decomposition mechanism of $LiPF_6$-based electrolytes for lithium-ion batteries. *J. Electrochem. Soc.* **2013**, *160*, A404–A409. [CrossRef]
38. Lee, H.; Wan, C.; Wang, Y. Thermal stability of the solid electrolyte interface on carbon electrodes of lithium batteries. *J. Electrochem. Soc.* **2004**, *151*, A542–A547. [CrossRef]

polymers MDPI

Article

Flexible PDMS-Based SERS Substrates Replicated from Beetle Wings for Water Pollutant Detection

Chen-Hsin Lu [1], Ming-Ren Cheng [1], Sheng Chen [1], Wei-Lin Syu [2], Ming-Yen Chien [2], Kuan-Syun Wang [3], Jeng-Shiung Chen [4], Po-Han Lee [1,*] and Ting-Yu Liu [2,5,6,*]

1 The Affiliated Senior High School of National Taiwan Normal University, Taipei 10658, Taiwan
2 Department of Materials Engineering, Ming Chi University of Technology, New Taipei City 243303, Taiwan
3 Institute of Polymer Science and Engineering, National Taiwan University, Taipei 10617, Taiwan
4 Yottadeft Optoelectronics Technology Co., Ltd., Taipei 10460, Taiwan
5 Research Center for Intelligent Medical Devices, Center for Plasma and Thin Film Technologies, Ming Chi University of Technology, New Taipei City 243303, Taiwan
6 Department of Chemical Engineering and Materials Science, Yuan Ze University, Taoyuan City 32003, Taiwan
* Correspondence: leepohan@gmail.com (P.-H.L.); tyliu0322@gmail.com (T.-Y.L.)

Abstract: The flexible surface-enhanced Raman scattering (SERS) sensor, which has the bionic 3D nanoarray structure of a beetle-wing substrate (BWS), was successfully prepared by replicated technology and thermal evaporation. The bionic structure was replicated with polydimethylsiloxane (PDMS) and then silver (Ag) nanoisland thin films were deposited by thermal evaporation. The deposition times and thicknesses (25–40 nm) of the Ag thin films were manipulated to find the optimal SERS detection capability. The Ag nanoisland arrays on the surface of the bionic replicated PDMS were observed by scanning electron microscope (SEM), X-ray diffraction (XRD), and contact angle, which can generate strong and reproducible three-dimensional hotspots (3D hotspots) to enhance Raman signals. The water pollutant, rhodamine 6G (R6G), was used as a model molecule for SERS detection. The results show that 35 nm Ag deposited on a PDMS-BWS SERS substrate displays the strongest SERS intensity, which is 10 times higher than that of the pristine BWS with 35 nm Ag coating, due to the excellent 3D bionic structure. Our results demonstrate that bionic 3D SERS sensors have the potential to be applied in wearable devices and sensors to detect biomolecules and environmental pollutants, such as industrial wastewater, in the future.

Keywords: surface-enhanced Raman scattering detection; water pollutants; bionic nanostructure; silver nanoislands; thermal evaporation

Citation: Lu, C.-H.; Cheng, M.-R.; Chen, S.; Syu, W.-L.; Chien, M.-Y.; Wang, K.-S.; Chen, J.-S.; Lee, P.-H.; Liu, T.-Y. Flexible PDMS-Based SERS Substrates Replicated from Beetle Wings for Water Pollutant Detection. *Polymers* **2023**, *15*, 191. https://doi.org/10.3390/polym15010191

Academic Editor: Wei Guo

Received: 12 November 2022
Revised: 14 December 2022
Accepted: 27 December 2022
Published: 30 December 2022

1. Introduction

Surface-enhanced Raman spectroscopy (SERS) detection is a powerful analytical tool for trace-level detection with high sensitivity to characteristic vibrational fingerprints. SERS enhancement is divided into two main mechanisms for improving the SERS enhancement effect of the analyte. Firstly, the charge transfer between the analyte and the SERS-active metal substrate improves the signal; this is called chemical enhancement (CE) [1,2]. When the metal nanoparticles and the molecules are in direct contact, an adsorbate–surface complex is formed due to the electronic coupling effect of the molecules and the metal nanoparticles. Adsorbed molecules in this complex display larger cross-sectional effects than free molecules in typical Raman experiments. For example, some researchers synthesized low-dimensional semiconductor substrates, such as ZnO and GaN, to study the chemical enhancement. Moreover, using molecules such as 4-mercaptopyridine (4-MPY), 4-mercaptobenzoic acid (4-MBA), and 4-adenosine triphosphate (4-ATP) as analytes to measure SERS spectra can elucidate the charge transfer mechanisms between substrate and analyte molecules, leading to chemical enhancement. Hence, the charge transfers responsible for chemical enhancement have a mainly enhancing effect in SERS. Secondly, the metallic surface irradiated by a specific laser will generate an electromagnetic field, which then induces localized surface plasmon resonance (LSPR) [3,4] to enhance the signal; this is called the electromagnetic effect [5,6]. The electromagnetic effect originates from the plasmonic resonance of the conductive electrons on the surface of noble metal nanoparticles and the lasing light at the nanoscale, thus forming a huge local electric field on the surface, i.e., the surface plasmon. Surface plasmons are surface electromagnetic waves that exist at the interface of metals and dielectrics. Surface plasmon modes are localized around the surface of metal nanoparticles, creating a highly enhanced near field. According to the formation characteristics of surface plasmons, they can be divided into two categories: surface plasmon polariton (SPP) and localized surface plasmon (LSP). For example, the gold nanorods exhibit two well-divided LSPR bands. These bands have optical characteristics and high stability in solution, so gold nanorods can be conveniently used for the detection and characterization of absorbed molecules. Since SERS was discovered in the 1970s [7,8], it has had constant improvement along with the advancement of nanotechnology. However, traditional SERS sensor platforms were based on robust materials, such as glass [9], FTO glass [10], and silicon wafers [11]. There are also some different procedures of SERS substrate, such as nanoimprint lithography and sylgard 184. Ding et al. used composite nanoimprint lithography of etched polymer/silica opal films with electron-beam evaporation to fabricate a high-performance sensing substrate for UV-SERS [12]. The UV-SERS performance of DNA base adenine had been enhanced, revealing that it matched well with the UV plasmonic optical signatures and simulations. Moreover, Anindita Das et al. applied nanoimprint lithography on a flexible SERS substrate through electron-beam lithography (EBL), which was utilized as a master template, and the mold was later replicated via a nanoimprinting process to prepare a gold-coated polymer nanopyramid three-dimensional (3D) SERS substrate [13]. The intense electric field confinement at nanotips and four edges of a gold-coated polymer nanopyramid enhanced the Raman signal of probe molecules, which meant that rhodamine 6G, with a limit of detection down to 3.3×10^{-9} M, was achieved. Furthermore, Manuel Gómez et al. presented the potential of combining a plasmonic Bragg grating structure, which obtained a rough Al layer to tune SERS enhancement in the visible region [14]. The potential structure with excellent analytical reproducibility and amazing enhancement factors (10^7–10^9) are produced by the scalable procedure, which means it corresponds to the development of producing sustainable SERS substrates. Abeer Alyami et al. also fabricated some flexible, sensitive, and on-site detection-enabled substrates via Ag NPs/PDMS composites [15]. The Ag NPs/PDMS composites were obtained by self-assembly of Ag nanoparticle solutions on flexible PDMS surfaces. As a result, thiram concentrations and crystal violet (CV) within 10^{-5} M and 10^{-7} M were measured on fish skin and orange peel, which revealed the analytical versatility of SERS substrates. These SERS sensors are usually accompanied by complex fabrication techniques and numerous complicated procedures that are disadvantageous to the manufacturing stability of the product. These SERS sensors generally have high performance and sensitivity. Nevertheless, there are strong demands for the development of SERS sensors with high performance via facile and efficient fabrication processes.

Recently, flexible substrates have attracted enormous attention in the SERS field because of various advantages over rigid substrates, such as facile fabrication and shape control [12]. In contrast to conventional rigid substrates, flexible materials, such as polymers [16,17], papers [18], and nanofiber [19] films, have excellent flexibility that allows them to be attached to arbitrary surfaces for in situ detection. This property could further collect analytes, by pasting and peeling off from irregular surfaces, for rapid detection that avoids complex pretreatment. Moreover, bionic materials, such as butterfly wings, cicada wings, mantis wings, lotus leaves, and rose petals, have been widely considered to be excellent SERS substrates because of their 3D periodic microstructures. Moreover, there are studies with SERS substrates synthesized using a simple, low-cost, and environment-friendly method, wherein the component chitosan/chitin was utilized as an in situ reducer to synthesize gold nanoparticles in natural 3D photonic architectures [20–22]. The results show that a SERS substrate could detect 10^{-9} M of 4-ATP and exhibited the lowest relative standard deviation (RSD) with a moderate SNR [23]. Zhang et al. reported that the tip-based continuous, different dimensions of micro/nanostructure arrays were fabricated by the overlap of pyramidal cavities with different adjacent distances [24]. The 1362 cm^{-1} peak of Raman intensity of Au-coated PDMS substrate is about eight times higher than that of the commercial substrate. The SERS enhancement factor achieved an ideal level by using the nanostructured Au-coated PDMS surface. In the procedure of the PDMS substrate, the PDMS-BWS substrate displays a convenient way to replicate the nanostructure from the beetle wings, making the commercial biosensor possible in the future. Additionally, polymers such as PDMS have plasticity, and polyamide is most suitable for transfer to a bionic structure. Some studies have successfully transferred the structure of shark skin onto the outer layer of PDMS membranes to manufacture an ultrahydrophilic exterior for inhibiting bacterial adhesion. As a result, the use of PDMS not only inhibited bacteria and protected the wound, but also improved the hydrophilicity and biocompatibility of the wound repair [25].

In this study, polydimethylsiloxane (PDMS) was used to replicate the 3D structure of beetle wings to fabricate PDMS-BWS SERS substrates. The morphology, surface energy, and roughness could be tuned by depositing various thicknesses (25–45 nm) of silver (Ag) nanoisland thin films, and the SERS performance of the PDMS-BWS SERS substrates was evaluated by the water dye, rhodamine 6G (R6G). A novel, flexible, and reproducible bionic-nanostructure SERS substrate is demonstrated for the rapid determination of water pollutants, such as heavy metal, toxic molecules, and pesticides.

2. Experimental Section

2.1. Material

Beetle wings from *Anomalocera olivacea insularis* were purchased from a specimen store in Taipei. Ethanol and rhodamine 6G (R6G) were purchased from Aldrich Chemical Co (Milwaukee, WI, USA). Sliver slug was purchased from the Gredmann Group. SYLGARD™ 184 was purchased from Dow Corning.

2.2. Preparation of Beetle Wings

First, the beetle wings were immersed in acetone, ethanol, and deionized (DI) water, respectively, for 10 min under the ultrasonicator to remove surface impurities. Subsequently, the beetle wings were washed with DI water to remove the redundant solvents and dried in a circulator oven at 60 °C for 30 min.

2.3. Fabrication of PDMS-BWS-Ag SERS Substrate

PDMS-BWS-Ag was fabricated as a SERS substrate to rapidly detect water pollutant molecules through replication of the bionic 3D structure and deposition of Ag nanoislands. First, the silicone elastomer base reagent and the curing reagent were mixed to homogenization in a ratio of 10:1. The mixed reagent was poured on the beetle wings, removing bubbles in a vacuum, and then heated for 2 h at 60 °C until completely cured. After finishing the curing process, PDMS gel was split from the beetle wing to obtain the PDMS-BWS. As illustrated in Scheme 1, PDMS-BWS-Ag was prepared from the PDMS-BWS by depositing Ag nanoislands with thermal evaporation at a rate of 0.5 Å/s and rotating at 10 rpm under high vacuum (10^{-6} Torr) to ensure homogenous deposition. Various Ag thicknesses (25, 30, 35, and 40 nm) were deposited onto the PDMS-BWS substrate, termed PDMS-BWS-Ag25, PDMS-BWS-Ag30, and PDMS-BWS-Ag35, respectively.

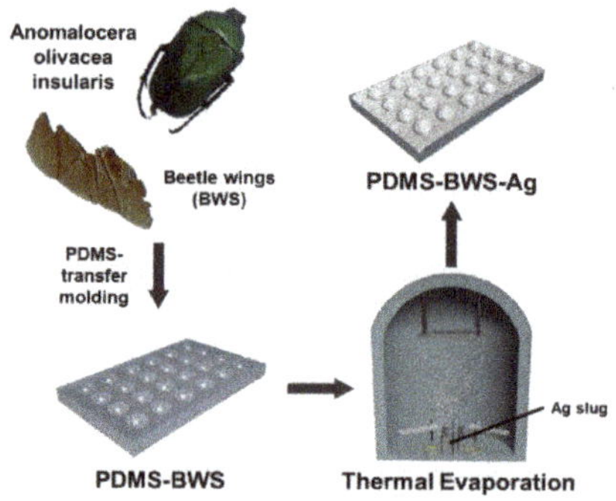

Scheme 1. Preparation of PDMS-replicated beetle-wing substrate with Ag deposition (PDMS-BWS-Ag) SERS substrates.

2.4. Characterization and SERS Detection

The 10 μL of R6G solution, with the concentrations of 10^{-4}, 5×10^{-5}, 10^{-5}, 5×10^{-6}, and 10^{-6} M, was dropped onto the PDMS-BWS-Ag substrate to measure the SERS spectra. The SERS measurements (632.8 nm He–Ne laser and 0.1 mW laser beam) were acquired by using a confocal Raman microscope (LabRAM HR Evolution, HORIBA France SAS, Kyoto, Japan) with a 50× objective lens with a detection range of 400–2000 cm^{-1}. The contact-angle measurements were obtained using a contact-angle goniometer (DSA 100, Krüss GmbH, Hamburg, Germany). The XRD pattern was measured using a PANalytical-X'Pert PRO MPD operated at a scanning speed of 2°/min with a Cu Kα X-ray radiation (λ = 1.5405 Å). Scanning electron microscopy (SEM) (JEOL JSM-6701F) and an atomic force microscope (AFM) (Dimension Edge, Bruker, Berlin, Germany) were used to determine the morphology of the PDMS-BWS-Ag substrates.

3. Results

3.1. XRD Spectra

The crystalline structure of the PDMS-BWS-Ag SERS substrate was examined using XRD spectra. Figure 1 shows the XRD pattern of the PDMS-BWC-Ag SERS substrate, in which five diffraction peaks can be observed, corresponding to the 38.2° (111), 44.3° (200), 64.5° (220), 77.6° (311), and 83.3° (222) given by the Joint Committee on Powder Diffraction Standards (JCPDS, file numbers 04-0783 and 84-0713). These diffraction peaks were assigned to the crystalline nature of silver nanoparticles and demonstrate that the silver nanoislands had been successfully deposited onto the surface of the PDMS-BWC substrates.

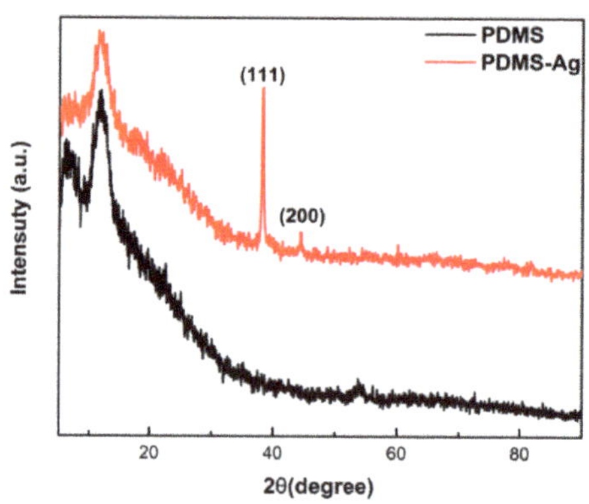

Figure 1. XRD spectra of the PDMS-BWS substrate before and after Ag deposition by thermal evaporation.

3.2. Wettability of SERS Substrate

Figure 2 demonstrates the surface energies of the BWS-Ag, PDMS-BWS-Ag, and PDMS-Ag SERS substrates (Ag deposited thickness is 35 nm) using the contact-angle goniometer. The Ag deposited onto the pristine BWS (BWS-Ag) substrate was more hydrophilic, with a contact angle of 72.22° (Figure 2a). After transferring the BWS nanostructure to PDMS and then depositing the Ag (PDMS-BWS-Ag), the contact angle of the PDMS-BWS-Ag SERS substrate increased to 89.26° (Figure 2b), which was attributed to the more hydrophobic PDMS and the replica of the bionic nanostructure. Compared with the contact angle of the Ag deposited onto the pristine PDMS (PDMS-Ag, 81.72°, Figure 2c), the contact angle of the PDMS-BWS-Ag SERS substrate increased by 7.54°, which confirmed that the bionic (beetle wing) nanostructure has been replicated on the PDMS polymer gels.

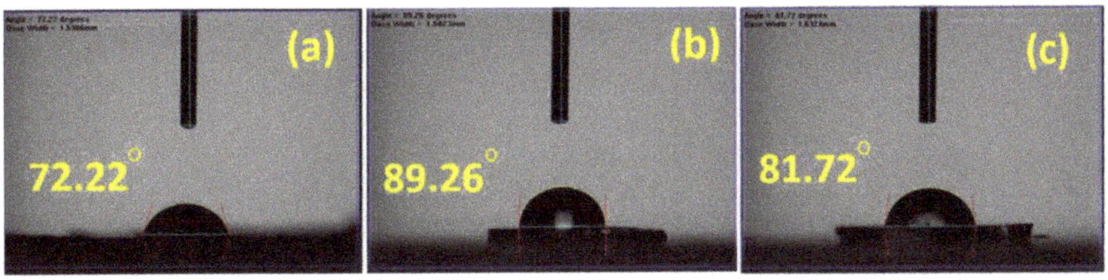

Figure 2. Contact angles of (**a**) BWS-Ag, (**b**) PDMS-BWS-Ag, and (**c**) PDMS-Ag SERS substrates (thickness = 35 nm).

3.3. Surface Topography Analysis

Figure 3 shows the SEM and AFM images of various thicknesses of a PDMS-BWS-Ag SESR substrate. Ag deposited onto the bionic PDMS-BWS by thermal evaporation forms a metal nanoisland film. The roughness of the PDMS-BWS-Ag SERS substrate increased with increasing thickness, which is similar to our previous findings [16]. According to the SEM image (Figure 3), Ag with a thickness of 25 nm deposited onto PDMS-BWS (abbreviated as PDMS-BWS-Ag25) demonstrated a relatively smooth surface, the root-mean-square roughness of which (Rq) was 0.44 nm (Figure 3d). In contrast, PDMS-BWS-Ag30 and PDMS-BWS-Ag35 exhibited a rougher surface by thickness increase, showing Rqs of 1.56 nm (Figure 3e) and 1.86 nm (Figure 3f), respectively. The results show that the Ag nanoislands successfully immobilized on the bionic beetle-wing nanostructure. The rougher nanostructure was beneficial to generating higher localized surface plasmon resonance (LSPR) effects, which obtain a higher Raman enhancement factor. The Raman enhancement analysis is discussed in detail in Figure 4.

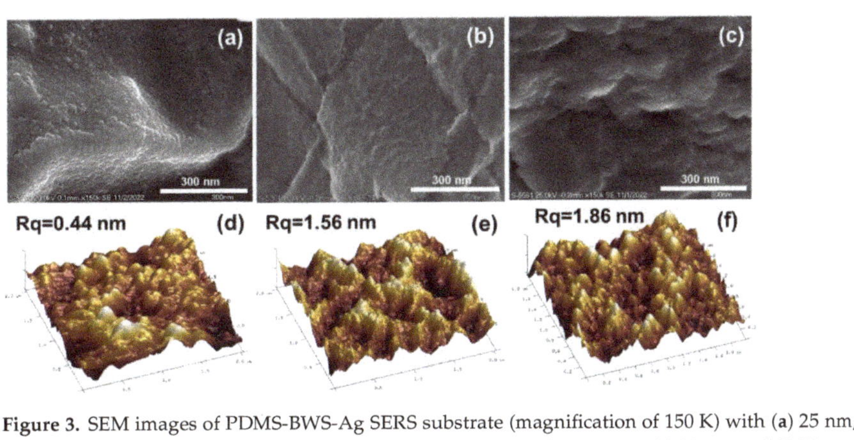

Figure 3. SEM images of PDMS-BWS-Ag SERS substrate (magnification of 150 K) with (**a**) 25 nm, (**b**) 30 nm, and (**c**) 35 nm. AFM 3D images of PDMS-BWS-Ag with (**d**) 25 nm, (**e**) 30 nm, and (**f**) 35 nm.

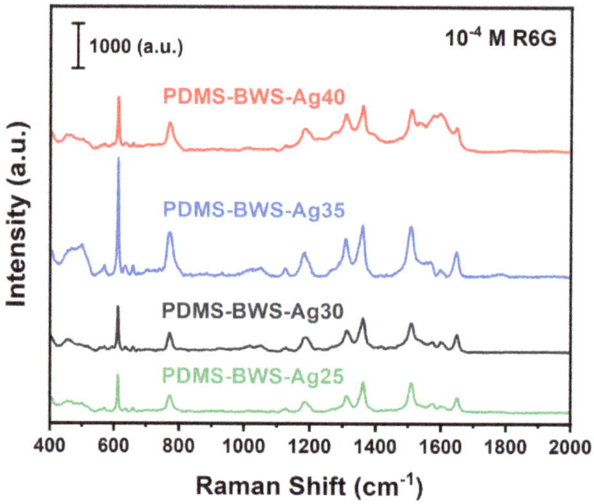

Figure 4. SERS spectra of rhodamine 6G (R6G) molecules (10^{-4} M) measured by PDMS-BWS-Ag25, PDMS-BWS-Ag30, PDMS-BWS-Ag35, and PDMS-BWS-Ag40 SERS substrates.

3.4. SERS Spectra

The various thicknesses (25–40 nm) of Ag deposition by thermal evaporation influence the extent of the SERS enhancement. Here, the water dye of R6G was used as a model molecule. According to a previous report [26–28], the characteristic SERS peaks of R6G are 612, 1361, 1510, and 1649 cm^{-1}, which are assigned to C-C-C ring in-plane bending, C-H in-plane bending, and aromatic C-C stretching, respectively [29]. Figure 4 shows the SERS spectra of R6G with varied thicknesses of the PDMS-BWS-Ag SERS substrate. The R6G characteristic peak located at 612 cm^{-1} was used as the characteristic indicator to evaluate the SERS enhancement. The results show that the SERS intensity increased in relation to the increasing thickness of Ag deposition from 25 to 35 nm (Figure 5). However, the SERS intensity was significantly reduced when the thickness was 40 nm. This suggests that the saturated Ag nanoparticles were coated on the PDMS-BWS substrate, causing a reduction in SERS intensity. Comparing the results, it could be inferred that 35 nm of Ag coating (PDMS-BWS-Ag35) the SERS substrate demonstrates the highest SERS signals. Therefore, the following experiment utilized PDMS-BWS-Ag35 as a SERS substrate to detect the different concentrations of R6G. The relation between the Ag thin film thickness and the Raman intensity of R6G can also be verified in the reference [30].

Herein, the change in SERS intensity was investigated with various SERS substrates after the deposition of 35 nm Ag nanoislands, i.e., PDMS-BWS-Ag35, BWS-Ag35, PDMS-Ag35, and glass-Ag35 (Figure 6). A 10^{-5} M R6G solution was dropped onto the different SERS substrates to evaluate their SERS enhancement. The results indicate that the order of SERS intensity is PDMS-BWS-Ag35 > BWS-Ag35 > PDMS-Ag35 > pristine glass-Ag35 SERS substrates. The pristine glass slide with 35 nm Ag nanoislands (glass-Ag35 SERS substrate) showed a very weak (almost absent) SERS signal. However, a stronger SERS signal with R6G was displayed using PDMS-Ag-35 due to the rougher surface structure compared to the glass slide. Furthermore, it was also possible to detect an obvious SERS signal with a 35 nm Ag coating on natural beetle-wing substrates (BWS-Ag-35 SERS substrate). Nevertheless, the SERS intensity increased ~10 times with the PDMS-replicated bionic beetle-wing substrate coated with 35 nm Ag nanoislands (PDMS-BWS-Ag35 SERS substrate). This means that PDMS with a beetle-wing nanostructure could create the optimal hotspot effect to induce the strongest SERS intensity.

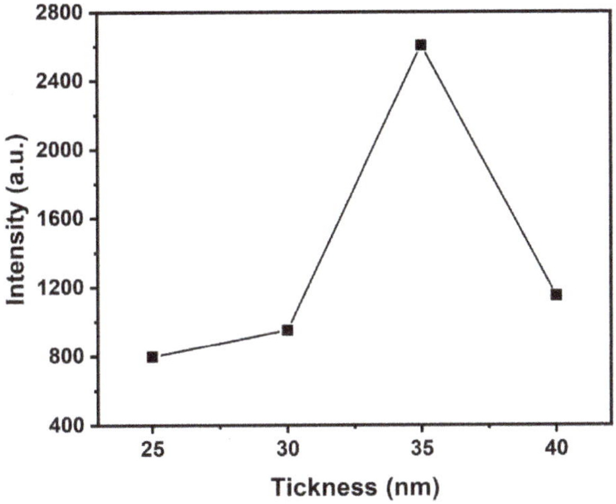

Figure 5. The relationship between Ag thickness and SERS intensity, measured with R6G molecules (10^{-4} M).

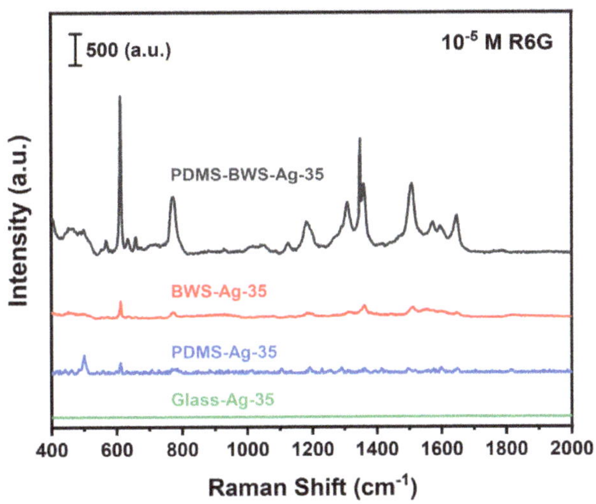

Figure 6. SERS spectra of PDMS-BWS-Ag-35 nm, BWS-Ag-35 nm, pristine PDMS-Ag-35 nm, and glass with R6G (10^{-5} M).

For limit of detection (LOD) measurement, various concentrations of the R6G solution were measured using the PDMS-BWS-Ag35 SERS substrate. The SERS intensity of the characteristic peaks (612 cm^{-1}) progressively decreased until it reached a concentration of 10^{-6} M (Figure 7). It should be noted that the SERS intensity decreased when the concentration approached 10^{-4} M, which can be attributed to the decreasing induction of the SERS signals due to R6G molecular aggregation. We tried to calculate a calibration curve for the PDMS-BWS-Ag35 SERS substrate using various concentrations (5×10^{-5}~10^{-6} M) of R6G molecules as a function of the SERS intensity (integrating the area of the characteristic peaks, 612 cm^{-1}) in Figure 8. The result shows a good linear relationship, suggesting that the calibration curve exhibits high reliability. Although the result of the LOD measurement compared with other similar research [31] was not good enough, this unique 3D bionic structure demonstrates the potential possibility of SERS detection. In addition, the PDMS-

BWS substrate has more flexibility within the advantages of being an environment-friendly and easy procedure than the cicada wings.

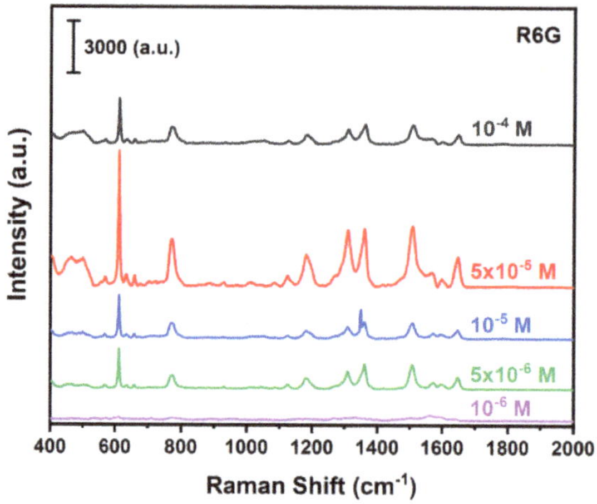

Figure 7. SERS spectra of PDMS-BWS-Ag35 with various concentrations of R6G molecules (from 10^{-4} to 10^{-6} M).

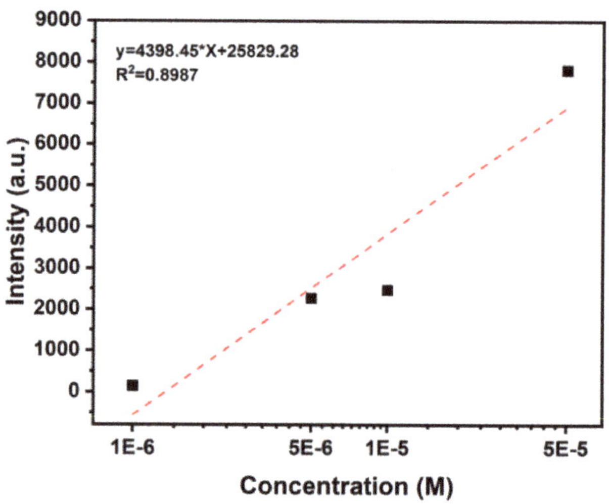

Figure 8. Calibration curve of PDMS-BWS-Ag35 for R6G at different concentrations.

In addition to using LOD analysis to evaluate the SERS performance of the PDMS-BWS-Ag35 SERS substrate, we utilized the following (Equation (1)) to calculate the enhancement factor (EF) of Raman spectroscopy:

$$EF = (I_{SERS}/N_{SERS})/(I_{Raman}/N_{Raman}) \tag{1}$$

where I_{SERS} and I_{Raman} are the intensity of the SERS spectra and Raman spectra, respectively, for R6G molecules at 612 cm^{-1}. N_{SERS} and N_{Raman} are the numbers of R6G adsorbed on the SERS and blank substrates, respectively. According to Figures 6 and 7, the EF of the PDMS-BWS-Ag-35 SERS substrate was 1.9×10^4 in optimized conditions.

Therefore, the proposed technique of PDMS replication and Ag deposition by thermal evaporation allows us to achieve SERS intensification through the utilization of LSPR plasmonic coupling phenomena and the generation of a huge electric field; this enhancement is attributed to the nanostructure of the bionic beetle wing replica.

4. Conclusions

We successfully developed a novel bionic nanostructure replication technique from beetle wings, with deposited Ag nanoislands, as a SERS substrate (PDMS-BWS-Ag) to detect water pollutants. Comparing with other references of nanoisland substrates (the dragonfly wings [32]), the beetle wings display unique and natural nanostructures to produce enormous hotspots for Raman enhancing. In addition, the BWS-PDMS-based substrate exhibits a flexible, environment-friendly, and easier to produce procedure, which makes the chance for the commercialization of the products. Subsequently, the SERS performance is optimized through the deposition of different thicknesses of Ag nanoislands by facile thermal evaporation methods. The results indicate that PDMS-BWS-Ag35 demonstrates excellent SERS enhancement performance, which is 10 times stronger than pristine BWS-Ag35 (without the replication process). The LOD of PDMS-BWS-Ag35 SERS is lower than 10^{-6} M and displays a linear calibration curve from 5×10^{-5}~10^{-6} M. These results illustrate that the PDMS-BWS SERS substrate has the potential to be applied in detecting biomolecules and water pollutants, such as adenine from DNA, bacteria, viruses, water toxicity, and water heavy-metal ions.

Author Contributions: Conceptualization, C.-H.L. and T.-Y.L.; methodology, C.-H.L., M.-R.C., S.C. and W.-L.S.; software, C.-H.L., M.-R.C., S.C., W.-L.S. and J.-S.C.; validation, T.-Y.L., W.-L.S. and M.-Y.C.; formal analysis, W.-L.S., M.-Y.C., K.-S.W., J.-S.C. and P.-H.L.; investigation, C.-H.L., M.-R.C. and S.C.; resources, T.-Y.L.; data curation, T.-Y.L.; writing—original draft preparation, C.-H.L., M.-Y.C., K.-S.W., J.-S.C., P.-H.L. and T.-Y.L.; writing—review and editing, C.-H.L., M.-Y.C., K.-S.W., P.-H.L. and T.-Y.L.; visualization, C.-H.L.; supervision, T.-Y.L.; project administration, P.-H.L. and T.-Y.L.; funding acquisition, T.-Y.L. All authors have read and agreed to the published version of the manuscript.

Funding: This research was funded by NSTC 111-2221-E-131-020-MY3, NSTC 110-2221-E-131-009, NSTC 111-2123-M-001-005 and NSTC 110-2622-E-131-005.

Institutional Review Board Statement: Not applicable.

Informed Consent Statement: Not applicable.

Data Availability Statement: The data presented in this study are available on request from the corresponding author.

Acknowledgments: Facilities are supported by Yuh-Lin Wang from the Institute of Atomic and Molecular Sciences, Academia Sinica and is gratefully acknowledged.

Conflicts of Interest: The authors declare no conflict of interest.

References

1. Sun, C.; Chen, T.; Ruan, W.; Jung, Y.M.; Cong, Q.; Zhao, B. A simple strategy to improve the sensitivity of probe molecules on SERS substrates. *Talanta* **2019**, *195*, 221–228. [CrossRef] [PubMed]
2. Wu, D.Y.; Liu, X.M.; Duan, S.; Xu, X.; Ren, B.; Lin, S.H.; Tian, Z.Q. Chemical Enhancement Effects in SERS Spectra: A Quantum Chemical Study of Pyridine Interacting with Copper, Silver, Gold and Platinum Metals. *J. Phys. Chem. C* **2008**, *112*, 4195–4204. [CrossRef]
3. Gabudean, A.M.; Biro, D.; Astilean, S. Localized surface plasmon resonance (LSPR) and surface-enhanced Raman scattering (SERS) studies of 4-aminothiophenol adsorption on gold nanorods. *J. Mol. Struct.* **2011**, *993*, 420–424. [CrossRef]
4. Liu, G.; Li, Y.; Duan, G.; Wang, J.; Liang, C.; Cai, W. Tunable Surface Plasmon Resonance and Strong SERS Performances of Au Opening-Nanoshell Ordered Arrays. *ACS Appl. Mater. Interfaces* **2012**, *4*, 1–5. [CrossRef] [PubMed]
5. Camden, J.P.; Dieringer, J.A.; Wang, Y.; Masiello, D.J.; Marks, L.D.; Schatz, G.C.; Van Duyne, R.P. Probing the Structure of Single-Molecule Surface-Enhanced Raman Scattering Hot Spots. *J. Am. Chem. Soc.* **2008**, *130*, 12616–12617. [CrossRef]
6. Pilot, R.; Signorini, R.; Durante, C.; Orian, L.; Bhamidipati, M.; Fabris, L. A Review on Surface-Enhanced Raman Scattering. *Biosensors* **2019**, *9*, 57. [CrossRef] [PubMed]
7. Fleischmann, M.; Hendra, P.J.; McQuillan, A.J. Raman spectra of pyridine adsorbed at a silver electrode. *Chem. Phys. Lett.* **1974**, *26*, 163–166. [CrossRef]
8. Singh, R. CV Raman and the Discovery of the Raman Effect. *Phys. Perspect.* **2002**, *4*, 399–420. [CrossRef]

9. Liu, J.W.; Wang, J.L.; Huang, W.R.; Yu, L.; Ren, X.F.; Wen, W.C.; Yu, S.H. Ordering Ag nanowire arrays by a glass capillary: A portable, reusable and durable SERS substrate. *Sci. Rep.* **2012**, *2*, 987. [CrossRef]
10. Xue, X.; Chen, L.; Zhao, C.; Qiao, Y.; Wang, J.; Shi, J.; Lin, Y.; Chang, L. Tailored FTO/Ag/ZIF-8 structure as SERS substrate for ultrasensitive detection. *Spectrochim. Acta Part A Mol. Biomol. Spectrosc.* **2022**, *282*, 121693. [CrossRef]
11. Deng, Y.-L.; Juang, Y.-J. Black silicon SERS substrate: Effect of surface morphology on SERS detection and application of single algal cell analysis. *Biosens. Bioelectron.* **2014**, *53*, 37–42. [CrossRef] [PubMed]
12. Ding, T.; Sigle, D.O.; Herrmann, L.O.; Wolverson, D.; Baumberg, J.J. Nanoimprint lithography of Al nanovoids for deep-UV SERS. *ACS Appl. Mater. Interfaces* **2014**, *6*, 17358–17363. [CrossRef]
13. Das, A.; Pant, U.; Cao, C.; Moirangthem, R.S.; Kamble, H.B. Fabrication of plasmonic nanopyramidal array as flexible SERS substrate for biosensing application. *Nano Res.* **2022**, *2022*, 1–9. [CrossRef]
14. Gómez, M.; Kadkhodazadeh, S.; Lazzari, M. Surface enhanced Raman scattering (SERS) in the visible range on scalable aluminum-coated platforms. *Chem. Commun.* **2018**, *54*, 10638–10641. [CrossRef]
15. Alyami, A.; Quinn, A.; Iacopino, D. Flexible and transparent Surface Enhanced Raman Scattering (SERS)-Active Ag NPs/PDMS composites for in-situ detection of food contaminants. *Talanta* **2019**, *201*, 58–64. [CrossRef]
16. Wu, M.; Zhang, C.; Ji, Y.; Tian, Y.; Wei, H.; Li, C.; Li, Z.; Zhu, T.; Sun, Q.; Man, B.; et al. 3D Ultrasensitive Polymers-Plasmonic Hybrid Flexible Platform for In-Situ Detection. *Polymers* **2020**, *12*, 392. [CrossRef] [PubMed]
17. Cheng, Y.W.; Hsiao, C.W.; Zeng, Z.L.; Syu, W.L.; Liu, T.Y. The interparticle gap manipulation of Au-Ag nanoparticle arrays deposited on flexible and atmospheric plasma-treated PDMS substrate for SERS detection. *Surf. Coat. Technol.* **2020**, *389*, 125653. [CrossRef]
18. Siebe, H.S.; Chen, Q.; Li, X.; Xu, Y.; Browne, W.R.; Bell, S.E. Filter paper based SERS substrate for the direct detection of analytes in complex matrices. *Analyst* **2021**, *146*, 1281–1288. [CrossRef] [PubMed]
19. Roque-Ruiz, J.H.; Martínez-Máynez, H.; Zalapa-Garibay, M.A.; Arizmendi-Moraquecho, A.; Farias, R.; Reyes-López, S.Y. Surface enhanced Raman spectroscopy in nanofibers mats of SiO2-TiO2-Ag. *Results Phys.* **2017**, *7*, 2520–2527. [CrossRef]
20. Huang, Y.; Wu, Y.; Tao, H.; Yuan, B. Bio-Based Porous Aerogel with Bionic Structure and Hydrophobic Polymer Coating for Efficient Absorption of Oil/Organic Liquids. *Polymers* **2022**, *14*, 4579. [CrossRef]
21. Jiang, S.; Xia, L. Bioinspired High-Performance Bilayer, pH-Responsive Hydrogel with Superior Adhesive Property. *Polymers* **2022**, *14*, 4425. [CrossRef] [PubMed]
22. Lin, Y.T.; Wu, C.H.; Syu, W.L.; Ho, P.C.; Tseng, Z.L.; Yang, M.C.; Lin, C.C.; Chen, C.C.; Chen, C.C.; Liu, T.Y. Replica of Bionic Nepenthes Peristome-like and Anti-Fouling Structures for Self-Driving Water and Raman-Enhancing Detection. *Polymers* **2022**, *14*, 2465. [CrossRef] [PubMed]
23. Mu, Z.; Zhao, X.; Xie, Z.; Zhao, Y.; Zhong, Q.; Bo, L.; Gu, Z. In situ synthesis of gold nanoparticles (AuNPs) in butterfly wings for surface enhanced Raman spectroscopy (SERS). *J. Mater. Chem. B* **2013**, *1*, 1607–1613. [CrossRef] [PubMed]
24. Zhang, J.; Yan, Y.; Miao, P.; Cai, J. Fabrication of gold-coated PDMS surfaces with arrayed triangular micro/nanopyramids for use as SERS substrates. *Beilstein J. Nanotechnol.* **2017**, *8*, 2271–2282. [CrossRef] [PubMed]
25. Lin, Y.T.; Ting, Y.S.; Chen, B.Y.; Cheng, Y.W.; Liu, T.Y. Bionic shark skin replica and zwitterionic polymer brushes functionalized PDMS membrane for anti-fouling and wound dressing applications. *Surf. Coat. Technol.* **2020**, *391*, 125663. [CrossRef]
26. Hunyadi, S.E.; Murphy, C.J. Bimetallic silver–gold nanowires: Fabrication and use in surface-enhanced Raman scattering. *J. Mater. Chem.* **2006**, *16*, 3929–3935. [CrossRef]
27. Wolf, M.P.; Salieb-Beugelaar, G.B.; Hunziker, P. PDMS with designer functionalities—Properties, modifications strategies, and applications. *Prog. Polym. Sci.* **2018**, *83*, 97–134. [CrossRef]
28. Zong, C.; Ge, M.; Pan, H.; Wang, J.; Nie, X.; Zhang, Q.; Zhao, W.; Liu, X.; Yu, Y. In situ synthesis of low-cost and large-scale flexible metal nanoparticle–polymer composite films as highly sensitive SERS substrates for surface trace analysis. *RSC Adv.* **2019**, *9*, 2857–2864. [CrossRef]
29. Chiang, C.Y.; Liu, T.Y.; Su, Y.A.; Wu, C.H.; Cheng, Y.W.; Cheng, H.W.; Jeng, R.J. Au nanoparticles immobilized on honeycomb-like polymeric films for surface-enhanced Raman scattering (SERS) detection. *Polymers* **2017**, *9*, 93. [CrossRef]
30. Yi, Z.; Niu, G.; Luo, J.; Kang, X.; Yao, W.; Zhang, W.; Yi, Y.; Yi, Y.; Ye, X.; Duan, T.; et al. Ordered array of Ag semishells on different diameter monolayer polystyrene colloidal crystals: An ultrasensitive and reproducible SERS substrate. *Sci. Rep.* **2016**, *6*, 1–11. [CrossRef]
31. Zhao, N.; Li, H.; Tian, C.; Xie, Y.; Feng, Z.; Wang, Z.; Yan, X.; Wang, W.; Yu, H. Bioscaffold arrays decorated with Ag nanoparticles as a SERS substrate for direct detection of melamine in infant formula. *RSC Adv.* **2019**, *9*, 21771–21776. [CrossRef]
32. Shi, G.C.; Wang, M.L.; Zhu, Y.Y.; Shen, L.; Ma, W.L.; Wang, Y.H.; Li, R.F. Dragonfly wing decorated by gold nanoislands as flexible and stable substrates for surface-enhanced Raman scattering (SERS). *Sci. Rep.* **2018**, *8*, 1–11. [CrossRef]

polymers

MDPI

Article

A Novel Approach to Increase the Oxygen Permeability of Soft Contact Lenses by Incorporating Silica Sol

Nguyen-Phuong-Dung Tran, Chuan-Cheng Ting, Chien-Hong Lin and Ming-Chien Yang *

Department of Materials Science and Engineering, National Taiwan University of Science and Technology, Taipei 10607, Taiwan; thaonguyeng89@gmail.com (N.-P.-D.T.); jeffting2926198619@gmail.com (C.-C.T.); chlin0805@gmail.com (C.-H.L.)

* Correspondence: myang@mail.ntust.edu.tw; Tel.: +886-2-2737-6528; Fax: +886-2-2737-6544

Received: 26 August 2020; Accepted: 11 September 2020; Published: 14 September 2020

Abstract: This study presents a novel approach to increase the oxygen permeability of hydrogel by the addition of silica sol. Herein, 2-hydroxyethyl methacrylate (HEMA) was copolymerized with N-vinyl-2-pyrrolidone (NVP) after mixing with silica sol. The resultant hydrogel was subject to characterizations including Fourier-transform infrared (FTIR), equilibrium water content (EWC), contact angle, optical transmittance, oxygen permeability (Dk), tensile test, anti-deposition of proteins, and cytotoxicity. The results showed that with the increase of silica content, the Dk values and Young's moduli increased, the optical transmittance decreased slightly, whereas the EWC and contact angle, and protein deposition were not much affected. Moreover, the cytotoxicity of the resultant poly(HEMA-*co*-NVP)-SNPs indicated that the presence of silica sol was non-toxic and caused no effect to the growth of L929 cells. Thus, this approach increased the Dk of soft contact lenses without affecting their hydrophilicity.

Keywords: silica sol; soft contact lens; poly(HEMA-*co*-NVP); oxygen permeability; hydrophilicity

1. Introduction

Currently, contact lenses are manufactured from hydrogels of hydrophilic monomers such as 2-hydroxyethylmethacrylate (HEMA) and 1-vinyl-2-pyrrolidinone (NVP). Hydrogel lenses were required to supply sufficient oxygen to the eyes during walking and sleeping. In general, for soft contact lenses, there are two possible avenues to increase oxygen permeability: develop the materials with higher water content or develop the hydrophobic materials. Oxygen can be absorbed and permeate through the soft lens by water phase or its matrix [1,2]. In conventional hydrogel, oxygen permeability is essentially governed by the equilibrium water content (EWC). Higher water content in soft lens can achieve higher oxygen transmissibility. However, the permeability of water (Dk_{water}) was approximate 80 barrer, restricting the delivery of oxygen to the eyes through hydrogel lenses [1]. In addition, high water content in hydrogel matrix may reduce the mechanical properties and increase protein deposition of contact lenses [3]. In the other permeation path based on the matrix, hydrophobic silicon-containing ingredients, such as polydimethylsiloxane (PDMS) and 3-(methacryloyloxy) propyltris(trimethylsiloxy) silane (TRIS), are incorporated into hydrogels to deliver sufficient oxygen to the eyes. Combining hydrophilic and hydrophobic components often incurs phase separation that impairs the transparency of the lens. Feasible approaches were developed in the manufacturing of commercial silicone hydrogel lenses [4,5]. Other than the lens's transparency, silicone materials exhibit some problems because of their characteristics. High oxygen supply with high silicone content is frequently in the expense of water absorbability as well as wetting surface [3,6–9]. High silicone content may also reduce resistance to protein deposition and mechanical properties [3,6,10–12].

A new approach to solve problems associated with silicone and to achieve high water content may be developed based on tetraethyl orthosilicate (TEOS) via Stöber process. Silica nanoparticles (SNPs) synthesized from TEOS were reported when blending into polymers could increase the gas permeability [13,14]. The oxygen permeability was enhanced to about 40% higher than the theoretical prediction. Based on this finding, we proposed a novel approach to increase the oxygen permeability by blending silica nanoparticles while keeping the hydrophilicity of the base hydrogels. In this study, SNPs were synthesized from TEOS through sol-gel process before mixing with the matrix of hydrophilic materials such as HEMA and NVP. All the resultant SNPs-loaded poly(HEMA-*co*-NVP) were characterized by dynamic light scattering (DLS), Fourier-transform infrared spectrometer (FTIR), equilibrium water content (EWC), oxygen permeability (Dk), contact angle, optical transparency, mechanical properties, protein adsorption, and in vitro cytotoxicity.

2. Materials and Methods

2.1. Materials

Tetraethyl orthosilicate (TEOS), 1-vinyl-2-pyrrolidinone (NVP), phosphate buffered saline powder (PBS, 0.1 M, pH 7.4), and 2-hydroxy-2-methylbenzene acetone (D-1173) were purchased from Sigma-Aldrich, St. Louis, MO, USA. Glycerol, 2-hydroxyethylmethacrylate (HEMA), and ethylene glycol dimethacrylate (EGDMA) were obtained from Acros Organics (Morris Plains, NJ, USA). Phosphate buffered saline solution (1 wt%, pH 7.4) was prepared in our laboratory.

2.2. Preparation of Silica Nanoparticles and Hydrogels

Silica nanoparticles (SNPs) were prepared through the sol-gel process. Briefly, 1 g of TEOS was slowly added into 5 mL of 0.5 M HCl at room temperature and stirred for 24 h. Afterward, the size of the SNPs in the sol was analyzed through dynamic light scattering (DLS-DKSH, Nano-ZS90, Malvern Instruments Ltd., UK). After verifying the synthesis of SNPs, an aliquot of the silica sol, ranging from 0 to 2 g, was added to 4 g of HEMA followed by 1 g of NVP, 0.5 g of glycerol, 0.1 g of EDGMA (crosslinking agent), and 0.1 g of D-1173 (photo-initiator). The solution was mixed for 1 h and molded before exposing to UV light (365 nm) at 5 mW/cm^2 for 20 min. After demolding, the resultant lenses were soaked in 50% ethanol for 20 h at 50 °C to remove un-reacted monomers and photo initiator. Then, the lenses were immersed in distilled water for 4 h at 50 °C to wash out ethanol. Finally, the lenses were preserved in PBS (pH 7.4) at room temperature. Table 1 presents the compositions of the samples. Figure 1 shows the synthesis of the hydrogels.

Table 1. Hydrophilicity, oxygen permeability, wettability and protein deposition of SNP-loaded contact lenses.

Sample	SNP Content (wt%)	EWC (%)	Dk (barrer)	Dk/t (barrer/mm)	Contact Angle (°)	Protein Deposition (nmol/cm^2)	
						HSA	Lysozyme
S0	0	52.5 ± 1.6	20.6 ± 1.0	198 ± 9	60.5 ± 2.3	0.76 ± 0.01	3.10 ± 0.05
S1	0.18	51.5 ± 2.1	26.6 ± 1.2	257 ± 11	60.1 ± 3.1	0.78 ± 0.01	3.21 ± 0.05
S2	0.46	53.5 ± 1.7	36.7 ± 1.5	355 ± 19	61.1 ± 2.9	0.75 ± 0.01	3.27 ± 0.04
S3	0.92	52.9 ± 1.9	42.8 ± 1.7	414 ± 11	59.5 ± 4.1	0.76 ± 0.01	3.16 ± 0.05
S4	1.37	51.9 ± 2.0	50.3 ± 1.9	494 ± 13	60.1 ± 3.3	0.75 ± 0.01	3.23 ± 0.06
S5	1.82	53.1 ± 1.8	54.3 ± 2.2	534 ± 15	61.2 ± 4.5	0.76 ± 0.01	3.29 ± 0.05

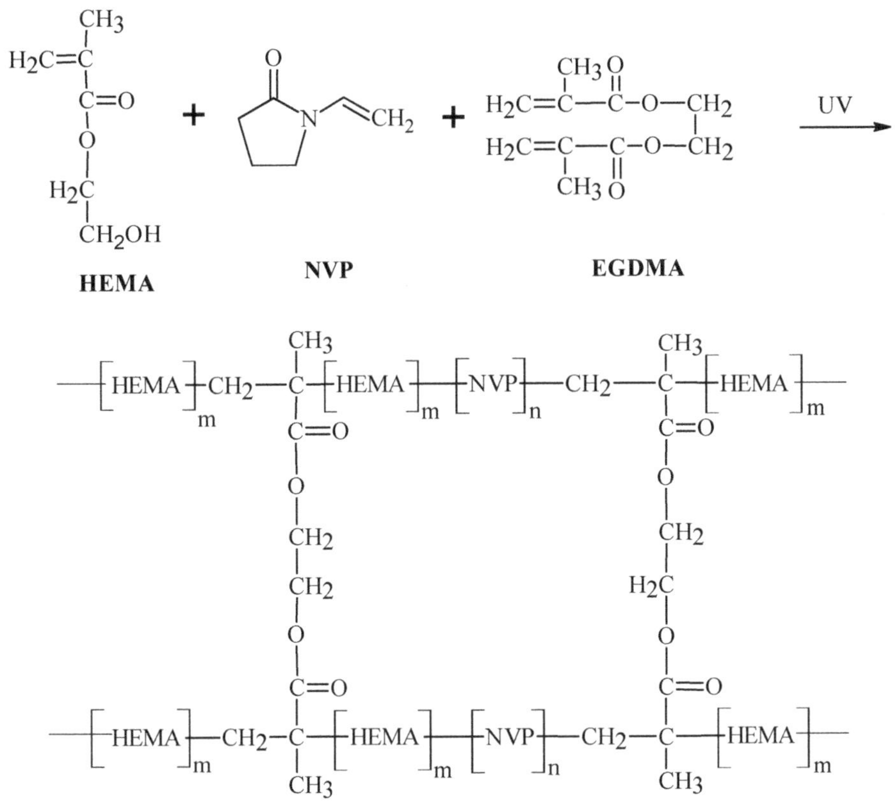

Figure 1. The reaction of poly(HEMA-*co*-NVP) lenses.

2.3. *Characterization of SNPs and Hydrogel Lenses*

2.3.1. Size of Particles and Chemical Structure

The size of the SNPs in the sol was analyzed through dynamic light scattering (DLS-DKSH, Malvern Instruments Ltd., Malvern, UK). The chemical structure of SNPs was examined using Raman Spectroscopy (iHR550, Horiba Scientific, Kyoto, Japan) via the wavenumber range of 400–4000 cm^{-1}.

2.3.2. Fourier-Transform Infrared Spectrometer

The structure of the hydrogels was characterized in the wavenumber range of 600–4000 cm^{-1} by using a Fourier-transform infrared spectrometer (FTIR, Nicolet 170 SX, Thermo Fisher Scientific, Madison, WI, USA).

2.3.3. Equilibrium Water Content

Equilibrium water content (EWC) of lenses was calculated as follows:

$$EWC\ (\%) = \frac{W_w - W_d}{W_w} \times 100 \quad (1)$$

where W_d and W_w are, respectively, the weights of the dry and rehydrated samples. The rehydrated lenses were immersed in distilled water for one day at room temperature before being weighed.

2.3.4. Oxygen Permeability

The oxygen permeability (Dk) of poly(HEMA-*co*-NVP)-SiO_2 was determined using an oxygen permeometer (Model 201T, Createch, CA, USA) based on polarographic method.

2.3.5. Contact Angle

The contact angle was determined using a contact angle goniometer (DSA 100, Krüss GmbH, Hamburg, Germany) at 37 °C.

2.3.6. Optical Transparency

The optical transparency of the contact lens was measured based on light transmittance (T%) in a wavelength range of 400–700 nm using a UV-Vis spectrometer (Cary 300, Agilient Technologies, Santa Clara, CA, USA).

2.3.7. Mechanical Properties

The mechanical properties of all specimens were measured using a tensile tester (MTS 810, Material Test System, MN, USA). The hydrated specimens were cut into dog bone shape before measuring the modulus at a crosshead speed of 50 mm/min.

2.3.8. Protein Deposition

Protein deposition on the contact lenses was studied against human serum albumin (HSA) and lysozyme using a UV-Vis spectrometer. Specimens (1 × 1 cm) were immersed in 1 mL PBS containing 2 mg/mL of either HSA or lysozyme. After being incubated at 37 °C for 24 h, specimens were rinsed 3 times with PBS. Then, specimens were placed into 2 mL of 1 wt% sodium dodecyl sulfate (SDS), and shaken at 100 rpm at 37 °C for 1 h to remove the protein adsorbed on the lenses surface. Finally, the protein deposition on the lenses were analyzed based on BCA assay after being incubated in 2 mL of bicinchoninic acid at 37 °C for 1 h.

2.3.9. Cytotoxicity Test

The cytotoxicity test of the soft lenses was analyzed according to ISO 10993-5. The cell culture medium for determining cytotoxicity was a mixture of 94% Dulbecco's Modified Eagle's medium (DMEM), 1% penicillin antibiotic (PNC), and 5% fetal bovine serum (FBS). After being sterilized under the UV light for 4 h, specimens (1 × 1 cm) were soaked in cell culture media for 24 h at 37 °C. Then, the extracted media were filtered by passing through a 0.22 μm filter. L929 cells were cultured with the extracted media, positive control, and negative control. Thiazolyl blue tetrazolium bromide (MTT) reagent was mixed into the cells at 37 °C for 4 h after being incubated for 48 h at 37 °C in a humidified atmosphere containing 5% carbon dioxide. Afterwards, dimethyl sulfoxide (DMSO) was added into the media to dissolve the purple product. Finally, the cytotoxicity test was analyzed at 570 nm using an Elisa reader (M965, Accu reader, Taiwan).

3. Results

3.1. Dynamic Light Scattering and Raman

Figure 2 shows that the mean diameter of the resultant SNPs was 13.5 nm determined by DLS measurement. In general, the hydrolysis and condensation of TEOS in water would first result in a sol, which is composed of SiO_2 nanoparticles suspending in water. After a period of time, these nanoparticles would further condense to become gel. In this present study, the condensation was terminated before the gelation by adding the sol directly to the monomer solution. Glycerol was added to improve the miscibility of the silica sol and the monomers. The solution was polymerized under UV (365 nm) followed by rinsing with ethanol and water to remove glycerol and unreacted monomers.

Figure 3 presents the Raman spectrum of SNPs through the sol-gel process. The peak of Si–O group was detected at 805 cm^{-1} while the peaks of C_2H_5OH were respectively observed at 430 cm^{-1} and 1086 cm^{-1} (C–O groups) and at 1456 cm^{-1} (CH_2 groups).

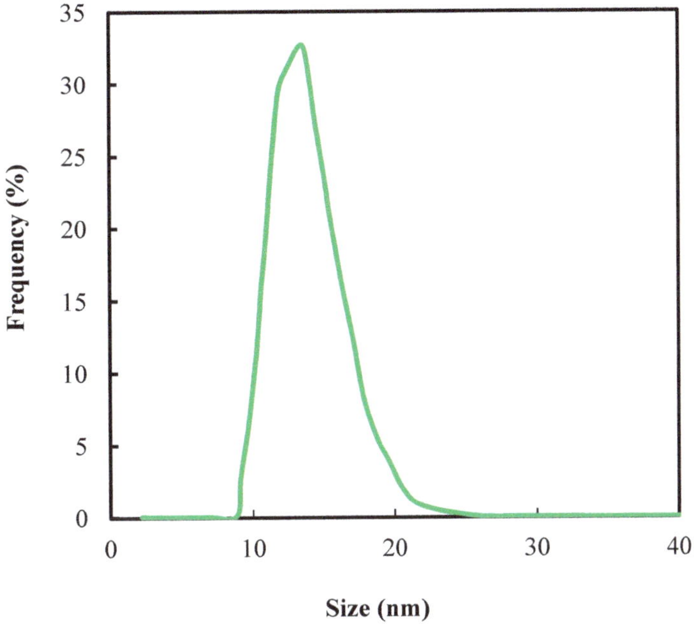

Figure 2. DLS of SNPs.

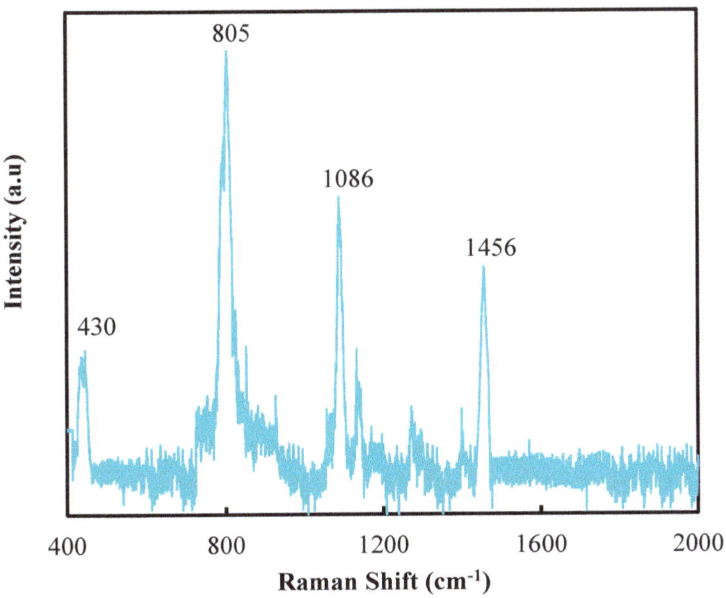

Figure 3. Raman spectrum of SNPs.

3.2. Fourier-Transform Infrared Spectrometer

Figure 4 shows the FTIR spectra of HEMA, NVP, and poly(HEMA-*co*-NVP)-SNPs, where the peaks of C–H groups (2878 cm^{-1}) and C=O groups (1720 cm^{-1}) were observed. Additionally, in SNP-loaded lenses, the peaks of Si–O–Si group in SiO_2 were appeared at 720 and 1068 cm^{-1} while the C–N group of NVP were observed in 1150 cm^{-1} [6,15]. The characteristic peaks of C–C double bond were absent from the infrared spectrum of the resultant lens, indicating the successful synthesis of the hydrogel.

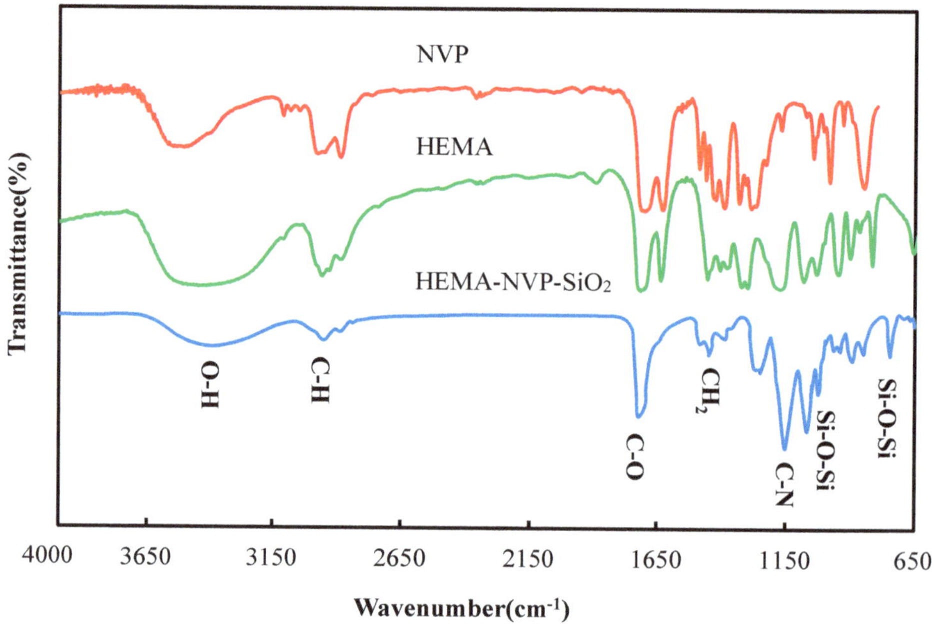

Figure 4. FTIR spectra of SNPs lenses.

3.3. Equilibrium Water Content

Table 1 shows that the equilibrium water content (EWC) of SNPs lenses was influenced by loading of SNPs various concentrations. The EWC value of lenses varied slightly around 52%, regardless of the content of SNPs. The quantity of nanoparticles added was small, and that SNPs had little interaction with the matrix of HEMA and NVP Thus EWC was maintained for the hydrogel of hydrophilic HEMA and NVP [3,16].

3.4. Contact Angle

Table 1 presents that contact angle was basically independent on the SNP content. The contact angle was varied around 60° while the SNPs concentration increased from 0 wt% to 1.8 wt%. The possible reason was that the SNPs exhibit content was less than 1.8%hydroxyl groups on the surface, making them hydrophilic and was embedded in the hydrogel matrix, thus affected little the hydrophilicity wettability of the hydrogel. This result also has the same tendency as stated in literature. The combination of small amount of SiNPs silicone nanoparticlescontent with hydrophilic monomers had the slight influence on the wetting surface of soft lenses [17].

3.5. Optical Transparency

Figure 5 presents the photos of SNPs hydrated contact lenses. Figure 6 shows that the light transmittance decreased with the increase of SNP content. The transparency of commercial contact lenses is usually above 90% [3,17], thus in this work, the maximum SNP content was limited to 1.8 wt%. The decrease in the light transmittance is probably caused by the formation of nodules due to the interaction between SNPs and functional groups in poly(HEMA-*co*-NVP), which scattered the light. A higher SNP content leads to more nodules, and thus impairs light transmission.

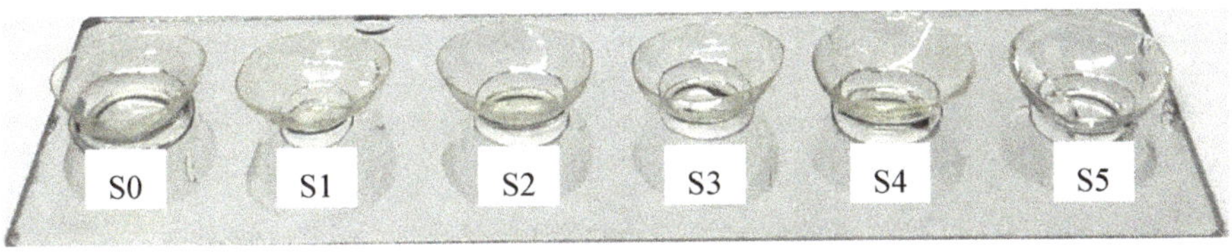

Figure 5. Photos of SNPs hydrated contact lenses.

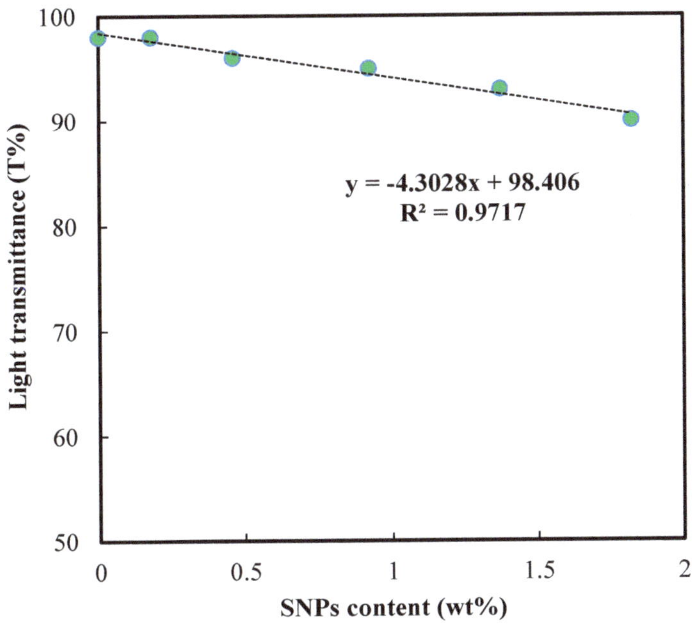

Figure 6. Effect of SNPs content on optical transparency.

3.6. Oxygen Permeability

Increasing Dk for hydrogel lenses using hydrophobic groups is important to restrict the corneal damage such as hypoxia, corneal edema, and red eyes [18,19]. Figure 7 shows that the oxygen permeability of SNP-loaded hydrogel increased with the increase of SNP content. The highest Dk reached at 54.3 barrer corresponding to the highest SNPs ratio (1.8 wt%). This trend is quite unusually because SiO_2 exhibits an oxygen permeability much lower than polymers and water. The presence of silica should hinder the diffusion of oxygen through the hydrogel. However, in SNPs-loaded hydrogel, these particles were entrapped in a hydrated environment. This would confer SNPs mobility, facilitating the creation of free volumes for oxygen to pass through, thus accelerating the diffusion of oxygen. This finding would be useful for developing novel materials for contact lens.

In general, the inverted correlation frequently appears between Dk and EWC in soft lenses whereby the increase of hydrophobic ingredients such as PDMS and TRIS will affect the water absorbability of soft lenses as reported previously [7,20,21]. However, in this study, the inverted relationship of Dk and EWC was absent with the addition of SNPs. Therefore, the result of this study proposed the new approach to solve the frequent trouble of hydrogel lenses.

3.7. Mechanical Properties

Generally, lower modulus lenses can provide more comfort and fit for wearers [6,22]. Figure 7 shows that Young's modulus of SNPs-loaded hydrogel increased almost linearly with the SNP content. The Young's modulus of lenses increased from 0.41 MPa to 1.02 MPa with increasing SNPs ratio (0 wt% for S0 and 1.8 wt% for S5). The reinforcement of the hydrogel suggests that SNPs did serve as the crosslinker for the poly(HEMA-*co*-NVP) molecular chains. It is also reported in the literature that silica nanoparticles can increase the mechanical properties [14].

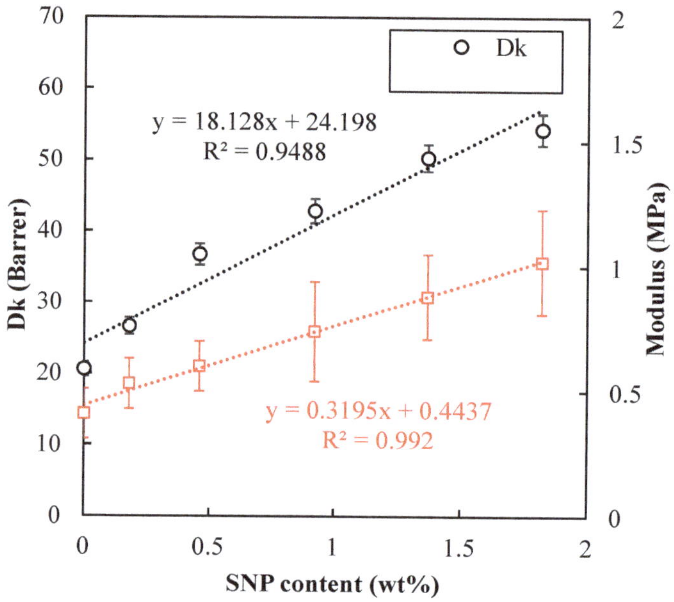

Figure 7. Effect of SNPs content on oxygen permeability and Young' modulus.

3.8. Protein Deposition

Basically, the deposition of protein has the close relationship with the wettability of hydrogel lenses. Because the wettability was not affected by the addition of SNPs, the deposition of proteins was not affected significantly either, as shown in Table 1. Comparing with 0 wt% of SNPs, the apparent amount of protein adsorption slightly varied around 0.75 and 0.78 nmol/cm^2 for HSA and between 3.1 and 3.3 nmol/cm^2 for lysozyme. The high HSA and lysozyme adsorption may be explained based on the properties of hydrogel materials, especially NVP monomers [3,17,23–25].

3.9. Cytotoxicity Test

The biocompatibility of an ophthalmic device, especially contact lenses, is significant factor determined via cellular behavior [6,21]. The in vitro cytotoxicity test was applied to observe the growth of L929 cells based on sample's extracted medium. In comparison to positive and negative controls, the L929 cells using extracted medium could growth normally after three days culture as shown in Figure 8. The result determined that the addition of SNPs content in hydrogel lenses did not influence the growth of L929 cells through cytotoxicity test.

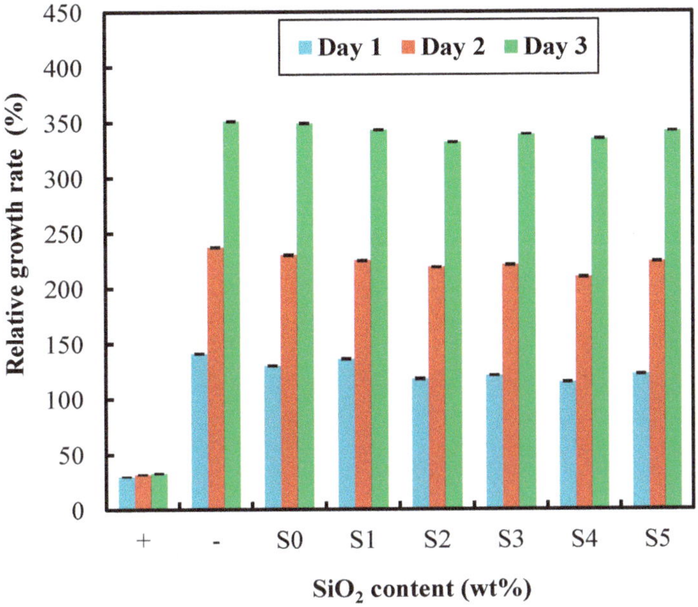

Figure 8. Effect of SNPs content on cytotoxicity.

3.10. Comparison with Commercial Contact Lenses

The Dk and EWC of SNPs hydrogel lenses were comparable or higher than those non-silicone commercial products such as Biomedics XC, Biomedics 38, Acuvue 2, and Acuvue Advance (Table 2) [3,26]. Although the Dk values of these SNP-loaded lenses were lower than those silicone commercial lenses such as Air Optix Night & Day, Air Optix, Acuvue Oasys, and PureVision, the EWC values were higher and the moduli were similar. The results of this research indicated that a small amount of SNPs would confer conventional hydrogel lenses improved ophthalmic properties to non-silicone commercial contact lenses including Biomedics XC and Acuvue Advance. These SNPs lenses can be considered to apply for ophthalmology materials as contact lenses.

Table 2. Properties comparison of this study and commercial lenses.

Product	Manufacturer	Dk (barrer)	EWC (%)	Contact angle (°)	Modulus (MPa)	Principle Monomers
Air Optix Night & Day	CIBA Vision	140	24		1.52	DMA, TRIS, siloxane monomer
Air Optix		110	33	44.4	1.00	DMA, TRIS, siloxane monomer
Acuvue Oasys	Johnson & Johnson Vision Care	103	38	78.7	0.72	MPDMS, DMA, HEMA, siloxane macromer, TEGDMA, PVP
Acuvue Advance		60	47	65.6	0.43	MPDMS, DMA, HEMA, EGDMA, siloxane macromer, PVP
Acuvue 2		19	58			HEMA, MAA, EGDMA
Pure Vision	Bausch & Lomb	91	36	93.6	1.10	TEGDMA, NVP, TPVC, NCVE, PBVC
Biomedics XC	CooperVision	44	60			HEMA, MAA, PC, TEGDMA
Biomedics 38		8.4	38	30	0.81	HEMA, EGDMA
S4	This work	50.3	51.9	60.1	0.88	HEMA, NVP, SNPs
S5		54.3	53.1	61.2	1.02	

[PVP] polyvinyl pyrrolidone; [MPDMS] monofunctional polydimethylsiloxane; [DMA] N,M-dimethylacrylamide; TEGDMA tetraethyleneglycol dimethacrylate; [TRIS] trimethyl siloxysilyl; [TPVC] tris-(trimethyl siloxysilyl)propylvinyl carbamate; [NCVE] N-carboxyvinyl ester; [PBVC] poly-(dimethysiloxy) di-(silylbutanol) bis-(vinyl carbamate).

4. Conclusions

Silica sol was prepared through the hydrolysis of tetraethoxysilane (TEOS) followed by blending with hydroxyethyl methacrylate (HEMA) and N-vinylpyrrolidone (NVP) before photoinitiated polymerization. With the silica content increasing to 1.8 wt%, the oxygen permeability of the resultant hydrogel was increased from 20 barrer to 54 barrer, while the equilibrium water content and contact angle remained around 52% and 60°, respectively. In addition, the Young's modulus increased from 0.4 MPa to 1 MPa when the content of silica nanoparticles increased to 1.8 wt%.

Author Contributions: N.-P.-D.T., C.-C.T., and C.-H.L. prepared experiments. N.-P.-D.T. analyzed the data and wrote the original draft. C.-H.L. analyzed the data. M.-C.Y. supervised the research project and revised the manuscripts. All authors have read and agreed to the published version of the manuscript.

Funding: This work was supported by the Ministry of Science and Technology, Taiwan through Grant No. MOST 106-2622-E-011-004-CC2.

Conflicts of Interest: The authors declare no conflict of interest.

References

1. Alvord, L.; Davis, T.; Morgan, C.F.; Schindhelm, K.; Vogt, J.; Winterton, L. Oxygen permeability of a new type of high Dk soft contact lens material. *Optom. Vis. Sci.* **1998**, *75*, 30–36. [CrossRef] [PubMed]
2. Maldonado, C.; Efron, N. Hydrogel lenses–Materials and manufacture: A review. *Optom. Pract.* **2003**, *4*, 101–115.
3. Tran, N.P.D.; Yang, M.C. Synthesis and characterization of silicone contact lenses based on TRIS-DMA-NVP-HEMA hydrogels. *Polymers* **2019**, *11*, 944. [CrossRef] [PubMed]
4. Lin, C.H.; Cho, H.L.; Yeh, Y.H.; Yang, M.C. Improvement of the surface wettability of silicone hydrogel contact lenses via layer-by-layer self-assembly technique. *Colloids Surf. B* **2015**, *136*, 735–743. [CrossRef] [PubMed]
5. Efron, N.; Morgan, P.B.; Cameron, I.D.; Brennan, N.A.; Goodwin, M. Oxygen permeability and water content of silicone hydrogel contact lens materials. *Optom. Vis. Sci.* **2007**, *84*, E328–E337. [CrossRef]
6. Lin, C.H.; Yeh, Y.H.; Lin, W.C.; Yang, M.C. Novel silicone hydrogel based on PDMS and PEGMA for contact lens application. *Colloids Surf. B* **2014**, *123*, 986–994. [CrossRef] [PubMed]
7. Song, M.; Shin, Y.H.; Kwon, Y. Synthesis and properties of siloxane-containing hybrid hydrogels: Optical transmittance, oxygen permeability and equilibrium water content. *J. Nanosci. Nanotechnol.* **2010**, *10*, 6934–6938. [CrossRef]
8. Zhao, Z.; Xie, H.; An, S.; Jiang, Y. The relationship between oxygen permeability and phase separation morphology of the multicomponent silicone hydrogels. *J. Phys. Chem. B* **2014**, *118*, 14640–14647. [CrossRef]
9. Zhao, Z.B.; An, S.S.; Xie, H.J.; Han, X.L.; Wang, F.H.; Jiang, Y. The relationship between the hydrophilicity and surface chemical composition microphase separation structure of multicomponent silicone hydrogels. *J. Phys. Chem. B* **2015**, *119*, 9780–9786. [CrossRef]
10. Garrett, Q.; Laycock, B.; Garrett, R.W. Hydrogel lens monomer constituents modulate protein sorption. *Investig. Ophthalmol. Vis. Sci.* **2000**, *41*, 1687–1695.
11. Paterson, S.M.; Liu, L.; Brook, M.A.; Sheardown, H. Poly (ethylene glycol)-or silicone-modified hyaluronan for contact lens wetting agent applications. *J. Biomed. Mater. Res. A* **2015**, *103*, 2602–2610. [CrossRef] [PubMed]
12. Kim, J.; Somorjai, G.A. Molecular packing of lysozyme, fibrinogen, and bovine serum albumin on hydrophilic and hydrophobic surfaces studied by infrared– visible sum frequency generation and fluorescence microscopy. *J. Am. Chem. Soc.* **2003**, *125*, 3150–3158. [CrossRef] [PubMed]
13. Ahn, J.; Chung, W.-J.; Pinnau, I.; Guiver, M.D. Polysulfone/silica nanoparticle mixed-matrix membranes for gas separation. *J. Membr. Sci.* **2008**, *314*, 123–133. [CrossRef]
14. Ahn, J.; Chung, W.-J.; Pinnau, I.; Song, J.; Du, N.; Robertson, G.P.; Guiver, M.D. Gas transport behavior of mixed-matrix membranes composed of silica nanoparticles in a polymer of intrinsic microporosity (PIM-1). *J. Membr. Sci.* **2010**, *346*, 280–287. [CrossRef]
15. Tran, N.P.D.; Yang, M.C. The ophthalmic performance of hydrogel contact lenses loaded with silicone nanoparticles. *Polymers* **2020**, *12*, 1128. [CrossRef]

16. Wang, J.; Li, X. Preparation and characterization of interpenetrating polymer network silicone hydrogels with high oxygen permeability. *J. Appl. Polym. Sci.* **2010**, *116*, 2749–2757. [CrossRef]
17. Tran, N.P.D.; Yang, M.C. Synthesis and characterization of soft contact lens based on the combination of silicone nanoparticles with hydrophobic and hydrophilic monomers. *J. Polym. Res.* **2019**, *26*, 143. [CrossRef]
18. Fonn, D.; Dumbleton, K.; Jalbert, I.; Sivak, A. Benefits of silicone hydrogel lenses. *Contact Lens Spectr.* **2006**, *21*, 38.
19. Awasthi, A.; Meng, F.; Künzler, J.; Linhardt, J.; Papagelis, P.; Oltean, G.; Myers, S. Ethylenically unsaturated polycarbosiloxanes for novel silicone hydrogels: Synthesis, end-group analysis, contact lens formulations, and structure–property correlations. *Polym. Adv. Technol.* **2013**, *24*, 557–567. [CrossRef]
20. Ghoreishi, S.; Abbasi, F.; Jalili, K. Hydrophilicity improvement of silicone rubber by interpenetrating polymer network formation in the proximal layer of polymer surface. *J. Polym. Res.* **2016**, *23*, 115. [CrossRef]
21. Korogiannaki, M.; Guidi, G.; Jones, L.; Sheardown, H. Timolol maleate release from hyaluronic acid-containing model silicone hydrogel contact lens materials. *J. Biomater. Appl.* **2015**, *30*, 361–376. [CrossRef]
22. Nicolson, P.C.; Vogt, J. Soft contact lens polymers: An evolution. *Biomaterials* **2001**, *22*, 3273–3283. [CrossRef]
23. Luensmann, D.; Jones, L. Protein deposition on contact lenses: The past, the present, and the future. *Cont. Lens Anterior Eye* **2012**, *35*, 53–64. [CrossRef] [PubMed]
24. Silva, D.; Fernandes, A.; Nunes, T.; Colaço, R.; Serro, A. The effect of albumin and cholesterol on the biotribological behavior of hydrogels for contact lenses. *Acta Biomater.* **2015**, *26*, 184–194. [CrossRef] [PubMed]
25. Subbaraman, L.N.; Glasier, M.-A.; Senchyna, M.; Sheardown, H.; Jones, L. Kinetics of in vitro lysozyme deposition on silicone hydrogel, PMMA, and FDA groups I, II, and IV contact lens materials. *Curr. Eye Res.* **2006**, *31*, 787–796. [CrossRef]
26. French, K.; Jones, L. A decade with silicone hydrogels: Part 1. *Optom. Today* **2008**, *48*, 42–46.

polymers

MDPI

Article

Synergistic and Regulatable Bioremediation Capsules Fabrication Based on Vapor-Phased Encapsulation of *Bacillus* Bacteria and its Regulator by Poly-*p*-Xylylene

Yen-Ching Yang [1], Wei-Shen Huang [2], Shu-Man Hu [1], Chao-Wei Huang [1], Chih-Hao Chiu [3,4,*] and Hsien-Yeh Chen [1,5,6,*]

1 Department of Chemical Engineering, National Taiwan University, Taipei 10617, Taiwan; youngeddie.tw@gmail.com (Y.-C.Y.); sofiahu870628@gmail.com (S.-M.H.); jordanmagic0712@gmail.com (C.-W.H.)
2 Institute of Oceanography, National Taiwan University, Taipei 10617, Taiwan; lowridertrent@gmail.com
3 Department of Orthopedic Surgery, Chang Gung Memorial Hospital, Taoyuan 33378, Taiwan
4 Bone and Joint Research Center, Chang Gung Memorial Hospital, Linkou 33305, Taiwan
5 Molecular Imaging Center, National Taiwan University, Taipei 10617, Taiwan
6 Advanced Research Center for Green Materials Science and Technology, National Taiwan University, Taipei 10617, Taiwan
* Correspondence: joechiu0115@gmail.com (C.-H.C.); hsychen@ntu.edu.tw (H.-Y.C.); Tel.: +886-2-33669476 (H.-Y.C.)

Abstract: A regulatable bioremediation capsule material was synthesized with isolated single-strain bacteria (*Bacillus* species, *B. CMC1*) and a regulator molecule (carboxymethyl cellulose, CMC) by a vapor-phased encapsulation method with simple steps of water sublimation and poly-*p*-xylylene deposition in chemical vapor deposition (CVD) process. Mechanically, the capsule construct exhibited a controllable shape and dimensions, and was composed of highly biocompatible poly-*p*-xylylene as the matrix with homogeneously distributed bacteria and CMC molecules. Versatility of the encapsulation of the molecules at the desired concentrations was achieved in the vapor-phased sublimation and deposition fabrication process. The discovery of the fabricated capsule revealed that viable living *B. CMC1* inhabited the capsule, and the capsule enhanced bacterial growth due to the materials and process used. Biologically, the encapsulated *B. CMC1* demonstrated viable and functional enzyme activity for cellulase activation, and such activity was regulatable and proportional to the concentration of the decorated CMC molecules in the same capsule construct. Impressively, 13% of cellulase activity increase was realized by encapsulation of *B. CMC1* by poly-*p*-xylylene, and a further 34% of cellulase activity increase was achieved by encapsulation of additional 2.5% CMC. Accordingly, this synergistic effectiveness of the capsule constructs was established by combining enzymatic *B. CMC1* bacteria and its regulatory CMC by poly-*p*-xylylene encapsulation process. This reported encapsulation process exhibited other advantages, including the use of simple steps and a dry and clean process free of harmful chemicals; most importantly, the process is scalable for mass production. The present study represents a novel method to fabricate bacteria-encapsulated capsule for cellulose degradation in bioremediation that can be used in various applications, such as wastewater treatment and transforming of cellulose into glucose for biofuel production. Moreover, the concept of this vapor-phased encapsulation technology can be correspondingly used to encapsulate multiple bacteria and regulators to enhance the specific enzyme functions for degradation of various organic matters.

Keywords: parylene; CVD; bioremediation; polymer; cellulase; *Bacillus*

Citation: Yang, Y.-C.; Huang, W.-S.; Hu, S.-M.; Huang, C.-W.; Chiu, C.-H.; Chen, H.-Y. Synergistic and Regulatable Bioremediation Capsules Fabrication Based on Vapor-Phased Encapsulation of *Bacillus* Bacteria and its Regulator by Poly-*p*-Xylylene. *Polymers* **2021**, *13*, 41. https://dx.doi.org/10.3390/polym13010041

Received: 30 November 2020
Accepted: 22 December 2020
Published: 24 December 2020

Publisher's Note: MDPI stays neutral with regard to jurisdictional claims in published maps and institutional affiliations.

1. Introduction

Bioremediation provides promising solutions for the removal of environmental pollutants, toxic elements, and poisoning management for clinical purposes [1–3]. Various remedial approaches were developed by precision screening of efficient bacteria from contaminated environments [4,5] to isolate genetically engineered bacteria with enhancement of catabolic pathways and microbial physiology [6–8]. To protect bacteria/cells from acidic environments and to elongate the treatment length, circulation time, and shelf life, different techniques, including spray drying/coating, freeze drying, emulsification, coacervation, extrusion, and microfluidics, have been developed for probiotic bacterial encapsulation [9,10]. However, the issue of the low vitality of the encapsulated bacteria used for bioremediation remains [11]. Imperatively, the technique used to create a microenvironment that can provide solid protection for bacteria, whose growth may be suppressed by parasites or predators [12] as well as control the release of either enzymes and/or bacteria to target pollution [13], is very important in the encapsulation of bacteria for bioremediation use. Therefore, the development of prospective encapsulation techniques for bioremediation has been pursued to provide additional properties that include (1) a favorable habitation microenvironment for functional bacteria with enhanced growth activities and (2) a regulatory mechanism to customize the enzymatic functions of bacteria, allowing increased efficiency and adjustment to the required bioremediation conditions.

We herein provide a vapor-phased encapsulation method for bacterial capsules with an elegantly equipped mechanism for increasing bacterial populations and the corresponding enzymatic functions. A vapor-deposited polymeric matrix system of poly-*p*-xylylenes of United States Pharmacopeia (USP) class VI with high biocompatibility and chemical resistance to strong acids, bases, and solvents was used for encapsulation during the fabrication process. The fabrication is performed in one step with a dry and clean vapor phase, which is desirable for sensitive biological substances such as cells, enzymes, growth factors, and other functional peptides and proteins [14,15]. The overall encapsulation process was realized based on our previously reported mechanism to deposit a vapor-phased poly-*p*-xylylene polymer on a template substrate that is eventually sublimated, and by manipulating the mass transport during the processing conditions, diverse species with distinct thermodynamic properties were subjected to sublimation and deposition within the confined space of the templates. Finally, transformation of the template resulted in the construction of composite materials with defined physical properties in terms of porosity, bulk size and geometry, and chemical functionality by the compartmentalized functional entities and the devised interfacial chemistries [16]. With respect to the vapor sublimation and deposition process, which has been reported to result in the precise localization and distribution of substances such as metals, molecules, and liquids to form composites with controlled homogeneity or anisotropy from various materials [15,17], we hypothesized that the benign and versatile vapor-phased process was able to encapsulate (i) unmodified native bacteria of the *Bacillus* species CMC1 (hereafter referred to as *B. CMC1*) with specific enzymatic function and the same precision to localize and distribute *B. CMC1* due to the encapsulation technique, enabling well-controlled bacterial growth activities and functions, and (ii) a regulator molecule, carboxymethyl cellulose (CMC), with a customizable encapsulation concentration to provide the same customizable stimulation dosages to regulate the enzymatic functions of the neighboring *B. CMC1* bacteria. The fabricated and encapsulated capsules were composed of inhabitant bacteria from (i) and the surrounding conditioning regulator molecules from (ii), and the synergistic activities exhibited by (i) and (ii) were able to deliver a combination of controlled enzymatic factors for bioremediation in the devised microenvironments (Figure 1a). The synergistic conditioning mechanism was effective in regulating the bacterial number and the corresponding enzymatic reactivity, and was proportional to the CMC regulator composition. The encapsulation process is versatile and is completed in simple steps with mass production scalability (Figure 1a,e), and the resultant bacteria/CMC capsules represent an advanced bacteria/polymer combination for bioremediation models and have potential for unlimited combinatorial composting configurations for bioremediation applications.

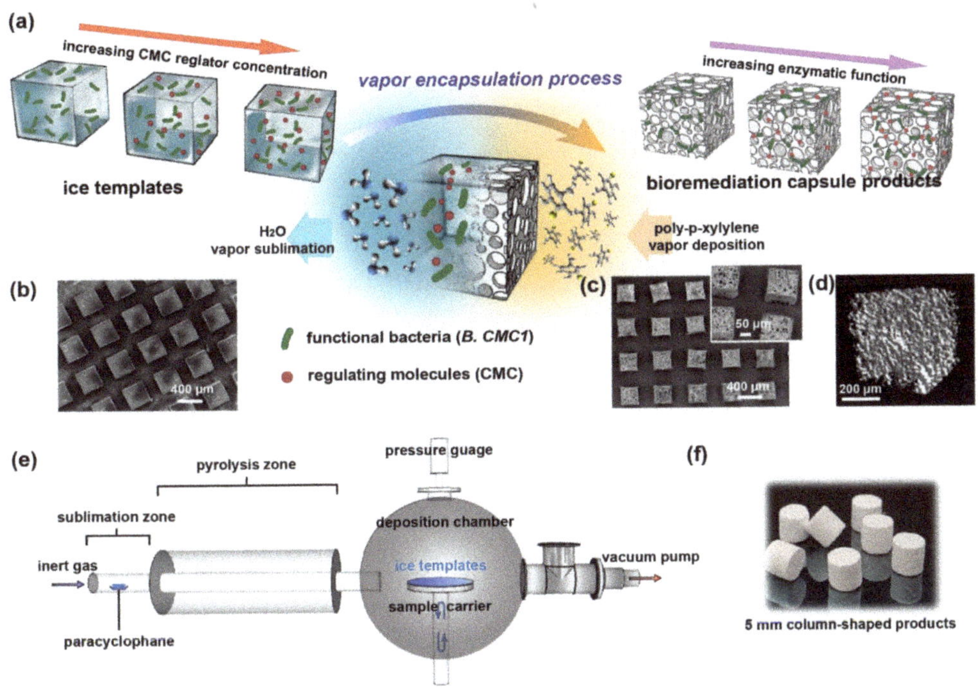

Figure 1. Fabrication of regulatable bioremediation capsules via vapor-phased encapsulation of bacteria. (**a**) Schematic illustration of the capsule fabrication process by vapor sublimation and deposition to encapsule (i) functional bacteria and (ii) regulatory molecules within a poly-*p*-xylylene polymer matrix. (**b**) Image by cryo SEM showing the iced templates of array cubes with dimensions of 400 μm × 400 μm × 400 μm used for encapsulation. (**c**) SEM images of the capsule products by the encapsulation process; the products showed the same replicated dimension and shape and comprised poly-*p*-xylylene as the matrix and encapsulated bacteria and CMC molecules. Panels (**b**,**c**) show the scalability of mass production potentials. (**d**) A 3-D image by micro-CT examination of the capsule product. (**e**) An illustration of the vapor deposition system, containing a sublimation zone, a pyrolysis zone, and a deposition chamber in the main body. (**f**) The fabricated column-shaped bioremediation capsule products with the dimension of 5 mm in diameter and 5 mm in height.

2. Materials and Methods

2.1. Bacterial Strain Isolation

Wastewater and activated sludge were collected from the sewage of a pig farm in Taoyuan, Taiwan. All tools and reservoirs were sterilized before sampling. Activated sludge was stored immediately at 4 °C after collection. The wastewater and activated sludge were used for further culturing of environmental bacteria in liquid nutrient medium. In one liter of culture medium, 3 g peptone (BD biosciences, San Jose, CA, USA), 1 g yeast extract (BD biosciences, USA), 5 g NaCl (Sigma-Aldrich, St. Louis, MO, USA), 0.2 g $MgSO_4 \cdot H_2O$ (Sigma-Aldrich, USA), 0.2 g $CaCl_2$ (Sigma-Aldrich, USA), 0.2 g KNO_2 (Sigma-Aldrich, USA), 2 g NH_4NO_3 (Sigma-Aldrich, USA), 0.3 g Na_2HPO_4 (Sigma-Aldrich, USA), 5 g glucose (Sigma-Aldrich, USA), and 2.5 mL tween 80 (Sigma-Aldrich, USA) were added and then pH with NaOH to 7.0. The genomic DNA of the isolated bacterial strain was extracted using a commercial genomic DNA extraction kit (Qiagen, Germantown, MD, USA). Then, the sequence of its 16S rRNA was confirmed by polymerase chain reaction by using the primers F8 (5′GAGAGTTTGATCCTGGCTCAG3′) and R1492 (5′GGTTACCTTGTTACGACTT3′). The obtained 16S rRNA sequence from the isolated bacteria was

TGCTATACATGCAGTCGAGCGGACAGATGGGAGCTTGCTCCCTGATGTTAGCG GCGGACGGGTGAGTAACACGTGGGTAACCTGCCTGTAAGACTGGGATA ACTCCGGG AAACCGGGGCTAATACCGGATGGTTGTTTGAACCGCATGGTTCAAACATAAAAGGT GGCTTCGGCTACCACTTACAGATGGACCCGCGGCGCATTAGCTAGTTGGTGAGGTA ACGGCTCACCAAGGCAACGATGCGTAGCCGACCTGAGAGGGTGATCGGCCACACT GGGACTGAGACACGGCCCAGACTCCTACGGGAGGCAGCAGTAGGGAATCTTCCGC AATGGACGAAAGTCTGACGGAGCAACGCCGCGTGAGTGATGAAGGTTTTCGGATC GTAAAGCTCTGTTGTTAGGGAAGAACAAGTACCGTTCGAATAGGGCGGTACCTTGA CGGTACCTAACCAGAAAGCCACGGCTAACTACGTGCCAGCAGCCGCGGTAATACGT AGGTGGCAAGCGTTGTCCGGAATTATTGGGCGTAAAGGGCTCGCAGGCGGTTTCTT AAGTCTGATGTGAAAGCCCCCGGCTCAACCGGGGAGGGTCATTGGAAACTGGGGA ACTTGAGTGCAGAAGAGGAGAGTGGAATTCCACGTGTAGCGGTGAAATGCGTAGA GATGTGGAGGAACACCAGTGGCGAAGGCGACTCTCTGGTCTGTAACTGACGCTGA GGAGCGAAAGCGTGGGGAGCGAACAGGATTAGATACCCTGGTAGTCCACGCCGTA AACGATGAGTGCTAAGTGTTAGGGGGTTTCCGCCCCTTAGTGCTGCAGCTAACGCAT TAAGCACTCCGCCTGGGGAGTACGGTCGCAAGACTGAAACTCAAAGGAATTGACGG GGGCCCGCACAAGCGGTGGAGCATGTGGTTTAATTCGAAGCAACGCGAAGAACCT TACCAGGTCTTGACATCCTCTGACAATCCTAGAGATAGGACGTCCCCTTCGGGGGCA GAGTGACAGGTGGTGCATGGTTGTCGTCAGCTCGTGTCGTGAGATGTTGGGTTAAGT CCCGCAACGAGCGCAACCCTTGATCTTAGTTGCCAGCATTCAGTTGGGCACTCTAAG GTGACTGCCGGTGACAAACCGGAGGAAGGTGGGGATGACGTCAAATCATCATGCCC CTTATGACCTGGGCTACACACGTGCTACAATGGACAGAACAAAGGGCAGCGAAACC GCGAGGTTAAGCCAATCCCACAAATCTGTTCTCAGTTCGGATCGCAGTCTGCAACTC GACTGCGTGAAGCTGGAATCGCTAGTAATCGCGGATCAGCATGCCGCGGTGAATACG TTCCCGGGCCTTGTACACACCGCCCGTCACACCACGAGAGTTTGTAACACCCGAAGT CGGTGAGGTAACCTTTTAGGAGCCAGC.

The sequence was further confirmed according to the 16S rRNA database at https://www.ezbiocloud.net/. The isolated bacteria were the *Bacillus* species with the top ranking of *B. subtilis* and was named *Bacillus* species CMC1 (*B. CMC1*) in the study.

2.2. Encapsulation Process

B. CMC1 was cultured for 30 h at 30 °C and then washed twice with diH_2O by centrifugation and resuspended into final solutions (either diH_2O, 0.25% CMC (Showa, Tokyo, Japan), 1.5% CMC, or 2.5% CMC) to prepare samples with the same concentration of bacteria (~1×10^8 bacteria/mL) on the same day of capsule fabrication. The prepared bacterial solution with/without CMC was solidified by liquid nitrogen for later encapsulation. A home-built vapor deposition system was used for the encapsulation and fabrication process in this study (illustrated in Figure 1e), and the specifications and detailed installations of the system are described elsewhere [18,19]. First, the precursor dichloro-[2,2]-paracyclophane (Galxyl C, Galentis, Venice, Italy) was transformed into di-radical monomers by pyrolysis at 670 °C for the subsequent vapor deposition of poly-*p*-xylylenes. Subsequently, the monomers started both polymerization and deposition under the conditions of 100 mTorr and 25 °C. Simultaneously, the solid ice in the bacterial ice templates was transferred into water vapor via a sublimation process with respect to the same conditions of 100 mTorr and 25 °C. The deposition of poly-*p*-xylylenes and sublimation of solid ice into water vapor produced porous poly-*p*-xylylene bacteria capsules of the desired size and shape.

2.3. Enzyme Activity Assays

Cellulase activity was examined on 9 cm agar plates by quantifying the size of the enzyme functioning zones. To one liter of cellulase visualizing agar, 1 g yeast extract (BD Biosciences, USA), 1 g $NH_4H_2PO_4$ (Sigma-Aldrich, USA), 0.2 g KCl (Sigma-Aldrich, USA), and 1 g $MgSO_4{\cdot}7H_2O$ (Sigma-Aldrich, USA) were added, and then the pH was adjusted to 6.3 with NaOH. Then, 9 g agar (Sigma-Aldrich, USA) and 10 mg trypan blue (Sigma-Aldrich, USA) were added. A single bacterial capsule (100 μL in volume) was seeded in the center of one 9 cm plate and left at 30 °C for up to 140 h for subsequent enzyme functioning zone quantification. For the quantification of DNA and RNA, a single bacterial capsule (100 µL in volume) was seeded in a 50 mL centrifuge tube with 30 mL of growth culture medium inside, and the culture was left at 30 °C for 50 h on an orbital shaker with a 100 rpm shaking speed. The same volume of the culture (5 mL) from encapsulated bacteria was centrifuged for genomic DNA extraction using a commercial extraction kit (Zymo Research, Irvine, CA, USA) for the following DNA quantification. The same number of bacterial cells (2×10^8 cells) was used for total RNA extraction by using a commercial extraction kit (Zymo Research, USA) for total RNA quantification. The concentrations of DNA and RNA were measured by a NanoDrop Spectrophotometer (ThermoFisher, Waltham, MA, USA). For this, 1 µL of purified DNA and RNA were used in the analysis. Note that only the DNA and RNA with the 260/280 ratio between 1.8 and 2.0 were used for concentration measurements, ensuring the sufficient purity of DNA and RNA for further quantification analysis.

2.4. Characterizations

A Nikon ECLIPSE 80i fluorescence microscope (Nikon, Tokyo, Japan) with a visible light source was used to visualize the crystal violet-stained *B. CMC1* in the porous parylene structure. A VK-9510 3D profile microscope (Keyence, Osaka, Japan) was used to analyze the external architecture of the bacterial capsules and the existence of the bacteria inside the structure. SEM images were recorded using a NovaTM NanoSEM (FEI, Hillsboro, OR, USA) operated at a primary energy of 10 keV and a pressure of 5×10^{-6} Torr to detailed the internal structure of the bacterial capsule. EDS elemental point analyses were captured for the quantification of the studied elements. 3D analysis of the interior structure was performed by using Bruker Skyscan 2211 (Bruker micro-CT, Kontich, Belgium) at 2.0 µm/pixel resolution. The setting of the voltage was 40 kVp, whereas the current was 700 µA at 8 Watt output with microfocus mode. Image reconstruction, ring artifacts, and beam-hardening correction were performed using reconstruction software, Instarecon (Bruker micro-CT, Belgium). FT-IR spectra of the fabricated capsules were recorded by a Spectrum 100 spectrometer equipped with an ATR detector (PerkinElmer, Waltham, MA, USA). The recorded spectra ranged from 600 to 4500 cm^{-1} with 4 scan times at 4 cm^{-1} resolution.

3. Results and Discussion

3.1. The Fabrication of Regulatable Bioremediation Capsules

The bacteria-encapsulated capsule was fabricated by first preparing ice templates by directly transforming the liquid-phased bacteria-cultured media into a solidified ice template in a liquid nitrogen-conditioned bath. Subsequently, the fabrication exploited the previously described sublimation of ice and deposition of poly-*p*-xylylene in one continuous step [16,17] to construct a bacterially encapsulated polymeric capsule. A second transformation was performed with vapor-deposited poly-*p*-xylylene molecules to replace the resulting space when the sublimated ice/water molecules evaporated from the ice templates. The resultant construct consisted of a three-dimensional porous poly-*p*-xylylene matrix with encapsulated *B. CMC1* bacteria in the matrix (Figure 1). The shape of the final capsule construct replicated the shape of the transformed ice template, which was theoretically shaped by molding, sculpting, or solidifying droplets to obtain various sizes and geometries [15]. Ideally, shrinkage or dislocation of the construct is avoided owing to the continuous sublimation and deposition process [16]. As a result, the dimension of the fabricated capsule is depended on the size and shape of the mold used in the ice template preparation. The present technology supports to fabricate the bacteria-encapsulated capsule as little as 400 µm cubic shape to a larger 5mm columnal shape. As shown in Figure 1b, a cubic shape with dimensions of 400 µm × 400 µm × 400 µm was used for the preparation of the ice templates and the fabrication of the construct of the bacterial-encapsulated capsule. The subsequent CVD process replaced ice with poly-*p*-xylylene to form a three-dimensional porous structure (Figure 1c). Characterization by microcomputed tomography (micro-CT, Figure 1d) also indicated the overall homogeneous porous structure of the polymer matrix. By using the column-shaped mold with the dimension of 5 mm in diameter and 5 mm in height, the larger size of bacteria-encapsulated capsule was obtained (Figure 1f).

In order to characterize the internal structure of the bacteria encapsulated capsule, the 400 μm cubic capsule was examined by using a combination of optical microscopy, scanning electron microscopy (SEM), and confocal microscopy to verify the localization and distribution of the bacteria (Figure 2). The image recorded by optical microscopy clearly showed that the crystal violet-stained bacteria were found only in the polymer matrix but not in the void pores, verifying the successful encapsulation of bacteria by the fabrication process (Figure 2a). Both confocal microscopy and SEM further confirmed that rod-shaped *B. CMC1* with dimensions of 0.5–1 μm existed in the fabricated capsule structures (Figure 2a). The anticipated homogeneity of the distributed bacteria was also observed and was believed to be due to the controlled mass transport of solidified ice templates and the continued sublimation and deposition process that prevented dislocation of the encapsulated substances. As the same concentration of bacteria was used to prepare the ice template, the quantity of the bacteria was depended on the volume of the ice template prepared for the capsule fabrication. Therefore, the larger column-shaped bacteria-encapsulated capsules which contained sufficient number of bacteria in the fabricated capsules were used for cellulase activities examination on 9 cm agar plates.

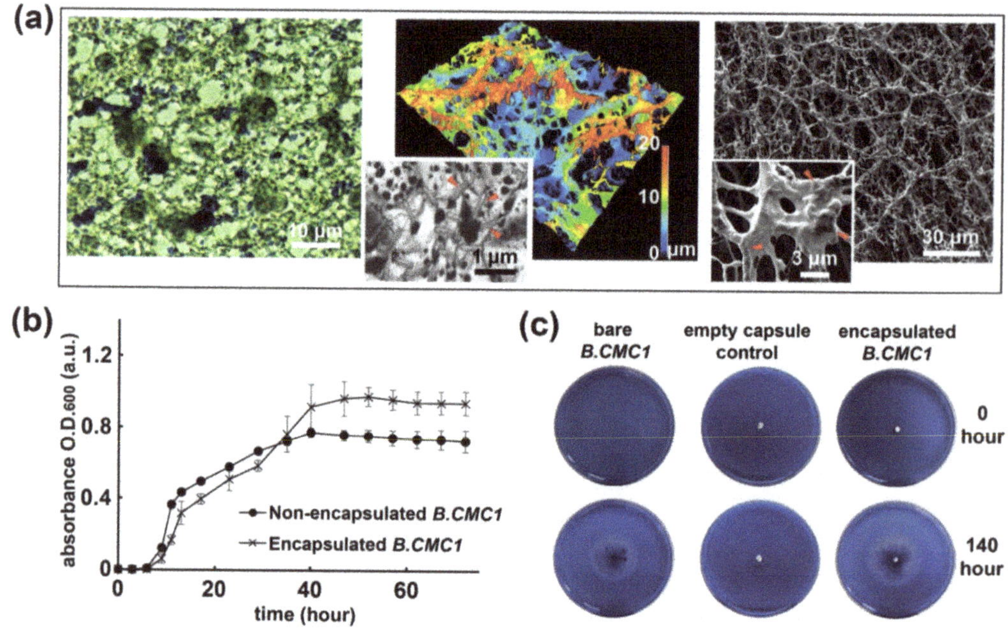

Figure 2. Viability analysis of *B. CMC1* activities of the fabricated capsule. (**a**) Micrographs showing stained (crystal violet, purple color), 3-D structural, and SEM topological images of encapsulated *B. CMC1* within the capsule products. The yellow marks indicate the presence of *B. CMC1* within the polymer structure. (**b**) Bacterial growth curves and (**c**) enzymatic functions were compared in parallel for pure bacteria to verify viability.

3.2. *Viability and Cellulase Activities of Encapsulated B. CMC1*

The cell viability of the encapsulated bacteria was further analyzed, and due to the natural self-protective mechanism of the cell wall in *B. CMC1*, we hypothesized that the bacteria were resistant to low temperatures during iced template preparation and low-pressure conditions (approximately 10^{-3} Torr) during the sublimation and deposition process. Cultured samples of encapsulated *B. CMC1* were compared to samples of *B. CMC1* without encapsulation (positive control), and their cell densities were monitored by OD_{600} measurement. The results showed comparable bacterial growth patterns and activities in liquid medium for both types of bacteria during the culture time frame of 10 h (Figure 2b), indicating the viability of cells and negligible disturbance of *B. CMC1* by the encapsulation process. In addition, by comparing the enzymatic function for cellulase decomposition on 9.0 cm agar plates for both groups of samples, the data at 140 h showed a 4.3 cm enzyme functional zone diameter for the encapsulated *B. CMC1* compared to a 3.8 cm diameter for samples without encapsulation (Figure 2c). Unambiguously, these results were found to support the hypothesis that the *B. CMC1* bacteria were preserved during the encapsulation and fabrication process. Interestingly, it was discovered that a higher population and a larger enzyme effective zone size were found for the *B. CMC1*-encapsulated samples, which indicated their enhanced bacterial growth and cellulase enzymatic activities compared to those of the nonencapsulated sample at the studied time point of 140 h, and it is believed that the highly biocompatible poly-*p*-xylylene [20–22] played an important role and provided an agreeable microenvironmental niche for *B. CMC1*.

3.3. Regulation of the Fabricated Capsules by CMC

CMC was found to enhance the cellulase activity of *Bacillus* species bacteria [23]. Here, in this study, the proposed regulation approach exploited the cellulase self-induction mechanism by supplying a various number of CMC molecules for *B. CMC1* in the same controlled volume in the capsule construct, and our second hypothesis was thus raised to validate whether increasing the concentration of CMC would proportionally increase the resultant enzymatic activity of *B. CMC1* in the same capsule construct. While the versatility of the sublimation and deposition process was able to encapsulate/include multiple ingredients with a precisely administered composition and without molecular phase separation [16], the fabrication was straightforward and conducted by preparing iced templates containing both (i) *B. CMC1* and (ii) various concentrations of CMC (0.25%, 1.5%, or 2.5% were selected for the demonstration in the study). Subsequently, the same sublimation and deposition process was performed to resolve the capsule construct that comprised a poly-*p*-xylylene matrix with simultaneous encapsulation (i) and (ii) the same matrix structure. In addition to the already confirmed encapsulation of *B. CMC1*, the use of FT-IR analysis indicated that the intensity of the characteristic –C=O and $–CH–O–CH_2$ peaks from CMC observed at 1558 and 1050 cm^{-1}, respectively, gradually increased from a low concentration to higher concentrations of CMC content in the fabricated capsules (Figure 3a). Moreover, quantification of the peak area for –C=O and $–CH–O–CH_2$, which were normalized with respect to nonshifted peaks (–C–H at approximately 2800 cm^{-1}) from poly-*p*-xylylene, additionally showed consistency by stoichiometry of the encapsulated CMC concentrations (Figure 3b). On the other hand, the consistent increases in elemental concentrations of oxygen (0% to 27.98%) and sodium (0% to 4.73%) and decreases in carbon (95.22% to 65.25%) and chloride (4.78% to 2.03%) in SEM-EDS analysis were also in close value to the theory, which also verified the proposed encapsulation and regulation of CMC compositions (Figure 3c). The SEM and optical microscopic images of bioremediation capsules containing various concentrations of CMC (0.25%, 1.5%, 2.5%) did not show any obvious difference in the porous structure within the capsules. This is because the poly-*p*-xylylene matrix replaced sublimated water to constitute the main structure of the capsule during the vapor encapsulation process, the poly-*p*-xylylene matrix might deposited directly on the CMC molecules when water sublimated. The function of CMC molecular in these bioremediation capsules was only to upregulate the function of cellulase activity [23]. However, the pore size of the capsule can be controlled by the deposition rate of the poly-*p*-xylylene [16].

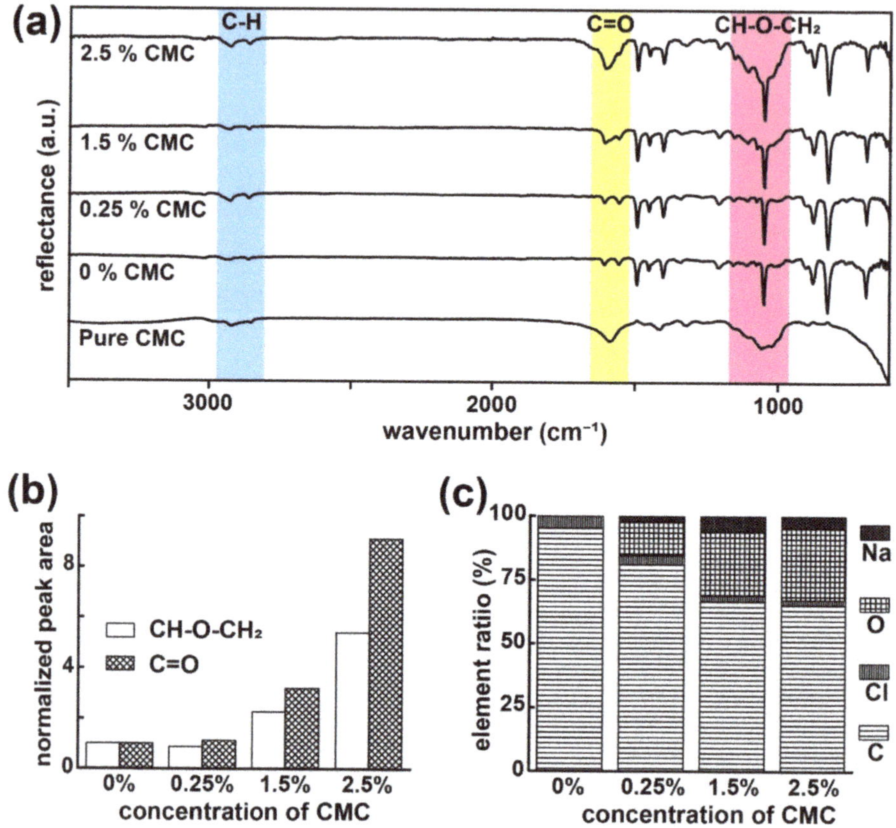

Figure 3. Regulation and encapsulation of CMC molecules. (**a**) FT-IR analysis showed regulated concentrations of CMC in the capsule products. (**b**) Quantified and normalized peak intensity of C=O and CH-O-CH_2 in panel (**a**) indicated a consistent increase according to the increased stoichiometric concentrations of CMC. (**c**) EDS elemental analysis revealed similar trends in the carbon and oxygen composition variations corresponding to the regulations of CMC concentrations.

With respect to the effectiveness of increasing CMC compositions toward the growth activities of *B. CMC1* in the same fabricated capsules, separate experiments were performed by culturing *B. CMC1* in liquid medium for samples with varied CMC compositions and were compared to the samples without CMC encapsulation. As shown in Figure 4a, the cultured regulation capsule samples (both *B. CMC1* and CMC were encapsulated) exhibited approximately 10 h of bacterial growth in the lag phase and showed consistency with the samples with only *B. CMC1* encapsulation (Figure 2b), indicative of negligible disturbance by the surrounded and encapsulated CMC molecules, and the bacteria adapted themselves well to the engineered microenvironments of the capsules. Interestingly, and surprisingly, the encapsulation of CMC molecules showed a positive influence on the growth of *B. CMC1* in the exponential phase; the presence of CMC resulted in the enhancement of *B. CMC1* growth activities, and such an enhancement was found to be proportional to the increase in CMC composition. Two stages in the exponential phase of showing a steep growth curve in the early exponential growth phase from approximately 10–15 h and a second gradual incline stage from 15–40 h were found. The bacterial growth finally reached a stationary phase after 40 h, and due to the regulation and enhancement of varied CMC compositions, increasing the final bacterial number was found accordingly for the studied groups.

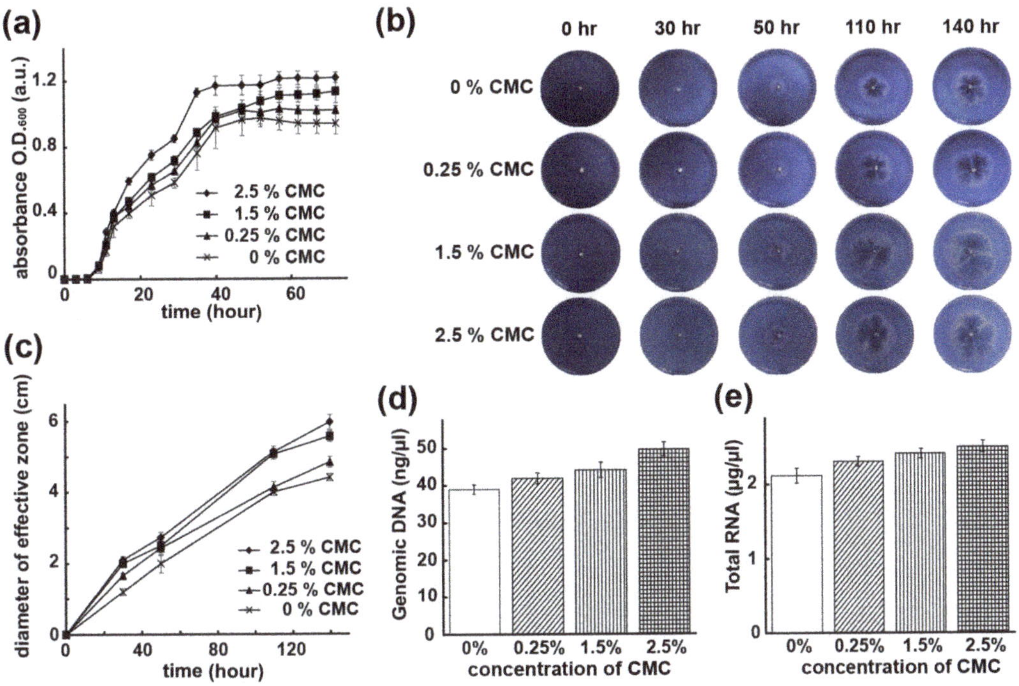

Figure 4. Regulation and quantification of bacterial biochemical activities of the bacterial capsules. (**a**) Regulation to enhance bacterial growth with respect to increasing CMC concentrations was verified in a studied culture time frame. (**b**) Regulated and enhanced enzyme activities of *B. CMC1* were discovered with increasing CMC concentrations on agar plates for up to 140 h. (**c**) Quantification of the effective zones in panel (**b**) was compared. (**d**) Analysis of the genomic DNA for the studied groups confirmed the upregulated expression of the corresponding DNA according to the increments of CMC concentrations. (**e**) The same upregulated expression of total RNA unambiguously verified the proposed regulation mechanism by increasing CMC concentrations in the bacterial capsules. Statistical data were expressed as the mean value and the standard deviation based on three independent samples.

The important expressions of regulated enzyme activities were finally analyzed for the synergistic capsules. Capsule samples containing both *B. CMC1* and CMC (with the same varied compositions of 0.25%, 1.5%, and 2.5%) were studied in parallel to bare samples with unregulated *B. CMC1* (0% CMC), and their enzymatic cellulase activities were first examined on agar plate results for up to 140 h (enzyme functioning zone measuring collected from three independent experiments were listed in Table 1). As shown in Figure 4b,c, comparisons of the size of the enzyme functioning zones were measured and quantified for these plates, and anticipated results indicated that increased zone sizes were found with increasing CMC compositions. More specifically, the quantification of the zone size revealed a 9.7% increase from 0% to 0.25% of CMC composition variation and a 26.5% and 35.6% increase for 1.5% and 2.5%, respectively, demonstrating a regulatable and enhanced enzymatic cellulase activity of the synergistic capsules in accordance with the encapsulated variation of CMC compositions.

Table 1. The diameter of enzyme functioning zones recorded in the enzymatic cellulase examination in Figure 4b,c.

	Time (h)	Diameter of the Functioning Zone (cm)				Time (h)	Diameter of the Functioning Zone (cm)		
		Exp. 1	Exp. 2	Exp. 3			Exp. 1	Exp. 2	Exp. 3
0% CMC	0	0	0	0	0.25% CMC	0	0	0	0
	30	1.2	1.1	1.3		30	1.6	1.7	1.7
	50	1.9	1.8	2.3		50	2.6	2.2	2.5
	110	3.9	4.1	4		110	4	4.1	4.3
	140	4.3	4.5	4.4		140	5	4.8	4.7
	Time (h)	Diameter of the Functioning Zone (cm)				Time (h)	Diameter of the Functioning Zone (cm)		
		Exp. 1	Exp. 2	Exp. 3			Exp. 1	Exp. 2	Exp. 3
1.5% CMC	0	0	0	0	2.5% CMC	0	0	0	0
	30	2	2.1	1.9		30	2	2.2	2.1
	50	2.5	2.4	2.6		50	2.7	2.9	2.6
	110	5.1	4.9	5.2		110	5.1	5.3	5
	140	5.6	5.4	5.7		140	5.9	6.2	5.8

More evidence showing the enhancement of the enzyme activities was further verified through the analysis of the genomic information, including the amount of genomic DNA and total RNA from the cultured samples of the capsules at a time point of 50 h, which was in the early stage of stationary phase in the bacterial growth curve. Calculation and quantification of the genomic DNA was performed based on the same volume of bacteria obtained from the capsule samples, and as shown in Figure 4d, an increase by 7.5% of the DNA concentration from unregulated *B. CMC1* (0% CMC) to a 0.25% regulation was found, and an anticipated enhancement of the expression with increasing CMC regulation compositions was accordingly discovered with a 13.1% increase for 1.5% CMC and 27.4% for 2.5% CMC of the studied capsule samples. On the other hand, based on the same bacteria number to extract their total RNA, the results in Figure 4e also indicated a similar trend with anticipation, with a 9% increase for 0.25% CMC, 14% increase for the 1.5% CMC sample, and 18.4% increase for the 2.5% CMC capsules. The genomic information specifically verified the enzymatic ability of *B. CMC1* was regulated for the synergistic capsules, and the synergy was found to be consistent with the same regulatable and enhancement of the bacterial population in the aforementioned study. Collectively, the results unambiguously verified the hypothesis that synergic effectiveness was achieved by vapor encapsulation of (i) *B. CMC1* and (ii) the CMC regulator in the same fabricated capsule constructs, and the regulation was achievable through the versatile use of the encapsulated CMC compositions.

4. Conclusions

Synergistic and regulatable bioremediation of cellulose by encapsulated *B. CMC1* was achieved in the current study. Because of the versatility of various types of remedial bacteria and regulator molecules, unlimited applications are expected beyond those shown in the report. In addition, the use of water/ice templates and a dry and clean vapor-phased process preserved the sensitive biomolecules and their biological functions, and the final fabricated capsule construct was composed of a USP Class VI compatible poly-*p*-xylylene matrix. Due to this specific production process, this novel fabrication process can not only encapsulate functional bacteria, regulating molecules, but also other potential absorbing material to concentrate the organic matters for degradation. It was reported that the encapsulation of phenol metabolic bacteria in microfiltration membrane capsules alone can create a confined environment to enhance the bioremediation of efficiency of encapsulated bacteria [24]. The poly-*p*-xylylene fabrication technology can constitute a defined inner porous structure to capture different size of organic particles to further concentrate organic matters and a controllable size and shape of the products to fit to different situations. Until now, most of the present studies still focused on identifying the bacteria with specific enzyme functions to use in the bioremediation process [5,25,26]. We foresee the application of capsule products to the degradation of organic compounds, wastewater, and environmental pollution, as well as the removal of potential hazardous chemicals.

Author Contributions: Conceptualization, Y.-C.Y., C.-H.C. and H.-Y.C.; methodology, Y.-C.Y. and W.-S.H.; validation, Y.-C.Y. and W.-S.H.; formal analysis, S.-M.H.; investigation, Y.-C.Y.; resources, C.-H.C.; data curation, C.-W.H.; writing—original draft preparation, Y.-C.Y.; writing—review and editing, Y.-C.Y. and H.-Y.C.; visualization, C.-H.C.; supervision, H.-Y.C.; project administration, H.-Y.C.; funding acquisition, C.-H.C. and H.-Y.C. All authors have read and agreed to the published version of the manuscript.

Funding: This research was funded by Ministry of Science and Technology of Taiwan (MOST 107-2314-B-182A-150-MY3; 108-2221-E-002-169-MY3; 108-2218-E-007-045; 109-2314-B-002-041-MY3; 109-2634-F-002-042). This research was also funded by Chang Gung Medical Foundation in Taiwan (CMRP 202007-202206). In addition, this work was further supported by the "Advanced Research Center for Green Materials Science and Technology" from The Featured Area Research Center Program within the framework of the Higher Education Sprout Project by the Ministry of Education (109L9006).

Institutional Review Board Statement: Not applicable.

Informed Consent Statement: Not applicable.

Data Availability Statement: Data available on request due to privacy.

Conflicts of Interest: The authors declare no conflict of interest.

References

1. Abatenh, E.; Gizaw, B.; Tsegaye, Z.; Wassie, M. The role of microorganisms in bioremediation. *Open J. Environ. Biol.* **2017**, *1*, 38–46. [CrossRef]
2. Giovanella, P.; Vieira, G.A.L.; Otero, I.V.R.; Pellizzer, E.P.; Fontes, B.J.; Sette, L.D. Metal and organic pollutants bioremediation by extremophile microorganisms. *J. Hazard. Mater.* **2020**, *382*, 121024. [CrossRef] [PubMed]
3. Ojuederie, O.B.; Babalola, O.O. Microbial and plant-assisted bioremediation of heavy metal polluted environments. *Int. J. Environ. Res. Pub. Health* **2017**, *14*, 1504. [CrossRef] [PubMed]
4. Lovley, D.R. Cleaning up with genomics: Applying molecular biology to bioremediation. *Nat. Rev.* **2003**, *1*, 35–44. [CrossRef] [PubMed]
5. Viesser, J.A.; Sugai-Guerios, M.H.; Malucelli, L.C.; Pincerati, M.R.; Karp, S.G.; Maranho, L.T. Petroleum-tolerant rhizospheric bacteria: Isolation, characterization and bioremediation potential. *Sci. Rep.* **2020**, *10*, 2060. [CrossRef]
6. Singh, J.S.; Abhilash, P.C.; Singh, H.B.; Singh, R.P.; Singh, D.P. Genetically engineered bacteria: an emerging tool for environmental remediation and future research perspectives. *Gene* **2011**, *480*, 1–9. [CrossRef]
7. Pieper, D.H.; Reineke, W. Engineering bacteria for bioremediation. *Curr. Option Biotechnol.* **2000**, *11*, 262–270. [CrossRef]
8. Chen, W.; Brühlmann, F.; Richins, R.D.; Mulchandani, A. Engineering of improved microbes and enzymes for bioremediation. *Curr. Option Biotechnol.* **1999**, *10*, 137–141. [CrossRef]
9. Martín, M.J.; Lara-Villoslada, F.; Ruiz, M.A.; Morales, M.E. Microencapsulation of bacteria: A review of different technologies and their impact on the probiotic effects. *Innov. Food Sci. Emerg. Technol.* **2015**, *27*, 15–25. [CrossRef]
10. Huq, T.; Khan, A.; Khan, R.A.; Riedl, B.; Lacroix, M. Encapsulation of probiotic bacteria in biopolymeric system. *Crit. Rev. Food Sci. Nutr.* **2013**, *53*, 909–916. [CrossRef]
11. Keskin, N.O.S.; Celebioglu, A.; Sarioglu, O.F.; Uyar, T.; Tekinay, T. Encapsulation of living bacteria in electrospun cyclodextrin ultrathin fibers for bioremediation of heavy metals and reactive dye from wastewater. *Coll. Surf. B Biointerfaces* **2018**, *161*, 169–176. [CrossRef] [PubMed]
12. Ramírez-García, R.; Gohil, N.; Singh, V. *Phytomanagement of Polluted Sites*; Elsevier Inc.: Amsterdam, The Netherlands, 2019; pp. 517–568.
13. Rathore, S.; Desai, P.M.; Liew, C.V.; Chan, L.W.; Heng, P.W.S. Microencapsulation of microbial cells. *J. Food Eng.* **2013**, *116*, 369–381. [CrossRef]
14. Tsai, Y.T.; Huang, C.W.; Liu, H.Y.; Huang, M.C.; Sun, T.P.; Chen, W.C.; Wu, C.Y.; Ding, S.T.; Chen, H.Y. Enhanced bone morphogenic property of parylene-C. *Biomater. Sci.* **2016**, *4*, 1754–1760. [CrossRef] [PubMed]

15. Chiu, Y.-R.; Hsu, Y.-T.; Wu, C.-Y.; Lin, T.-H.; Yang, Y.-Z.; Chen, H.-Y. Fabrication of asymmetrical and gradient hierarchy structures of poly-p-xylylenes on multiscale regimes based on a vapor-phase sublimation and deposition process. *Chem. Mater.* **2020**, *32*, 1120–1130. [CrossRef]
16. Tung, H.-Y.; Sun, T.-P.; Sun, H.-Y.; Guan, Z.-Y.; Hu, S.-K.; Chao, L.; Chen, H.-Y. Construction and control of 3D porous structure based on vapor deposition on sublimation solids. *Appl. Mater. Today* **2017**, *7*, 77–81. [CrossRef]
17. Tung, H.-Y.; Guan, Z.-Y.; Liu, T.-Y.; Chen, H.-Y. Vapor sublimation and deposition to build porous particles and composites. *Nat. Commun.* **2018**, *9*, 2564. [CrossRef]
18. Chen, H.-Y.; Lahann, J. Designable biointerfaces using vapor-based reactive polymers. *Langmuir* **2011**, *27*, 34–48. [CrossRef]
19. Chen, H.-Y.; Lin, T.-J.; Tsai, M.-Y.; Su, C.-T.; Yuan, R.-H.; Hsieh, C.-C.; Yang, Y.-J.; Hsu, C.-C.; Hsiao, H.-M.; Hsu, Y.-C. Vapor-based tri-functional coatings. *Chem. Commun.* **2013**, *49*, 4531–4533. [CrossRef]
20. Wu, C.-Y.; Huang, C.-W.; Guan, Z.-Y.; Wu, J.-T.; Yeh, S.-Y.; Su, C.-T.; Chang, C.-H.; Ding, S.-T.; Chen, H.-Y. Vapor-based coatings for antibacterial and osteogenic functionalization and the immunological compatibility. *Mater. Sci. Eng. C* **2016**, *69*, 283–291. [CrossRef]
21. Chang, C.-H.; Yeh, S.-Y.; Lee, B.-H.; Hsu, C.-W.; Chen, Y.-C.; Chen, C.-J.; Lin, T.-J.; Hung-Chih Chen, M.; Huang, C.-T.; Chen, H.-Y. Compatibility balanced antibacterial modification based on vapor-deposited parylene coatings for biomaterials. *J. Mater. Chem. B* **2014**, *2*, 8496–8503. [CrossRef]
22. Chen, S.-T.; Wu, C.-Y.; Chen, H.-Y. Enhanced growth activities of stem cell spheroids based on a durable and chemically defined surface modification coating. *ACS Appl. Mater. Interfaces* **2018**, *10*, 31882–31891. [CrossRef] [PubMed]
23. Deka, D.; Bhargavi, P.; Sharma, A.; Goyal, D.; Jawed, M.; Goyal, A. Enhancement of cellulase activity from a new strain of *Bacillus subtilis* by medium optimization and analysis with various cellulosic substrates. *Enzyme Res.* **2011**, *2011*, 151656. [CrossRef] [PubMed]
24. Kurzbaum, E.; Raizner, Y.; Kuc, M.E.; Kulikov, A.; Hakimi, B.; Kruh, L.I.; Menashe, O. Phenol biodegradation by bacterial cultures encapsulated in 3D microfiltration-membrane capsules. *Environ. Technol.* **2020**, *41*, 2875–2883. [CrossRef] [PubMed]
25. Sun, Z.; Pang, B.; Xi, J.; Hu, H.-Y. Screening and characterization of mixotrophic sulfide oxidizing bacteria for odorous surface water bioremediation. *Biores. Technol.* **2019**, *290*, 121721. [CrossRef] [PubMed]
26. Kalaimurugan, D.; Balamuralikrishnan, B.; Durairaj, K.; Vasudhevan, P.; Shivakumar, M.; Kaul, T.; Chang, S.; Ravindran, B.; Venkatesan, S. Isolation and characterization of heavy-metal-resistant bacteria and their applications in environmental bioremediation. *Int. J. Environ. Sci. Technol.* **2020**, *17*, 1455–1462. [CrossRef]

polymers

MDPI

Article

Clinical Efficacy of Polycaprolactone β-Calcium Triphosphate Composite for Osteoconduction in Rabbit Bone Defect Model

Chiu-Ming Chen [1], Shen-Mao Chen [1], Shiou-Fu Lin [2], Huang-Chien Liang [3] and Chia-Chun Wu [1,*]

[1] Department of Orthopaedics, Tri-Service General Hospital, National Defense Medical Center, Taipei City 11490, Taiwan; ccm20349@gmail.com (C.-M.C.); smchen1113@gmail.com (S.-M.C.)
[2] Department of Pathology, Shuang-Ho Hospital, Taipei Medical University, Taipei City 23561, Taiwan; 18011@s.tmu.edu.tw
[3] Department of Materials Engineering, Ming Chi University of Technology, New Taipei City 24301, Taiwan; 930747@gmail.com
* Correspondence: doc20281@gmail.com

Abstract: The combination of β-tricalcium phosphate (β-TCP) with polycaprolactone (PCL) has been considered a promising strategy for designing scaffolds for bone grafting. This study incorporated PCL with commercially available β-TCP (Osteocera^{TM}) to fabricate an injectable bone substitute and evaluate the effect of PCL on compressive strength and setting time of the hydraulic cement. The mechanical testing was compliant with the ASTM D695 and ASTM C191-13 standards. Results showed that PCL-TCP composite presented a well-defined architecture with uniform pore distribution and a significant increase in compressive strength compared with β-TCP alone. Eighteen rabbits, each with two surgically created bone defects, were treated using the PCL-TCP composites. The composite materials were resorbed and replaced by newly formed bone tissue. Both PCL-TCP and β-TCP demonstrated equivalent clinical effects on osteoconduction property in terms of the percentage of newly formed bone area measured by histomorphometric analysis. PCL-TCP was proven to be as effective as the commercially available β-TCP scaffold (Osteocera^{TM}).

Keywords: β-tricalcium phosphate; biocomposites; osteoconduction; polycaprolactone

Citation: Chen, C.-M.; Chen, S.-M.; Lin, S.-F.; Liang, H.-C.; Wu, C.-C. Clinical Efficacy of Polycaprolactone β-Calcium Triphosphate Composite for Osteoconduction in Rabbit Bone Defect Model. *Polymers* **2021**, *13*, 2552. https://doi.org/10.3390/polym13152552

Academic Editor: Yu-Wei Cheng

Received: 8 June 2021
Accepted: 29 July 2021
Published: 31 July 2021

1. Introduction

Repair of large bone defects remains an unmet clinical need in modern orthopedics. These defects have resulted in poor quality of life for aging populations and have become a growing socioeconomic concern around the world [1]. Bone regeneration occurs with a balance of signaling events in osteoblasts, bone-forming cells and osteoclasts to restore bone structure and function [2]. Natural bone is made of inorganic and organic components such as extracellular matrix (ECM) and calcium phosphate ceramics [3]. To address the continuous resorption of bone because of age or disease, use of biomaterials that are compatible with bone and are inert to the immune system has been developed.

Biomaterials such as polymers, metals and ceramics used in clinical applications of orthopedics and dentistry procedures have been known to contain several disadvantages. Synthetic and biodegradable polymers with tunable properties used as bone scaffolds have been reported to raise the risk of immunogenicity and toxicity [4,5]. Metals such as titanium (Ti), magnesium and stainless steel possess great mechanical strength and excellent fatigue resistance. However, they can have poor integration properties with surrounding tissues and may release metal ions [6,7]. Currently, use of calcium phosphate-based biomaterials is considered as the gold standard [8].

Depending on osteoinduction, an ability to induce osteoblastic differentiation and osteoconduction, different types of calcium phosphate were developed. Based on the Ca/P atomic ratios of 1.5 to 2, calcium phosphates such as hydroxyapatite, tricalcium phosphate and Whitlockite are widely used. Further, a variety of applications of these materials are in use to enhance bioactivity as coatings, to fill and heal bone defects as cements and to control the porosity and biocompatibility as scaffolds [9]. β-phase form of tricalcium phosphate (β-TCP; $Ca_3(PO_4)_2$) has been extensively investigated and widely used clinically because of its similar chemical composition to the apatite naturally present in bone tissue [10]. Although β-TCP demonstrates various favorable characteristics for its clinical use, it is difficult to deliver to the target site and hard to compact adequately. To overcome the difficulty in shaping the material, a combination of different biomaterials has become a major area of research. Several polymers such as poly(L-lactic acid), PCL, poly(lactide-co-glycolide) and poly(3-hydroxybutyrate) have been reported for their potential to improve the handling properties of β-TCP [11–14].

PCL is a semicrystalline linear aliphatic polyester with a high degree of crystallinity and hydrophobicity, and it has been widely studied in tissue engineering due to its high level of biocompatibility and biodegradability [15]. It has already been approved for use, along with a range of medical and drug delivery devices. In this study, we evaluated the use of 25% PCL as an additive to improve the mechanical and osteoconductive properties of the β-TCP in terms of physicochemical analysis; static compression test, in vivo study and bone histomorphometric were performed. The formulated combination of ceramic scaffold PCL–TCP is considered a promising strategy for designing useful scaffolds for bone grafting in clinical practice.

2. Materials and Methods

2.1. Preparation of PCL–TCP Composites

The PCL–TCP composites were prepared from the combination of PCL (inherent viscosity: 1.0–1.3 dL/g, Sigma-Aldrich, St. Louis, MO, USA) and β-TCP (0.25–0.5 mm, Wiltrom Co. Ltd, Hsinchu County, Taiwan) with a biocomposite weight proportion of 1:3. PCL was stirred vigorously in dichloromethane (CH_2Cl_2, Sigma-Aldrich, St. Louis, MO, USA) for 2 h, followed by mixing with TCP for 2 min. Sodium chloride particles (Sigma-Aldrich, St. Louis, MO, USA) were then incorporated into the suspension. The ratio of TCP/PCL/sodium chloride was 3/1/17. Finally, the dispersion was cast into a Ti mold and air-dried for 24 h to allow the solvent to evaporate. Subsequently, the TCP/PCL/sodium chloride powders were immersed in deionized water for 6 h, and the water was changed approximately every 2 h at room temperature in order to leach the salt out. The PCL–TCP powder was dried for 24 h in an oven at 30 °C.

2.2. Physicochemical Analysis

The physicochemical properties of the scaffolds were evaluated based on the porosity, swells, flow rate and pore size analysis. SEM (SEM, JEOL 5565, Tokyo, Japan) was used to examine the microstructure of PCL–TCP powder. Pore size distribution and porosity were determined by analyzing the SEM images using the ImageJ software (NIH, Bethesda, MD, USA). Ten fields were randomly selected for each SEM image. In total, five replicates were conducted.

2.3. Setting Time and Static Compression Test

PCL–TCP was heated to 70 °C in a water bath. The setting time of PCL–TCP was then measured by Vicat Needle (HCH-122, Hsinchu County, Taiwan) on a Material Test Equipment (Instron®ElectroPuls™ E10000, Norwood, MA, USA) at 37 °C. The mean setting time was recorded from three individual batches.

Compression test was performed to examine the compressive strength (MPa) in response to an applied compression load using a universal testing system (Instron®ElectroPuls™ E10000, Norwood, MA, USA). A constant extension of 1.3 ± 0.3 mm/min was applied up to a strain value of 0.1 mm/min. The resulting force displacement response was recorded continuously.

2.4. Surgical Procedures and Implantation

Eighteen New Zealand rabbits (BioLASCO Taiwan Co., Ltd., Taipei, Taiwan) with a weight of at least 2.8 kg were used in this study. The investigation was approved by Institutional Animal Care and Use Committee (IACUC) of MASTER LABORATORY Co., Ltd. (IACUC approval number: MS20161101). All experimental procedures were conducted in accordance with National Institutes of Health Guidelines. The rabbits were divided into three groups according to the implantation time (Table 1). The rabbits were anesthetized by intramuscular injection of a mixture of 0.5 mL/kg Zoletil (Virbac, France) and 0.5 mL/kg Rompun (BAYER KOREA LTD., Seoul, KOREA) and 2% Xylocaine (Recipharm Monts, Monts, France) subcutaneously. Both hind legs of each rabbit were shaved and a 2.5 cm skin incision was made (Figure 1A,B). The skin was then retracted laterally to allow lateral arthrotomy on the stifle joints (Figure 1C). A surgical drill bit and a stopper set were used to create a bony defect of 5 mm in diameter and 10 mm in depth in the femur at the midpoint of the lateral condyle from the lateral fabella to the anterior portion of lateral trochlea (Figure 1D,E). The target sites were then filled with appropriate bone substitute cylindrical filaments, followed (Figure 1F,G) by endodermis and muscle tissue being stitched using absorbable suture and skin stitches with nylon suture (Figure 1H). The PCL-TCP bone substitutes were prepared by a heating device and injected into the irregular bone damaged during the surgery to facilitate the process. All rabbits were monitored before the designated implantation observation time at 4, 12 and 24 weeks when the left and right femora were collected. Tissues were placed in 10% formalin until further analyses.

Table 1. Grouping, treatment details and implantation time of each rabbit.

Group	No.	Treatment		Implantation Time (Week)
		Left	Right	
A	1	PCLTCP	β-TCP	4
	2	β-TCP	PCLTCP	4
	3	PCLTCP	Empty	4
	4	β-TCP	Empty	4
	5	Empty	PCLTCP	4
	6	Empty	β-TCP	4
B	7	PCLTCP	β-TCP	12
	8	β-TCP	PCLTCP	12
	9	PCLTCP	Empty	12
	10	β-TCP	Empty	12
	11	Empty	PCLTCP	12
	12	Empty	β-TCP	12
C	13	PCLTCP	β-TCP	24
	14	β-TCP	PCLTCP	24
	15	PCLTCP	Empty	24
	16	β-TCP	Empty	24
	17	Empty	PCLTCP	24
	18	Empty	β-TCP	24

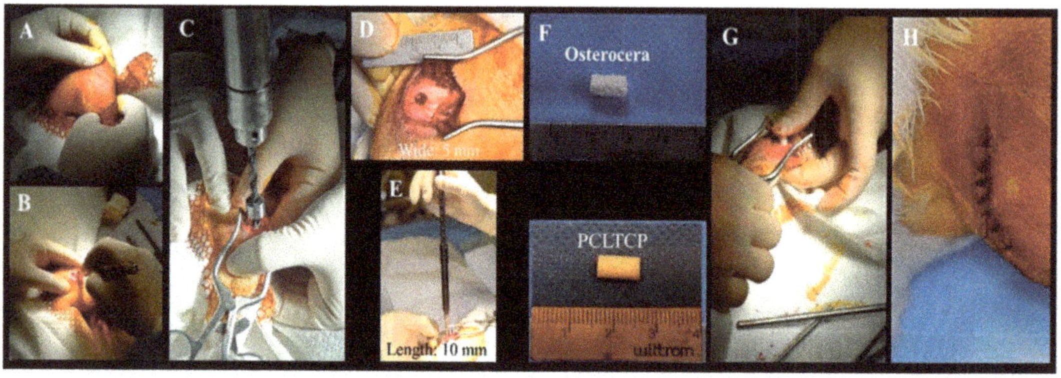

Figure 1. Surgical procedure for creating lateral femur lesion site and implantation. Both hind legs of each rabbit were shaved and a 2.5 cm skin incision was made (**A**,**B**). The skin was then retracted laterally to allow lateral arthrotomy on the stifle joints (**C**). A surgical drill bit and a stopper set were used to create a bony defect of 5 mm in diameter and 10 mm in depth in the femur at the midpoint of the lateral condyle from the lateral fabella to the anterior portion of lateral trochlea (**D**,**E**). The target sites were then filled with appropriate bone substitute cylindrical filaments, followed (**F**,**G**) by endodermis and muscle tissue being stitched using absorbable suture and skin stitches with nylon suture (**H**).

2.5. Radiological and Histomorphometric Analyses

The X-ray of the femoral condyles obtained at the designated implantation time were taken using the Siemens Arcadis Varic C-arm system (SOMA TECH INTL, Bloomfield, CT, USA). The femurs for undecalcified sectioning were dehydrated and embedded in polymethyl methacrylate (PMMA) (Merck, Germany). A section cut was made in the sagittal plane of the target region and surrounding bone with 5–10 mm thickness and 5 mm from the top of the defect at the lateral condyle for each specimen. Sections with a final thickness of 5 μm were obtained and stained with Masson–Goldner trichrome for histomorphometric analyses. The sections were examined under a light microscope (Olympus BX43) to identify the newly formed bone and the implant. Area measurements were made using the "Tile Overlapping Image" function in Media Cybernetics' Image Pro Plus (IPP) program (Image Pro Plus, Media Cybernetics, Inc., Rockville, MD, USA).

2.6. Statistical Analysis

Data are presented as mean ± standard deviation. Statistical differences were determined by one-way analysis of variance with Tukey's HSD post hoc tests to compare the mean changes between difference composite scaffolds. p-values of <0.05 were considered statistically significant.

3. Results

3.1. Morphology and Mechanical Analysis

The physicochemical characterization of the β-TCP (Osteocera^{TM}) and PCL–TCP focused on the pore size, porosity, compressive strength and the material setting time. Results showed that the designed scaffolds presented a well-defined architecture with uniform pore distribution. The pore size of the PCL–TCP scaffolds was 235.28 ± 113.50 μm, which was smaller than the β-TCP (463 ± 88.75 μm, $p < 0.001$), while the porosity percentage was 43.00 ± 15.98% and 83.05 ± 1.52%, respectively ($p < 0.001$) (Table 2). A significant increase in the compression load of PCLTCP was observed at a strain of 0.1 (PCL–TCP: 15.10 ± 0.53 MPa vs β-TCP: 0.85 ± 0.19 MPa, $p < 0.001$). PCLTCP was further tested for a maximum setting time of 85 s, and the final curing time was recorded as 72.6 ± 8.5 s (Table 2 and Figure 2B). Figure 2A showed SEM images (top and cross section views) of PCL–TCP scaffolds with a uniform pore distribution.

Table 2. The physicochemical characterization of the composite scaffolds.

	PCL-TCP	β-TCP
Product name	"Wiltrom" Bitrans Bone Graft Substitute	"Wiltrom" Osteocera Bone Graft Substitute
Composite	β-tricalcium phosphate, β-TCP 75% polycaprolactone, PCL 25%	β-tricalcium phosphate, β-TCP > 95%
Pore size From SEM	235.28 ± 113.50 (μm)	463 ± 88.75 (μm)
Porosity From SEM	43.00% ± 15.98%	83.05% ± 1.53%
Compressive strength	15.104 ± 0.530 (MPa)	0.85 ± 0.19 (MPa)
Setting time	72.6 ± 8.5 (s)	NA

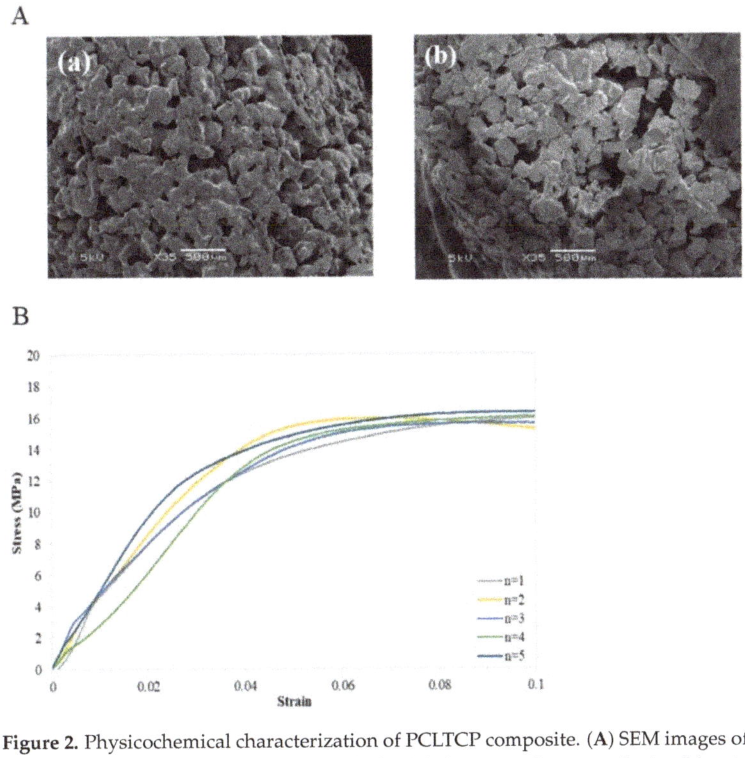

Figure 2. Physicochemical characterization of PCLTCP composite. (**A**) SEM images of: (**a**) top and (**b**) cross section view of PCL-TCP scaffolds. (**B**) Stress–strain curve obtained by static compression test.

3.2. Clinical Observation and Gross Bone Morphology

All animals were carefully monitored throughout the study period with body weights taken daily. The average body weight increases were 249.5 ± 53.8 g, 490.2 ± 74.8 g and 608.3 ± 45.3 g in groups A, B and C, respectively. There were no differences in intragroup body weight, suggesting that different treatments did not affect the growth of the animals. All surgical wounds were closed and healed with no signs of infections at the time of the final evaluation. There were no complications observed throughout the study period. The appearances of the femurs obtained from all eighteen rabbits are shown in Figure 3. All implant sites were successfully healed without any abnormalities.

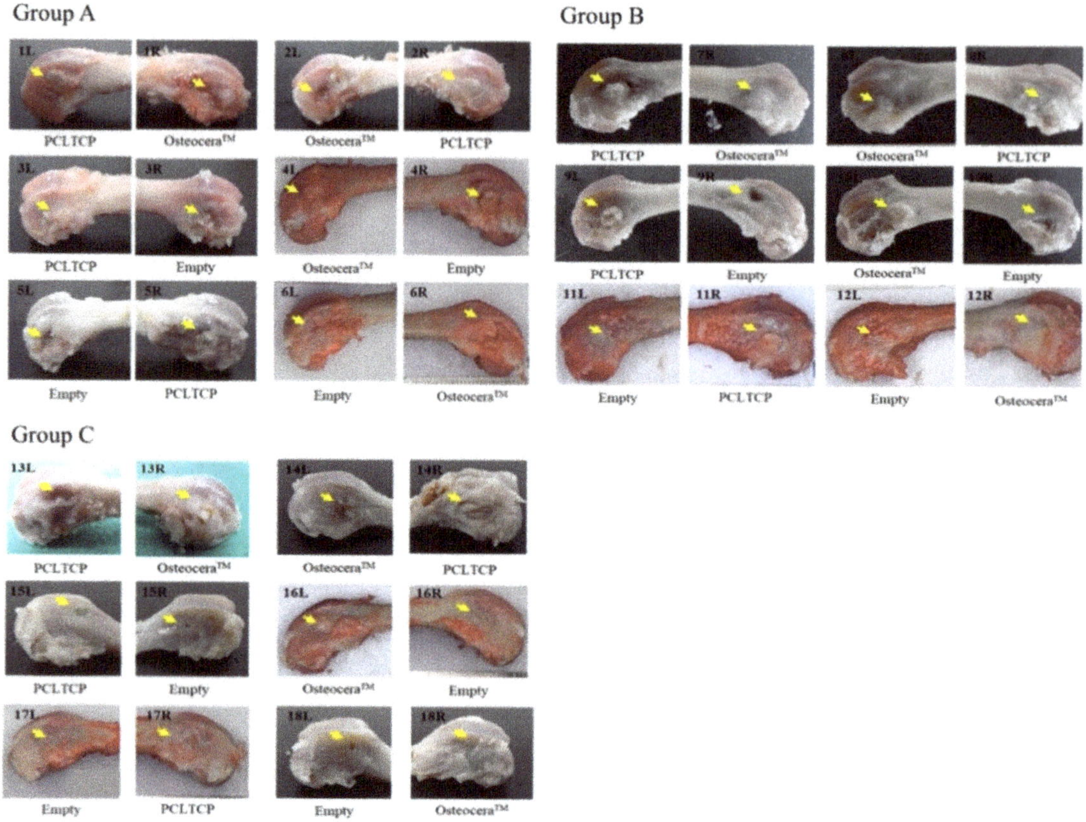

Figure 3. Gross bone morphology at 4-, 12- and 24-weeks post-implantation with PCL–TCP. The femurs with appropriate treatments of all 18 rabbits at designated implantation observation time points are shown. Yellow arrows indicate the implant sites.

3.3. *Radiographic Analyses*

X-ray images were taken to examine bone regeneration at different implantation time points (Figure 4). Cylindrical filaments of PCLTCP and β-TCP were clearly seen at the implant sites at 4 weeks post-implantation, while part of the boundaries between implant and surrounding bones were indistinct in β-TCP (2/4 specimen) but not in PCLTCP filled implant sites at 12 weeks. We observed smeared and less distinguishable boundaries in all the implant sites from PCL–TCP and β-TCP at 24 weeks post-implantation, indicating bone graft degradation of PCL–TCP and β-TCP after 24 weeks. The injury in the control group remained throughout the study period.

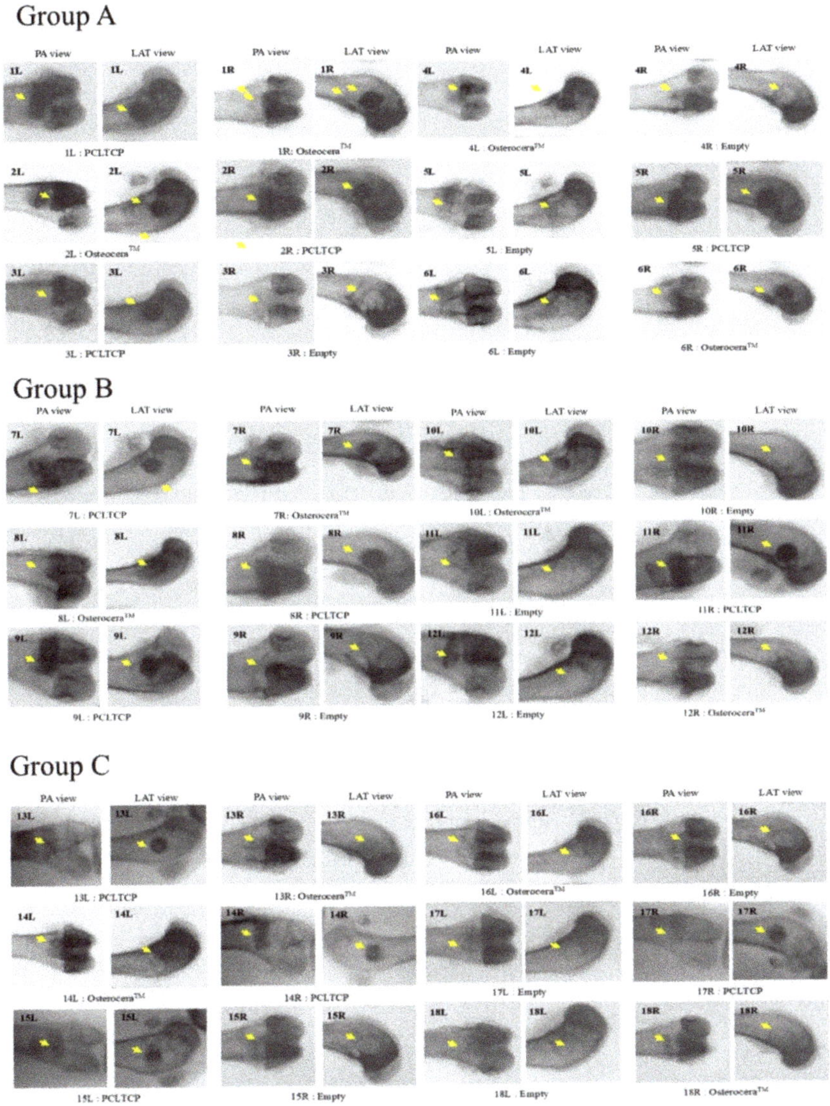

Figure 4. Radiographic images at 4- 12-, and 24-weeks post-implantation with PCLTCP. X-ray was taken from all the femurs with appropriate treatment of eighteen rabbits at designated implantation observation time points.

3.4. Histomorphometric Examinations

The clinical effectiveness of PCLTCP and β-TCP were harvested at each implant site for histomorphometric evaluations (Figure 5 and Table 3). The results showed that PCLTCP possessed the same biological affinity as β-TCP without adverse effects one month after rabbit femur condyle implantation. The newly formed bone area (NFBA) of the PCLTCP was significantly increased at 24 weeks compared to 4 weeks ($p < 0.001$). However, no statistical significance in percentage of NFBA between the PCLTCP and β-TCP was observed at 4 (p = 0.219), 12 (p = 0.139) and 24 weeks (p = 0.331). The boundary of the implants showed no surrounding soft tissue. The growth of newly formed bone was observed inside the materials. After three to six months, bone cavities were filled with new bone tissue and then reconstructed into secondary bone structure with a cancellous-like bone structure. Data suggested that the composite materials were resorbed and replaced by newly formed bone tissue. Both PCL-TCP and β-TCP demonstrated equivalent effects on biological affinity and osteoconduction property at 4-, 12- and 24-weeks post-implantation.

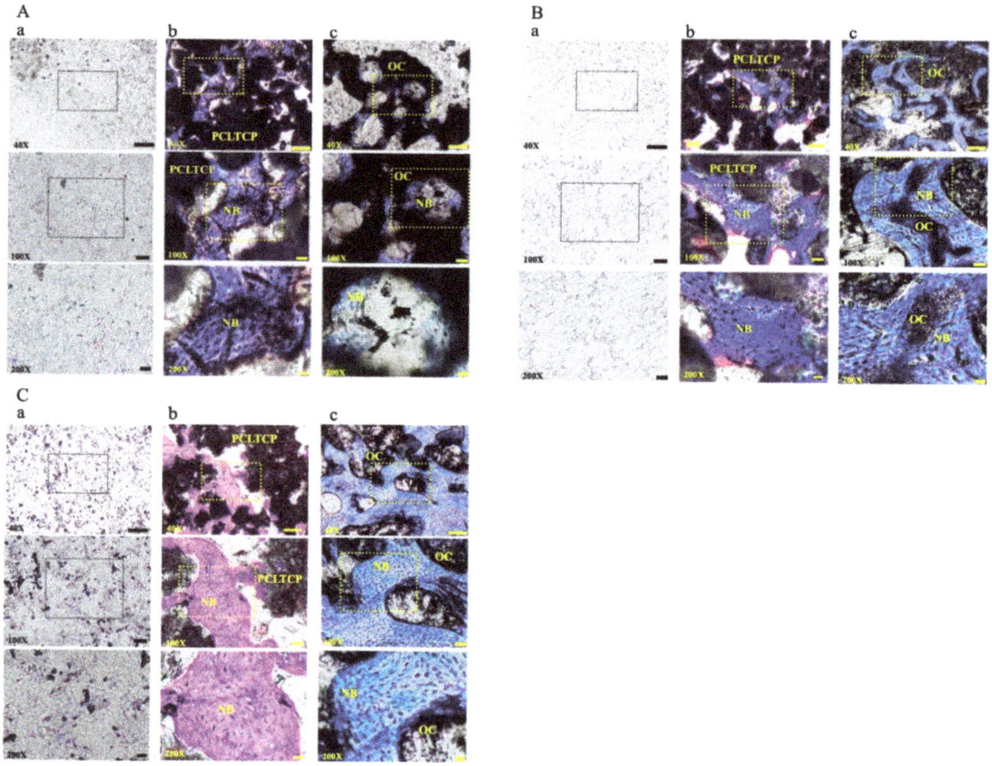

Figure 5. Representative photomicrographs from PCL–TCP implants at (**A**) 4, (**B**) 12 and (**C**) 24 weeks post-implantation. The treatments in each implantation observation time group are labelled as (**a**) empty, (**b**) PCL–TCP and (**c**) β-TCP (OC). NB represents newly formed bone. (Scale bar = 200 μm in 40×, 50 μm in 100×, 20 μm in 200×).

Table 3. Summary of histomorphometric analysis. Values are presented as mean ± standard deviation.

Group/Treatment		Remaining Implant (%)	Void Space Area (%)	Newly Formed Bone Area (%)
A	PCLTCP	50.22 ± 7.61	31.06 ± 7.36	18.72 ± 4.48
	β-TCP	51.73 ± 8.22	26.57 ± 6.92	21.70 ± 7.44
	Empty	NA	98.00 ± 0.92	2.00 ± 0.92
B	PCLTCP	53.72 ± 4.61	18.25 ± 5.66	28.03 ± 4.82 *
	β-TCP	32.94 ± 6.96 *	35.41 ± 10.72	31.32 ± 5.91 *
	Empty	NA	98.05 ± 0.92	1.95 ± 0.92
C	PCLTCP	36.51 ± 6.84 *	28.80 ± 4.04	34.69 ± 6.06 *
	β-TCP	23.02 ± 7.22 *	39.87 ± 11.14	37.11 ± 10.65 *
	Empty	NA	98.10 ± 0.76	1.90 ± 0.76

Asterisks (*) indicate significance ($p < 0.05$) compared with Group A.

4. Discussion

The aim of our study was to fabricate an injectable and mechanically strong bone substitute that can be better applied to irregular bone defects during surgery. We have successfully demonstrated the injectability of PCL–TCP composite material, and characterized the mechanical properties of a PCL–TCP composite and its osteoconduction property following the implantation in rabbit bone injury model. The fabricated PCL–TCP provides higher compressive strength and better shaping characteristics than β-TCP, and enhances new bone growth from residual bone into the cancellous bone-like structure. The newly formed bone that connects both ends of the injury could compensate for the loss of mechanical strength due to degradation of β-TCP scaffolds [16]. The void left by β-TCP dissolution additionally increases the available surface area. Physiologically, β-TCP degrades quickly through both osteoclastic resorption and material dissolution [10,17], hence promoting the bone regeneration process [18]. Although spontaneous bone regeneration has been reported in rabbit bone defect model [19], a superior new bone formation was observed in the groups treated with filling composites compared to the control group.

The mechanical property and setting time of the hardened materials are also important indexes for bone repair in clinical practice, as setting time gives an advantage for the surgical operation with more time. Huang et al. incorporated different concentrations of β-TCP into PCL, with the aim of developing injectable and highly flexural strength materials [20]. They have shown that PCL containing a higher ratio of β-TCP resulted in declining mechanical strength but longer setting time. Ramay et al. combined β-TCP and hydroxyl apatite (HA) to fabricate a biodegradable nanocomposite porous scaffold that contained a compressive strength of 9.8 ± 0.3 MPa and stiffness of 1.72 ± 0.02 kN/m [21]. Recently, a composite of poly (propylene fumarate) (PPF) and β-TCP as a potential bone repair material with a maximum compressive strength of 133 ± 6 MPa and curing temperature of 54.7 ± 1.69 °C was also developed [22]. In yet another study, a composite of hydroxyapatite and β-TCP was mixed with polycaprolactone (PCL) to improve mechanical properties [23]. PCL-based β-TCP has been reported to have high mechanical strength and low degradation rate, which is considered a promising strategy for long-term hard tissue engineering [24].

The quality of bone integration is related to pore size and porosity as a function of structural permeability and biomechanics [25]. β-TCP is more hydrophilic than PCL by acting as a conductor to facilitate water diffusion [26]. The pore size and porosity percentage determine the solvent convective flow and describe the capacity of composites for bone substitution in terms of solutes and nutrient transportation [27]. The reported minimum pore size for a bone substitute is 100 μm [28]; however, the recommended pore size for great osteogenesis is over 300 μm [29,30]. Although PCL–TCP used in our study had a relatively small pore size and porosity, our data demonstrated equivalent effectiveness on osteoconduction property compared to the commercially available β-TCP scaffold (Osteocera™). Results implied the potential of 25% PCL with heterogeneous pore size β-TCP scaffolds for rapid bone ingrowth.

The primary limitation in this study was the sample size of animals and the assessment of the PCLTCP degradation. The remaining implants of PCLTCP were 36% while β-TCP showed 23% left after 24 weeks implantation in rabbit femur. Previous in vivo study demonstrated 33% degradation of PCL-based β-TCP scaffold from 6 to 9 months after implantation in canine mandible [31]. More investigation is necessary for long-term observation of degradation of scaffolds and associated inflammatory response. Even with these limitations, the findings showed favorable and acceptable potential for further clinical implementation on patients using PCL–TCP scaffolds.

5. Conclusions

This study aimed to develop an injectable and mechanically strong bone substitute, and investigate the efficacy of PCL–TCP scaffolds for bone regeneration in rabbit defect model. Results were compared against β-TCP (Osteocera™), a commercially available composite, which is commonly used in the clinical setting. PCL–TCP was proven to be as effective as the commercially available β-TCP scaffold in terms of the percentage of NFBA measured by histomorphometric analysis. The incorporation of 25% PCL with β-TCP leads to a relatively smaller pore size and porosity in scaffolds and higher compression load with good shaping characteristics, which also contributes to the clinical effectiveness of the in vivo model. Future work should further explore the bio-affinity, adaptability and the rate of desired degradation of PCL–TCP composite.

Author Contributions: C.-C.W. and H.-C.L. conceived and designed the experiments. S.-M.C., H.-C.L. performed the experiments. C.-C.W. and S.-F.L. analyzed the data. C.-M.C. and S.-M.C. contributed materials/analysis tools. C.-C.W. and C.-M.C. wrote the paper. C.-C.W., C.-M.C. and S.-F.L. revised the manuscript critically for important intellectual content. All authors have read and agreed to the published version of the manuscript.

Funding: This research received no external funding.

Institutional Review Board Statement: Not applicable.

Informed Consent Statement: Not applicable.

Data Availability Statement: The data presented in this study are available on request from the corresponding author.

Conflicts of Interest: The authors declare no conflict of interest.

References

1. Chapekar, M.S. Tissue engineering: Challenges and opportunities. *J. Biomed. Mater. Res.* **2000**, *53*, 617–620. [CrossRef]
2. Einhorn, T.A. The cell and molecular biology of fracture healing. *Clin. Orthop. Relat. Res.* **1998**. [CrossRef] [PubMed]
3. Bose, S.; Tarafder, S. Calcium phosphate ceramic systems in growth factor and drug delivery for bone tissue engineering: A review. *Acta Biomater.* **2012**, *8*, 1401–1421. [CrossRef] [PubMed]
4. Gunatillake, P.A.; Adhikari, R. Biodegradable synthetic polymers for tissue engineering. *Eur Cell Mater.* **2003**, *5*, 1–16. [CrossRef]
5. Williams, C.G.; Malik, A.N.; Kim, T.K. Variable cytocompatibility of six cell lines with photoinitiators used for polymerizing hydrogels and cell encapsulation. *Biomaterials* **2005**, *26*, 1211–1218. [CrossRef]
6. Glenske, K.; Donkiewicz, P.; Köwitsch, A. Applications of metals for bone regeneration. *Int. J. Mol. Sci.* **2018**, *19*, 826. [CrossRef]
7. Pobloth, A.M.; Checa, S.; Razi, H. Mechanobiologically optimized 3D titanium-mesh scaffolds enhance bone regeneration in critical segmental defects in sheep. *Sci. Transl. Med.* **2018**. [CrossRef]
8. LeGeros, R.Z. Biodegradation and bioresorption of calcium phosphate ceramics. *Clin. Mater.* **1993**, *14*, 65–88. [CrossRef]
9. Jeong, J.; Kim, J.H.; Shim, J.H. Bioactive calcium phosphate materials and applications in bone regeneration. *Biomater. Res.* **2019**, *23*, 1–11. [CrossRef]
10. Yamada, S.; Heymann, D.; Bouler, J.M. Osteoclastic resorption of calcium phosphate ceramics with different hydroxyapatite/β-tricalcium phosphate ratios. *Biomaterials* **1997**, *18*, 1037–1041. [CrossRef]
11. Kung, F.C.; Lin, C.C.; Lai, W.F.T. Osteogenesis of human adipose-derived stem cells on hydroxyapatite-mineralized poly (lactic acid) nanofiber sheets. *Mater. Sci. Eng. C* **2014**, *45*, 578–588. [CrossRef]
12. Shim, J.H.; Yoon, M.C.; Jeong, C.M. Efficacy of rhBMP-2 loaded PCL/PLGA/β-TCP guided bone regeneration membrane fabricated by 3D printing technology for reconstruction of calvaria defects in rabbit. *Biomed. Mater.* **2014**, *9*, 065006. [CrossRef]
13. Hsu, S.; Hsieh, C.T.; Sun, Y.M. Synthesis and characterization of waterborne polyurethane containing poly (3-hydroxybutyrate) as new biodegradable elastomers. *J. Mater. Chem.* **2015**, *3*, 9089–9097. [CrossRef]
14. Zhang, E.; Zhu, C.; Yang, J. Electrospun PDLLA/PLGA composite membranes for potential application in guided tissue regeneration. *Mater. Sci. Eng. C* **2016**, *58*, 278–285. [CrossRef]
15. Cooke, S.L.; Whittington, A.R. Influence of therapeutic radiation on polycaprolactone and polyurethane biomaterials. *Mater. Sci. Eng. C* **2016**, *60*, 78–83. [CrossRef]
16. Lee, S.; Choi, D.; Shim, J.H. Efficacy of three-dimensionally printed polycaprolactone/beta tricalcium phosphate scaffold on mandibular reconstruction. *Sci. Rep.* **2020**, *10*, 1–9. [CrossRef]
17. Horch, H.H.; Sader, R.; Pautke, C. Synthetic, pure-phase beta-tricalcium phosphate ceramic granules (Cerasorb®) for bone regeneration in the reconstructive surgery of the jaws. *Int. J. Oral. Maxill. Surg.* **2006**, *35*, 708–713. [CrossRef]
18. Liu, S.; Jin, F.; Lin, K. The effect of calcium silicate on in vitro physiochemical properties and in vivo osteogenesis, degradability and bioactivity of porous β-tricalcium phosphate bioceramics. *Biomed. Mater.* **2013**, *8*, 025008. [CrossRef]
19. Delgado-Ruiz, R.A.; Calvo-Guirado, J.L.; Romanos, G.E. Critical size defects for bone regeneration experiments in rabbit calvariae: Systematic review and quality evaluation using ARRIVE guidelines. *Clin. Oral Implant. Res.* **2015**, *26*, 915–930. [CrossRef]
20. Huang, S.H.; Hsu, T.T.; Huang, T.H. Fabrication and characterization of polycaprolactone and tricalcium phosphate composites for tissue engineering applications. *J. Dent. Sci.* **2017**, *12*, 33–43. [CrossRef]

21. Ramay, H.R.R.; Zhang, M. Biphasic calcium phosphate nanocomposite porous scaffolds for load-bearing bone tissue engineering. *Biomaterials* **2004**, *25*, 5171–5180. [CrossRef]
22. Ma, C.; Ma, Z.; Yang, F. Poly (propylene fumarate)/β-calcium phosphate composites for enhanced bone repair. *Biomed. Mater.* **2019**, *14*, 045002. [CrossRef]
23. Huang, B.; Caetano, G.; Vyas, C.; Blaker, J.J.; Diver, C.; Bártolo, P. Polymer-ceramic composite scaffolds: The effect of hydroxyapatite and β-tri-calcium phosphate. *Materials* **2018**, *11*, 129. [CrossRef]
24. Chen, C.H.; Lee, M.Y.; Shyu, V.B.; Chen, Y.C.; Chen, C.T.; Chen, J.P. Surface modification of polycaprolactone scaffolds fabricated via selective laser sintering for cartilage tissue engineering. *Mater. Sci. Eng. C Mater. Biol. Appl.* **2014**, *40*, 389–397. [CrossRef] [PubMed]
25. Hannink, G.; Arts, J.J.C. Bioresorbability, porosity and mechanical strength of bone substitutes: What is optimal for bone regeneration? *Injury* **2011**, *42*, S22–S25. [CrossRef] [PubMed]
26. Yeh, C.H.; Chen, Y.W.; Shie, M.Y. Poly (dopamine)-assisted immobilization of Xu Duan on 3D printed poly (lactic acid) scaffolds to up-regulate osteogenic and angiogenic markers of bone marrow stem cells. *Materials* **2015**, *8*, 4299–4315. [CrossRef]
27. Liu, F.; Liu, C.; Zheng, B.; He, J.; Liu, J.; Chen, C.; Lee, I.S.; Wang, X.; Liu, Y. Synergistic Effects on Incorporation of beta-Tricalcium Phosphate and Graphene Oxide Nanoparticles to Silk Fibroin/Soy Protein Isolate Scaffolds for Bone Tissue Engineering. *Polymers* **2020**, *12*, 69. [CrossRef]
28. Fisher, J.; Vehof, J.; Dean, D.; van der Waerden, J.; Holland, T.; Mikos, A. Soft and hard tissue response to photocross-linked poly(propylene fumarate) scaffolds in a rabbit model. *J. Biomed. Mater. Res.* **2002**, *59*, 547–556. [CrossRef] [PubMed]
29. Kuboki, Y.; Jin, Q.; Kikuchi, M.; Mamood, J.; Takita, H. Geometry of artificial ECM: Sizes of pores controlling phenotype expression in BMP-induced osteogenesis and chondrogenesis. *Connect. Tissue Res.* **2002**, *43*, 529–534. [CrossRef] [PubMed]
30. Kujala, S.; Ryhanen, J.; Danilov, A.; Tuukkanen, J. Effect of porosity on the osteointe-gration and bone ingrowth of a weight-bearing nickel-titanium bone graft substitute. *Biomaterials* **2003**, *24*, 4691–4697. [CrossRef]
31. Rai, B.; Ho, K.H.; Lei, Y.; Si-Hoe, K.M.; Teo, C.M.J.; Yacob, K.B.; Chen, F.; Ng, F.C.; Teoh, S.H. Polycaprolactone-20% tricalcium phosphate scaffolds in combination with platelet-rich plasma for the treatment of critical-sized defects of the mandible: A pilot study. *J. Oral Maxillofac. Surg.* **2007**, *65*, 2195–2205. [CrossRef]

polymers

Article

Nanocomposites of Chitosan/Graphene Oxide/Titanium Dioxide Nanoparticles/Blackberry Waste Extract as Potential Bone Substitutes

Carlos Humberto Valencia-Llano [1], Moisés A. Solano [2] and Carlos David Grande-Tovar [2,*]

[1] Grupo Biomateriales Dentales, Escuela de Odontología, Universidad del Valle, Calle 4B # 36-00, Cali 76001, Colombia; carlos.humberto.valencia@correounivalle.edu.co
[2] Grupo de Investigación de Fotoquímica y Fotobiología, Facultad de Ciencias, Programa de Química, Universidad del Atlántico, Carrera 30 Número 8-49, Puerto Colombia 081008, Colombia; madolfosolano@mail.uniatlantico.edu.co
* Correspondence: carlosgrande@mail.uniatlantico.edu.co; Tel.: +57-5-3599-484

Abstract: New technologies based on nanocomposites of biopolymers and nanoparticles inspired by the nature of bone structure have accelerated their application in regenerative medicine, thanks to the introduction of reinforcing properties. Our research incorporated chitosan (CS) covalently crosslinked with glutaraldehyde (GLA) beads with graphene oxide (GO) nanosheets, titanium dioxide nanoparticles (TiO_2), and blackberry processing waste extract (BBE) and evaluated them as partial bone substitutes. Skullbone defects in biomodels filled with the scaffolds showed evidence through light microscopy, scanning electron microscopy, histological studies, soft tissue development with hair recovery, and absence of necrotic areas or aggressive infectious response of the immune system after 90 days of implantation. More interestingly, newly formed bone was evidenced by elemental analysis and Masson trichromacy analysis, which demonstrated a possible osteoinductive effect from the beads using the critical size defect experimental design in the biomodels. The results of this research are auspicious for the development of bone substitutes and evidence that the technologies for tissue regeneration, including chitosan nanocomposites, are beneficial for the adhesion and proliferation of bone cells.

Keywords: bone tissue regeneration; chitosan nanocomposites; critical size defect; graphene oxide; titanium dioxide nanoparticles

Citation: Valencia-Llano, C.H.; Solano, M.A.; Grande-Tovar, C.D. Nanocomposites of Chitosan/Graphene Oxide/Titanium Dioxide Nanoparticles/Blackberry Waste Extract as Potential Bone Substitutes. *Polymers* **2021**, *13*, 3877. https://doi.org/10.3390/polym13223877

Academic Editor: Ting-Yu Liu

Received: 20 October 2021
Accepted: 8 November 2021
Published: 10 November 2021

1. Introduction

Bone connective tissue is crucial to the human body, protecting vital organs, housing the bone marrow and minerals such as calcium and phosphate, supporting the body, and allowing its locomotion [1]. The global collective lawsuit for bone grafting grows daily due to many accidents and increased human osteoporosis, neoplasm, and osteoarthritis. Autografts are usually a common surgical practice for bone defects, removing material from the patient's iliac crest, femur, or tibia [2].

Safety due to the absence of an immune response, histocompatibility, and the stimulation of the growth of new bone cells and their development (osteogenesis), make this technique attractive. On the other side, allografts include using bone from a donor (usually a cadaver) to repair and implant the bone-critical size defect [3].

Both autographs and allografts present osteoconduction, since their three-dimensional structure and bone morphogenetic proteins, together with various growth factors, facilitate the incorporation of vascularity and other osteoprogenitor cells through the differentiation of mesenchymal stem cells, stimulating bone renewal [4]. However, all these benefits are clouded by the report of numerous cases of chronic pain, bleeding, infection, and even patient morbidity [5]. A decrease in availability in conjunction with the previously reported problems of the gold standard techniques for bone grafting has generated a growing demand for safer and more practical solutions with sufficient availability.

On the other hand, the use of biomaterials to mimic the biological, mechanical, and structural properties of connective tissues has grown significantly in the last decade by leaps and bounds [6]. Thanks to their similarity to the bone, many of these materials showed that they could stimulate genes for cell differentiation, extracellular matrix growth, and osteogenesis [3]. The design of three-dimensional and porous materials, organized at the microscopic and even the molecular level, allowed many investigations to demonstrate that biomaterials, including nanoparticles, can stimulate the growth of new bone and other connective tissues [7–9].

Even though promising biomaterials such as polysaccharides and proteins have been used for bone regeneration [10], there is still a great need to develop materials that meet all the requirements to be considered scaffolds for bone regeneration due to their low mechanical resistance. Biocompatible synthetic polymers such as polylactic acid (PLA) [11], poly ε-caprolactone (PCL) [12], polyvinyl alcohol (PVA) [13], among others, have even been used for tissue regeneration, with promising results. However, the selection of synthetic polymers has concerns related to environmental contamination and low-rate biodegradability. For this reason, the use of biopolymers reinforced with nanomaterials and nanofillers has been introduced in the last few years [14].

Chitosan is a derivative obtained from the chemical modification of chitin, a polysaccharide abundant in nature (mycelium of fungi, exoskeleton of crustaceans and insects) as cellulose or starch [15]. Chitosan has biological properties that meet the criteria required to prepare scaffolds that stimulate bone cells and new tissue growth and differentiation [16]. Many investigations reported excellent results in tissue regeneration using chitosan as a material for the construction of scaffolds. However, chitosan reinforcement with nanomaterials and natural extracts such as the blackberry extract (BBE) would guarantee biocompatibility, excellent durability, and better mechanical resistance, which are essential requirements in the regeneration of bone tissue not fulfilled yet. In this sense, incorporating graphene oxide (GO) and titanium dioxide (TiO_2) nanoparticles would provide a greater contact surface and more excellent mechanical resistance without affecting biocompatibility, as we previously demonstrated in subcutaneous tissue studies [17]. Motivated by the observed biocompatibility in the previously prepared chitosan beads [18], we decided to study the possible stimulating effect for bone tissue regeneration of the chitosan beads implanted in critical bone defects of Wistar rats during 90 days of implantation. The objective was to determine by histological and SEM-EDS studies whether the CS/GLA/GO/TiO_2/BBE beads can promote bone neoformation under critical size defect experimental design, which to the best of our knowledge, has not been reported yet with chitosan/graphene oxide/titanium dioxide beads. The present work results demonstrated that CS/GLA/GO/TiO_2/BBE beads are potential bone tissue substitutes and are an open a window for new nanocomposites, including nanoparticles and natural extracts with durability and stability under physiological conditions.

2. Materials and Methods

2.1. Materials

Chitosan (CS) (deacetylation $\geq$ 75%; medium molecular weight, Merck KGaA, Darmstadt, Germany), graphite (325 mesh, Alfa-Aesar, Tewksbury, MA, USA) for GO synthesis following Marcano's procedure [19], Titanium isopropoxide (TTIP; reagent grade, 99%, Aldrich, Palo Alto, CA, USA) for TiO_2 nanoparticles synthesis as reported previously [20]. GLA for crosslinking CS was provided by (Fischer Chemical, Philadelphia, PA, USA). The blackberry residues were obtained from a fruit and vegetable pulper distributor located in the Santa Elena neighborhood of the city of Cali, in the southern region of Valle del Cauca, at coordinates 3°27′00″ N, 76°32′00″ W. The residues consisted of a mixture of seeds, peel, and pulp of the fruit, preserved in a −20 °C freezer (Barnstead/Lab-line, Waltham, MA,

USA) until use. The extraction and characterization procedures were presented in a recent publication [21].

2.2. Synthesis of Chitosan Beads

CS/GLA; CS/GLA/GO; CS/GLA/TiO_2; CS/GLA/GO/TiO_2, and CS/GLA/GO/TiO_2/BBE beads (Table 1) were synthesized and characterized elsewhere [18]. The different formulations used were:

Table 1. Composition of chitosan beads for the critical bone defect study.

Name	System	Formulation
F1	CS/GLA	Chitosan, glutaraldehyde
F2	CS/GLA/GO	Chitosan, glutaraldehyde, graphene oxide
F3	CS/GLA/TiO_2	Chitosan, glutaraldehyde, titanium dioxide nanoparticles
F4	CS/GLA/GO/TiO_2	Chitosan, glutaraldehyde, graphene oxide, titanium dioxide nanoparticles
F5	CS/GLA/GO/TiO_2/BBE	Chitosan, glutaraldehyde, graphene oxide, titanium dioxide nanoparticles, blackberry waste extract

2.3. Critical Bone Defect Studies

For the study, 15 biomodels (male Wistar rats, three months old and 380 g of average weight) were used in an experimental design of critical size bone, with defects of 5 mm × 0.8 mm (diameter × deep) in parietal bones. According to the five formulations of beads used (Table 1), the biomodels were organized into five groups of three animals. Two intraosseous preparations were made for each biomodel; in one, the corresponding bead formulation was implanted, and in the other, it was left as an empty defect control. The preparations were made with a trephine bur with an external diameter of 5 mm at low speed and constant irrigation with physiological saline. Anesthetic medications used were an intramuscular solution of Ketamine 70 mg/kg (Holliday Scott S Laboratory, Buenos Aires, Argentina) and Xylazine 30 mg/kg (Laboratorios ERMA, Celta, Colombia).

After 90 days of implantation, the biomodels were euthanized by an intraperitoneal application of 0.3 mL of Eutanex™ (Euthanex-INVET, Medellín, Colombia) consisting of 390 mg Pentobarbital sodium and 50 mg of diphenylhydantoin sodium/mL of saline solution. Subsequently, the samples were recovered and fixed in buffered formalin for 48 h. They were then decalcified in five days with TBD-2™ Decalcifier, Epredia™ and processed with the Tissue Processor™ (Leica Microsystems, Mannheim, Germany) and Thermo Scientific™ Histoplast Paraffin™ equipment. Finally, for histology studies, the samples were cut to 5 μm with the Leica microtome equipment (Leica Microsystems Mannheim, Germany). The sections were stained with Masson's Trichrome (MT) and Hematoxylin-Eosin (H&E) stains to be analyzed by a Leica microscope with the Leica suite application for imaging (Leica Microsystems, Mannheim, Germany).

The surface morphology of the samples and their respective elemental analysis was assessed using a scanning electron microscope (SEM) JEOL Model JSM 6490 LV coupled to an X-ray Energy Dispersion Spectrometer (EDS) (Akishima, Tokyo, Japan) after metalized coating on its surface with gold using the Denton Vacum Model Desk IV equipment. Finally, the guidelines for Animal Research: Reporting in vivo experiments (ARRIVE) were considered [22]. The protocols applied in this research were based on the ethical principles for animal research adopted by the LABBIO Laboratory of the Universidad del Valle (Cali-, Colombia). The Ethics Committee approved them with Biomedical Experimental Animals (CEAS) from the Universidad del Valle (CEAS 012-019).

3. Results

3.1. Characterization of the Synthesized Chitosan Beads

We previously discussed the preparation and characterization of the CS beads (Table 1) by Fourier Transformed Infrared spectroscopy (FT-IR), X-ray diffraction (XRD), scanning electron microscopy (SEM), and thermogravimetric analysis (TGA) [18]. The characterization of the ethanolic extract used to prepare the F5 presented the main component, cyanidin-3-rutinoside, as shown by ultra-high-performance liquid chromatography (UHPLC, Figure S1) [21]. Briefly, the FT-IR spectrum confirmed the crosslinking reaction of CS with GLA through the band at 1641 cm^{-1}, usually attributed to the imine-like bond from the amino groups of CS and the carbonyl groups of GLA. On the other hand, the XRD analysis for the formulations including TiO_2 showed characteristic angles $2\theta = 25°$, 38°, 48°, 54°, 62°, 68°, 74°, and 82°, of the anatase phase of TiO_2. The CS/GLA/GO/TiO_2 beads presented a small peak at $2\theta = 13°$ attributed to GO intercalation. SEM microstructure studies demonstrated alterations in the surfaces of the CS/GLA beads, increasing the roughness when GO and TiO_2 were added due to the intercalation of the nanomaterials with the polymeric chains.

3.2. Critical Bone Defect Studies

In orthopedic, trauma, and maxillofacial surgery procedures, bone tissue grafts (autografts) or bone substitutes in patients to regenerate or rebuild lost tissue [23]. Most of the bone substitutes currently available commercially come from mainly dead human donors (allograft) or from animals (xenograft), which tend to be rejected by patients because of their origin or because of the belief that they can transmit diseases [24]. In our research, the possibility of manufacturing bone substitutes of a synthetic origin (polyvinyl alcohol) combined with a safe and biocompatible polysaccharide (chitosan), which was also reinforced with graphene oxide and titanium dioxide nanoparticles, was explored to improve durability without sacrificing the biocompatibility of the beads. Blackberry residue extract was introduced to provide anthocyanins that improve biocompatibility and provide antioxidant characteristics. This system has not been reported so far for applications as a bone substitute in critical bone defects. This approach, if successful, will make it possible to end the dependence on grafts of human origin, which depend on the donation made by patients or family members before they die. Likewise, it could be a good alternative to grafts of animal origin.

Under the ISO 10993-6 standard, a macroscopic inspection of the biomodels and samples was carried out in search of alterations related to inflammation or necrosis. Figure 1 corresponds to the recovery of the samples after 90 days of implantation. The figure shows how the hair in the cranial area was completely recovered. After the trichotomy, the skin presented a healthy appearance without scars or inflamed areas. Besides, after tissue separation accessing the intervened area, it was easy to appreciate that all the beads were covered with soft transparent tissues in the bone area without necrosis.

The macroscopic observation of hair recovery and the absence of inflammation or necrotic tissues is the first indication of biocompatibility, because the implanted materials allowed the tissues to heal generally without interfering in this process. The macroscopic observation of the tissues is a recommendation of the iso 10993-6 standard, and the finding of good skin conditions after implantation of compatible materials has been reported by other authors [25].

On the other hand, the microscopic images of the samples recovered with bone fragments are observed in Figure 2. In all cases, the samples were covered by a soft scar tissue of a transparent appearance that seems to integrate with the periosteum. The periosteum is a part of the soft tissue (skin, gum, or oral mucosa) that covers the bone tissue and can repair it when injured. Our investigation's histological results showed that the intraosseous preparation healing is carried out at the expense of a fibrous tissue composed of bundles of type I collagen fibers that seem to continue with the periosteum, which seems to be to indicate typical intraosseous scarring. What is interesting is that type I collagen is one of the main components of the extracellular bone matrix, and the process is occurring in a critical size defect, which would not be expected to regenerate unless the implanted material has the property of activating this process.

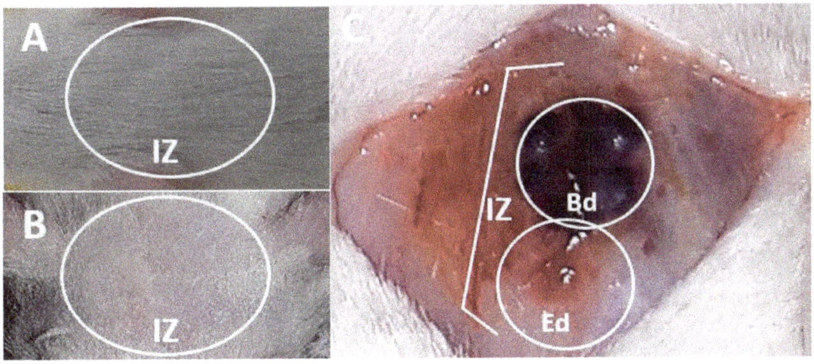

Figure 1. Macroscopic image of the skull intervened area. (**A**): Presence of hair. (**B**): Trichotomy. (**C**): Exposed bone area. IZ: Implantation zone. Bd: Beads. Ed: Empty defect.

However, little reabsorption of the implanted materials is observed without the presence of a severe or necrotic immune response. The minimum reabsorption observed in the implanted beads is due to the material's stability, and this guarantees its presence during the creeping substitution process, which ranges between six and twelve months to have a partial replacement of the material [26], and years for a complete replacement, depending on the type of material used [27]. In the case of the critical size design in parietal bone, it would be expected that the reabsorption of the implanted materials will be difficult due to the characteristic of the parietal bone, which is very cortical and therefore not very vascularized [28].

In this work, the critical size defect experimental design was applied. The critical size of the defect has been defined as an intraosseous preparation that did not heal spontaneously during the experiment [29] and is considered an ideal scenario to evaluate biomaterials that may have utility in bone regeneration [30]. In Figure 3, it is possible to observe how in the control samples (empty defect), the preparation is also covered by a soft tissue with a transparent appearance, without evidence of regeneration of the bone defect. The absence of regeneration of the control defect agrees with what was expected because this type of preparation should not show an appreciable regeneration during the duration of the experiment, which in this case was 90 days [31].

Using the SEM technique (Figure 4), it was identified that the covering soft tissue is made up of fibrous bundles, and in some places, the integration between the tissue that covers the beads and the periosteum that covers the surrounding bone tissue, is appreciated. Other authors have already reported the bundles of collagen fibers that cover the surgical preparations, but especially in histological studies with trichrome staining [32], which are also reported as a fibrous membrane covering the preparations [33].

Histological and histochemical studies also confirmed soft tissue covering the implanted areas and individually surrounding each bead. The first column of images in Figure 5 shows the five types of beads surrounded by soft tissue that appears to integrate with the periosteum of the adjacent bone. Utilizing Masson's Trichromacy staining confirmed that this tissue corresponds to collagen type I and surrounds each bead individually. However, it joins the bone tissue, having continuity with the periosteum, which has also been reported in implanting other compatible materials [34].

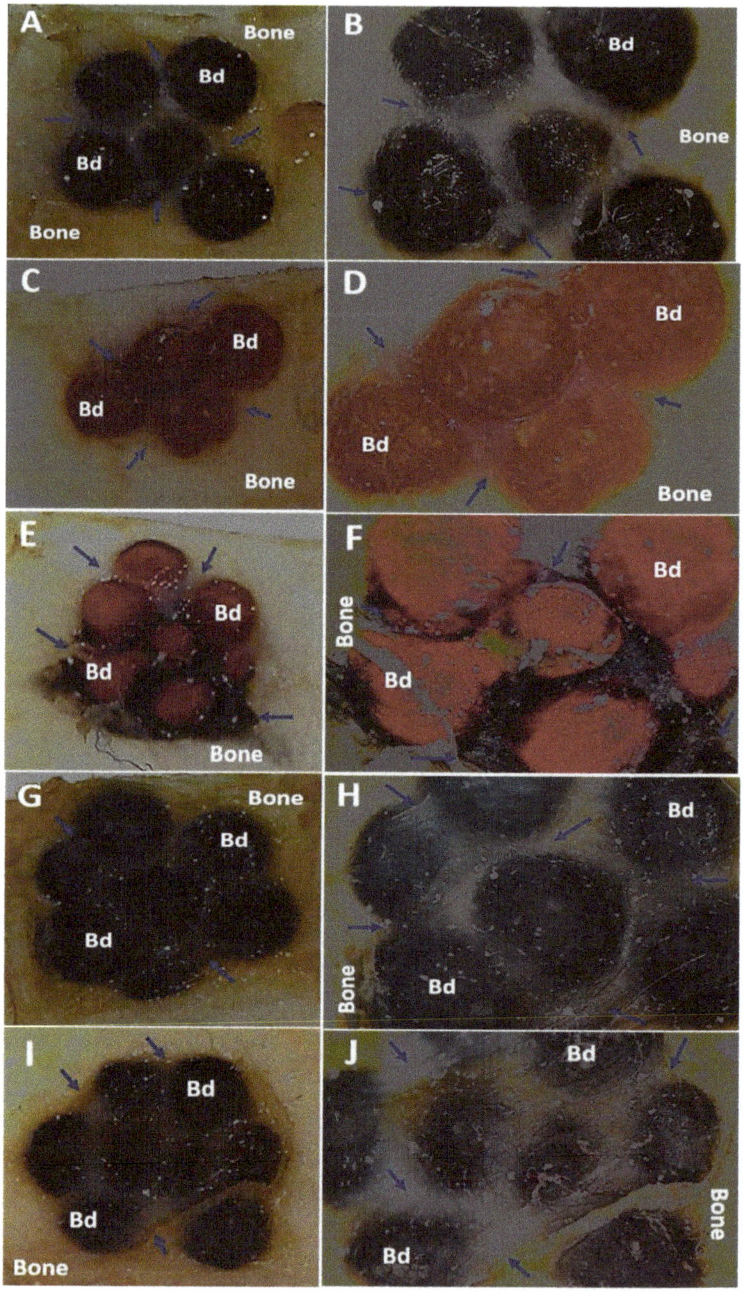

Figure 2. Cranial implantation areas. (**A**): F1 at 2×. (**B**): F1 at 3×. (**C**): F2 at 1×. (**D**): F2 at 3×. (**E**): F3 at 1×. (**F**): F3 at 3×. (**G**): F4 at 1×. (**H**): F4 at 3×. (**I**): F5 at 1×. (**J**): F5 at 3×. Bd: Bead. Blue arrows: soft tissue surrounding intervened areas. Stereoscopic microscope technique.

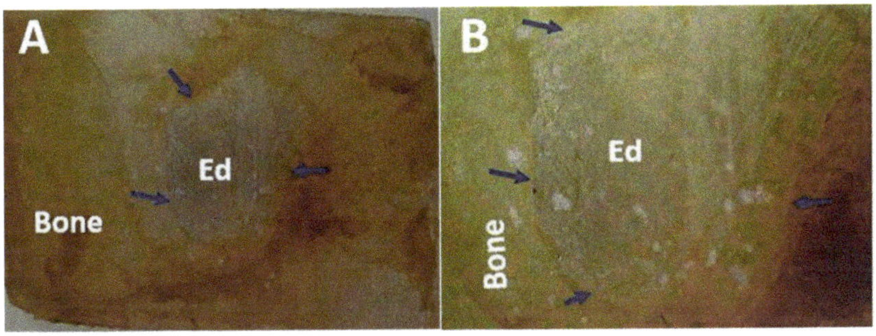

Figure 3. Cranial implantation area of the empty defect. (**A**): Preparation at 1×. (**B**): Preparation at 3×. Blue arrows: soft tissue covering. Stereoscopic microscope technique.

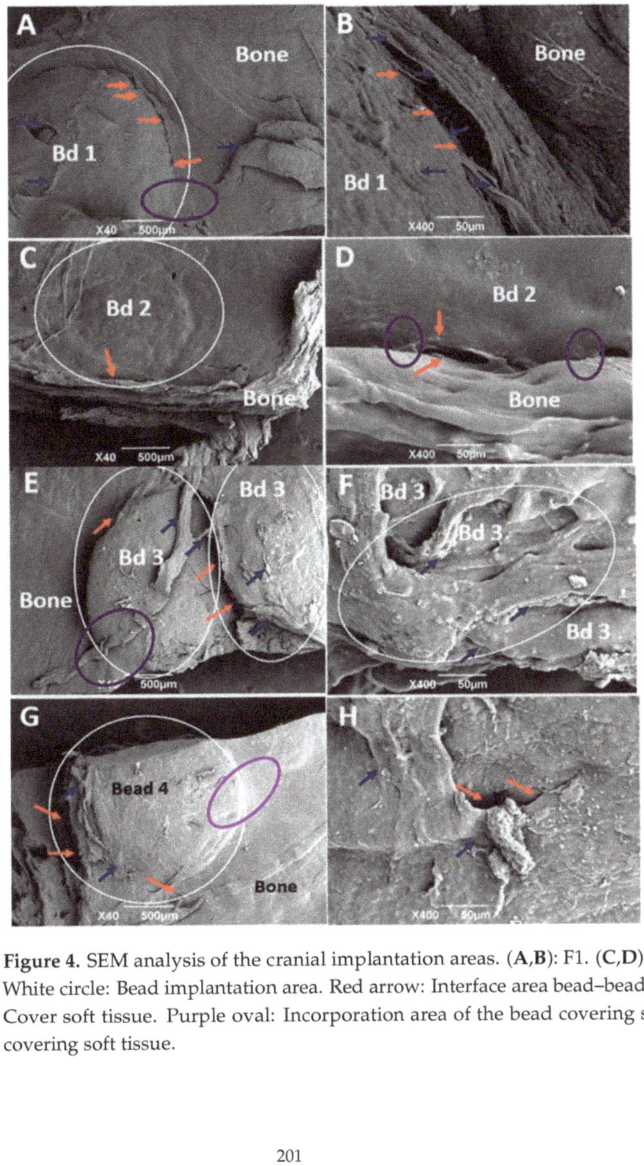

Figure 4. SEM analysis of the cranial implantation areas. (**A**,**B**): F1. (**C**,**D**): F2. (**E**,**F**): F3. (**G**,**H**): F4.. White circle: Bead implantation area. Red arrow: Interface area bead–bead, bead–bone. Blue arrow: Cover soft tissue. Purple oval: Incorporation area of the bead covering soft tissue with the bone covering soft tissue.

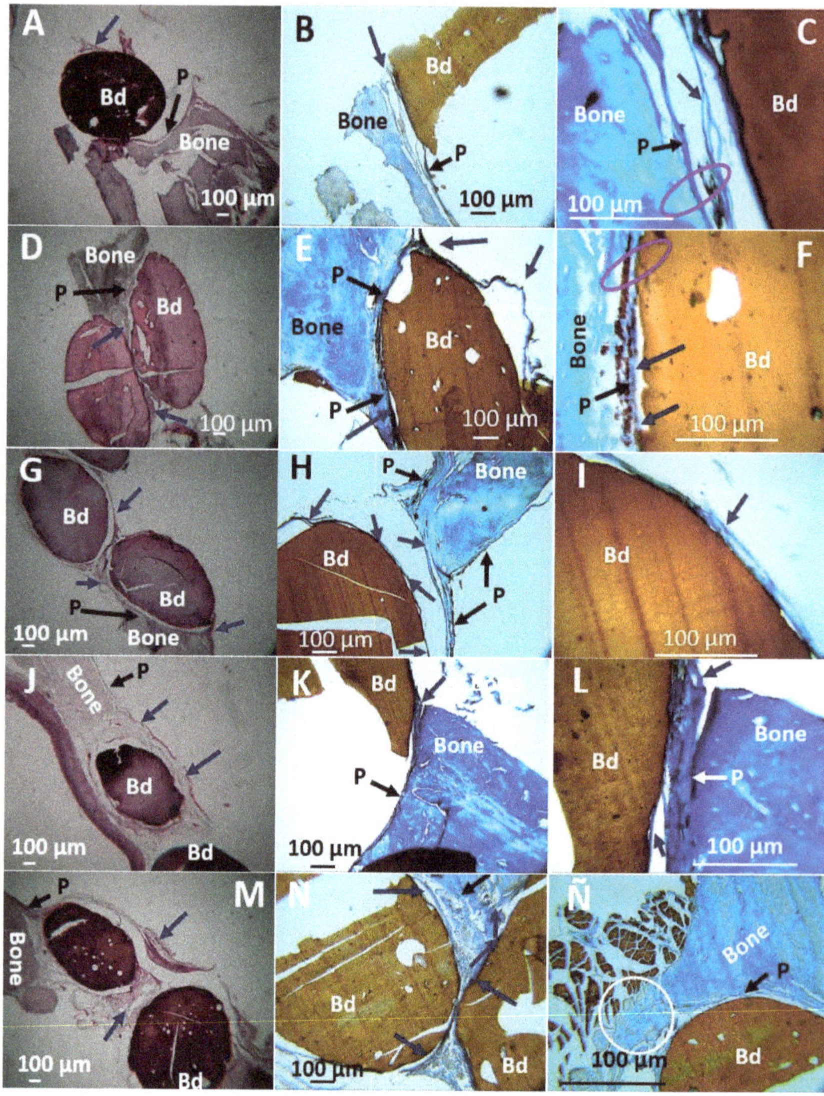

Figure 5. Samples implanted in the cranial bone defect. (**A–C**): F1. (**D–F**): F2. (**G–I**): F3. (**J–L**): F4. (**M–Ñ**): F5. (**A**,**D**,**G**,**J**,**K**) at 4 × H-E technique. (**B**,**E**,**H**,**K**,**N**) at 10 × MT technique. (**C**,**F**,**I**,**L**,**Ñ**) at 40 × MT technique. Bd: Bead. Black arrow P: Periosteum. Blue arrow: Soft tissue that covers the beads. Purple Oval: Areas where there is continuity of bone-bead soft tissue. White Circle: area of new bone formation.

In a critical size defect, it would not be expected to find bone tissue formation unless the material that is implanted has an osteogenic potential as has been reported in other studies in which materials such as calcium phosphates have been implanted [35], or as we reported previously with graphene oxide [36]. In this investigation, the presence of newly formed bone tissue was found in the interface area between the grains and the surrounding bone tissue, corresponding to the area of F5 implantation (Figure 5Ñ), where it can be seen that these structures have a slightly different appearance compared with the tissue surrounding the bone, indicating a lower level of maturity (Figure 6).

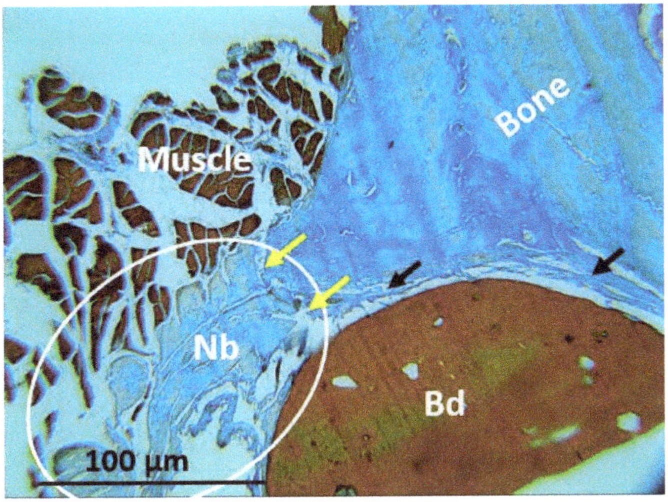

Figure 6. Implantation area for F5. White circle: Area of newly formed bone. NB: Newly formed bone. Yellow arrow: Interface between the newly formed bone and the surrounding bone. Bd: Bead.

Figure 6 corresponds to the image of F5, where newly formed bone tissue was evidenced. The image clearly shows that the newly formed bone tissue is not fully mature and occupies the interface area, filling the space between the bead and the muscle and continuing with the peripheral bone tissue. This finding contrasts with the results obtained with the other formulations with which bone tissue formation was not evidenced in the areas of the preparations. The presence of bone tissue in the implantation area of this specific material would indicate that the implanted material has stimulated the regeneration of the critical size defect, since it is accepted that this type of preparation does not regenerate spontaneously [37].

Masson's trichromacy technique is accepted in histology as specific staining to identify collagen fibers in connective tissues [38]. Figure 5, for example, shows solid blue staining that is explained because the organic component of the bone extracellular matrix mainly is collagen Type I [39].

The presence of collagen fibers in the bone defect, surrounding the beads and occupying the space between them, is a finding that could be related to mineralization if the relationship between fibrillogenesis and mineralization is taken into account, because the collagen fibers present in the extracellular bone matrix are the scaffold used by osteoblasts as a support for mineralization [40].

Size defects are an experimental design widely used to test materials that show potential bone regeneration applications. This design is characterized by the preparation of intraosseous defects that, due to their dimensions, do not allow spontaneous regeneration through new bone formation [31]. In this way, everything observable in the experiment can be attributed to the properties of the material. In this work, the empty defect did not show evidence of regeneration, remaining as a preparation covered by tissue similar to the periosteum.

The periosteum is the tissue that lines the bone surfaces and is considered to have great osteogenic potential. Here, the periosteum is removed to perform intraosseous preparations; however, it recovers and recoats the surfaces when healing occurs. This can be observed when reviewing histological images of similar works, in which the periosteum covers both the neighboring bone tissue and the intervening areas [41].

Despite the tremendous osteogenic capacity of the periosteum and its role in the regeneration of other types of bone defects [42], the potential of this tissue is not sufficient to stimulate the regeneration of the control preparations (empty defects), which are the main characteristic of this experimental design. Usually, when a material with osteogenic potential is implanted, periosteum cells can contribute to the healing of the defect [43].

In this work, the control preparations remained empty, while, of the experimental defects, only F5 showed new bone formation at the preparation site (Figure 6). The formation of new bone seems to be stimulated by the additional presence of the anthocyanin cyanidin-3-rutenoside that could function as an elicitor to produce growth factors or as a gene activator for cell differentiation. Anthocyanins have previously been shown to benefit human health (anticancer activity, anti-inflammatory activity, neuroprotective activity, prevention of cardiovascular disease, anti-obesity, and anti-diabetic activities) [44]. Anthocyanins stimulate nuclear reprogramming through the increased transcription factor expression, which could also stimulate cell differentiation [45]. For this reason, this study is of great interest, adding value to waste from a pulping company that does not re-use this waste.

SEM-EDS analysis was performed for all the formulations used in the present study. Figure 7 shows the elemental analysis, confirming the presence of calcium in F3 and F5. However, F5 also has phosphorus, which is an essential component in the mineralization process. As previously stated, the mineralization process might be stimulated for the anthocyanin present in the blackberry waste extract.

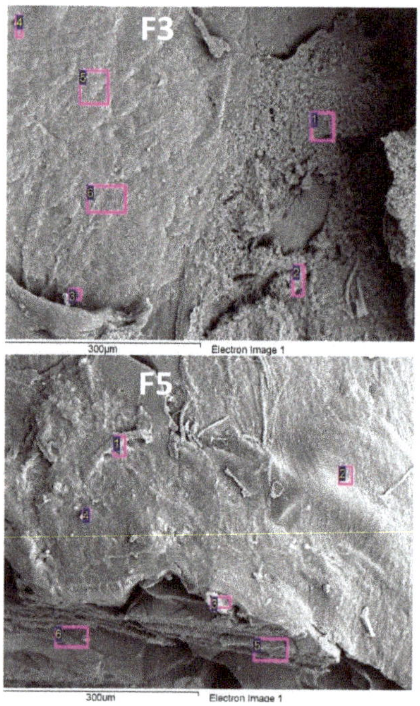

Spectrum	In stats.	C	O	Al	Si	K	Ca	Total
Spectrum 1	Yes	58.01	41.99					100.00
Spectrum 2	Yes	68.26	31.74					100.00
Spectrum 3	Yes	45.50	40.48		3.72	3.15	7.14	100.00
Spectrum 4	Yes	37.81	36.03	2.26	10.81	1.93	11.17	100.00
Spectrum 5	Yes	61.43	38.57					100.00
Spectrum 6	Yes	59.17	40.83					100.00
Max.		68.26	41.99	2.26	10.81	3.15	11.17	
Min.		37.81	31.74	2.26	3.72	1.93	7.14	

Spectrum	In stats.	C	O	Na	P	Cl	K	Ca	Ti	Total
Spectrum 1	Yes	59.41	31.46	1.41	1.06		6.66			100.00
Spectrum 2	Yes	61.16	38.84							100.00
Spectrum 3	Yes	25.92	25.81				0.76	1.12	46.39	100.00
Spectrum 4	Yes	51.86	45.85	0.97		0.42	0.90			100.00
Spectrum 5	Yes	51.24	48.76							100.00
Spectrum 6	Yes	40.08	47.60						12.32	100.00
Max.		61.16	48.76	1.41	1.06	0.42	6.66	1.12	46.39	
Min.		25.92	25.81	0.97	1.06	0.42	0.76	1.12	12.32	

Figure 7. SEM-EDS analysis for F3 and F5 formulations inside after bone implantation studies.

4. Conclusions and Future Perspectives

This research found that the materials behaved as compatible in the five formulations, allowing soft tissue healing with hair recovery and the absence of necrotic areas. At the level of intraosseous implantations, the materials were covered by soft tissue with a transparent appearance that seems to come from the periosteum and histologically is made up of bundles of collagen type I.

Histologically, the presence of newly formed bone tissue could only be identified in the sample corresponding to F5. The presence of calcium (~10%) in F3 can be explained by the beginning of mineralization and extracellular bone matrix formation. Moreover, calcium and phosphorus (~1%) in F5 indicates a maturation process of the bone matrix, stimulated by the anthocyanin from the BBE.

The results obtained with the formulations F3 and F5 concerning the new extracellular material deposited observed with SEM indicate potential use as a bone substitute. However, it is imperative to assess a study with a more significant number of biomodels to quantify the newly bone matrix formed area and the percentage of calcium deposited.

The results of this research are auspicious for the development of bone substitutes and evidence that the technologies for tissue regeneration, including chitosan nanocomposites, are beneficial for the adhesion and proliferation of bone cells.

Supplementary Materials: The following are available online at https://www.mdpi.com/article/10.3390/polym13223877/s1. Figure S1. UHPLC chromatogram of anthocyanin cyanidin-3-rutinoside.

Author Contributions: Conceptualization, C.D.G.-T.; Data curation, C.H.V.L., M.A.S. and C.D.G.-T.; Formal analysis, C.H.V.L. and C.D.G.-T.; Investigation, C.H.V.L., M.A.S. and C.D.G.-T.; Methodology, C.H.V.L., M.A.S. and C.D.G.-T.; Supervision, C.D.G.-T.; Writing—original draft, C.H.V.L. and C.D.G.-T.; Writing—review and editing, C.D.G.-T. All authors have read and agreed to the published version of the manuscript.

Funding: This research and the APC charge were funded by Universidad del Atlántico through the Project 111685270984 of Minciencias.

Institutional Review Board Statement: Finally, the guidelines for Animal Research: Reporting in vivo experiments (ARRIVE) were considered. The protocols applied in this research were based on the ethical principles for animal research adopted by the LABBIO Laboratory of the Universidad del Valle (Cali-, Colombia). The Ethics Committee approved them with Biomedical Experimental Animals (CEAS) from the Universidad del Valle (CEAS 012-019).

Informed Consent Statement: Not applicable.

Data Availability Statement: Data available under request to the corresponding author.

Acknowledgments: The authors thanks to the Ministerio de Ciencia Tecnología e Innovación (Minciencias) for the economic support of this research through the Project 111685270984 of Minciencias.

Conflicts of Interest: The authors declare no conflict of interest.

References

1. Florencio-Silva, R.; da Sasso, G.R.; Sasso-Cerri, E.; Simões, M.J.; Cerri, P.S. Biology of bone tissue: Structure, function, and factors that influence bone cells. *BioMed Res. Int.* **2015**, *2015*, 421746. [CrossRef]
2. Bharadwaz, A.; Jayasuriya, A.C. Recent trends in the application of widely used natural and synthetic polymer nanocomposites in bone tissue regeneration. *Mater. Sci. Eng. C* **2020**, *110*, 110698. [CrossRef] [PubMed]
3. Pina, S.; Oliveira, J.M.; Reis, R.L. Natural-based nanocomposites for bone tissue engineering and regenerative medicine: A review. *Adv. Mater.* **2015**, *27*, 1143–1169. [CrossRef] [PubMed]
4. Alonzo, M.; Alvarez Primo, F.; Anil Kumar, S.; Mudloff, J.A.; Dominguez, E.; Fregoso, G.; Ortiz, N.; Weiss, W.M.; Joddar, B. Bone tissue engineering techniques, advances, and scaffolds for treatment of bone defects. *Curr. Opin. Biomed. Eng.* **2021**, *17*, 100248. [CrossRef] [PubMed]
5. Dinçel, Y.M. Bone Graft Types. In *Bone Grafting-Recent Advances with Special References to Cranio-Maxillofacial Surgery*; IntechOpen: London, UK, 2018; ISBN 1789848830.
6. Silva, M.; Ferreira, F.N.; Alves, N.M.; Paiva, M.C. Biodegradable polymer nanocomposites for ligament/tendon tissue engineering. *J. Nanobiotechnol.* **2020**, *18*, 23. [CrossRef]
7. Yaqoob, A.A.; Ahmad, H.; Parveen, T.; Ahmad, A.; Oves, M.; Ismail, I.M.I.; Qari, H.A.; Umar, K.; Mohamad Ibrahim, M.N. Recent advances in metal decorated nanomaterials and their various biological applications: A review. *Front. Chem.* **2020**, *8*, 341. [CrossRef] [PubMed]
8. Erol-Taygun, M.; Unalan, I.; Idris, M.I.B.; Mano, J.F.; Boccaccini, A.R. Bioactive Glass-Polymer Nanocomposites for Bone Tissue Regeneration Applications: A Review. *Adv. Eng. Mater.* **2019**, *21*, 1900287. [CrossRef]
9. Idumah, C.I. Progress in polymer nanocomposites for bone regeneration and engineering. *Polym. Polym. Compos.* **2021**, *29*, 509–527.
10. Biswal, T. Biopolymers for tissue engineering applications: A review. *Mater. Today Proc.* **2020**, *41*, 397–402. [CrossRef]
11. Hassanajili, S.; Karami-Pour, A.; Oryan, A.; Talaei-Khozani, T. Preparation and characterization of PLA/PCL/HA composite scaffolds using indirect 3D printing for bone tissue engineering. *Mater. Sci. Eng. C* **2019**, *104*, 109960. [CrossRef]

12. Dwivedi, R.; Kumar, S.; Pandey, R.; Mahajan, A.; Nandana, D.; Katti, D.S.; Mehrotra, D. Polycaprolactone as biomaterial for bone scaffolds: Review of literature. *J. Oral Biol. Craniofac. Res.* **2020**, *10*, 381–388. [CrossRef] [PubMed]
13. Kumar, A.; Han, S.S. PVA-based hydrogels for tissue engineering: A review. *Int. J. Polym. Mater. Polym. Biomater.* **2017**, *66*, 159–182. [CrossRef]
14. Hasan, A.; Morshed, M.; Memic, A.; Hassan, S.; Webster, T.J.; Marei, H.E.-S. Nanoparticles in tissue engineering: Applications, challenges and prospects. *Int. J. Nanomed.* **2018**, *13*, 5637. [CrossRef]
15. Yaqoob, A.A.; Ahmad, A.; Ibrahim, M.N.M.; Rashid, M. Chitosan-based nanocomposites for gene delivery: Application and future perspectives. In *Polysaccharide-Based Nanocomposites for Gene Delivery and Tissue Engineering*; Elsevier: Amsterdam, The Netherlands, 2021; pp. 245–262.
16. Islam, M.M.; Shahruzzaman, M.; Biswas, S.; Nurus Sakib, M.; Rashid, T.U. Chitosan based bioactive materials in tissue engineering applications-A review. *Bioact. Mater.* **2020**, *5*, 164–183. [CrossRef] [PubMed]
17. Valencia, C.; Valencia, C.; Zuluaga, F.; Valencia, M.; Mina, J.; Grande-Tovar, C. Synthesis and Application of Scaffolds of Chitosan-Graphene Oxide by the Freeze-Drying Method for Tissue Regeneration. *Molecules* **2018**, *23*, 2651. [CrossRef]
18. Grande Tovar, D.C.; Castro, I.J.; Valencia, H.C.; Zapata, A.P.; Solano, A.M.; Florez López, E.; Chaur, N.M.; Valencia Zapata, E.M.; Mina Hernandez, H.J. Synthesis of Chitosan Beads Incorporating Graphene Oxide/Titanium Dioxide Nanoparticles for In Vivo Studies. *Molecules* **2020**, *25*, 2308. [CrossRef]
19. Marcano, D.C.; Kosynkin, D.V.; Berlin, J.M.; Sinitskii, A.; Sun, Z.; Slesarev, A.; Alemany, L.B.; Lu, W.; Tour, J.M. Improved Synthesis of Graphene Oxide. *ACS Nano* **2010**, *4*, 183–191. [CrossRef]
20. Mahshid, S.; Askari, M.; Ghamsari, M.S. Synthesis of TiO2 nanoparticles by hydrolysis and peptization of titanium isopropoxide solution. *J. Mater. Process. Technol.* **2007**, *189*, 296–300. [CrossRef]
21. Grande-Tovar, C.; Araujo-Pabón, L.; Flórez-López, E.; Aranaga-Arias, C. Determinación de la actividad antioxidante y antimicrobiana de residuos de mora (*Rubus glaucus* Benth). *Inf. Técnico* **2021**, *85*, 64–82. [CrossRef]
22. McGrath, J.C.; Drummond, G.B.; McLachlan, E.M.; Kilkenny, C.; Wainwright, C.L. Guidelines for reporting experiments involving animals: The ARRIVE guidelines. *Br. J. Pharmacol.* **2010**, *160*, 1573–1576. [CrossRef]
23. Deev, R.V.; Drobyshev, A.Y.; Bozo, I.Y.; Isaev, A.A. Ordinary and activated bone grafts: Applied classification and the main features. *Biomed Res. Int.* **2015**, *2015*, 365050. [CrossRef] [PubMed]
24. Bucchi, C.; Del Fabbro, M.; Arias, A.; Fuentes, R.; Mendes, J.M.; Ordonneau, M.; Orti, V.; Manzanares-Céspedes, M.-C. Multicenter study of patients' preferences and concerns regarding the origin of bone grafts utilized in dentistry. *Patient Prefer. Adherence* **2019**, *13*, 179. [CrossRef] [PubMed]
25. Zanchetta, P.; Lagarde, N.; Uguen, A.; Marcorelles, P. Mixture of hyaluronic acid, chondroitin 6 sulphate and dermatan sulphate used to completely regenerate bone in rat critical size defect model. *J. Cranio-Maxillofac. Surg.* **2012**, *40*, 783–787. [CrossRef] [PubMed]
26. Bouaicha, S.; von Rechenberg, B.; Osterhoff, G.; Wanner, G.A.; Simmen, H.-P.; Werner, C.M.L. Histological remodelling of demineralised bone matrix allograft in posterolateral fusion of the spine—An ex vivo study. *BMC Surg.* **2013**, *13*, 58. [CrossRef] [PubMed]
27. Wang, W.; Yeung, K.W.K. Bone grafts and biomaterials substitutes for bone defect repair: A review. *Bioact. Mater.* **2017**, *2*, 224–247. [CrossRef]
28. Dupoirieux, L.; Pourquier, D.; Picot, M.C.; Neves, M. Comparative study of three different membranes for guided bone regeneration of rat cranial defects. *Int. J. Oral Maxillofac. Surg.* **2001**, *30*, 58–62. [CrossRef]
29. Senos, R.; Hankenson, K.D. Calvaria critical-size defects in rats using piezoelectric equipment: A comparison with the classic trephine. *Injury* **2020**, *51*, 1509–1514. [CrossRef]
30. Li, Y.; Chen, S.-K.; Li, L.; Qin, L.; Wang, X.-L.; Lai, Y.-X. Bone defect animal models for testing efficacy of bone substitute biomaterials. *J. Orthop. Transl.* **2015**, *3*, 95–104. [CrossRef]
31. Vajgel, A.; Mardas, N.; Farias, B.C.; Petrie, A.; Cimões, R.; Donos, N. A systematic review on the critical size defect model. *Clin. Oral Implants Res.* **2014**, *25*, 879–893. [CrossRef]
32. Nagata, M.J.H.; Santinoni, C.S.; Pola, N.M.; De Campos, N.; Messora, M.R.; Bomfim, S.R.M.; Ervolino, E.; Fucini, S.E.; Faleiros, P.L.; Garcia, V.G. Bone marrow aspirate combined with low-level laser therapy: A new therapeutic approach to enhance bone healing. *J. Photochem. Photobiol. B Biol.* **2013**, *121*, 6–14. [CrossRef]
33. Farnezi Bassi, A.P.; Bizelli, V.F.; de Mendes Brasil, L.F.; Pereira, J.C.; Al-Sharani, H.M.; Momesso, G.A.C.; Faverani, L.P.; de Almeida Lucas, F. Is the Bacterial Cellulose Membrane Feasible for Osteopromotive Property? *Membranes* **2020**, *10*, 230. [CrossRef]
34. Ghiacci, G.; Graiani, G.; Ravanetti, F.; Lumetti, S.; Manfredi, E.; Galli, C.; Cacchioli, A.; Macaluso, G.M.; Sala, R. "Over-inlay" block graft and differential morphometry: A novel block graft model to study bone regeneration and host-to-graft interfaces in rats. *J. Periodontal Implant Sci.* **2016**, *46*, 220–233. [CrossRef]
35. Luvizuto, E.R.; de Oliveira, J.C.S.; Gomes-Ferreira, P.H.S.; Pereira, C.C.S.; Faverani, L.P.; Antoniali, C.; Okamoto, R. Immunohistochemical response in rats of beta-tricalcium phosphate (TCP) with or without BMP-2 in the production of collagen matrix critical defects. *Acta Histochem.* **2017**, *119*, 302–308. [CrossRef]
36. López Tenorio, D.; Valencia, H.C.; Valencia, C.; Zuluaga, F.; Valencia, E.M.; Mina, H.J.; Grande Tovar, D.C. Evaluation of the Biocompatibility of CS-Graphene Oxide Compounds In Vivo. *Int. J. Mol. Sci.* **2019**, *20*, 1572. [CrossRef] [PubMed]

37. Tan, R.; She, Z.; Wang, M.; Yu, X.; Jin, H.; Feng, Q. Repair of rat calvarial bone defects by controlled release of rhBMP-2 from an injectable bone regeneration composite. *J. Tissue Eng. Regen. Med.* **2012**, *6*, 614–621. [CrossRef]
38. Chen, Y.; Yu, Q.; Xu, C.-B. A convenient method for quantifying collagen fibers in atherosclerotic lesions by ImageJ software. *Int. J. Clin. Exp. Med.* **2017**, *10*, 14904–14910.
39. Sroga, G.E.; Vashishth, D. Effects of bone matrix proteins on fracture and fragility in osteoporosis. *Curr. Osteoporos. Rep.* **2012**, *10*, 141–150. [CrossRef] [PubMed]
40. Vuong, J.; Hellmich, C. Bone fibrillogenesis and mineralization: Quantitative analysis and implications for tissue elasticity. *J. Theor. Biol.* **2011**, *287*, 115–130. [CrossRef] [PubMed]
41. Ezirganlı, Ş.; Kazancıoğlu, H.O.; Mihmanlı, A.; Aydın, M.Ş.; Sharifov, R.; Alkan, A. The effect of local simvastatin application on critical size defects in the diabetic rats. *Clin. Oral Implants Res.* **2014**, *25*, 969–976. [CrossRef] [PubMed]
42. Zhao, L.; Zhao, J.; Tuo, Z.; Ren, G. Repair of long bone defects of large size using a tissue-engineered periosteum in a rabbit model. *J. Mater. Sci. Mater. Med.* **2021**, *32*, 105. [CrossRef]
43. Torquato, L.C.; Suárez, E.A.C.; Bernardo, D.V.; Pinto, I.L.R.; Mantovani, L.O.; Silva, T.I.L.; Jardini, M.A.N.; Santamaria, M.P.; De Marco, A.C. Bone repair assessment of critical size defects in rats treated with mineralized bovine bone (Bio-Oss®) and photobiomodulation therapy: A histomorphometric and immunohistochemical study. *Lasers Med. Sci.* **2021**, *36*, 1515–1525. [CrossRef] [PubMed]
44. Li, D.; Wang, P.; Luo, Y.; Zhao, M.; Chen, F. Health benefits of anthocyanins and molecular mechanisms: Update from recent decade. *Crit. Rev. Food Sci. Nutr.* **2017**, *57*, 1729–1741. [CrossRef] [PubMed]
45. You, J.; Kim, J.; Lim, J.; Lee, E. Anthocyanin stimulates in vitro development of cloned pig embryos by increasing the intracellular glutathione level and inhibiting reactive oxygen species. *Theriogenology* **2010**, *74*, 777–785. [CrossRef] [PubMed]

polymers

MDPI

Article

Self-Assembled Supramolecular Micelles Based on Multiple Hydrogen Bonding Motifs for the Encapsulation and Release of Fullerene

Cheng-Wei Huang [1,*], Ya-Ying Chang [2], Chih-Chia Cheng [3], Meng-Ting Hung [1] and Mohamed Gamal Mohamed [4,5]

1 Department of Chemical and Materials Engineering, National Kaohsiung University of Science and Technology, Kaohsiung 80778, Taiwan
2 Department of Applied Chemistry, National Chiao Tung University, Hsinchu 30010, Taiwan
3 Graduate Institute of Applied Science and Technology, National Taiwan University of Science and Technology, Taipei 10607, Taiwan
4 Department of Materials and Optoelectronic Science, College of Semiconductor and Advanced Technology Research, Center for Functional Polymers and Supramolecular Materials, National Sun Yat-Sen University, Kaohsiung 804, Taiwan
5 Chemistry Department, Faculty of Science, Assiut University, Assiut 71515, Egypt
* Correspondence: cwhuangs@nkust.edu.tw

Abstract: Living creatures involve several defense mechanisms, such as protecting enzymes to protect organs and cells from the invasion of free radicals. Developing antioxidant molecules and delivery systems to working with enzymes is vital. In this study, a supramolecular polymer PNI-U-DPy was used to encapsulate C_{60}, a well-known antioxidant that is hard to dissolve or disperse in the aqueous media. PNI-U-DPy exhibits characteristics similar to PNIPAM but could form micelles even when the environment temperature is lower than its LCST. The U-DPy moieties could utilize their strong complementary hydrogen bonding–interaction to create a physically crosslinked network within PNIPAM micelles, thus adjusting its LCST to a value near the physiological temperature. Morphological studies suggested that C_{60} could be effectively loaded into PNI-U-DPy micelles with a high loading capacity (29.12%), and the resulting complex PNI-C_{60} is stable and remains temperature responsive. A series of measurements under variable temperatures was carried out and showed that a controlled release process proceeded. Furthermore, PNI-C_{60} exhibits hydroxyl radicals scavenging abilities at a low dosage and could even be adjusted by temperature. It can be admitted that the micelle system can be a valuable alternative for radical scavengers and may be delivered to the desired position with good dispersibility and thermo-responsivity. It is beneficial to the search progress of scientists for drug delivery systems for chemotherapeutic treatments and biomedical applications.

Keywords: supramolecular micelle; C_{60}; hydroxyl radical scavenger

Citation: Huang, C.-W.; Chang, Y.-Y.; Cheng, C.-C.; Hung, M.-T.; Mohamed, M.G. Self-Assembled Supramolecular Micelles Based on Multiple Hydrogen Bonding Motifs for the Encapsulation and Release of Fullerene. *Polymers* **2022**, *14*, 4923. https://doi.org/10.3390/polym14224923

Academic Editor: Leonard Ionut Atanase

Received: 20 October 2022
Accepted: 10 November 2022
Published: 15 November 2022

1. Introduction

Medical technology is constantly improving to combat the harm caused by diseases to human beings. Among these techniques, the drug delivery system is critical in treating various conditions. Scientists have worked hard to develop drug carriers that are simple, accessible to process, have high stability, good biocompatibility, and are sensitive to environmental stimuli. As a proven effective drug carrier, micelles have been reported in quite a few works in the literature [1,2]. Generally, micelles are formed using multi-block copolymers or amphiphilic macromolecules as their main structures. These polymers can self-assemble in aqueous media and include various morphologies, such as core–shell type micelles, rods, cylinders, and vesicles [3,4]. Among these microstructures, micelles are very suitable for entrapping hydrophobic compounds such as drugs since their inner hydrophobic core can isolate them from the external medium [5,6]. It was also reported that when environmental factors (such as temperature, pH, etc.) were changed, the micelles changed their structure so that the drug could be released [7]. However, the synthesis procedure of multi-block copolymers is usually more complicated. The polymeric micelles formed via covalent bonds often exhibit inefficient stability, low stimuli response, and poor drug release rates [8,9], limiting their practical usage in clinical treatments.

On the other hand, polymers with supramolecular interactions, as another way to form micelles, have attracted extensive attention in recent years [10]. This kind of material can be assembled into micelles of different sizes, and the stimuli-responsivity can be manipulated by the strength of interactions [11]. The physical crosslinking network offered by supramolecular functionalities may contribute to a more stable system in the aqueous media compared to traditional materials. The continued dedication to developing supramolecular micelles is of considerable value for drug delivery and chemotherapy research.

Free radicals, a critical species in nature, are involved in many functions of the organism to maintain good health. Reactions involving free radicals can also be observed in other fields, such as atmospheric and polymer chemistry. Even though our bodies produce some antioxidants, excessive free radicals may still affect the biosystem. Reactive oxygen species (ROS), such as hydrogen peroxide and hydroxyl radicals, may cause severe and irreversible damage to tissues and organs [12]. Efficient and safe free radical scavengers were persuaded. Fullerenes, such as C_{60} and C_{70}, exhibited high electron affinities and reactive exteriors and are the ideal candidate for quenching ROS [13]. However, the poor water solubility strictly limited the applications of fullerenes in the biomedical field. Chemical modification is one of the common approaches to improving their solubility. Nevertheless, the derivatives and isomers produced after the change may lead to uncertain toxicity. There were some strategies to encapsulate C_{60} to fabricate fullerene-involved composites, such as through carbohydrate polymers [14], silver nanoparticles [15], and metal-organic frameworks (MOFs) [16]. The hybrid materials and encapsulation techniques can be used for drug release, electrode fabrication, and guest molecule adoption [14–16]. Another way is toward supramolecular complexation, but this still suffers from low complex stability and a low drug loading content (DLC)(~1 wt%) [13]. Wang and co-workers presented a thermo-responsive polymer P(NIPAM-co-CD), which could form a supramolecular complex by self-assembling β–cyclodextrin and adamantane. This idea could incorporate C_{60} into an aqueous medium [17]. This complex can scavenge hydroxyl radicals, and the antioxidative property can be controlled by temperature. Eom and co-workers recently prepared fullerene-containing nanoparticles through solvophobic and aromatic interactions, followed by a core crosslinking process, which led to a high DLC (~8.12 wt%) of C_{60}. However, even if covalent cross-linking provides more stability in the core, it still brings some practical limitations. On the other side, physical cross-linking via non-covalent interactions, such as H–bonding or π–π interactions, are spontaneous and can be accomplished under relatively mild conditions [18]. Therefore, establishing a strategy to improve the water solubility and to reach the target of the controlled release of C_{60}, especially through non-covalent interactions, has a significant value in expanding the application of fullerenes.

Polymers with supramolecular interactions have been developed as an ideal strategy to form physically crosslinked micelles. It was proposed that *N*-(6-(3-(2,4-dioxo-3,4-dihydropyrimidin-1(2H)-yl)propanamido)pyridin-2-yl)undec-10-enamide (U-DPy) was a suitable functionality to form supramolecular polymers due to its sextuple complementary hydrogen–bonding interaction. The strong H–bonding interaction can be proved by ^{1}H NMR varian temperature experiment and the ^{1}H NMR titration experiment, which is stronger than other base pairs such as adenine-thymine (two-point hydrogen bonding system) and diaminopyridine-uracil (three-point hydrogen bonding system) [19]. It was also mentioned in another work that the Polyethylene glycol (PEG) oligomer with the U-DPy end group can form telechelic supramolecular polymers through the U-DPy:U-DPy interaction, suggesting that the recognition units possessing an extremely high K_a could lead to new supramolecular polymers in bulk as well as in solution [20]. In the previous study, we proposed two supramolecular poly(N-isopropylacrylamide) (PNIPAAm), PNI-

DAP and PNI-U-DPy, which contained quadruple and sextuple hydrogen bonding groups on the polymer side chain, respectively [9]. The association constant (K_a) of the DAP:DAP complex was 170 M^{-1} [21], and U-DPy:U-DPy was over 10^6 M^{-1} [19]. The significant difference of K_a between the two hydrogen bonding pairs led to totally different properties of two PNIPAAm derivatives, such as the value of a lower critical solution temperature (LCST), particle sizes (PNI-DAP: 38 nm; PNI-U-DPy: 164 nm), kinetic stabilities under surfactants, and micelle morphologies [9]. The strong hydrogen–bonding interaction can also be evidenced by the increase in the polymer's T_g, the supramolecular aggregation in the GPC trace, the decrease in LCST, and the increased stability against long-time SDS treatment. Owing to the strong self-complementary hydrogen–bonding interaction, PNI-U-DPy exhibits excellent micelle stability and drug loading capacities, which are suitable for application as a drug carrier. We have studied PNI-U-DPy as a carrier to load planar compounds such as doxorubicin (DOX). It was proved that the loaded micelle could successfully endocytose cancer cells [9]. However, to our knowledge, there are few examples of using supramolecular polymeric micelles to encapsulate 3D molecules such as fullerene. Herein, we proposed an efficient and rapid route to use PNI-U-DPy to load C_{60}, which is difficult to load and hard to use without chemical modification. The self-assembly of hydrogen bonding moieties and phase transition behaviors of PNI-U-DPy lead to stable micellar structures in aqueous solutions. These micelles show good C_{60} encapsulation capabilities and can effectively release C_{60} upon heating. The loaded micelles (PNI-C_{60}) exhibit hydroxyl radicals and scavenging activities at a low dosage and could even be enhanced by ramping the temperature. Such results suggest the micelle system may have the potential to use as a vehicle for C_{60} radical scavenger and provide the possibility of fullerene dispersion and delivery, which expand its application on biomedical applications.

2. Materials and Methods

2.1. Materials

The supramolecular polymer PNI-U-DPy was synthesized and characterized in the previous literature [9]. C_{60} (purity 99.5%) was purchased from Alfa Aesar (Ward Hill, MA, USA). Organic solvents were purchased from TEDIA (Fairfield, OH, USA) and used as received. DMF was distilled over CaH_2, and THF was refluxed with sodium lumps and distilled to remove water. All other reagents not mentioned were obtained from Sigma–Aldrich (St. Louis, MO, USA), Alfa Aesar (Ward Hill, MA, USA), Showa (Osaka, Japan), or TCI (Tokyo, Japan) without further purifications.

2.2. Characterization

The molecular weight information, such as the number of the average molecular weight (M_n), weight average molecular weight (M_w), and polydispersity (PDI) was investigated by Waters 510 gel permeation chromatography (GPC) system equipped with three Ultrastyragel^{TM} columns (100, 500, and 1000 Å), and Dimethylformamide (DMF) was used as the eluent at a flow rate of 1.0 mL min^{-1} at 50 °C. Polystyrene standards with a narrow PDI were used to calibrate the GPC system. UV-Vis spectra were recorded by HP 8453 spectrometer in the concentration described in the following paragraph. The measurements proceeded with a scanning speed of 200 nm/min and UV-Vis bandwidth of 1.5 nm. The recording wavelength ranged from 190 nm to 1100 nm with a sampling interval of 0.9 nm. Each sample was measured three times to make sure the result was consistent. Dynamic light scattering (DLS) measurements were performed by Brookhaven Instrument Corporation 90-plus at a scattering angle of 90°. C_{60}-loaded micelle samples for DLS measurement were prepared in 1,1,2,2-tetrachloroethane (TCE) and first dialyzed against *N*,*N*-Dimethylacetamide (DMAC), then dialyzed against DI water. Transmission electron microscopy (TEM) images were recorded using an FEI Tecnai T12 microscope operating at 120 kV, and the Image J program was utilized to calculate the number and diameter of micelles from TEM micrographs. Samples for the TEM observation were prepared using micelle dispersant in DI water, which was dropped onto the carbon-coated copper grids and dried under 10 °C for 72 h. The sample of the PNI-U-DPy micelle was stained with RuO4, and the C_{60}-loaded micelle samples were observed without stain. The micelle morphologies were observed with Hitachi S-4200 field-emission scanning electron microscopy (FE-SEM, Hitachi, Tokyo, Japan), and the samples were sputtered with Pt before imaging. Wide-angle-X-ray scattering (WAXS) patterns were performed using the wiggler beamline BL17A1 of the National Synchrotron Radiation Research Center (NSRRC), Taiwan, and the radiation with a wavelength of 1.33 Å was carried out under room temperature. Atomic force microscopy (AFM) images were recorded in tapping mode with Dimension 3100 (Digital Instrument) equipped with silicon cantilevers (PPP-NCH-50, 204–497 kHz, 10–130 N/m) at room temperature. The scan rate was 0.5 Hz with a tip velocity of 5 µm/s. The resonance frequency was 394.8 kHz, and the spring constant was 42 N/m. All images were subjected to a first-order plane-fitting procedure to compensate for the sample tilt.

2.3. Preparation of C_{60}-Loaded Micelles

To prepare C_{60}-loaded micelles, PNI-U-DPy were first dissolved, and C_{60} was suspended in TCE with a concentration of 6 mg mL^{-1}, respectively, then mixed in a 5:1 ratio and stirred for 10 min. The solution was transferred to an MWCO 6–8 kDa dialysis bag and dialyzed against *N,N*-Dimethylacetamide (DMAC) for 48 h, then dialyzed against DI water for another 48 h to remove the organic solvent. The resulting dispersion was centrifuged at 1000 rpm for 10 min to give a transparent supernatant. The supernatant was lyophilized for 24 h, then redispersed and diluted with DI water or a pH 7.4 Phosphate buffered saline (PBS) solution.

2.4. Determination of Fullerene Loading

The lyophilized product was dissolved in TCE with sonication. After 10 min, the absorbance of the mixture was recorded by the UV-Vis spectrometer at 410 nm and compared to the calibration curve for C_{60} in TCE (Figure S2 and Table S1).

The drug loading (DL) and encapsulation efficiency (EE) of the micelles were determined using Equations (1) and (2):

$$DL = \frac{mass\ of\ drug\ encapsulated\ in\ micelles}{mass\ of\ drug - loaded\ micelles} \times 100\% \quad (1)$$

$$EE = \frac{mass\ of\ drug\ encapsulated\ in\ micelles}{the\ initial\ mass\ of\ drug\ before\ dialysis} \times 100\% \quad (2)$$

2.5. Hydroxyl Radical ($^{\bullet}OH$) Scavenging Activity

To determine the interception of hydroxyl radicals by the C_{60}-loaded micelles, we measured the UV absorption of Rhodamine-B (Rh-B) in the Fenton reaction. Based on Equation (3), hydrogen peroxide could dissociate into hydroxide ions (OH^-) and hydroxyl radicals ($^{\bullet}OH$) at low pH in the presence of ferrous ions (Fe^{2+}) [22,23].

$$Fe^{2+} + H_2O_2 \rightarrow Fe^{3+} + OH^- + {}^{\bullet}OH \quad (3)$$

First, we prepared the Rh-B solution with different concentrations, then mixed it with 25 mM $FeSO_4$ solution with the ratio of 2.8:0.2 (*v/v*). The calibration curve was set up by measuring the UV absorption at 554 nm with different concentrations. The radical scavenging experiment was carried out by mixing 0.012 mM Rh-B solution (2.8 mL), PNC_{60}, 25 mM $FeSO_4$ (0.1 mL), and 6.25 mM H_2O_2 (0.1 mL) together. The UV absorption was measured and compared to the calibration curve described above to calculate the scavenging activity. The molar concentration of PNC_{60} is defined by C_{60}. The percentage inhibition was calculated with Equation (4) using the following expression [24]:

$$Inhibition\% = \frac{|A_{control} - A_{sample}|}{A_{control}} \times 100\% \quad (4)$$

where $A_{control}$ = absorbance of the control (containing all reagents except the test compound) and A_{sample} = absorbance of samples (containing all reagents including the test compound). Hydroxyl radical ($^{\bullet}$OH) scavenging activity under different temperatures was measured with a similar procedure. However, the mixture's temperature was kept at a specific value (26, 28, 30, 32, 34, 36, 38, 40, 44 °C) for 5 min before the H_2O_2 solution was added. The scavenging activity was calculated with the method mentioned above.

3. Results and Discussion

3.1. Preparation and Characterization of PNI-U-DPy and Supramolecular Micelles

The preparation of the supramolecular polymer PNI-U-DPy and micelles were briefly depicted in Scheme 1. A precursor polymer with the propargyl group (PNIPAAm-PA) was "clicked" with UPy-N_3, a compound containing the azide group and U-DPy hydrogen bonding motif, to give a PNIPAAm derivative with a complementary sextuple hydrogen bonding group at the side chain [9]. The PDI of PNI-U-DPy is 5.97, and M_n is 18,073. The relatively broad PDI is caused by the multimodal distribution observed at the GPC measurement, which was attributed to the formation of aggregation by the strong hydrogen–bonding interaction and has been described before [9]. The pendant group containing the U-DPy moiety only occupied 5.1% of the whole amount of the repeat unit. This is different from PNIPAM, which can only form an extended coil in water when the temperature is lower than its LCST. PNI-U-DPy was proved to form stable micelles with a low critical micelle concentration (CMC) in the aqueous media due to its strong hydrogen–bonding interaction (Scheme 1c). The LCST of PNI-U-DPy was investigated by observing the turbidity under different temperatures in the DI water and PBS solution, respectively (Figure S1). After calculating the 50% transmittance value from each curve (Figure 1a), the obtained LCST of PNI-U-DPy was 37 °C in DI water and 31 °C in the PBS buffer, which is lower than PNIPAAm's LCST due to the strong supramolecular network contributed by U-DPy moities [9]. During the temperature ramping process, the forming PNI-U-DPy micelles still exhibited a phase transition ability at temperatures higher than LCST. The transition of PNI-U-DPy in PBS was sharper and faster than in DI water, which may explain the different hydration behaviors between water molecules and PNI-U-DPy in the two media [25]. It also suggested that PNI-U-DPy can form a more stable physically crosslinked network in PBS.

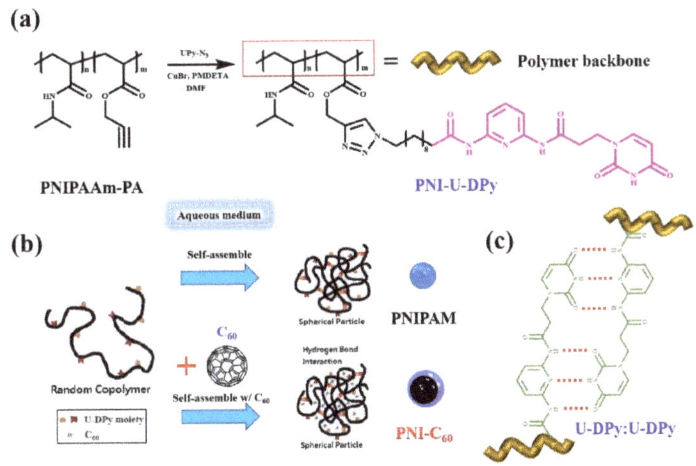

Scheme 1. (**a**) Synthesis of PNI-U-DPy; (**b**) Schematic illustration of supramolecular micelles formed by the strong U-DPy:U-DPy complementary multiple hydrogen bonding–interactions and the formation of C_{60}-loaded micelles; (**c**) Illustration of complementary hydrogen–bonding interaction within PNI-U-DPy.

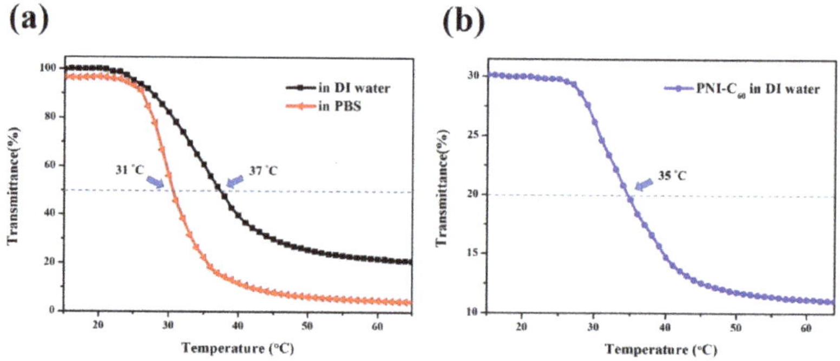

Figure 1. Temperature-dependent transmittance curves at 600 nm of (**a**) PNI-U-DPy in DI water and PBS; (**b**) C_{60}-loaded PNI-U-DPy micelles (PNI-C_{60}). The blue dotted line represented the 50% transmittance value in the UV-Vis spectra.

To investigate the thermal responsivity after drug encapsulation, we used the same method to monitor the release profile of C_{60}-loaded PNI-U-DPy micelles (PNI-C_{60}) (Figure 1b). The lyophilized PNI-C_{60} was redispersed in DI water with 5 mg mL^{-1}. There was no precipitate during the redisperse process, suggesting that the structure of the PNI-U-DPy micelle was strong enough to sustain the C_{60} inside the micelles. As shown in Figure 1b, the change in the transmittance profile of PNI-C_{60} is similar to the PNI-U-DPy obtained in the DI water. However, it should be mentioned that after loading, the transmittance of PNI-C_{60} becomes relatively low (~30%) at 600 nm because C_{60} could absorb the radiation at visible light [26,27]. PNI-C_{60} releases half of C_{60} at 35 °C, near the original LCST of PNI-U-DPy, suggesting its suitability and non-selectivity as a drug loading and release system.

3.2. C_{60} Encapsulation

PNI-U-DPy can form stable micelles due to the dimmerizable side-chain U-DPy functionality. It also exhibits suitable LCST near the physiological temperature, which is especially important for drug delivery systems. Here, we proceeded with the C_{60} encapsulation experiment by mixing the C_{60} solution with PNI-U-DPy solution, then removing the organic solvent and transferring the mixture to DI water through dialysis, as mentioned above. Because C_{60} has an intrinsic hydrophobic nature, which may aggregate into the PNI-U-DPy micelles' hydrophobic core, we successfully encapsulated C_{60} during the PNI-U-DPy self-assembly process in water. The comparison of the size of the PNI-U-DPy micelles before and after C_{60} loading is shown in Figure 2. When C_{60} was loaded into PNI-U-DPy, the resulting micelle composite PNI-C_{60} exhibited a broad single distribution with a PDI of 0.177 ± 0.013, which was significantly different from the result obtained from the PNI-U-DPy micelles. Additionally, the particle size increased from 170 nm to 345 nm after loading, which suggests that C_{60} was successfully encapsulated into the micelles. To calculate the C_{60}-loading efficiency, we first set up the calibration curve by dissolving C_{60} in TCE under different concentrations and measuring the absorbances with a UV-Vis spectrometer (Figure S2). The lyophilized PNI-C_{60} was then dissolved in TCE and measured for UV-Vis absorbance, followed by calculations with Equations (1) and (2). The DLC is 29.12%, and EE is 5.83%, comparable with the proposed PNIPAM micelles [28] loading with the drug in a planar structure such as doxorubicin (DOX). The moderate loading efficiency compared to other carrier systems, such as MOFs, may be due to the nature of the smaller size and surface area of the micelle carrier [29]. However, it should be mentioned there is relatively little in the literature on unmodified C_{60} encapsulating with micelles. It is difficult to find comparison standards from other studies. Thus, the excellent DLC and EE value for micelles may lead to easier control of the loading content actually needed.

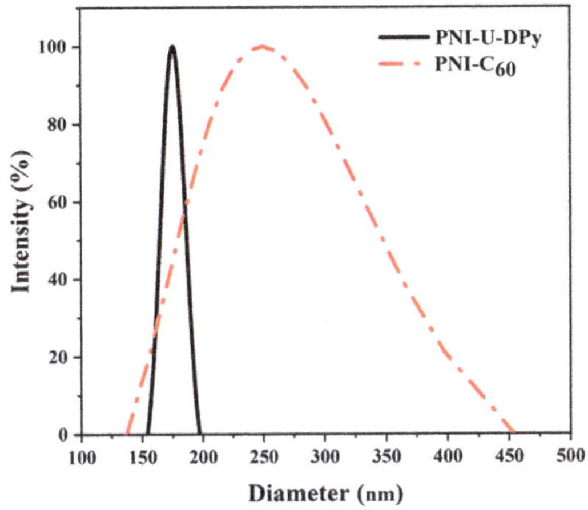

Figure 2. DLS analyses results of PNI-U-DPy micelles and PNI-C_{60} micelles.

To investigate the morphology of PNI-C_{60}, we prepared samples by redispersing the lyophilized PNI-C_{60} in the water and dropping it onto silicon wafers and carbon-coated copper grids for SEM and TEM measurements, respectively. PNI-U-DPy micelles were also used to compare in TEM observation, which was stained with RuO_4. The corresponding images were collected, as shown in Figure 3. The TEM image of C_{60} is shown in Figure S3. PNI-U-DPy micelles exhibit vesicle-like structures with average diameters near the DLS measurement values (Figure 3a,b). In contrast, PNI-C_{60} micelles show different morphologies from the PNI-U-DPy micelles (Figure 3c,d). The interior of the micelles became extremely dark compared with the PNI-U-DPy micelles, which suggested that the high electron density species, C_{60}, were encapsulated. The mean size of the complex calculated by the 76 micelles in the TEM image (Figure 3c) is 342 $\pm$ 7 nm, which was in good agreement with the DLS results. Additionally, we can observe that several PNI-C_{60} micelles may gather closely to form large aggregates, but the structure did not collapse to form huge micelles, indicating that the PNI-U-DPy micelles were stable enough to maintain the network even in the dialysis process due to the strong–hydrogen bonding interaction. SEM images of PNI-C_{60} also show the aggregate formed from tiny micelles (Figure 3e,f). AFM images were recorded with a drop-coated sample, as shown in Figure 4. The height of the complex is 79 $\pm$ 20 nm, which was calculated by the AFM z-scale measure results. The mean size calculated from the AFM image is 212 $\pm$ 125 nm. It presents similar results to SEM, which contains particles with a diameter ranging from 150 nm to 350 nm. The combined results of DLS, TEM, SEM, and AFM prove the formation of the supramolecular micelle complex.

C_{60} was reported to show strong crystallinity even in the solvated state [30–32]. Therefore, we decided to use the wide-angle X-ray technique to check the crystalline behavior of the samples discussed in this study. As shown in Figures 5 and S4, PNIPAM-PA displays a classic amorphous pattern without any crystalline peaks. In contrast, C_{60} reveals a series of crystalline peaks that contributes to its FCC crystalline nature [32]. Interestingly, PNI-C_{60} exhibits a mixed state containing both amorphous and crystalline peaks. The broadband should be offered by the PNI-U-DPy, which is an amorphous polymer, while the intense peak is from C_{60}. The 2D WAXS images also imply crystalline in the interior of the PNI-C_{60} micelles (Figure 5d and Figure S4). When the C_{60} was encapsulated by the micelles, it still maintained a partial crystallinity since there were several peaks in the WAXS pattern. However, it contains not only FCC but other crystalline structures, as shown in Figure 5b. It may be explained by the close aggregation of C_{60} inside the PNI-U-DPy micelles, which was indicated through the TEM observation results. Accordingly, the encapsulation of C_{60}

through the PNI-U-DPy micelles remained crystalline, which may offer a new strategy for fabricating the bottom-up fullerene materials [33].

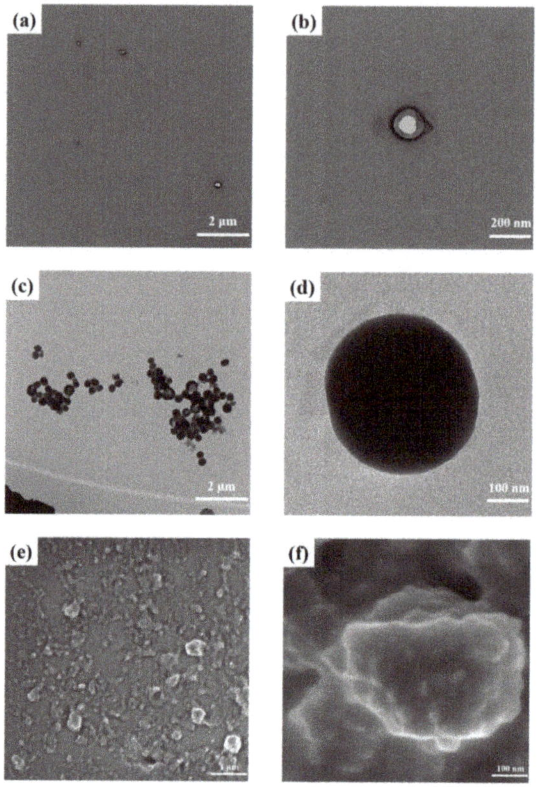

Figure 3. TEM and SEM images of the micelles: (**a**) TEM of PNI-U-DPy micelles; (**b**) TEM of PNI-U-DPy micelles (magnified); (**c**) TEM of PNI-C_{60} micelles; (**d**) TEM of PNI-U-DPy micelles (magnified); (**e**) SEM of PNI-C_{60} micelles; (**f**) SEM of PNI-C_{60} micelles (magnified).

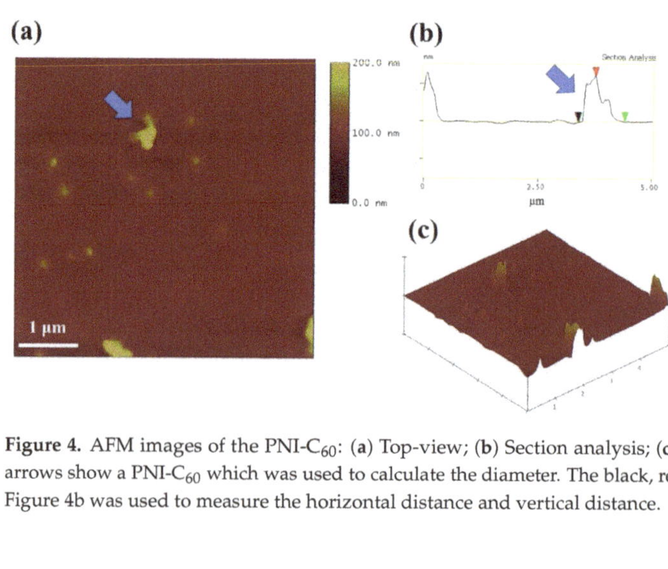

Figure 4. AFM images of the PNI-C_{60}: (**a**) Top-view; (**b**) Section analysis; (**c**) Side-view. The blue arrows show a PNI-C_{60} which was used to calculate the diameter. The black, red, and green marks in Figure 4b was used to measure the horizontal distance and vertical distance.

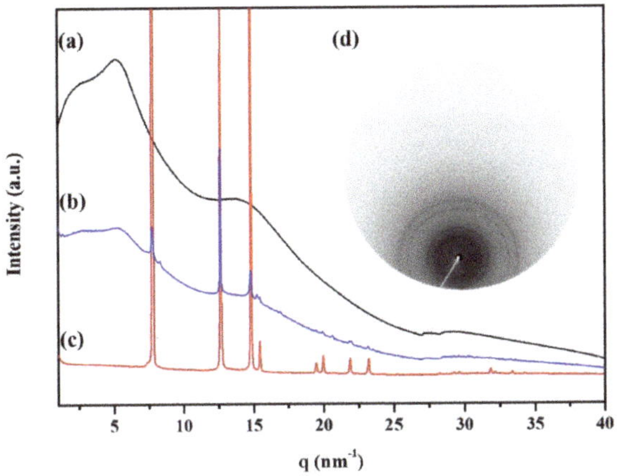

Figure 5. WAXS profiles of (**a**) PNIPAM-PA; (**b**) PNI-C_{60}; (**c**) C_{60} and (**d**) Two-dimensional WAXS image of PNI-C_{60}.

3.3. Drug Release Study of Micelles

A fast and controlled release behavior is quite essential for a drug delivery system. We have confirmed that PNI-U-DPy exhibits LCST near the physiological temperature above. Here, we continued to discuss the release behavior of PNI-C_{60} through morphological analyses. The variable-temperature of the DLS measure result was carried out from 25 °C to 65 °C (Figure 6). As described before, PNI-C_{60} exhibits a diameter of 345 nm at room temperature, with a relatively broad polydispersity (Table 1). The diameter dramatically changed to 268 nm when the temperature increased to 35 °C, near the LCST of PNI-U-DPy. While the temperature became higher, the diameter continued to decrease, though only slightly. The diameter seems to be gradually tending to a value of about 220~230 nm when the temperature is over 45 °C. This indicates that most encapsulated C_{60} may be released between the temperature range of 35 °C to 45 °C. On the other hand, TEM observation of the samples prepared under 25 °C and heated at 45 °C is collected in Figure 7. The sample prepared with an anneal process shows a collapsed structure that did not contain a dark interior as the sample prepared at 25 °C. Furthermore, the size of the micelles decreased significantly, and some shadow surrounded the released micelles. It further proves that the loading and controlled release of PNI-U-DPy micelle system is successfully proceeding.

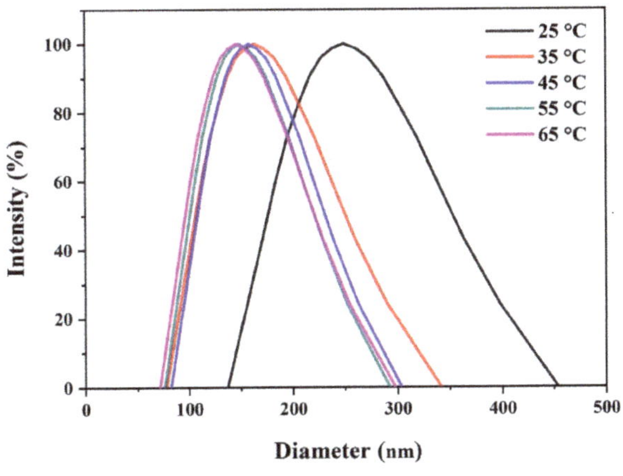

Figure 6. DLS measurement of PNI-C_{60} micelles under different temperatures.

Table 1. DLS measure results of PNI-C_{60} under different temperatures.

Temperature	25 °C	35 °C	45 °C	55 °C	65 °C
Dh (nm)	345 ± 3	268 ± 2	247 ± 4	235 ± 2	234 ± 2
polydispersity	0.277 ± 0.01	0.247 ± 0.01	0.227 ± 0.01	0.198 ± 0.01	0.194 ± 0.01

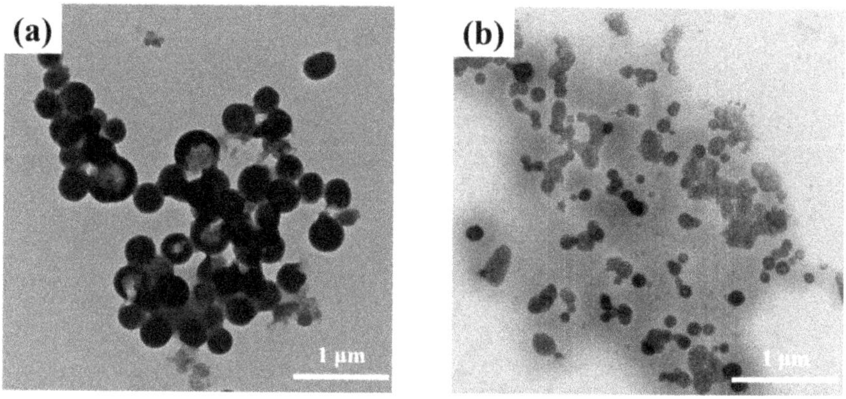

Figure 7. TEM images of (**a**) PNI-C_{60} micelles and (**b**) PNI-C_{60} micelles after heating.

3.4. *Hydroxyl Radical Scavenging Ability*

The hydroxyl radical scavenging activity of PNI-C_{60} was evaluated through the Fenton reaction [34]. According to the literature, because fullerene species could quench the fluorescence of Rh-B, the analyses were recorded through a UV-Vis spectrometer [17]. Rh-B shows a strong absorption at 554 nm in the aqueous medium [35], and the absorption will decrease when the structure of Rh-B has been reacted by the hydroxyl radicals. PNI-C_{60} was used as an antioxidant to protect Rh-B against oxidation. A PBS solution with 0.1 M EDTA was used as a solvent in the experiment. Stock solutions of 0.012 mM Rh-B, 25 mM $FeSO_4$, and 500 mM H_2O_2 were prepared. The Rh-B solution, PNI-C_{60} dispersion, $FeSO_4$ solution, and H_2O_2 solution were added step by step in sequence into a quartz cell and were stirred for 10 s, followed by UV-Vis measurement. The absorption of Rh-B is very strong for control sample 1(without H_2O_2 and PNI-C_{60}) since the hydroxyl radical did not generate (Figure 8a). When the H_2O_2 was added to the cell (control sample 2), the absorption decreased seriously within 10 s, suggesting that when the hydroxyl radicals were produced and oxidized, the Rh-B molecules immediately decreased (Figure 8e). However, if PNI-C_{60} is present in the mixture, the absorption at 554 nm could be higher than the control sample 2, indicating that PNI-C_{60} may inhibit the oxidation of hydroxyl radicals. Furthermore, we also found that with the increase in the PNI-C_{60} addition amount, the inhibition effect became more evident (Figure 8b–d). The hydroxyl radical scavenging efficiency of different PNI-C_{60} was calculated, as shown in Figure 9. Surprisingly, the scavenging efficiency achieved 40% when only 0.1 mM PNI-C_{60} was added to the assay. The addition is much lower than the previous study [17], signifying that PNI-C_{60} could be used as an effective inhibitor against hydroxyl radicals.

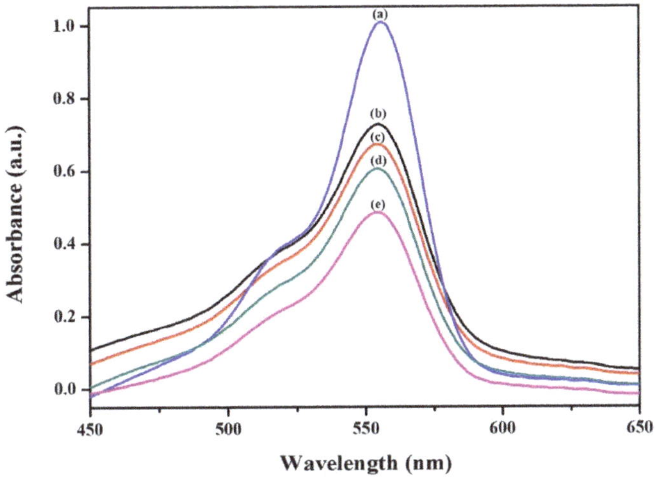

Figure 8. UV-Vis spectra of mixture with different conditions: (**a**) Control sample 1, 0.012 mM Rh-B + 25 mM $FeSO_4$; (**b**) 0.012 mM Rh-B + 0.15 mM PNI-C_{60} + 25 mM $FeSO_4$ + 500 mM H_2O_2; (**c**) 0.012 mM Rh-B + 0.10 mM PNI-C_{60} + 25 mM $FeSO_4$; (**d**) 0.012 mM Rh-B + 0.05 mM PNI-C_{60} + 25 mM $FeSO_4$ + 500 mM H_2O_2; (**e**) Control sample 2, 0.012 mM Rh-B + 25 mM $FeSO_4$ + 500 mM H_2O_2.

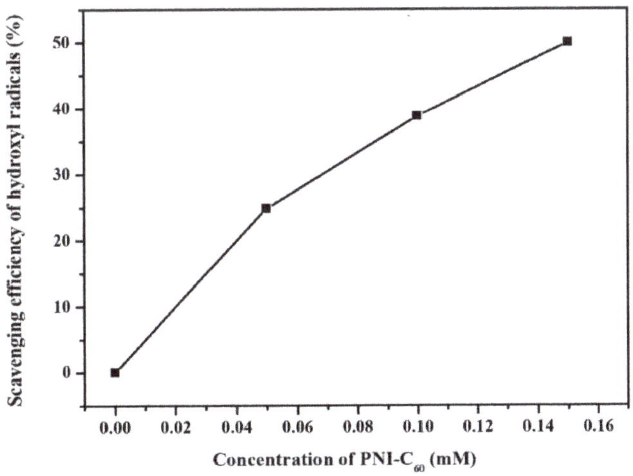

Figure 9. Hydroxyl radical scavenging efficiency of PNI-C_{60} in Fenton reaction at different concentrations.

As described in the above paragraph, PNIPAM is a thermo-responsive material, and its LCST could be adjusted by introducing U-DPy functionalities into the polymer side chain. Additionally, the C60-loaded micelles exhibit thermo-responsivity since the size of the micelles decreases after heating, which was proved by DLS and electronic microscopy observations. We continued to explore how the temperature change affected the scavenging property. The experiment was carried out at a PNI-C_{60} concentration of 0.1 mM, and the temperature was changed from 26 to 44 °C. The result was collected and calculated in Figure 10. While the environment temperature is lower than PNI-U-DPy's LCST, it maintains a scavenging efficiency of around 40%. In contrast, when the temperature exceeds the LCST, the scavenging efficiency starts to increase. This may be explained by the change in morphologies upon heating. As discussed, PNI-C_{60} undergoes the phase transition process, rapidly releasing the encapsulated C_{60} into the mixture and then triggering the fullerene's inhibition mechanism. Therefore, the results mentioned above give a conclusion that PNI-C_{60} has the potential to act as an antioxidant with thermo-responsivity and a controlled release manner.

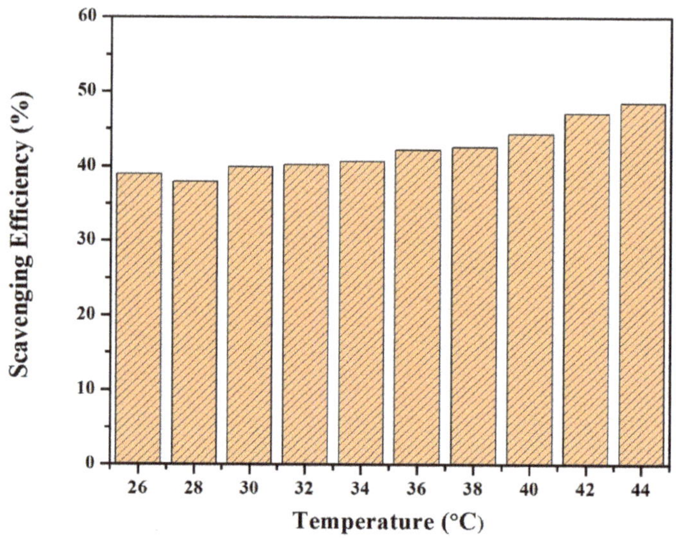

Figure 10. Hydroxyl radical scavenging efficiency of PNI-C_{60} in Fenton reaction from 26 to 44 °C with 0.10 mM PNI-C_{60}.

4. Conclusions

In this study, it was demonstrated that the supramolecular micelles were obtained from a designed polymer PNI-U-DPy. This PNIPAM derivative contains a strong self-complementary hydrogen bonding interaction at the side chain, which constructs a physically crosslinked network and increases the stability of the micelles. The intermolecular interaction leads to an apparent change in LCST and makes it closer to the physiological temperature. It also enables PNI-U-DPy micelles to encapsulate C_{60} with high loading content and entrapment stability, which is seldom reported by unmodified fullerene and polymeric micelle systems. DLS, TEM, SEM, and WAXS observation results suggest that C_{60} has been effectively loaded into a micelle to form a complex hybrid PNI-C_{60}. The encapsulated C_{60} exhibits a crystalline behavior inside the supramolecular micelles. The behavior of PNI-C_{60} was investigated under several variable temperature measurements and gave practical proof for the controlled release progress. Due to the existence of C_{60}, the PNI-C_{60} complex exhibit hydroxyl radicals and scavenging activity at a low dosage and could even be enhanced by ramping the temperature. These findings show that supramolecular polymers could be a potential candidate as a C_{60} carrier without destroying its structure. This strategy provides the possibility for designing an efficient drug delivery system for those drugs that are difficult to encapsulate. It may also contribute to understanding the chemistry and biological behavior of fullerene in aqueous media, therefore offering a potential route for the fabrication of effective and eco-friendly assays for biomedical applications.

Supplementary Materials: The following supporting information can be downloaded at: https://www.mdpi.com/article/10.3390/polym14224923/s1, Table S1. The transmittance of C_{60} in TCE; Figure S1. Temperature-dependent transmittance curves of PNI-U-DPy in (**a**) DI water; (**b**) PBS; Figure S2. Calibration curve set up by C_{60} in TCE; Figure S3. TEM image of C_{60}; Figure S4. Two-dimensional WAXS image of (**a**) PNIPAM-PA; (**b**) C_{60}. [9,19]

Author Contributions: Conceptualization, C.-W.H.; methodology, C.-W.H. and C.-C.C.; formal analysis, C.-W.H., Y.-Y.C., C.-C.C., M.-T.H. and M.G.M.; investigation, C.-W.H. and Y.-Y.C.; data curation, C.-W.H. and Y.-Y.C.; writing—original draft preparation, C.-W.H.; writing—review and editing, C.-W.H., M.-T.H. and M.G.M.; supervision, C.-W.H.; project administration, C.-W.H.; funding acquisition, C.-W.H. All authors have read and agreed to the published version of the manuscript.

Funding: This work was financially supported by the Ministry of Science and Technology, Taiwan, under contracts MOST 102-2815-C-009-001 m and 111-2222-E-992-002-MY3.

Institutional Review Board Statement: Not applicable.

Informed Consent Statement: Not applicable.

Data Availability Statement: Data will be available from the first author on demand.

Acknowledgments: The authors extend their appreciation to Chao-yuan Chu for funding and technical support.

Conflicts of Interest: The authors declare no conflict of interest.

References

1. Nishiyama, N.; Matsumura, Y.; Kataoka, K. Development of polymeric micelles for targeting intractable cancers. *Cancer Sci.* **2016**, *107*, 867–874. [CrossRef]
2. Read, E.S.; Armes, S.P. Recent advances in shell cross-linked micelles. *Chem. Commun.* **2007**, *29*, 3021–3035. [CrossRef] [PubMed]
3. Lv, Y.; Wang, L.; Wu, F.; Gong, S.; Wei, J.; Lin, S. Self-assembly and stimuli-responsive behaviours of side-chain liquid crystalline copolymers: A dissipative particle dynamics simulation approach. *Phys. Chem. Chem. Phys.* **2019**, *21*, 7645–7653. [CrossRef] [PubMed]
4. Zhang, L.F.; Eisenberg, A. Multiple Morphologies of Crew-Cut Aggregates of Polystyrene-B-Poly(Acrylic Acid) Block-Copolymers. *Science* **1995**, *268*, 1728–1731. [CrossRef] [PubMed]
5. Tong, R.; Cheng, J. Anticancer Polymeric Nanomedicines. *Polym. Rev.* **2007**, *47*, 345–381. [CrossRef]
6. Cagel, M.; Tesan, F.C.; Bernabeu, E.; Salgueiro, M.J.; Zubillaga, M.B.; Moretton, M.A.; Chiappetta, D.A. Polymeric mixed micelles as nanomedicines: Achievements and perspectives. *Eur. J. Pharm. Biopharm.* **2017**, *113*, 211–228. [CrossRef] [PubMed]
7. Ganta, S.; Devalapally, H.; Shahiwala, A.; Amiji, M. A review of stimuli-responsive nanocarriers for drug and gene delivery. *J. Control. Release* **2008**, *126*, 187–204. [CrossRef]
8. Sui, M.; Liu, W.; Shen, Y. Nuclear drug delivery for cancer chemotherapy. *J. Control. Release* **2011**, *155*, 227–236. [CrossRef]
9. Cheng, C.C.; Chang, F.C.; Kao, W.Y.; Hwang, S.M.; Liao, L.C.; Chang, Y.J.; Liang, M.C.; Chen, J.K.; Lee, D.J. Highly efficient drug delivery systems based on functional supramolecular polymers: In vitro evaluation. *Acta Biomater.* **2016**, *33*, 194–202. [CrossRef]
10. Abebe Alemayehu, Y.; Tewabe Gebeyehu, B.; Cheng, C.-C. Photosensitive Supramolecular Micelles with Complementary Hydrogen Bonding Motifs To Improve the Efficacy of Cancer Chemotherapy. *Biomacromolecules* **2019**, *20*, 4535–4545. [CrossRef]
11. Cheng, C.-C.; Sun, Y.-T.; Lee, A.-W.; Huang, S.-Y.; Fan, W.-L.; Chiao, Y.-H.; Chiu, C.-W.; Lai, J.-Y. Hydrogen-bonded supramolecular micelle-mediated drug delivery enhances the efficacy and safety of cancer chemotherapy. *Polym. Chem.* **2020**, *11*, 2791–2798. [CrossRef]
12. Hu, J.; Yang, L.; Yang, P.; Jiang, S.; Liu, X.; Li, Y. Polydopamine free radical scavengers. *Biomater. Sci.* **2020**, *8*, 4940–4950. [CrossRef] [PubMed]
13. Eom, T.; Barat, V.; Khan, A.; Stuparu, M.C. Aggregation-free and high stability core-shell polymer nanoparticles with high fullerene loading capacity, variable fullerene type, and compatibility towards biological conditions. *Chem. Sci.* **2021**, *12*, 4949–4957. [CrossRef]
14. Sukhodub, L.; Sukhodub, L.; Kumeda, M.; Prylutska, S.; Deineka, V.; Prylutskyy, Y.I.; Ritter, U. C60 fullerene loaded hydroxyapatite-chitosan beads as a promising system for prolonged drug release. *Carbohydr. Polym.* **2019**, *223*, 115067. [CrossRef] [PubMed]
15. Matsuoka, K.-i.; Banya, S.; Hosoi, I.; Koyama, N.; Suzuki, K.; Jeyadevan, B.; Oku, T.; Fujita, K.; Yamada, S.; Akiyama, T. Preparation of silver-nanoparticle-loaded C60-ethylenediamine adduct microparticles and their application to photoelectric conversion. *Appl. Phys. Express* **2021**, *14*, 067003. [CrossRef]
16. Saura—Sanmartin, A.; Martinez-Cuezva, A.; Marin-Luna, M.; Bautista, D.; Berna, J. Effective Encapsulation of C60 by Metal—Organic Frameworks with Polyamide Macrocyclic Linkers. *Angew. Chem.* **2021**, *133*, 10909–10914. [CrossRef]
17. Wang, H.; Wang, L.; Wang, X.; Xu, J.; Luo, Q.; Liu, J. Self-assembled nanostructures from C60-containing supramolecular complex: Its stimuli-responsive reversible transition and biological antioxidative capacity. *New J. Chem.* **2011**, 35. [CrossRef]
18. Santra, S.; Kolay, S.; Sk, S.; Ghosh, D.; Mishra, A.; Roy, L.; Sarkar, K.; Molla, M.R. Supramolecularly cross-linked nanoassemblies of self-immolative polyurethane from recycled plastic waste: High encapsulation stability and the triggered release of guest molecules. *Polym. Chem.* **2022**, *13*, 3294–3303. [CrossRef]
19. Cheng, C.-C.; Yen, Y.-C.; Chang, F.-C. Hierarchical structures formed from self-complementary sextuple hydrogen-bonding arrays. *RSC Adv.* **2011**, *1*, 1190–1194. [CrossRef]

20. Cheng, C.-C.; Lin, I.H.; Yen, Y.-C.; Chu, C.-W.; Ko, F.-H.; Wang, X.; Chang, F.-C. New self-assembled supramolecular polymers formed by self-complementary sextuple hydrogen bond motifs. *RSC Adv.* **2012**, *2*, 9952–9957. [CrossRef]
21. Wang, J.-H.; Cheng, C.-C.; Yen, Y.-C.; Miao, C.-C.; Chang, F.-C. Block-copolymer-like supramolecules confined in nanolamellae. *Soft Matter* **2012**, *8*, 3747–3750. [CrossRef]
22. Anirudhan, T.S.; Sandeep, S. Synthesis and characterization of a novel pH-controllable composite hydrogel for anticancer drug delivery. *New J. Chem.* **2011**, *35*, 2869–2876. [CrossRef]
23. Morgan, B.; Lahav, O. The effect of pH on the kinetics of spontaneous Fe(II) oxidation by O2 in aqueous solution–basic principles and a simple heuristic description. *Chemosphere* **2007**, *68*, 2080–2084. [CrossRef] [PubMed]
24. Atere, T.G.; Akinloye, O.A.; Ugbaja, R.N.; Ojo, D.A.; Dealtry, G. In vitro antioxidant capacity and free radical scavenging evaluation of standardized extract of Costus afer leaf. *Food Sci. Hum. Wellness* **2018**, *7*, 266–272. [CrossRef]
25. Perkins, S.J. Protein volumes and hydration effects. The calculations of partial specific volumes, neutron scattering matchpoints and 280-nm absorption coefficients for proteins and glycoproteins from amino acid sequences. *Eur. J. Biochem.* **1986**, *157*, 169–180. [CrossRef]
26. Lu, H.-C.; Lin, M.-Y.; Peng, Y.-C.; Chou, S.-L.; Lo, J.-I.; Cheng, B.-M. Absorption, emission and photolysis of C60 with far-UV excitation. *Mon. Not. R. Astron. Soc.* **2015**, *452*, 2788–2793. [CrossRef]
27. Cataldo, F.; Iglesias-Groth, S. On The Molar Extinction Coefficients Of The Electronic Absorption Spectra Of C 60 And C 70 Fullerenes Radical Cation Introduction. *Eur. Chem. Bull* **2013**, *2*, 1013–1018.
28. Wang, H.; Ullah, A. Synthesis and Evaluation of Thermoresponsive Renewable Lipid-Based Block Copolymers for Drug Delivery. *Polymers* **2022**, *14*, 3436. [CrossRef]
29. Ibrahim, M.; Abuwatfa, W.H.; Awad, N.S.; Sabouni, R.; Husseini, G.A. Encapsulation, Release, and Cytotoxicity of Doxorubicin Loaded in Liposomes, Micelles, and Metal-Organic Frameworks: A Review. *Pharmaceutics* **2022**, *14*, 254. [CrossRef]
30. Wong, H.C.; Cabral, J.T. Mechanism and Kinetics of Fullerene Association in Polystyrene Thin Film Mixtures. *Macromolecules* **2011**, *44*, 4530–4537. [CrossRef]
31. Wang, X.-S.; Metanawin, T.; Zheng, X.-Y.; Wang, P.-Y.; Ali, M.; Vernon, D. Structure-Defined C60/Polymer Colloids Supramolecular Nanocomposites in Water. *Langmuir* **2008**, *24*, 9230–9232. [CrossRef] [PubMed]
32. Sathish, M.; Miyazawa, K.I. Synthesis and Characterization of Fullerene Nanowhiskers by Liquid-Liquid Interfacial Precipitation: Influence of C-60 Solubility. *Molecules* **2012**, *17*, 3858–3865. [CrossRef] [PubMed]
33. Sathish, M.; Miyazawa, K.; Hill, J.P.; Ariga, K. Solvent engineering for shape-shifter pure fullerene (C60). *J. Am. Chem. Soc.* **2009**, *131*, 6372–6373. [CrossRef]
34. Brands, S.; Schein, P.; Castro-Ochoa, K.F.; Galinski, E.A. Hydroxyl radical scavenging of the compatible solute ectoine generates two N-acetimides. *Arch. Biochem. Biophys* **2019**, *674*, 108097. [CrossRef] [PubMed]
35. Yu, F.; Xu, D.; Lei, R.; Li, N.; Li, K.A. Free-Radical Scavenging Capacity Using the Fenton Reaction with Rhodamine B as the Spectrophotometric Indicator. *J. Agric. Food. Chem.* **2008**, *56*, 730–735. [CrossRef]

polymers

MDPI

Article

Enhanced Thermal Conductivity of Silicone Composites Filled with Few-Layered Hexagonal Boron Nitride

Wei-Cheng Cheng, Yi-Ting Hsieh and Wei-Ren Liu *

Department of Chemical Engineering, R&D Center for Membrane Technology, Research Center for Circular Economy, Chung Yuan Christian University, 200 Chung Pei Road, Chung Li District, Taoyuan City 32023, Taiwan; z902082005@gmail.com (W.-C.C.); sunday52052@gmail.com (Y.-T.H.)

* Correspondence: wrliu1203@gmail.com; Tel.: +886-3-2654140; Fax: +886-3-2654199

Received: 26 August 2020; Accepted: 11 September 2020; Published: 12 September 2020

Abstract: In this study, we demonstrate the use of silicone/few-layered hexagonal boron nitride (FL-hBN) composites for heat dissipation applications. FL-hBN is synthesized via a green, facile, low-cost and scalable liquid exfoliation method using a jet cavitation process. The crystal structures, surface morphologies and specific surface areas of pristine h-BN and FL-hBN were characterized by XRD, SEM, TEM and AFM (atomic force microscopy). The results confirmed that FL-hBN with a thickness of ~4 nm was successfully obtained from the exfoliation process. In addition, we introduced both pristine h-BN and FL-hBN into silicone with different ratios to study their thermal properties. The results of the laser flash analysis indicate that the silicon/FL-hBN composite exhibited a higher thermal conductivity than that of the silicone/h-BN composite. With the optimal loading content of 30 wt.% FL-hBN content, the thermal conductivity of the composite could be enhanced to 230%, which is higher than that of silicone/h-BN (189%). These results indicate that jet cavitation is an effective and swift way to obtain few-layered hexagonal boron nitride that could effectively enhance the thermal conductivity of silicone composites.

Keywords: hexagonal boron nitride; jet cavitation; silicone; thermal conductivity; silicone

---

1. Introduction

In recent decades, heat dissipation has been one of the most critical challenges in current high-density and high-power electronic products due to rapid developments in the miniaturization of microelectronic devices. Silicone-based materials have been widely applied as the preferred matrix material, such as elastomeric thermal pads [1] and electronic packaging materials [2], due to their good electrical insulation, thermal stability, superior adhesion, good mechanical properties, ease of processing and low cost. However, it is well known that silicone has some disadvantages, such as a complex packaging process and low thermal conductivity, which might limit its application in the advanced microelectronic packaging field [3,4]. Hence, the development of highly thermally conductive silicone-based composites which maintain their low dielectric constant plays an important role in electronic packaging technologies.

It is well known that some potential fillers with high thermal conductivity material, such as graphene [5,6], metal particles [7,8], aluminum nitride (AlN) [9,10], alumina (Al_2O_3) [11,12] and silicon carbide (SiC) [13,14], have been reported. Mixing these fillers into polymer matrix composites could greatly increase the thermal conductivity of the polymer composites. In recent years, there have been many reports about improving the thermal conductivity of polymer matrices by incorporating hexagonal boron nitride (h-BN) due to its excellent properties, including high thermal conductivity, electrical insulation and excellent mechanical properties [15,16]. Hou's group grafted silane molecules onto the surface of BN particles to improve the wettability and homogeneous dispersion of BN in the polymer matrix with a strong interface interaction. A thermal conductivity of 1.178 W/m·K was obtained at 30 wt.% modified-BN loading [17]. Muratov et al. focused on the investigation of hexagonal boron nitride powder (h-BN) in combination with 3-amino-propyl-3-ethoxy-silane (APTES) as a thermal conductivity-increasing filler for polypropylene. They discovered that annealed h-BN powder before treatment by APTES had the highest –OH group. The maximum filler load increased from 11.6 to 33.7 wt.%, and the thermal conductivity of the composite samples produced using surface treated filler powder increased from 0.256 to 0.369 W/m·K at room temperature [18]. According to these references, we could understand that by introducing modified-BN as a filler into a polymer matrix, the thermal conductivity could be enhanced. When the filler in a polymer matrix has a larger aspect ratio, the thermal conductivity of the composite will be more enhanced based on the same weight percentage because fillers with a large aspect ratio easily form bridges between each other [19,20].

The aspect ratio is given by the ratio of the diameter of the platelet to its laminar thickness [21]. Thus, we can obtain a large aspect ratio if we decrease the thickness of filler. According to the existing research, the preparation of two-dimensional nanomaterials generally includes a micro-mechanical exfoliation method [22], ball milling [23,24], laser exfoliation [25], liquid exfoliation [26–28], a lithium intercalation method [29] and chemical vapor deposition (CVD) [30,31]. Fan et al. demonstrated a novel approach to fabricate boron nitride nanosheets (BNNS) via hypochlorite-assisted ball milling. This method involves the synergetic effects of chemical peeling and mechanical shear forces, which can improve the yield and dispersion [32]. However, the disadvantage of this process is that it would generate many defects due to the strong hit in the process of ball milling. Lithium intercalation is another approach to prepare few-layered materials. The drawbacks of this approach, however, include its complicated process and incomplete removal of lithium ions in the intercalation process. Thus, this process cannot be practically applied. Recently, liquid exfoliation methods have been reported as simple and solution-processable methods for preparing few-layered nanosheets. In this process, ultrasonic waves are applied to a mixture of bulk layered material and an appropriate solvent is chosen; the exfoliated few-layered nanosheets are produced by enough energy to overcome the Van der Waals forces between layers. Wu et al. obtained a thickness of BNNS (boron nitride nanosheets) between one to three layers using the liquid exfoliation method. Experimental results indicated that the BNNS could improve thermal stability and promote the curing of the matrix [33]. Recent studies have used liquid exfoliation techniques of two-dimensional materials, such as high pressure homogenization [34] and sonication [35], to delaminate these materials, or used a jet cavitation method to obtain few-layer two-dimensional materials [36]. According to the above research, these solvents are organic-based solvents, such as *N*-Methyl Pyrrolidone (NMP), Dimethylformamide (DMF) or acetone, which are used in the preparation process, and even the surfactants, including sodium carboxymethyl cellulose (CMC), non-ionic surfactant TWEEN®80 (TW80) or sodium dodecyl sulfate (SDS), are used. In our previous work, we synthesized few-layer graphene, Tungsten disulfide (WS_2) and Molybdenum diselenide ($MoSe_2$) by using a jet cavitation method [37–39].

In this study, few-layered hexagonal boron nitrides (FL-hBN) were firstly obtained via a jet cavitation method. The as-synthesized and the pristine FL-hBN were used as fillers in silicone to enhance its thermal conductivity. The jet cavitation method is a rapid, environmental-friendly, cost-effective and facile process. In order to prove that the FL-hBN was successfully obtained, h-BN and FL-hBN were characterized by using SEM, AFM (atomic force microscopy) and TEM to observe the thickness and testing BET (Brunauer–Emmett–Teller) to obtain the specific surface areas to calculate aspect ratios. In the thermal property analysis, the silicone composites with FL-hBN showed a much higher thermal conductivity than that of the h-BN/silicone composite in the same loading weight.

2. Experimental Section

2.1. Preparation of FL-hBN Powder

1 g hexagonal boron nitride (h-BN, purity > 99%, Asia Carbons & Technology Inc.®, Taoyuan, Taiwan) was mixed with 100 mL of deionized water to achieve a concentration of 1 wt.%. A dispersion of h-BN in H_2O was transferred into the tank of the low temperature ultra-high pressure continuous homemade flow cell disrupter (LTHPD). The solution was poured into the device and the process was operated three times at different pressures (800, 1300 and 1800 bar) in a circulation cooling water bath which kept the temperature at 14–16 °C. Afterwards, the few-layered h-BN suspension was produced by high pressure (Figure S1). Finally, the black particles were collected by centrifugation (10,000× *g* rpm) and washed with ethanol several times. After drying at 80 °C, FL-hBN powder was obtained.

2.2. Preparation of h-BN/silicone and FL-hBN/silicone Composites

A schematic diagram of the preparation of the h-BN/silicone and FL-hBN/silicone composites is shown in Figure 1. Firstly, 3 g silicone and different weight ratios of h-BN (10, 20 and 30 wt.%) and FL-hBN (10, 20 and 30 wt.%) were mixed well in the beaker. Appropriate amounts of ethyl ethanoate (Sigma Aldrich®, St. Louis, MO, USA, purity > 99.8%) were introduced into the beaker with magnetic stirring for 5 h. Secondly, the mixing solution was put in the furnace at 150 °C for 1 h to evaporate the solvent. Third, 3 g hardener (Sigma Aldrich®, KJC-1200) was added into the beaker and mixed for 2 h. Finally, we put the final product into the mold at 150 °C in a vacuum for thermal curing. The as-prepared bare silicone, h-BN/silicone and FL-hBN/silicone composites were obtained.

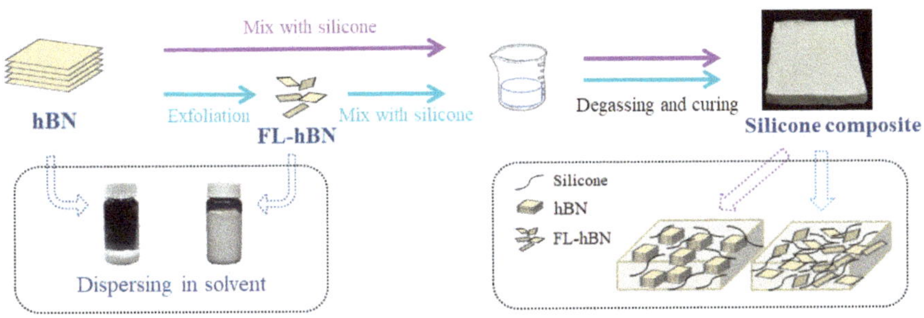

Figure 1. Schematic diagram of the preparation of hexagonal boron nitride (h-BN)/silicone and few-layered hexagonal boron nitride (FL-hBN)/silicone composites by our approach.

3. Characterizations

The thermal behaviors of the as-prepared materials were investigated using TGA analysis (thermogravimetric analysis). The ramp rate was maintained at 5 °C/min under air atmosphere. The crystal phase structure and purity were determined by an XRD (X-ray diffractometer) analysis using a D8 diffractometer (Bruker®, Billerica, MA, USA) with monochromatic CuKα radiation. The operating voltage, current and wavelength (λ) were 40 kV, 30 mA and 1.54060 Å, respectively. Diffraction data were recorded in the range (2θ) of 10°–80°. The morphological natures of the as-prepared samples were observed using SEM (scanning electron microscopy, Hitachi S-4100, Okinawa, Japan) with electron mapping (EDS, energy-dispersive X-ray spectroscopy, Okinawa, Japan). Atomic force microscope (AFM) images were captured by a Bruker Dimension Icon. The samples for AFM were prepared by dropping the dispersion directly onto freshly cleaved mica wafers with an injector. The thermal diffusivity (α) of the film was measured by a laser flash thermal analyzer (LFA457 Micro Flash, Netzsch, Germany). The sample size was 1 cm × 1 cm × 0.2 cm. Brunauer–Emmett–Teller (BET) specific surface area was determined from N_2 adsorption by using a Micromeritics TriStar 3000 (Norcross, GA, USA) analyzer at liquid nitrogen temperature.

4. Results and Discussion

The XRD patterns of pristine h-BN and few-layered h-BN (FL-hBN) are shown in Figure 2a,b. All the diffraction peaks at 2θ values of 26.76°, 41.59°, 43.87°, 50.15° and 55.17° corresponded to the diffraction planes of h-BN in (002), (100), (101), (102) and (004), respectively. All of the diffraction peaks matched well with the standard values and agreed with the hexagonal structure of the Bragg positions in JCPDS-34-0421, shown in Figure 2c. Compared with the pristine h-BN, the (002) and (004) peaks of few-layer BN showed a remarkable peak broadening, indicating the presence of thinner h-BN sheets and much less extended/ordered stacking in the c direction (Table S1). Figure 2d shows photo images of the h-BN and FL-hBN solutions with different settling times. Obviously, after the exfoliation process, the settling time of FL-hBN was much longer than that of pristine h-BN, which might be due to the fact that the thickness of h-BN may have become much thinner after the exfoliation process.

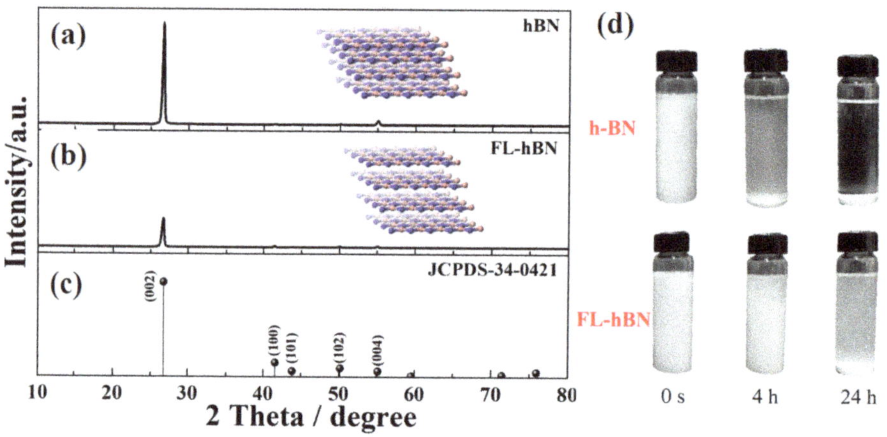

Figure 2. XRD patterns of: (**a**) h-BN; (**b**) FL-hBN; (**c**) standard data JCPDS-34-0421. Photo images (**d**) of h-BN and FL-hBN solutions with different settling times.

Figure 3a,d show the FE-SEM images of the h-BN and FL-hBN, respectively. The lateral size of h-BN was 4–9 μm. After the exfoliation process, the lateral size of the FL-hBN was smaller than that of the h-BN in the region of 2–5 μm. Figure 3a,b also demonstrated that the thickness of the FL-hBN prepared by a jet cavitation method was thinner than that of h-BN. The phenomenon proved that h-BN could be exfoliated successfully. Figure 3b,e display TEM images of the h-BN and FL-hBN. The FL-hBN became transparent due to the small thickness shown in Figure 3e. The HR-TEM images of these samples are also shown in Figure 3c,f. A lattice fringe with interplane spacing of 1.67 Å, corresponding to (004) plane, was revealed. These results could be seen due to the change in thickness with FL-hBN after delamination.

The surface morphology and thickness distribution of the as-synthesized FL-hBN was further examined by AFM characterization, as shown in Figure 4a, by 30 samples. The results indicate that the thickness of the h-BN was ~500 nm (Figure 4a), which is much thicker than that of the FL-hBN with a thickness in the range of 3.9 nm~4.3 nm, shown in Figure 4b. Figure 4c shows the histogram of the thickness distribution from AFM images of the FL-hBN from 30 samples. The results further confirm that more than 40% of the FL-hBN nanosheets' thicknesses ranged between 3.8 and 4.2 nm.

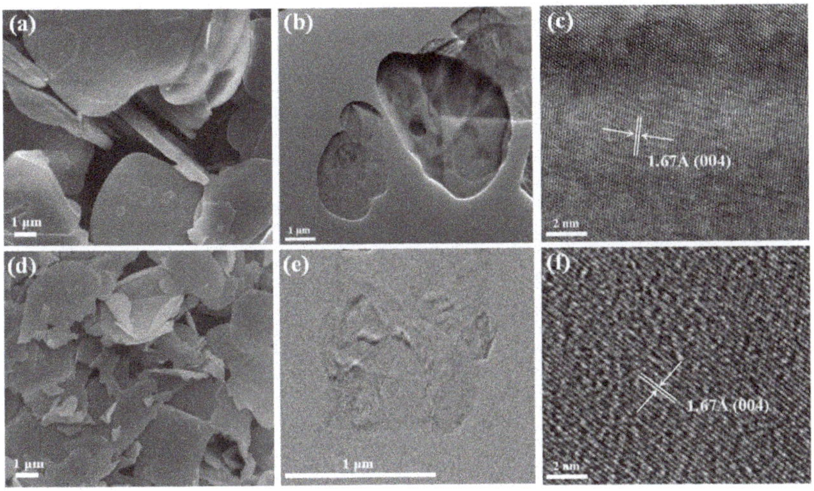

Figure 3. SEM images of (**a**) h-BN and (**d**) FL-hBN; TEM images of (**b**) h-BN and (**e**) FL-hBN; HR-TEM images of (**c**) h-BN and (**f**) FL-hBN. The insets show the selected area electron diffraction patterns.

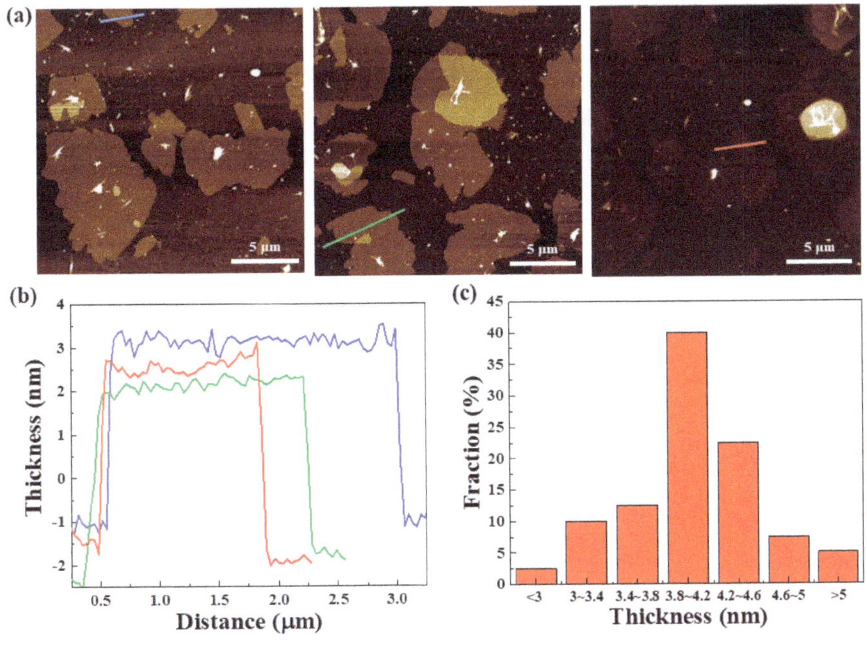

Figure 4. (**a**) AFM images (**b**) and the corresponding line scans of the FL-hBN on Si substrate; (**c**) histogram of the number visual observations of sheets as a function of the number of the layers of FL-hBN per sheet.

When the bulk material is peeled off layer by layer, it can be imagined that the specific surface area will change. Thus, we used N_2 adsorption and desorption isothermal curves of h-BN (black line + symbol) and FL-hBN (red line + symbol), shown in Figure 5, to understand the difference in specific surface areas. These two samples demonstrated type IV isotherms, as determined by the International Union of Pure and Applied Chemistry (IUPAC), which is usually associated with mesoporous materials (i.e., pore sizes between 2 and 50 nm). The specific surface areas of the h-BN and FL-hBN were about 4.456 and 11.19 m^2g^{-1}, respectively. In addition, from the data of pore size distribution analysis, the corresponding pore volumes of h-BN and FL-hBN were 0.01917 and 0.04241 cm^3g^{-1}, respectively. The higher specific surface area of FL-hBN could represent that the method we used to exfoliate materials was successful.

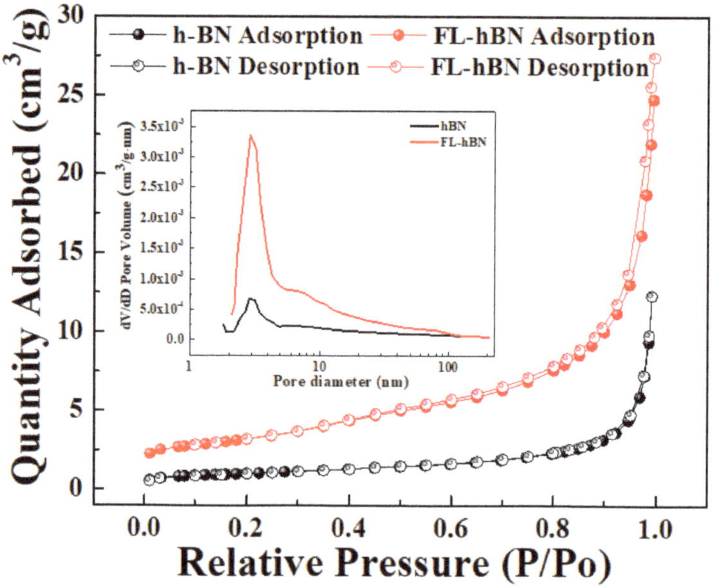

Figure 5. N_2 adsorption/desorption isotherms and pore size distribution analyses of h-BN (black line + symbol) and FL-hBN (red line + symbol). Inset: pore size distribution analyses of h-BN and FL-hBN.

Figure 6 shows the XRD analysis of the composites after the h-BN or FL-hBN were added into the silicone. The lattice orientation of the filler in the composites can be observed. It can be proved that h-BN exhibited high crystallinity in the composites (h-BN/silicone). Highly crystalline materials may have some advantages in some properties, such as their mechanical properties that are mentioned later. The diffraction intensity of FL-hBN on (002) and (004) decreased in composites with FL-hBN, indicating that the thickness of the filler (FL-hBN) was thinner in the composite. This trend was consistent with other reports [40–43].

Figure 7a,b show the thermal diffusivity and thermal conductivity of the composites of h-BN/silicone and FL-hBN/silicone composites with different loading ratios. Table 1 summarizes the physical properties of all of these composites. The results showed that both thermal diffusivity and thermal conductivity increased with the increasing content of the fillers. The composites containing FL-hBN exhibited much higher thermal conductivity than those with pristine h-BN at the same fraction in silicone. The thermal conductivity of 30% FL-hBN/silicone (30 wt.% FL-hBN into silicone) composites was 0.515 W/m·K, which was higher than that of 30% hBN/silicone composites (0.424 W/m·K). Compared to pristine silicone, the thermal conductivity of the FL-hBN/silicone composite enhanced by 230% (2.3 times) by introducing 30 wt.% FL-h-BN.

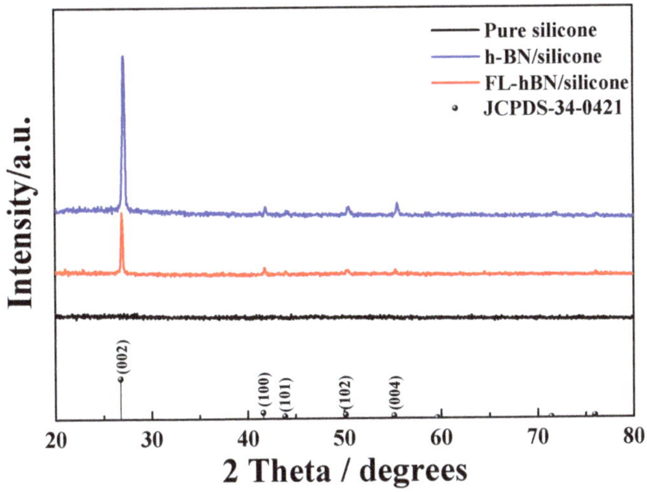

Figure 6. XRD patterns of pure silicone, (30%) h-BN/silicone and (30%) FL-hBN/silicone.

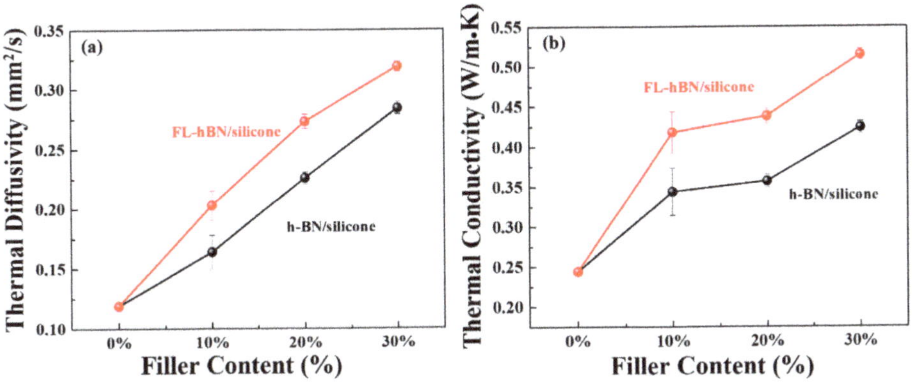

Figure 7. (**a**) Thermal diffusivity and (**b**) thermal conductivity of h-BN/silicone composites and FL-hBN/silicone composites.

Table 1. The physical properties of all the composites.

	Fraction (wt.%)	Density (cm^3 g^{-1})	Specific Heat Capacity (J g^{-1} K^{-1})	Thermal Diffusivity (mm^2 s^{-1})	Thermal Conductivity (W m^{-1} K^{-1})
Pure silicone	0	1.030	1.993	0.119	0.244
hBN/silicone	10	1.089	1.928	0.164	0.344
	20	1.141	1.386	0.226	0.357
	30	1.182	1.264	0.284	0.424
FL-hBN/silicone	10	1.060	1.944	0.203	0.418
	20	1.119	1.434	0.273	0.438
	30	1.170	1.381	0.319	0.515

The aspect ratio (Equation (1)) of the filler affects the thermal properties of composites [21]. Materials possessing a higher aspect ratio could disperse well in a polymer matrix and build a good thermal conductive network in the composite even if the filler content is low. We can obtain the aspect ratio from the measured specific surface area by BET and the particle size of the filler. The aspect ratios of h-BN and FL-hBN were 29.06 and 56.10, respectively. From these results, FL-hBN has a larger aspect ratio so it can enhance compatibility with the polymer matrix. In order to reveal the filler dispersing in these composites, the morphologies of the composites were characterized by SEM micrographs in Figure S2a–f. We can observe that FL-hBN had excellent dispersion in the composites of different filler loading. Compared with h-BN/silicone, h-BN could not connect well in the polymer matrix. The same results were shown in EDX mapping (Figure 8a,b).

$$AR = \frac{D}{t} = \frac{S\rho D}{2} - 2 \tag{1}$$

where

AR = Aspect ratio,
D = The diameter of the platelets (average particle size),
t = The thickness of the platelets,
S = The specific surface area of the particles,
ρ = The density of the platelets.

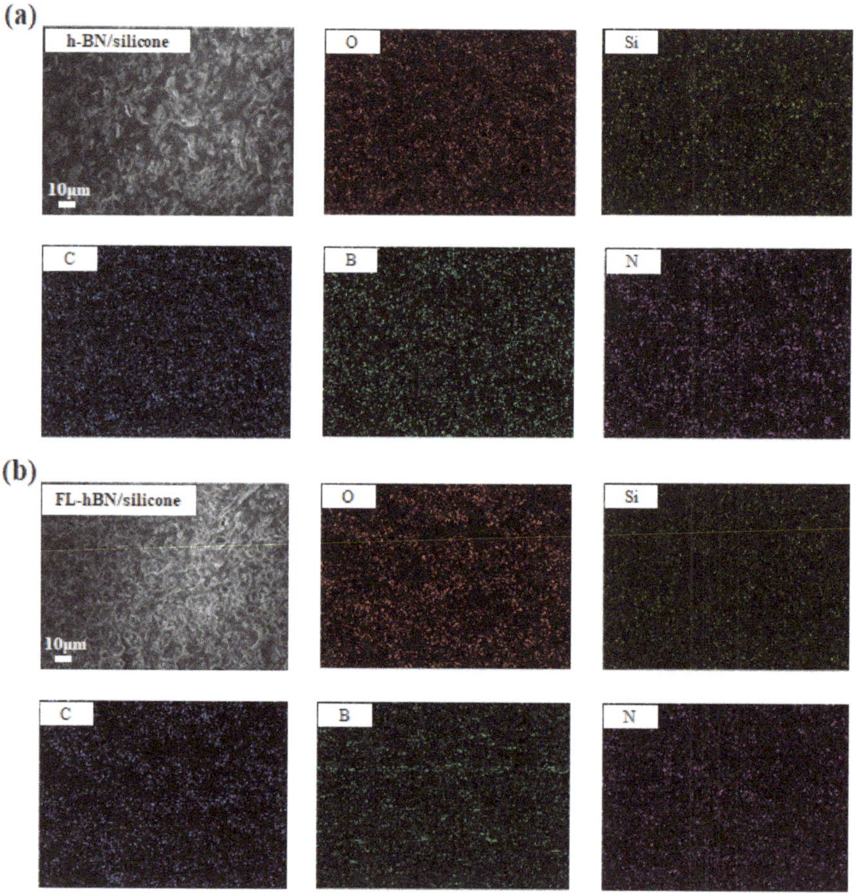

Figure 8. The Energy-dispersive X-ray spectroscopy (EDX) mapping of (**a**) h-BN/silicone (30 wt.%) and (**b**) FL-hBN/silicone (30 wt.%).

Figure 9 shows the TGA curves for pure silicone, h-BN/silicone (30 wt.%) and FL-hBN/silicone (30 wt.%) at 800 °C in nitrogen. The results showed that the temperature for 5% weight loss of these composites appeared at 410.1 °C (bare silicone), 483.1 °C (h-BN/silicone (30 wt.%)) and 488.4 °C (FL-hBN/silicone (30 wt.%)). At 600 °C, bare silicone, h-BN/silicone (30 wt.%) and FL-hBN/silicone (30 wt.%) exhibited 19.67%, 16.54% and 13.06% weight loss, respectively, indicating that the FL-hBN/silicone composite had lower weight loss than the other. If the fillers connect well with each other in the polymer matrix, heat will transfer quickly by thermal conduction. On the other hand, FL-hBN in silicone prevented heat from accumulating in the polymer matrix so that the composite will decompose more slowly.

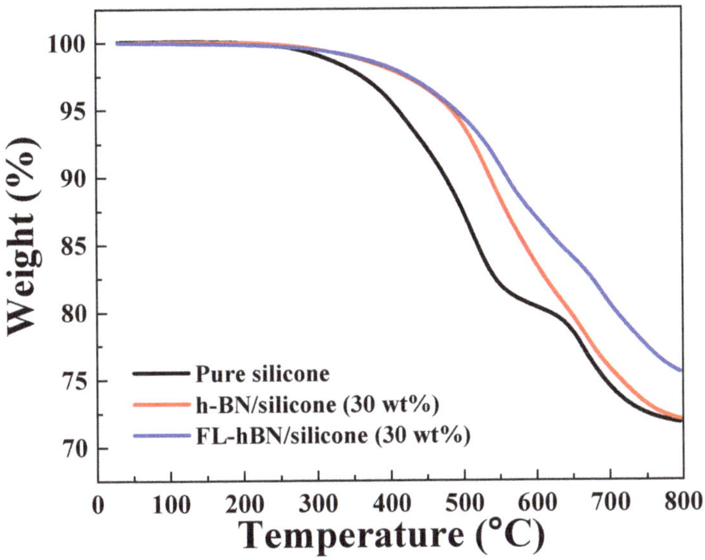

Figure 9. Thermal stabilities of pure silicone, hBN/silicone composites and FL-hBN/silicone.

Dynamic mechanical analysis (DMA) was performed to obtain the temperature dependent properties of composites, such as the storage modulus. The storage modulus is highly dependent on the dispersion between the filler and the silicone. Figure 10 shows the DMA curve of the three composites. The storage modulus of the composites increased after adding the fillers. For example, at 328 °C, the storage modulus of the pure silicone and the composites with hBN and FL-hBN were 9.19, 20.73 and 19.86 MPa, respectively. The main reason for the higher mechanical property is that the fillers will transfer some strength to each other when composite is stretched by external force.

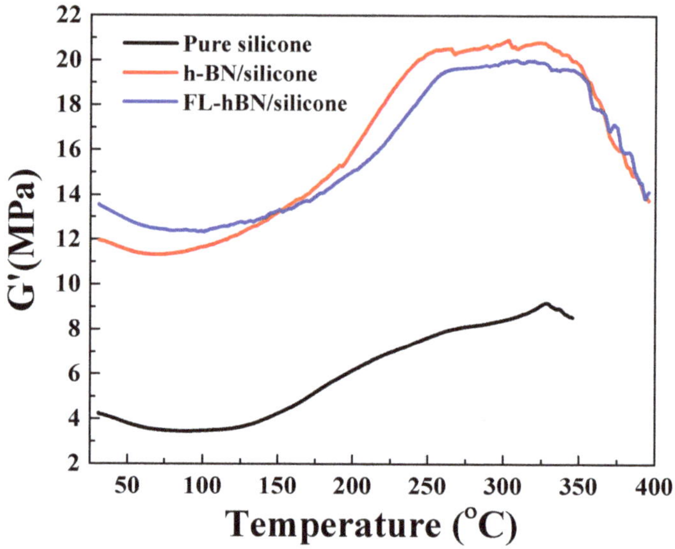

Figure 10. Dynamic mechanical analysis (DMA) curves of pure silicone, h-BN/silicone and FL-hBN/silicone composites.

5. Conclusions

We successfully obtained FL-hBN with an extremely facile and environmental-friendly jet cavitation method. According to AFM analysis, the thickness of all the FL-hBN layers were less than 5 nm. Moreover, the FL-hBN layers had smaller particle sizes with an increased surface area and a higher aspect ratio of 56.10. The thermal properties of both h-BN/silicone and FL-hBN/silicone composites improved, indicating that hexagonal boron nitride is an exceptional filler for silicone. Silicone composites containing FL-hBN exhibited both outstanding thermal conductivity and thermal stability—better than composites with h-BN. Compared to pure silicone, the thermal conductivity of the FL-hBN/silicone composite was 0.515 W/m·K at 30 wt.% filler loading. Silicone incorporated with FL-hBN also exhibited higher thermal stability at 800 °C. FL-hBN has a higher aspect ratio, indicating that it has good dispersion in silicone rubber, so the filler can connect well to form a thermal network. However, exfoliating h-BN to FL-hBN yielded superior thermal conductivity properties which can be used for industrial applications.

Supplementary Materials: The following are available online at http://www.mdpi.com/2073-4360/12/9/2072/s1, Figure S1: Photo images of (a) pristine h-BN and (b) FL-hBN solution after different resting time, Figure S2: (a–c) SEM images of h-BN/silicone composites with powder loading of 10 wt%, 20 wt%, 30 wt%; (d–f) SEM images of FL-hBN/silicone composites with powder loading of 10 wt%, 20 wt%, 30 wt%, Table S1: FWHM, Lc and La of h-BN and FL-hBN.

Author Contributions: Y.-T.H. wrote the paper with a contribution to the experiments and analysis; W.-C.C. contributed to the experiments and analysis; W.-R.L. supervised and edited the manuscript. All authors have read and agreed to the published version of the manuscript.

Funding: This research received no external funding.

Acknowledgments: The authors gratefully acknowledge the Ministry of Science and Technology, Taiwan, for the project grants 107-2811-M-033-506, 108EFD0500003, 108-2811-E-033-500, 109-2911-I-033-502, 108-E-033-MY3 and 109-3116-F-006 -018.

Conflicts of Interest: The authors declare no conflict of interest.

References

1. Sim, L.C.; Ramanan, S.R.; Ismail, H.; Seetharamu, K.N.; Goh, T.J. Thermal characterization of Al_2O_3 and ZnO reinforced silicone rubber as thermal pads for heat dissipation purposes. *Thermochim. Acta* **2005**, *43*, 155–165. [CrossRef]
2. Gao, N.; Liu, W.; Yan, Z.; Wang, Z. Synthesis and properties of transparent cycloaliphatic epoxy–silicone resins for opto-electronic devices packaging. *Opt. Mater.* **2013**, *35*, 567–575. [CrossRef]
3. Chen, C.; Tang, Y.; Ye, Y.S.; Xue, Z.; Xue, Y.; Xie, X.; Mai, Y.-W. High-performance epoxy/silica coated silver nanowire composites as underfill material for electronic packaging. *Compos. Sci. Technol.* **2014**, *105*, 80–85. [CrossRef]
4. Fang, L.; Wu, W.; Huang, X.; He, J.; Jiang, P. Hydrangea-like zinc oxide superstructures for ferroelectric polymer composites with high thermal conductivity and high dielectric constant. *Compos. Sci. Technol.* **2015**, *107*, 67–74. [CrossRef]
5. Song, H.; Park, K.H.; Kim, B.H.; Choi, Y.W.; Jun, G.H.; Lee, D.J.; Kong, B.S.; Paik, K.W.; Jeon, S. Enhanced thermal conductivity of epoxy–graphene composites by using non-oxidized graphene flakes with non-covalent functionalization. *Adv. Mater.* **2012**, *25*, 732–737. [CrossRef] [PubMed]
6. Teng, C.-C.; Ma, C.-C.M.; Lu, C.-H.; Yang, S.-Y.; Lee, S.-H.; Hsiao, M.-C.; Yen, M.-Y.; Chiou, K.-C.; Lee, T.-M. Thermal conductivity and structure of non-covalent functionalized graphene/epoxy composites. *Carbon* **2011**, *49*, 5107–5116. [CrossRef]
7. Bjorneklett, A.; Halbo, L.; Kristiansen, H. Thermal conductivity of epoxy adhesives filled with silver particles. *Int. J. Adhes. Adhes.* **1992**, *12*, 99–104. [CrossRef]
8. Papanicolaou, G.C.; Theocaris, P.S. Thermal properties and volume fraction of the boundary interphase in metal-filled epoxies. *Colloid Polym. Sci.* **1979**, *257*, 239–246. [CrossRef]
9. Huang, X.; Iizuka, T.; Jiang, P.; Ohki, Y.; Tanaka, T. Role of interface on the thermal conductivity of highly filled dielectric epoxy/AlN composites. *J. Phys. Chem. C* **2012**, *116*, 13629–13639. [CrossRef]
10. Yu, H.; Li, L.; Kido, T.; Xi, G.; Xu, G.; Guo, F. Thermal and insulating properties of epoxy/aluminum nitride composites used for thermal interface material. *J. Appl. Polym. Sci.* **2012**, *124*, 669–677. [CrossRef]
11. Im, H.; Kim, J. The effect of Al_2O_3 doped multi-walled carbon nanotubes on the thermal conductivity of Al_2O_3/epoxy terminated poly (dimethylsiloxane) composites. *Carbon* **2011**, *49*, 3503–3511. [CrossRef]
12. Gao, Z.; Zhao, L. Effect of nano-fillers on the thermal conductivity of epoxy composites with micro-Al_2O_3 particles. *Mater. Des.* **2015**, *66*, 176–182. [CrossRef]
13. Zhou, T.; Wang, X.; Liu, X.; Xiong, D. Improved thermal conductivity of epoxy composites using a hybrid multi-walled carbon nanotube/micro-SiC filler. *Carbon* **2010**, *48*, 1171–1176. [CrossRef]
14. Zhou, T.; Wang, X.; Gu, M.; Xiong, D. Study on mechanical; thermal and electrical characterizations of nano-SiC/epoxy composites. *Polym. J.* **2009**, *41*, 51. [CrossRef]
15. Wang, J.; Zhao, D.; Zou, X.; Mao, L.; Shi, L. The exfoliation and functionalization of boron nitride nanosheets and their utilization in silicone composites with improved thermal conductivity. *J. Mater. Sci. Mater. Electron.* **2017**, *28*, 12984–12994. [CrossRef]
16. Yuan, C.; Li, J.H.; Lindsay, L.; Cherns, D.; Pomeroy, J.W.; Liu, S.; Edgar, J.H.; Kuball, M. Modulating the thermal conductivity in hexagonal boron nitride via controlled boron isotope concentration. *Commun. Phys.* **2019**, *2*, 1–8. [CrossRef]
17. Hou, J.; Li, G.; Yang, N.; Qin, L.; Grami, M.E.; Zhang, Q.; Wang, N.; Qu, X. Preparation and characterization of surface modified boron nitride epoxy composites with enhanced thermal conductivity. *RSC Adv.* **2014**, *4*, 44282–44290. [CrossRef]
18. Muratov, D.S.; Kuznetsov, D.V.; Il'Inykh, I.A.; Burmistrov, I.N.; Mazov, I.N. Thermal conductivity of polypropylene composites filled with silane-modified hexagonal BN. *Compos. Sci. Technol.* **2015**, *111*, 40–43. [CrossRef]
19. Lee, G.-W.; Park, M.; Kim, J.; Lee, J.I.; Yoon, H.G. Enhanced thermal conductivity of polymer composites filled with hybrid filler. *Compos. Part A Appl. Sci. Manuf.* **2006**, *37*, 727–734. [CrossRef]
20. Liu, M.-S.; Lin, M.C.-C.; Huang, I.-T.; Wang, C.-C. Enhancement of thermal conductivity with carbon nanotube for nanofluids. *Int. Commun. Heat Mass Transf.* **2005**, *32*, 1202–1210. [CrossRef]

21. Gantenbein, D.; Schoelkopf, J.; Matthews, G.P.; Gane, P.A. Determining the size distribution-defined aspect ratio of platy particles. *Appl. Clay Sci.* **2011**, *53*, 544–552. [CrossRef]
22. Pacile, D.; Meyer, J.C.; Girit, Ç.Ö.; Zettl, A. The two-dimensional phase of boron nitride: Few-atomic-layer sheets and suspended membranes. *Appl. Phys. Lett.* **2008**, *92*, 133107. [CrossRef]
23. Li, L.H.; Glushenkov, A.M.; Hait, S.K.; Hodgson, P.; Chen, Y. High-efficient production of boron nitride nanosheets via an optimized ball milling process for lubrication in oil. *Sci. Rep.* **2014**, *4*, 7288.
24. Zhu, K.; Wang, W.; Meng, A.; Zhao, M.; Wang, J.; Zhao, M.; Zhang, D.; Jia, Y.; Xu, C.; Li, Z. Mechanically exfoliated g-C_3N_4 thin nanosheets by ball milling as high performance photocatalysts. *RSC Adv.* **2015**, *5*, 56239–56243. [CrossRef]
25. An, S.-J.; Kim, Y.H.; Lee, C.; Park, D.Y.; Jeong, M.S. Exfoliation of Transition Metal Dichalcogenides by a High-Power Femtosecond Laser. *Sci. Rep.* **2018**, *8*, 12957. [CrossRef] [PubMed]
26. Rafie-Sarmazdeh, Z.; Jafari, S.H.; Ahmadi, S.J.; Zahedi-Dizaji, S.M. Large-scale exfoliation of hexagonal boron nitride with combined fast quenching and liquid exfoliation strategies. *J. Mater. Sci.* **2016**, *51*, 3162–3169. [CrossRef]
27. Güler, Ö.; Güler, S.H. Production of graphene–boron nitride hybrid nanosheets by liquid-phase exfoliation. *Optik* **2016**, *127*, 4630–4634. [CrossRef]
28. Gerchman, D.; Alves, A.K. Solution-processable exfoliation and suspension of atomically thin WSe_2. *J. Colloid Interface Sci.* **2016**, *468*, 247–252. [CrossRef]
29. Mayorga-Martinez, C.C.; Ambrosi, A.; Eng, A.Y.S.; Sofer, Z.; Pumera, M. Transition metal dichalcogenides (MoS_2, $MoSe_2$, WS_2 and WSe_2) exfoliation technique has strong influence upon their capacitance. *Electrochem. Commun.* **2015**, *56*, 24–28. [CrossRef]
30. Jeon, J.; Jang, S.K.; Jeon, S.M.; Yoo, G.; Jang, Y.H.; Park, J.-H.; Lee, S. Layer-controlled CVD growth of large-area two-dimensional MoS_2 films. *Nanoscale* **2015**, *7*, 1688–1695. [CrossRef]
31. Perea-López, N.; Lin, Z.; Pradhan, N.R.; Iñiguez-Rábago, A.; Elías, A.L.; McCreary, A.; Lou, J.; Ajayan, P.M.; Terrones, H.; Balicas, L. CVD-grown monolayered MoS_2 as an effective photosensor operating at low-voltage. *2D Mater.* **2014**, *1*, 011004. [CrossRef]
32. Fan, D.; Feng, J.; Liu, J.; Gao, T.; Ye, Z.; Chen, M.; Lv, X. Hexagonal boron nitride nanosheets exfoliated by sodium hypochlorite ball mill and their potential application in catalysis. *Ceram. Int.* **2016**, *42*, 7155–7163. [CrossRef]
33. Wu, X.F.; Zhao, Z.-h.; Sun, Y.; Li, H.; Zhang, C.-X.; Wang, Y.-J.; Zheng, S.-S.; Zhang, H. Few-layer boron nitride nanosheets: Preparation; characterization and application in epoxy resin. *Ceram. Int.* **2017**, *43*, 2274–2278. [CrossRef]
34. Nacken, T.; Damm, C.; Walter, J.; Rüger, A.; Peukert, W. Delamination of graphite in a high pressure homogenizer. *RSC Adv.* **2015**, *5*, 57328–57338. [CrossRef]
35. Skaltsas, T.; Ke, X.; Bittencourt, C.; Tagmatarchis, N. Ultrasonication induces oxygenated species and defects onto exfoliated graphene. *J. Phys. Chem. C* **2013**, *117*, 23272–23278. [CrossRef]
36. Liang, S.; Shen, Z.; Yi, M.; Liu, L.; Zhang, X.; Cai, C.; Ma, S. Effects of Processing Parameters on Massive Production of Graphene by Jet Cavitation. *J. Nanosci. Nanotechnol.* **2015**, *15*, 2686–2694. [CrossRef]
37. Lin, P.-C.; Wu, J.-Y.; Liu, W.-R. Green and facile synthesis of few-layer graphene via liquid exfoliation process for Lithium-ion batteries. *Sci. Rep.* **2018**, *8*, 9766. [CrossRef]
38. Yeh, Y.-Y.; Chiang, W.-H.; Liu, W.-R. Synthesis of few-layer WS_2 by jet cavitation as anode material for lithium ion batteries. *J. Alloy. Compd.* **2019**, *775*, 1251–1258. [CrossRef]
39. Wu, Y.-C.; Liu, W.-R. Few-layered $MoSe_2$ ultrathin nanosheets as anode materials for lithium ion batteries. *J. Alloys Compd.* **2020**, *831*, 152074. [CrossRef]
40. Cai, W.; Hong, N.; Feng, X.; Zeng, W.; Shi, Y.; Zhang, Y.; Wang, B.; Hu, Y. A facile strategy to simultaneously exfoliate and functionalize boron nitride nanosheets via Lewis acid-base interaction. *Chem. Eng. J.* **2017**, *330*, 309–321. [CrossRef]
41. Lei, W.; Mochalin, V.N.; Liu, D.; Qin, S.; Gogotsi, Y.; Chen, Y. Boron nitride colloidal solutions; ultralight aerogels and freestanding membranes through one-step exfoliation and functionalization. *Nat. Commun.* **2015**, *6*, 8849. [CrossRef] [PubMed]

42. Yuan, C.; Duan, B.; Li, L.; Xie, B.; Huang, M.; Luo, X.B. Thermal Conductivity of Polymer-Based Composites with Magnetic Aligned Hexagonal Boron Nitride Platelets. *ACS Appl. Mater. Interfaces* **2015**, *7*, 13000–13006. [CrossRef] [PubMed]
43. Lin, Z.Y.; Liu, Y.; Raghavan, S.; Moon, K.-S.; Suresh, K.; Sitaraman, S.K.; Wong, C.-P. Magnetic Alignment of Hexagonal Boron Nitride Platelets in Polymer Matrix: Toward High Performance Anisotropic Polymer Composites for Electronic Encapsulation. *ACS Appl. Mater. Interfaces* **2013**, *5*, 7633–7640. [CrossRef] [PubMed]

polymers

MDPI

Article

Effect of Strain Rate and Silica Filler Content on the Compressive Behavior of RTM6 Epoxy-Based Nanocomposites

Ahmed Elmahdy [1,*,†], Aldobenedetto Zotti [2,†], Simona Zuppolini [2], Mauro Zarrelli [2,*], Anna Borriello [2] and Patricia Verleysen [1]

[1] Materials Science and Technology-DyMaLab Research Group, Department of Electromechanical Systems and Metals Engineering, Faculty of Engineering and Architecture, Ghent University, Tech Lane Ghent Science Park, Technologiepark 46, 9052 Zwijnaarde, Belgium; patricia.verleysen@ugent.be

[2] Institute of Polymers, Composites and Biomaterials, National Research Council of Italy, P.le Fermi, 1, 80055 Naples, Portici, Italy; aldobenedetto.zotti@unina.it (A.Z.); simona.zuppolini@cnr.it (S.Z.); borriell@unina.it (A.B.)

* Correspondence: ahmed.elmahdy@ugent.be (A.E.); mauro.zarrelli@cnr.it (M.Z.)

† These authors contributed equally to this work.

Abstract: The aim of this paper is to investigate the effect of strain rate and filler content on the compressive behavior of the aeronautical grade RTM6 epoxy-based nanocomposites. Silica nanoparticles with different sizes, weight concentrations and surface functionalization were used as fillers. Dynamic mechanical analysis was used to study the glass transition temperature and storage modulus of the nanocomposites. Using quasi-static and split Hopkinson bar tests, strain rates of 0.001 s^{-1} to 1100 s^{-1} were imposed. Sample deformation was measured using stereo digital image correlation techniques. Results showed a significant increase in the compressive strength with increasing strain rate. The elastic modulus and Poisson's ratio showed strain rate independency. The addition of silica nanoparticles marginally increased the glass transition temperature of the resin, and improved its storage and elastic moduli and peak yield strength for all filler concentrations. Increasing the weight percentage of the filler slightly improved the peak yield strength. Moreover, the filler's size and surface functionalization did not affect the resin's compressive behavior at different strain rates.

Keywords: epoxy resin; nanocomposites; silica nanoparticles; mechanical behavior; high strain rate; split Hopkinson bar

Citation: Elmahdy, A.; Zotti, A.; Zuppolini, S.; Zarrelli, M.; Borriello, A.; Verleysen, P. Effect of Strain Rate and Silica Filler Content on the Compressive Behavior of RTM6 Epoxy-Based Nanocomposites. *Polymers* **2021**, *13*, 3735. https://doi.org/10.3390/polym13213735

Academic Editors: Ting-Yu Liu and Yu-Wei Cheng

Received: 26 September 2021
Accepted: 25 October 2021
Published: 28 October 2021

1. Introduction

Epoxy resins are widely used as matrix material for high-performance composites in aeronautical applications. They are generally characterized by a high cross-linking density compared to other thermoset polymers. This gives epoxy resins and their composites many advantages such as high stiffness, good chemical resistance, good performance at high temperatures and excellent fatigue performance [1]. Additionally, their low curing shrinkage does not cause curing cracks in large aerospace components. However, because of the high cross-linking density, epoxy resins are generally very brittle with a very low fracture strain and have poor resistance to impact and crack propagation [2]. For this reason, efforts were made to improve the mechanical performance of the epoxy resins by the addition of different types of fillers, such as inorganic particles [3–5], elastomer particles [6,7], carbon nanotubes [8,9], hyperbranched polymers [10–12] and recently graphene nanoplatelets [2,13]. Compared to other filler types, silica nanoparticles are widely studied as fillers to epoxy resins. This is related to the marginal effect of the silica-based fillers on the glass transition temperature of the hosting epoxy matrix, and hence its curing temperature [14–17]. Moreover, the advancement in synthesis processes, particularly sol–gel and modified sol–gel techniques, allow the production of these nanoparticles either as precipitates or directly in the epoxy resin itself (in situ) [18], which can be considered for large scale manufacturing of epoxy nanocomposites with a relatively low cost [19]. In addition, these synthesis techniques allow a very high degree of control over the size and distribution of the formed nanoparticles [6]. The addition of silica nanoparticles up to a weight content of 25% generally improves the overall mechanical performance of epoxy resins such as tensile strength and stiffness [15,20], fracture toughness, compressive strength [21] and fatigue crack growth [22]. Additionally, when combined with carbon fibers, the silica nanoparticles can improve the overall toughness of the carbon epoxy composites by enhancing the interfacial adhesion with the fibers [23].

The extent of the improvement of the physical and the mechanical properties of epoxy nanocomposites is highly affected by the size and surface condition of the silica nanoparticles. Surface functionalization of the silica nanoparticles generally improves the compatibility of the particles with the hosting matrix [24] and improves the overall mechanical performance and glass transition temperature of the hosting epoxy resin up to silica particle sizes of 400 nm [25–28]. With regard to the size, the addition of silica nanoparticles of size ranging from 7 nm to 80 nm does not significantly affect glass transition temperature or the mechanical properties of the hosting epoxy resin [16,25,29]. However, at silica particle size of 100 nm or larger, no clear trends can be established. On the one hand, Dittanet et al. [29] showed that for a silica particle size range of 23 to 170 nm and up to 30% weight content, the mechanical properties and the glass transition temperature of the epoxy remained nearly constant regardless of the silica particle size. On the other hand, Bondioli et al. [30] showed that the elastic modulus of the epoxy resin increased by the addition of 1% weight content of 75 nm silica nanoparticle compared to 330 nm silica nanoparticles, which partially contradicts with the findings of Dittanet et al. [29]. Sun et al. [31] also reported a decrease in the glass transition temperature for epoxy filled with 100 nm silica nanoparticles and 10% weight content, compared to a constant glass transition temperature for the same epoxy filled with 3 μm silica particles at the same weight content, which also contradicts with the findings of Dittanet et al. [29].

As mentioned earlier, the main aim of improving epoxy resins is to enhance their use as matrix material and to improve the performance of aeronautical composite structures. These structures are typically subjected to extreme impact events, such as bird strike or fan blade-out events. Therefore, studying the effect of strain rate on the mechanical behavior of silica/epoxy nanocomposites is essential. While several studies report on the quasi-static mechanical behavior of silica/epoxy nanocomposites, few results are available regarding their high strain rate behavior. Miao et al. [32] showed that adding silica nanoparticles of size 20 nm and 10% weight content to epoxy only marginally improved its compressive yield strength at strain rates up to 5000 s^{-1}. Additionally, a significant strain rate sensitivity was reported, where the yield strength increased with increasing the strain rate. Tian et al. [33] reported that the addition of 30 nm silica nanoparticles with 10% weight content increased the compressive modulus and yield strength of epoxy resin with increasing strain rates up to 3000 s^{-1}. However, the improvement of the compressive performance was more pronounced in the low strain rate regime compared to the high strain rate regime. Contrary to these findings, Guo et al. [34] showed that the improvement in the compressive strength by the addition of 90 nm silica nanoparticles up to 7% weight content was more pronounced at high strain rates up to 10^4 s^{-1}, whereas no improvement could be observed at low strain rates. Ma et al. [35] showed that for epoxy filled with silica nanoparticles of size 50 nm and up to 15% weight content, the compressive failure strength increased at strain rates up to 200 s^{-1} and higher silica contents. The compressive stiffness, however, showed a reduction at higher silica weight contents. Yohanes [36] found that the addition 17 nm silica nanoparticles increased the dynamic stiffness of the epoxy at high strain rates, regardless of the weight content of the particles. However, when mixed with 34 μm silica particles, the dynamic stiffness is significantly reduced.

The previous literature overview clearly indicates that the experimental data are contradicting, and no clear trends can be established regarding the effect of strain rate and silica particles size and content on the mechanical behavior of epoxy resins. Despite the contradictions, part of the data still suggests that a lower weight content of silica particles, combined with a submicron size scale, has the potential to improve the mechanical properties of the epoxy resin without compromising its thermal or physical properties. This can be achieved by silica nanoparticle sizes of 300 nm up to 1 μm. However, to the best of the authors' knowledge, no data are available in the literature regarding the effect of strain rate on the compressive properties of epoxy resins filled with silica nanoparticles within this specific size range.

The aim of the present paper is to study the effect of strain rate and silica filler content on the compressive behavior of epoxy resin. The aeronautical grade RTM6 was considered, as it is suitable for low volume aircraft structures made by the resin transfer molding technique. Silica nanoparticles of sizes 300 nm and 800 nm with different surface functionalization conditions and weight percentages of 0.1%, 1% and 5% (5% wt. only for non-functionalized particles) and were investigated. High strain rate compression experiments were performed using a split Hopkinson pressure bar test (SHBT) setup. Reference quasi-static experiments were also performed in order to study the compressive behavior at a wide range of strain rates. In order to obtain accurate values of the multiaxial strain components, full-field strain measurements were performed using stereo digital image correlation techniques (DIC). The effect of strain rate on the compressive stiffness, Poisson's ratio and peak yield strength is discussed. In addition, the effect of the weight content, the size, and the surface functionalization conditions of the silica nanoparticles on the compressive behavior of epoxy nanocomposite at different strain rates is presented.

2. Materials and Methods

2.1. Matrix Material

The epoxy resin used in this study was the aeronautical grade RTM6, supplied by Hexcel Composites (Duxford, Cambrige, UK). It was made up of tetra-functional epoxy resin tetraglycidyl methylene dianiline (TGMDA) and two hardeners, namely 4,4′-methylenebis (2,6-diethylaniline) and 4,4′-methylenebis (2-isopropyl-6-methylaniline). The equivalent weight of the epoxy after mixing with the hardeners was 116 g/eq and the viscosity was 33 mPa.s at 120 °C. For the synthesis of the silica nanoparticles, tetraethyl orthosilicate (TEOS), 3-aminopropyl triethoxysilane (APTES) and other solvents supplied by Sigma-Aldrich (St. Louis, MO, USA) were used. All the chemicals were used as-received.

2.2. Nanoparticles Synthesis and Nanocomposite Preparation

Non-functionalized silica nanoparticles (NPsNF) were prepared using the Stöber method [37] with TEOS as precursors. TEOS (19.6 mL) was added drop by drop, while stirring to an alcoholic solution containing 50 mL of ethanol, 18 mL of water and 6.3 mL of ammonia. The mixture was then heated under reflux at 78 °C for 68 min. The solution was filtered and washed with deionized water, then dried in a vacuum oven (SALVIS VC20, Germany) overnight at 90 °C. The same procedure was employed for the synthesis of the functionalized silica nanoparticles (NPsF), however, an equimolar mixture of 9.8 mL TEOS and 10.3 mL APTES was employed instead of only TEOS [38]. Figure 1 shows a schematic illustration of the manufacturing process of the nanocomposites. The average diameter of the non-functionalized silica nanoparticles was 880 nm, whereas the average diameter of the functionalized silica nanoparticles was 300 nm, as depicted from the SEM images of the prepared nanoparticles (see Figure 2a,b). The Scanning Electron Microscopy (SEM) images were analyzed by ImageJ software (version 1.53m) and at least 15 particles were used to measure the average particle diameters. The reason for the size difference could be attributed to the functionality of the APTES precursor which is characterized by only three reactive functional groups (O-CH_2CH_3) compared to TEOS which has four reactive functional groups. The reduced functionality of APTES limits the nanoparticle growth, thus, explaining the smaller dimensions of the functionalized silica nanoparticles. Table 1 lists the composition of the manufactured nanocomposites.

Table 1. Composition of the manufactured nanocomposites.

Sample	Matrix	Filler	Filler Content [%]
RTM6 neat resin	RTM6	-	0
RTM6 + 0.1 wt% NPsNF	RTM6	NPsNF	0.1
RTM6 + 1 wt% NPsNF	RTM6	NPsNF	1
RTM6 + 5 wt% NPsNF	RTM6	NPsNF	5
RTM6 + 0.1 wt% NPsF	RTM6	NPsF	0.1
RTM6 + 1 wt% NPsF	RTM6	NPsF	1

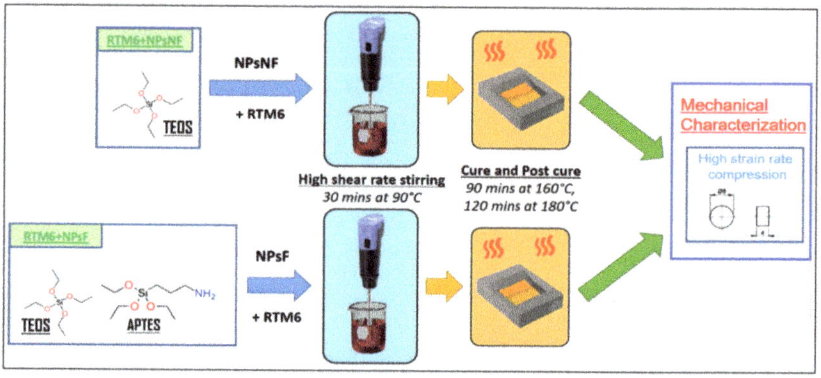

Figure 1. Manufacturing procedures of the silica/epoxy nanocomposites.

The RTM6 resin was prepared by first degassing the resin at 90 °C for 30 min in a vacuum oven, then the hardener was added and carefully mixed, according to the specified mixing ratio by the manufacturer. Different weight contents of nanoparticles were mixed in the resin using a high shear rate mixer (T25 digital ULTRA-TURRAX, from IKA, Staufen, Germany) to ensure a uniform dispersion, as depicted in Figure 2c. The weight contents of the non-functionalized silica nanoparticles were 0.1%, 1% and 5%, while the weight contents of the functionalized silica nanoparticles were 0.1% and 1%. The unreacted mixes of the resin and the silica nanoparticles were molded into long, hollow metallic cylinders which were coated with a release agent (FREKOTE 70 manufactured by Henckel, Rocky Hill, CT, USA) to facilitate the extraction of the samples. The resins were cured in an oven at 160 °C for 90 min, followed by a post-curing stage of 2 h at 180 °C, and left to cool to room temperature in the oven for 24 h.

The fully cured cylindrical rods of both neat and filled resins were finally cut into small cylindrical samples, having a diameter of 8 mm and a height of 4 mm. The selected height-to-diameter ratio of 0.5 helps to reduce the effects of interfacial friction during compression, which can give rise to significant sample barreling [39]. To eliminate any discrepancies related to the sample geometry, the same sample geometry and testing boundary conditions were used for both reference quasi-static and high strain rate tests.

Figure 2. SEM images of the synthesized silica nanoparticles: (**a**) non-functionalized, (**b**) functionalized and (**c**) non-functionalized nanoparticles (indicated by arrows) at 5% weight content in the RTM6 epoxy resin.

2.3. Dynamic Mechanical Analysis

Dynamic mechanical analysis (DMA) was performed by using the DMA Q800 system manufactured by TA Instruments. Double cantilever testing mode was employed on samples of neat and filled epoxy resins having nominal dimension of 60 mm × 12 mm × 2.5 mm. Samples were tested at an amplitude of 60 µm, a frequency of 1 Hz and a heating rate of 3 °C/min. Three tests for each silica nanoparticle concentrations were performed.

2.4. Quasi-Static Testing

Reference quasi-static experiments were carried out on the cylindrical samples using an Instron 5569 universal testing machine (supplied by Instron, Boechout, Belgium). The height of each sample was 4 mm and the diameter was 8 mm. Quasi-static compression tests were performed at speeds of 0.2, 2 and 20 mm/min, aiming at strain rates of 0.001, 0.01 and 0.1 s^{-1}, respectively, in the samples. Samples were placed between two flat ended steel bars, whose loading interfaces were polished to a mirror finish and lubricated with a PTFE lubricant to reduce friction. To eliminate any discrepancy related to sample geometry or boundary conditions, the same sample geometry and boundary conditions used in the quasi-static testing were used in the high strain rate testing.

Global displacements and strains were measured using 3 linear variable displacement transducers LVDTs (supplied by RDP Group, Le Spijkenisse, The Netherlands) fixed on the bars close to the sample. The measurements obtained from the LVDTs were later corrected for the bar compliance during compression. Local full-field displacements and strains were measured using a low speed 3D DIC setup. For this purpose, the lateral surfaces of the compression samples were painted with a black-on-white speckle pattern before testing. The deformation of the samples, from their speckle pattern, was recorded by two machine vision cameras (stingray 504b by Allied Vision, Stadtroda, Germany) positioned under a stereo angle of 18.85° (value obtained after a calibration procedure). Images were recorded at a resolution of 2452 × 2056 pixels and a rate of 1 images/s for the testing speeds of 0.2 and 2 mm/min, and 3 image/s for the testing speed 20 mm/min. Figure 3 shows the quasi-static setup used.

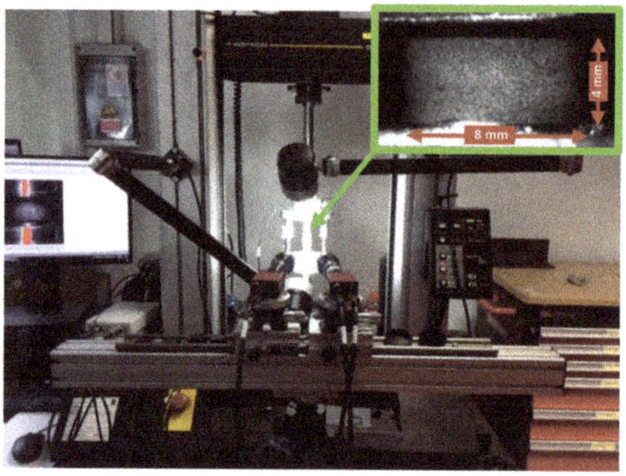

Figure 3. Quasi-static compression setup with a detail of the speckled sample (top right).

2.5. High Strain Rate Testing

The high strain rate compression experiments were performed using the split Hopkinson pressure bar (SHPB) facility available at MST-DyMaLab at Ghent University. The details of the setup were explained in previous work [11]. Figure 4 shows a schematic of the SHPB setup used. The cylindrical sample was placed between two long, aluminum bars, called input and output bars. The dynamic incident compressive loading wave was generated by accelerating the impactor towards a flange at the end of the input bar.

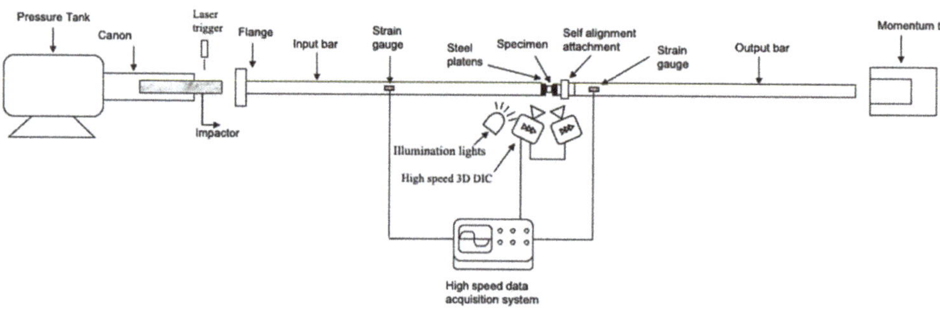

Figure 4. Schematic of the SHPB setup.

The interaction of the compressive incident loading wave with the sample resulted in the wave being partly reflected back to the input bar, and partly transmitted to the output bar. Using strain gauges attached to both bars at well-chosen locations, the strain histories corresponding with the reflected wave $\varepsilon_r(t)$ and transmitted wave $\varepsilon_t(t)$ were measured. A time shifting is applied to the measured waves to shift them from the strain gauge locations on the bars to the interfaces of the bars and sample [40]. The loading time of the incident wave was ~1.2 ms, which is long enough for the samples to reach yielding. The chosen sample geometry and small dimensions guaranteed that, from the early stage of deformation, the sample reached a state of quasi-static force equilibrium. Additionally, inertia effects were negligible and sufficiently high strain rates were achieved in the dynamic tests [41,42]. In that case, using the one dimensional wave propagation analysis developed by Kolsky [42], the time histories of the average axial strain rate $\dot{\varepsilon}^{Hop}$, strain ε^{Hop} and stress σ^{Hop} in the sample can be calculated as follows:

$$\dot{\varepsilon}^{Hop} = -2\frac{C_o}{H_s}\varepsilon_r(t) \tag{1}$$

$$\varepsilon^{Hop} = -2\frac{C_o}{H_s}\int_0^t \varepsilon_r(t)dt \tag{2}$$

$$\sigma^{Hop} = E_b\frac{A_b}{A_s}\varepsilon_t(t) \tag{3}$$

where C_0 is the elastic wave speed in the bar material; H_s is the height of the sample; E_b is the elastic modulus of the bar material; and A_b and A_s are the cross section areas of the bar and the sample, respectively. Assuming conservation of volume, the axial compressive true stress σ_t^{Hop} and axial true strains ε_t^{Hop} based on the Hopkinson analysis, i.e., Equations (2) and (3), can be calculated using the following relations:

$$\sigma_t^{Hop} = \sigma^{Hop}\left(1+\varepsilon^{Hop}\right) \tag{4}$$

$$\varepsilon_t^{Hop} = \ln\left(1+\varepsilon^{Hop}\right) \tag{5}$$

A high-speed 3D DIC technique was used to measure the local strains and strain rates on the surface of the sample. Similar to the quasi-static DIC setup, the lateral surfaces of the samples were painted with a thin black-on-white speckle pattern prior to testing. The deformation of the speckle pattern was recorded using two high-speed cameras (Photron Mini AX200) positioned at a stereo angle of 26.36° (value obtained after a calibration procedure). Images were recorded at a resolution of 384 × 265 pixels^2 and a rate of 54,000 images/s. Figure 5a shows the high-speed stereo DIC setup used.

Figure 5b shows an example of the incident, the reflected and the transmitted waves recorded in one of the dynamic compression experiments. To impose different strain rates to the sample, three impactor velocities were used: 8, 11 and 14 m/s. Note that some factors, such as sample indentation into the bars and the contact conditions between the bars and sample interfaces, could influence the accuracy of the split Hopkinson bar technique, especially in small strains measurement range [43]. To eliminate these sources of error, a self-alignment attachment was fixed to the output bar, as shown in Figure 4. This attachment was especially designed to reduce the impedance mismatch effects on the propagating waves, as evident in the strain signals of Figure 5b. Moreover, the self-alignment attachment ensured perfect contact between the specimen and the bar interfaces, even if the sample interfaces are slightly tilted. Additionally, two thin steel plates were attached to the loading interfaces of the sample to eliminate any indentation into the bars. Furthermore, the loading interfaces of the sample were lubricated with a PTFE based lubricant to minimize the interfacial friction.

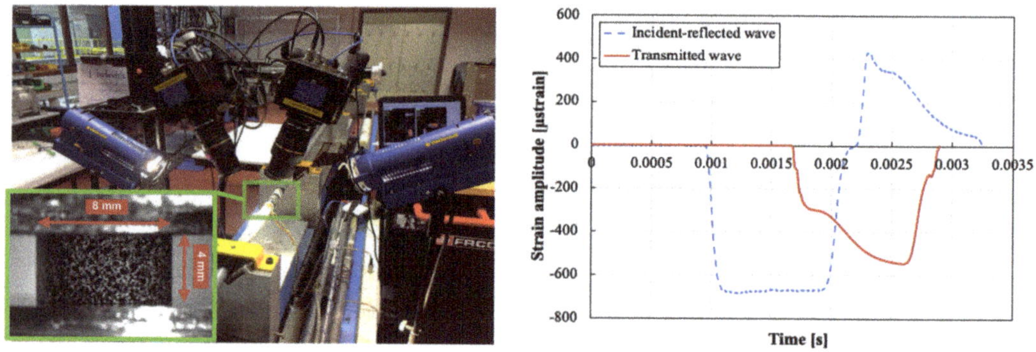

(a) (b)

Figure 5. (**a**) High-speed 3D DIC setup used, with detail of the speckled sample (bottom left). (**b**) An example of the incident, reflected and transmitted waves recorded by strain gauges on the Hopkinson bars during a dynamic compression experiment.

2.6. DIC Data Reduction and Processing Parameters

The MatchID commercial digital image correlation software (supplied by MatchID, Ghent, Belgium) was used to analyze and process the images of the deformed samples during the tests. Table 2 shows the processing parameters used for both quasi-static and dynamic tests. These parameters allowed to achieve a strain resolution of ~155 microstrains for quasi-static tests and ~400 microstrains for high strain rate tests. At each moment during the quasi-static and dynamic tests, the average full field in-plane strains and out-of-plane displacements were extracted from an area of 3.5 mm × 3.5 mm at the center of the sample. The axial engineering and true strains ε^{DIC} and ε_t^{DIC} were calculated based on the reference Biot and Hencky strain conventions, respectively. In order to increase the accuracy of the axial compressive true stress, its value was also calculated based on the transverse component of the strain ε_{hoop}^{DIC} (i.e., hoop strain) obtained using stereo DIC measurements. Therefore, it was possible to calculate the axial compressive true stress based on the instantaneous cross section, without the need to assume volume conservation as was the case for Equation (4). For this purpose, the following relation can be used:

$$\sigma_t^{DIC} = \frac{F}{A} = \frac{F}{\pi r^2} = \frac{F}{\pi r_0^2 \left(1 + \varepsilon_{hoop}^{DIC}\right)^2} = \frac{\sigma^{Hop}}{\left(1 + \varepsilon_{hoop}^{DIC}\right)^2} \tag{6}$$

Table 2. Processing parameters for DIC.

Parameter	Value
Correlation criterion	Zero normalized sum of square differences (ZNSSD)
Interpolation order	Bi-cubic spline
Shape function	Affine
Subset size (pixels x pixels)	55 × 55 (quasi-static) and 21 × 21 (high strain rate)
Step size (pixels)	10
Strain window	15

3. Results and Discussion

3.1. Dynamic Mechanical Analysis

The dynamic mechanical properties of silica-based nanocomposites were examined within the temperature range of 40 °C to 250 °C by DMA analysis. Figure 6 shows both the storage modulus and loss factor (tan delta) as a function of temperature for both nanocomposite types. It can be seen that the addition of the silica nanofillers improved the storage modulus of the RTM6 epoxy resin. The largest increase in the storage modulus, of ~11.6%, was associated with the addition of 5 wt% of non-functionalized silica nanoparticles to the neat resin. It is well known in the literature [44] that the addition of silica particles of micro and nanoscale sizes increase the storage modulus of the hosting matrix, both in the glassy and rubbery regions, due to the reduction in the free volume of the matrix. Moreover, the addition of the silica nanofillers also slightly increased the glass transition temperature of the RTM6 epoxy resin, as evident from the temperatures corresponding to the peak of the tan delta curves. This behavior is attributable to the higher surface area of the smaller nanoparticles. In fact, the greater surface of interaction between filler and matrix limits the thermal movements of the polymeric chains, causing the increase of the glass transition temperature [31]. Table 3 shows the results of the DMA.

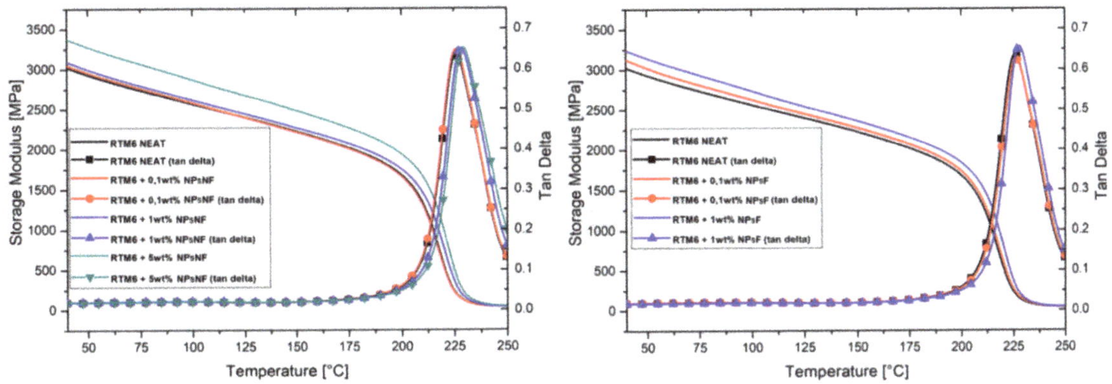

(a) (b)

Figure 6. DMA curves of the silica/RTM6 epoxy nanocomposites compared to the neat RTM6 epoxy: (**a**) NPsNF and (**b**) NPsF filled nanocomposites.

Table 3. Dynamical mechanical analysis results.

Sample Type	Storage Modulus at 40 °C (MPa)		Storage Modulus at 250 °C (MPa)		Glass Transition Temperature (°C)
	Mean	Std. Dev	Mean	Std. Dev	
RTM6 neat resin	3023	±23	38.4	±2.1	226.6 ± 0.2
RTM6 + 0.1 wt% NPsNF	3047	±43	38.8	±1.8	225.7 ± 0.3
RTM6 + 1 wt% NPsNF	3093	±17	39.4	±1.7	229.0 ± 0.2
RTM6 + 5 wt% NPsNF	3375	±33	44.5	±2.4	229.4 ± 0.4
RTM6 + 0.1 wt% NPsF	3123	±56	39.9	±6.2	226.0 ± 0.3
RTM6 + 1 wt% NPsF	3243	±31	41.0	±3.1	228.3 ± 0.4

3.2. Compressive Stress-Strain Response of RTM6 Epoxy Nanocomposites at Different Strain Rates

Figure 7 shows representative true stress–true strain curves for the RTM6 epoxy nanocomposites at different strain rates and particle weight contents. At least three experiments were performed for each testing condition. Note that the speckle pattern could not follow the deformation of the samples beyond ~30% true strain. Therefore, the quasi-static and high strain rate curves in Figure 7 are based on the LVDT and classical Hopkinson analysis (Equations (4) and (5)), respectively, as this information—unlike the DIC data—is available until fracture or unloading. For all the dynamic compression tests, the stresses at both bar interfaces were calculated using the 1D wave propagation theory, to confirm the achievement of quasi-static equilibrium. As equilibrium was established from the early stages of deformation, and the errors in the small deformation range were experimentally reduced, it was possible to calculate the elastic modulus and Poisson's ratio at high strain rates. Accurate values were obtained using the DIC strains. Moreover, the evolution of the true strain rate (i.e., true strain—time curve obtained by DIC) revealed a bilinear behavior, and consequently two stages with a different, yet relatively constant strain rate, separated with a transition point at ~0.07 true strain. Therefore, the strains rates corresponding to the elastic constants of the material were calculated in the first stage, i.e., from strain values of 0 to 0.07, while the strain rates corresponding to the yielding of the material were calculated in the second stage, i.e., from strain values 0.07 to 0.3. The strain rates indicated in Figure 7 correspond to the strain rate in the second stage. It can be seen that the compressive behavior of all tested epoxies, i.e., neat and filled, is highly strain rate sensitive. The true stress–true strain response for all materials follows 5 distinct stages, as depicted in Figure 8: (1) an initial, linear stage corresponding to the material's viscoelastic behavior; (2) a nonlinear stage corresponding to the yielding of the material [45], which reaches a maximum value at the peak yield point [45,46]; (3) a strain softening stage following the yielding and (4) further strain hardening; and (5) fracture for the quasi-static strain rates, or unloading for the high strain rates. Note that all the statically tested samples were loaded until fracture, whereas all the dynamically tested samples were not fractured at the end of loading and spring back during unloading was observed. Therefore, the strains at unloading cannot be considered a material property. All tested epoxies, filled and unfilled, showed an increase in strength with increasing strain rates. Both Gerlach et al. [41] and Morelle et al. [45] reported similar trends for RTM6 neat resin. Table 4 summarizes the results of all compression tests. The elastic modulus was calculated as the slope of the true stress–true strain curve between in the true axial strain range of 0 to 0.02, based on DIC strain measurements and Equation (6). The Poisson's ratio was calculated as the slope of the hoop strain–axial strain curve in the same strain range, based on DIC strain measurements. The peak yield strength was considered as the strength at the peak yield point [45,46], and was obtained based on the instantaneous cross section area, i.e., using Equation (6).

Table 4. Summary of the results of the neat RTM6 and RMT6/ silica epoxy nanocomposite (functionalized and non-functionalized) during compression at different strain rates.

Material Condition	Achieved True Strain Rates for Elastic Modulus and Poisson's Ratio (s^{-1})	Elastic Modulus (MPa)		Poisson's Ratio		Achieved True Strain Rates for Peak Yield Strength (s^{-1})	True Peak Yield Strength (MPa)	
		Mean	Std. Dev.	Mean	Std. Dev.		Mean	Std. Dev.
RTM6 neat resin	0.001	3095.653	46.804	0.3243	0.1256	0.001	136.403	6.2101
	0.01	3225.250	100.59	0.3208	0.0985	0.01	145.318	4.2153
	0.1	2953.969	105.521	0.2861	0.0178	0.1	155.196	7.8117
	478.638	3024.542	267.920	0.3683	0.09698	365.760	204.840	8.2448
	638.877	3166.459	52.097	0.3174	0.0017	527.334	211.603	11.305
	1034.972	3135.218	9.455	0.3608	0.0375	1017.751	212.295	1.774
RTM6 + 0.1% NPsNF	0.001	3332.913	69.506	0.3471	0.0675	0.001	142.063	3.574
	0.01	3333.440	162.954	0.3175	0.0978	0.01	153.812	7.4526
	0.1	3259.936	336.447	0.3259	0.0045	0.1	163.187	1.1321
	454.339	3264.105	1.148	0.3046	0.0531	318.821	207.955	0.994
	750.6	3233.544	109.158	0.3398	0.0073	640.054	213.176	0.717
	946.35	3174.352	58.153	0.3846	0.0189	891	213.292	0.9457
RTM6 + 1% NPsNF	0.001	3243.419	5.8576	0.3318	0.0325	0.001	146.242	0.873
	0.01	3310.858	26.257	0.3147	0.2677	0.01	152.376	0.0631
	0.1	3275.084	65.296	0.2677	0.0300	0.1	164.471	0.1899
	520.731	3264.376	79.535	0.3090	0.0013	369.592	208.867	3.8367
	786.096	3150.037	28.658	0.3172	0.0675	737.255	220.455	5.8581
	1093.333	3198.981	44.3911	0.3516	0.0306	1089.456	219.829	3.6738
RTM6 + 5% NPsNF	0.001	3282.311	53.296	0.2324	0.0038	0.001	139.690	6.1066
	0.01	3310.217	130.112	0.3299	0.0240	0.01	153.717	1.6009
	0.1	3378.446	73.952	0.2516	0.0597	0.1	170.631	0.4492
	431.035	3270.771	70.508	0.3174	0.0607	285.279	209.522	2.3349
	670.14	3180.850	83.435	0.3031	0.0809	596.957	214.067	0.5062
	1015.2	3179.854	107.427	0.3341	0.0766	1003.242	222.986	2.6408
RTM6 + 0.1% NPsF	0.001	3301.023	3.642	0.3573	0.0028	0.001	141.454	2.6183
	0.01	3327.148	66.729	0.3300	0.0792	0.01	149.692	1.6139
	0.1	3186.456	50.550	0.2572	0.0465	0.1	172.510	0.6441
	476.222	3261.809	25.714	0.3439	0.0621	342.424	203.886	1.0950
	782.692	3198.150	205.811	0.2900	0.0068	683.038	207.785	3.3535
	991.71	3273.225	45.430	0.3723	0.072	982.414	220.873	3.3542
RTM6 + 1% NPsF	0.001	3349.470	36.207	0.3013	0.0457	0.001	144.687	0.9358
	0.01	3342.377	10.823	0.3707	0.0112	0.01	152.181	3.4609
	0.1	3166.964	60.340	0.3152	0.0189	0.1	163.443	3.0203
	524.656	3202.538	132.026	0.2990	0.0818	355.420	205.585	1.7706
	711.72	3360.668	122.428	0.3875	0.0334	656.1	215.882	2.0841
	1048.6	3290.045	222.304	0.3576	0.1438	1004.4	230.579	3.1928

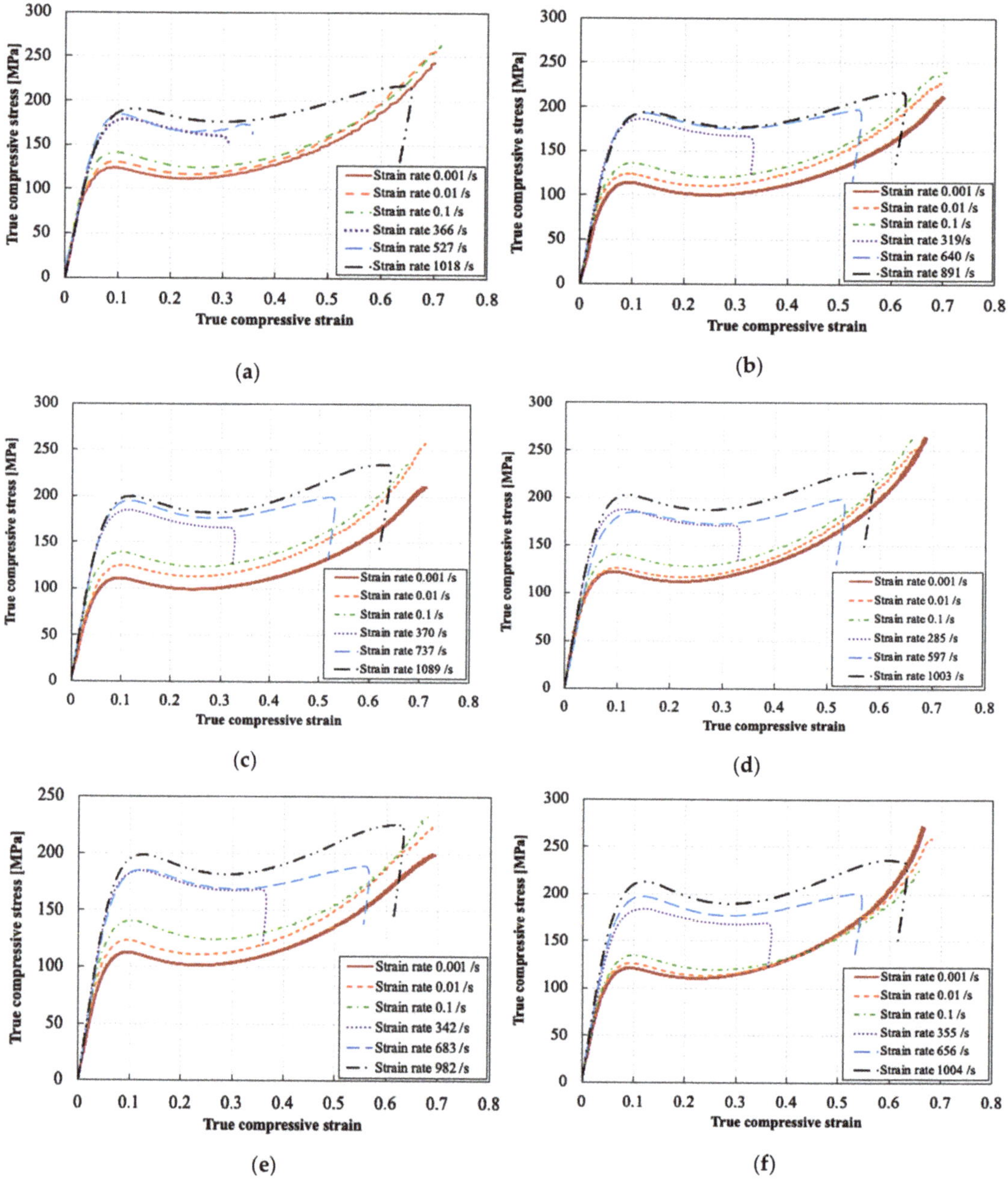

Figure 7. Representative compressive true stress-true strain curves at different strain rates for the RTM6 epoxy/silica nanocomposites: (**a**) unfilled neat resin, (**b**) non-functionalized 0.1%, (**c**) non-functionalized 1%, (**d**) non-functionalized 5%, (**e**) functionalized 0.1% and (**f**) functionalized 1%.

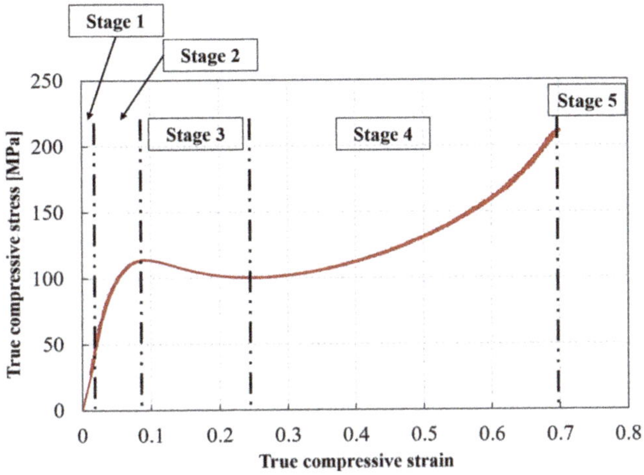

Figure 8. Stages of deformation of a silica nanoparticle sample (0.1% NPsNF) in compression.

3.3. *Effect of Strain Rate and Weight Content on the Elastic Modulus and Poisson's Ratio of the Silica Nanoparticles Filled RTM6 Epoxy Resin*

Figures 9 and 10 show the effect of strain rate on the elastic modulus and the Poisson's ratio of silica nanoparticle-filled epoxy resins at different particle weigh contents and functionalization conditions. The elastic modulus was slightly improved by the addition of the silica nanoparticles with different weight percentages of the particles across all strain rates. The elastic modulus of the neat resin was increased from ~3100 MPa to ~3300 MPa by the addition of silica nanoparticles. This corresponds to an increase of ~6.4%. Moreover, the elastic modulus was hardly affected by strain rate. Increasing the weight content of both particle types up to 5% also did not have a significant effect on the elastic modulus across all strain rates.

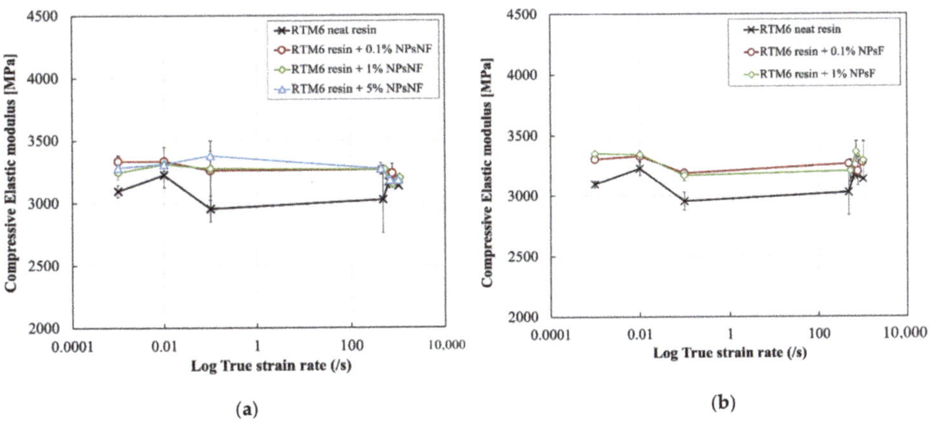

Figure 9. Effect of strain rate on the elastic modulus of the silica nanoparticle-filled epoxy at different particle weight contents and functionalization conditions: (**a**) non-functionalized and (**b**) functionalized.

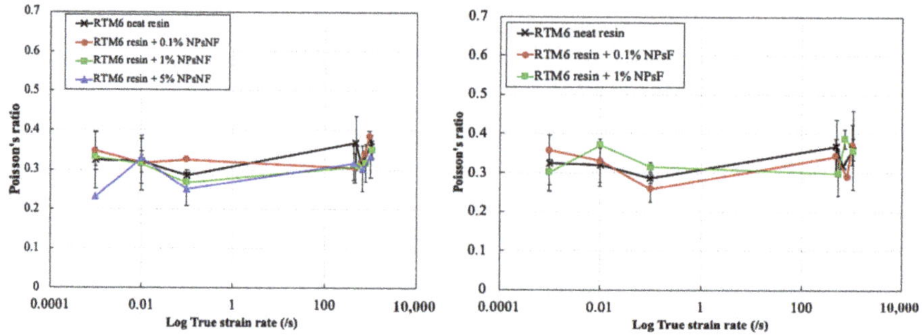

(a) (b)

Figure 10. Effect of strain rate on the Poisson's ratio of the silica nanoparticle-filled epoxy at different particle weight contents and functionalization conditions: (**a**) non-functionalized and (**b**) functionalized.

The Poisson's ratio also is independent of strain rate for all materials. The average value of the Poisson's ratio for the neat and silica nanoparticle filled epoxies was ~0.32. The slight increase in the elastic modulus by the addition of the silica nanoparticles is due to the transfer of forces from the matrix to the higher stiffness particles. As seen earlier in Section 3.1, the epoxy resin showed a predominantly elastic behavior rather than a viscoelastic behavior at the early stages of loading based on the DMA results. Consequently, the resin material is expected to show a strain rate independent response at small strains due to the lack of contribution of the damping component in the deformation behavior at high strain rates.

3.4. *Effect of Strain Rate and Weight Content on the Peak Yield Strength of the Silica Nanoparticle-Filled RTM6 Epoxy Resin*

Figure 11 shows the effect of strain rate on the true peak yield strength for functionalized and non-functionalized silica nanoparticle filled epoxy resins. The addition of the silica nanoparticles with different weight contents improved the true peak yield strength at different strain rates compared to the neat resin. Hardly any change was observed in the trend of the true peak yield strength with increasing strain rate as a result of the surface functionalization conditions of the silica nanoparticles. Compared to the neat resin at low strain rates, the peak yield strength of the 0.1% silica content increased from 116.4 MPa to 122.42 MPa at strain rate of 0.0008 s^{-1} and from 135.19 MPa to 143.18 MPa at strain rate of 0.08 s^{-1}. This corresponds to percentage increase of 5.2% and 5.9%, respectively. This increase was almost the same regardless of the filler content in the low strain rate range up to 0.08 s^{-1}.

Moreover, it can be seen that the improvement of the peak yield strength was much more significant in the low strain rate range compared to the high strain rate range. Indeed, compared to the neat resin, the peak yield strength of the 0.1% silica nanoparticle content increased from 184.84 MPa to 187.95 MPa at strain rate of 277.5 s^{-1} and 303.7 s^{-1}, respectively, and from 192.3 MPa to 193.39 MPa at strain rate of 1019.65 s^{-1} and 904.1 s^{-1}. This corresponds to percentage increases of 1.6% and 0.56%, respectively, which are much lower compared to the percentage increases in the low strain rate range. The most significant improvement of the true peak yield strength was observed at silica weight contents of 1% and 5%, especially at low strain rates. Similar results were also reported by Tian et al. [33] and Miao et al. [32], but for a different epoxy formulation and a different size of nanoparticles. The increase in the true peak yield strength of the epoxy/silica nanocomposite is attributed mainly to the viscoelastic nature of the resin. However, further research is still required to understand why the improvement of the yield strength is reduced at high strain rates compared to low strain rates. At low strain rates, the viscoelastic resin has enough time to deform. This allows the transfer of the forces from the matrix to the higher strength and higher stiffness silica nanoparticles, which further increase the yield strength. However, at high strain rates, the viscoelastic resin does not have enough time to fully deform due to the reduced molecular mobility of the polymer chains, as demonstrated by Chen et al. [47]. The reduced molecular mobility at high strain rates could reduce the interaction between the resin and the silica nanoparticles [48]. This behavior was also observed for carbon nanotube fillers, as reported by Del Rio et al. [9].

Another contributing factor is the adiabatic heating at high strain rates, which cannot be neglected. Indeed, Pan et al. [49] reported that the temperature rise in epoxy samples due to adiabatic heating at high strain rates can reach up to 90 °C. Furthermore, it was reported by Del-Rio et al. [50] that an increase of 40 °C can reduce the yield strength of epoxy by 23% at high strain rates. Note that Miao et al. [32] suggested that the strain softening of the epoxy matrix is the main contributing factor of the reduction in the yield strength at higher strain rates regardless of the nanoparticle weight percentage added. However, this conclusion was based on a simple model which was validated only for silica nanoparticle content of 10% and cannot be directly extended to other highly crosslinked epoxy resins and other weight percentages of silica nanoparticles.

The increasing trend of the true peak yield strength of the silica filled epoxies with increasing strain rate can be described by a least square power fit relation as follows (R^2 values are greater than 0.9):

$$\sigma_t^{Peak\ yield} = C\dot{\varepsilon}^d \tag{7}$$

where $\sigma_t^{Peak\ yield}$ is the true peak yield strength in compression, C is the compressive strength coefficient and d is the strain rate sensitivity exponent. For the non-functionalized silica filled epoxy resin, the compressive strength coefficients for the 0.1%, 1% and 5% weight contents were 154.85 MPa, 156.13 MPa and 157.04 MPa respectively, while the strain rate sensitivity exponents for the same consecutive weight contents were 0.0335, 0.0353 and 0.0354, respectively. For the functionalized silica-filled epoxy resin, the compressive strength coefficients for the 0.1% and 1%, weight contents were 155.09 MPa and 156.96 MPa, respectively, while the strain rate sensitivity exponents for the same consecutive weight contents were 0.0331 and 0.0355, respectively.

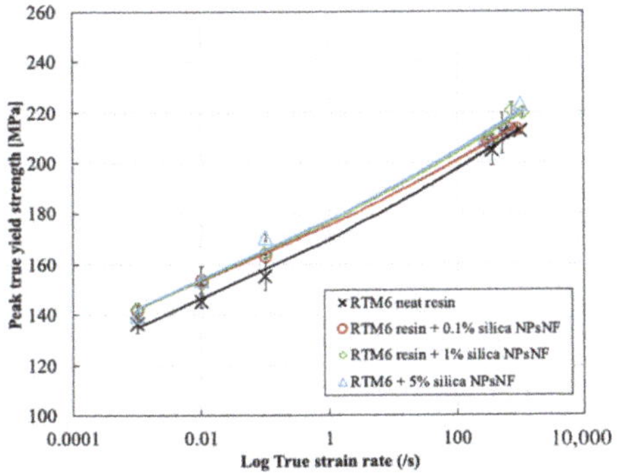

(a)

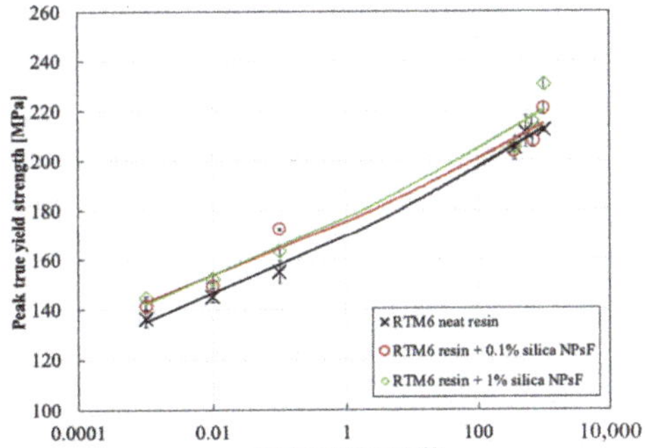

(b)

Figure 11. Effect of strain rate on the compressive true peak yield strength for the silica nanoparticle-filled epoxy at different particle weight contents and functionalization conditions: (**a**) non-functionalized and (**b**) functionalized.

3.5. *Effect of the Silica Nanoparticles Size and Surface Functionalization of on the Elastic Modulus, Poisson's Ratio and Peak Yield Strength of RTM6 Epoxy Nanocomposite*

Despite the different surface functionalization conditions of the silica nanoparticles used in this study, a rough estimate of the effect of the different sizes of these nanoparticles on the compressive behavior of the epoxy resin can still be studied. Figures 12–14 show the effect of the silica nanoparticles size and surface functionalization conditions on the peak true yield strength, elastic modulus, and Poisson's ratio, respectively, for weight percentages of 0.1% and 1% at different strain rates. It can be seen that for a silica nanoparticle content of 0.1%, the size of the particles and the surface functionalization conditions did not have a significant effect on the true peak yield strength and the elastic modulus all strain rates. Whereas for a silica nanoparticle content of 1%, a very slight increase in the true peak yield strength can be seen at the high strain rate range as a result of reducing the particle size from 880 nm to 300 nm and the functionalization of the particle surface. The size and the surface functionalization of the nanoparticles also did not show a significant effect on the elastic modulus and the Poisson's ratio at different strain rates for both filler contents. Similar results were reported by Dittanet et al. [29] for a similar epoxy system at quasi-static strain rates and silica nanoparticle size range from 23 nm to 170 nm. Here, again, further research is required to understand why the change of the silica nanoparticle sizes in the range 300 nm to 880 nm does not significantly affect the compressive properties of the epoxy resin, particularly at high strain rate. As explained earlier, the combined effect of viscoelasticity and adiabatic heating could be the main contributing factors in that case.

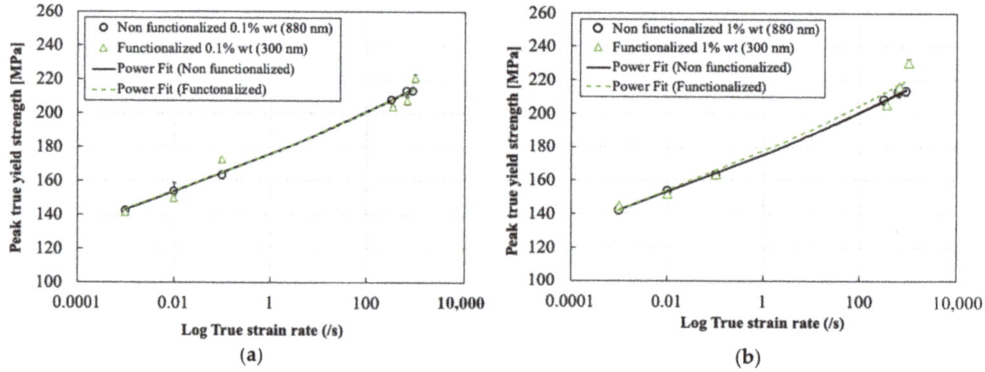

Figure 12. Effect of silica nanoparticles size and surface functionalization on the compressive true peak yield strength: (**a**) 0.1% and (**b**) 1%.

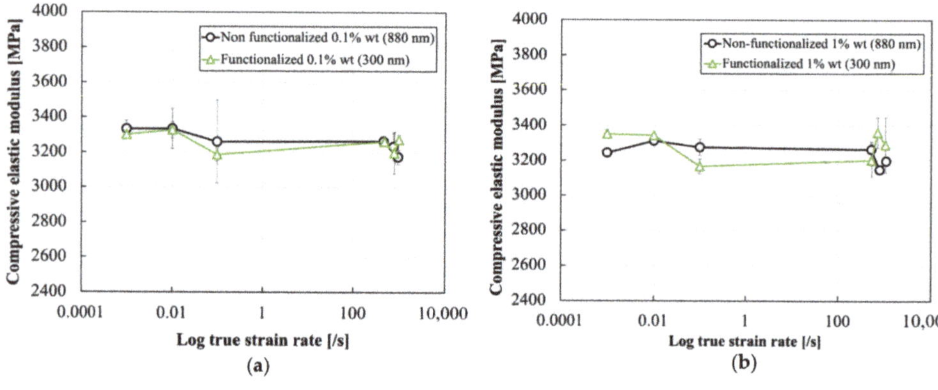

Figure 13. Effect of silica nanoparticles size and surface functionalization on the compressive elastic modulus for the silica nanoparticle-filled epoxy at different particle weight contents: (**a**) 0.1% and (**b**) 1%.

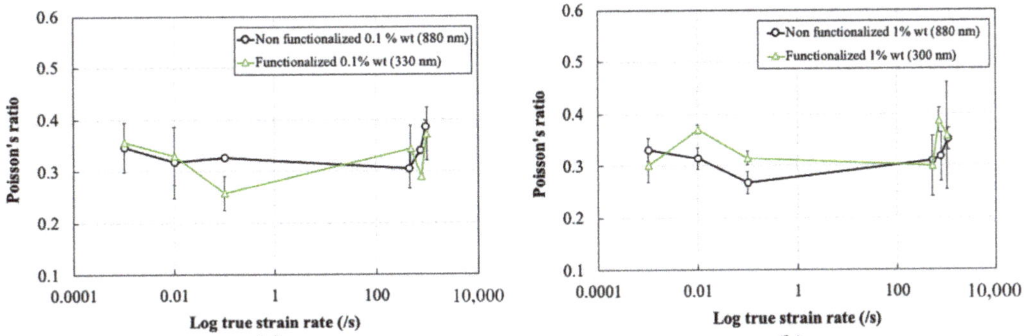

(a) (b)

Figure 14. Effect of silica nanoparticles size and surface functionalization on the Poisson's ratio for the silica nanoparticle: (**a**) 0.1% and (**b**) 1%.

4. Conclusions

Several experiments were performed to study the effect of strain rate and filler content on the compressive behavior of several epoxy-based nanocomposites. The aeronautical grade RTM6 epoxy was filled with silica nanoparticles of sizes 300 nm and 880 nm and different surface functionalization conditions. Three weight percentages were considered for the fillers: 0.1%, 1% and 5% (5% wt. only for non-functionalized particles). Quasi-static and high strain rate compression experiments were performed using a universal testing machine and a SHPB setup, respectively, to cover a strain rate range from 0.001 s^{-1} up to 1100 s^{-1}. Local displacements and strains in the sample were measured using the 3D digital image correlation technique. The effect of the strain rate, the size and the weight percentage of the silica nanoparticles on the elastic modulus, the Poisson's ratio and the true peak yield strength of the tested materials were discussed. Considering the tested materials, manufacturing techniques used, testing equipment and results, the following can be concluded:

1. The tested RTM6 neat and nanoparticle filled resins were all strain rate-sensitive in compression. All materials showed an increase in strength with increasing strain rates for all the weight percentages and sizes of the fillers.
2. The elastic modulus and Poisson's ratio of the tested epoxy nanocomposites were independent of the strain rate and showed a nearly constant behavior at different strain rates for all weight percentages and sizes of the particles. However, the true peak yield strength showed an increase with increasing strain rates for all weight percentages and sizes of the particles used.
3. The addition of silica nanoparticles to the RTM6 epoxy resin generally improved both its elastic modulus and its peak yield strength at different strain rates for all the weight percentages of the particles. Increasing the weight percentage of both types the silica nanoparticles from 0.1% to 5% did not yield any improvement in the elastic modulus and the Poisson's ratio but led to a slight increase in the peak yield strength. Additionally, it was found that the improvement in the peak yield strength due to the addition of silica nanoparticles was more prominent in the quasi-static strain rate regime compared to the high strain rate regime.
4. The dynamic mechanical analysis showed an increase in the storage modulus and a marginal increase in the glass transition temperature of the resin by the addition of silica nanoparticles of different weight percentages.
5. The sizes of the silica nanoparticles used (300 nm and 880 nm) did not significantly affect the compressive properties of the RTM6 epoxy resin, regardless of the weight percentages of the particles.

Author Contributions: Conceptualization, A.E., A.Z., M.Z., A.B. and P.V.; methodology, A.E., A.Z. and P.V.; formal analysis, A.E. and A.Z.; investigation, A.E., A.Z. and S.Z.; data curation, A.E. and A.Z.; writing—original draft preparation, A.E. and A.Z.; writing—review and editing, M.Z., A.B. and P.V.; visualization, A.E. and A.Z.; supervision, A.B., M.Z. and P.V.; funding acquisition, A.B. and P.V. All authors have read and agreed to the published version of the manuscript.

Funding: The project "EXTREME" leading to this publication has received funding from the European Union's Horizon 2020 research and innovation program under grant agreement No. 636549.

Institutional Review Board Statement: Not applicable.

Informed Consent Statement: Not applicable.

Data Availability Statement: The data presented in this study are available on request from the corresponding author. The data are not publicly available due to confidentiality.

Conflicts of Interest: The authors declare no conflict of interest. The funders had no role in the design of the study; in the collection, analyses or interpretation of data, in the writing of the manuscript; or in the decision to publish the results.

References

1. Mallick, P.K. *Composites Engineering Handbook*, 1st ed.; Mallick, P.K., Ed.; CRC Press: Boca Raton, FL, USA, 1997; ISBN 9780429175497.
2. Domun, N.; Hadavinia, H.; Zhang, T.; Liaghat, G.; Vahid, S.; Spacie, C.; Paton, K.R.; Sainsbury, T. Improving the fracture toughness properties of epoxy using graphene nanoplatelets at low filler content. *Nanocomposites* **2017**, *3*, 85–96. [CrossRef]
3. Duan, Z.; He, H.; Liang, W.; Wang, Z.; He, L.; Zhang, X. Tensile, quasistatic and dynamic fracture properties of nano-Al2O3-modified epoxy resin. *Materials* **2018**, *11*, 905. [CrossRef]
4. Naik, N.K.; Pandya, K.S.; Kavala, V.R.; Zhang, W.; Koratkar, N.A. Alumina nanoparticle filled epoxy resin: High strain rate compressive behavior. *Polym. Eng. Sci.* **2014**, *54*, 2896–2901. [CrossRef]
5. Gurusideswar, S.; Velmurugan, R.; Gupta, N.K. High strain rate sensitivity of epoxy/clay nanocomposites using non-contact strain measurement. *Polymer* **2016**, *86*, 197–207. [CrossRef]
6. Sprenger, S. Epoxy resins modified with elastomers and surface-modified silica nanoparticles. *Polymer* **2013**, *54*, 4790–4797. [CrossRef]
7. Kausar, A. Rubber toughened epoxy-based nanocomposite: A promising pathway toward advanced materials. *J. Macromol. Sci. Part A* **2020**. [CrossRef]
8. Bie, B.X.; Han, J.H.; Lu, L.; Zhou, X.M.; Qi, M.L.; Zhang, Z.; Luo, S.N. Dynamic fracture of carbon nanotube/epoxy composites under high strain-rate loading. *Compos. Part A Appl. Sci. Manuf.* **2015**, *68*, 282–288. [CrossRef]
9. Gómez-Del Río, T.; Rodríguez, J.; Pearson, R.A. Compressive properties of nanoparticle modified epoxy resin at different strain rates. *Compos. Part B Eng.* **2014**, *57*, 173–179. [CrossRef]
10. Buonocore, G.G.; Schiavo, L.; Attianese, I.; Borriello, A. Hyperbranched polymers as modifiers of epoxy adhesives. *Compos. Part B Eng.* **2013**, *53*, 187–192. [CrossRef]
11. Zotti, A.; Elmahdy, A.; Zuppolini, S.; Borriello, A.; Verleysen, P.; Zarrelli, M. Aromatic hyperbranched polyester/RTM6 epoxy resin for extreme dynamic loading aeronautical applications. *Nanomaterials* **2020**, *10*, 188. [CrossRef] [PubMed]
12. Boogh, L.; Pettersson, B.; Månson, J.A.E. Dendritic hyperbranched polymers as tougheners for epoxy resins. *Polymer* **1999**, *40*, 2249–2261. [CrossRef]
13. Wang, F.; Drzal, L.T.; Qin, Y.; Huang, Z. Enhancement of fracture toughness, mechanical and thermal properties of rubber/epoxy composites by incorporation of graphene nanoplatelets. *Compos. Part A Appl. Sci. Manuf.* **2016**, *87*, 10–22. [CrossRef]
14. Kinloch, A.J.; Mohammed, R.D.; Taylor, A.C.; Eger, C.; Sprenger, S.; Egan, D. The effect of silica nano particles and rubber particles on the toughness of multiphase thermosetting epoxy polymers. *J. Mater. Sci.* **2005**, *40*, 5083–5086. [CrossRef]
15. Hsieh, T.H.; Kinloch, A.J.; Masania, K.; Taylor, A.C.; Sprenger, S. The mechanisms and mechanics of the toughening of epoxy polymers modified with silica nanoparticles. *Polymer* **2010**, *51*, 6284–6294. [CrossRef]
16. Liang, Y.L.; Pearson, R.A. Toughening mechanisms in epoxy-silica nanocomposites (ESNs). *Polymer* **2009**, *50*, 4895–4905. [CrossRef]
17. Linec, M.; Mušič, B. The effects of silica-based fillers on the properties of epoxy molding compounds. *Materials* **2019**, *12*, 1811. [CrossRef] [PubMed]
18. Bondioli, F.; Darecchio, M.E.; Luyt, A.S.; Messori, M. Epoxy resin modified with in situ generated metal oxides by means of sol-gel process. *J. Appl. Polym. Sci.* **2011**, *122*, 1792–1799. [CrossRef]
19. Aliofkhazraei, M. *Handbook of Nanoparticles*; Springer International Publishing: Cham, Switzerland, 2015; ISBN 9783319153384.

20. Chen, C.; Justice, R.S.; Schaefer, D.W.; Baur, J.W. Highly dispersed nanosilica-epoxy resins with enhanced mechanical properties. *Polymer* **2008**, *49*, 3805–3815. [CrossRef]
21. Jumahat, A.; Soutis, C.; Jones, F.R.; Hodzic, A. Effect of silica nanoparticles on compressive properties of an epoxy polymer. *J. Mater. Sci.* **2010**, *45*, 5973–5983. [CrossRef]
22. Kothmann, M.H.; Bakis, G.; Zeiler, R.; Ziadeh, M.; Breu, J.; Altstädt, V. Fatigue crack growth behaviour of epoxy nanocomposites—Influence of particle geometry. *Superconductivity* **2017**, *247*, 23–32.
23. Tian, Y.; Zhang, H.; Zhang, Z. Influence of nanoparticles on the interfacial properties of fiber-reinforced-epoxy composites. *Compos. Part A Appl. Sci. Manuf.* **2017**, *98*, 1–8. [CrossRef]
24. Rahman, I.A.; Padavettan, V. Synthesis of silica nanoparticles by Sol-Gel: Size-dependent properties, surface modification, and applications in silica-polymer nanocompositesa review. *J. Nanomater.* **2012**, *2012*, 132424. [CrossRef]
25. Macan, J.; Paljar, K.; Burmas, B.; Špehar, G.; Leskovac, M.; Gajović, A. Epoxy-matrix composites filled with surface-modified SiO2 nanoparticles. *J. Therm. Anal. Calorim.* **2017**, *127*, 399–408. [CrossRef]
26. León, L.M.; Laza, M.; Puchol, V.; Torrens, F.; Abad, C.; Campos, A. Dynamic mechanical measurements of epoxy matrix-silica nanocomposites ii. *Polym. Polym. Compos.* **2009**, *17*, 313–324. [CrossRef]
27. Kamran-Pirzaman, A.; Rostamian, Y.; Babatabar, S. Surface improvement effect of silica nanoparticles on epoxy nanocomposites mechanical and physical properties, and curing kinetic. *J. Polym. Res.* **2020**, *27*, 13. [CrossRef]
28. Kang, S.; Hong, S.I.; Choe, C.R.; Park, M.; Rim, S.; Kim, J. Preparation and characterization of epoxy composites filled with functionalized nanosilica particles obtained via sol-gel process. *Polymer* **2001**, *42*, 879–887. [CrossRef]
29. Dittanet, P.; Pearson, R.A. Effect of silica nanoparticle size on toughening mechanisms of filled epoxy. *Polymer* **2012**, *53*, 1890–1905. [CrossRef]
30. Bondioli, F.; Cannillo, V.; Fabbri, E.; Messori, M. Epoxy-silica nanocomposites: Preparation, experimental characterization, and modeling. *J. Appl. Polym. Sci.* **2005**, *97*, 2382–2386. [CrossRef]
31. Sun, Y.; Zhang, Z.; Moon, K.S.; Wong, C.P. Glass transition and relaxation behavior of epoxy nanocomposites. *J. Polym. Sci. Part B Polym. Phys.* **2004**, *42*, 3849–3858. [CrossRef]
32. Miao, Y.G.; Liu, H.Y.; Suo, T.; Mai, Y.W.; Xie, F.Q.; Li, Y.L. Effects of strain rate on mechanical properties of nanosilica/epoxy. *Compos. Part B Eng.* **2016**, *96*, 119–124. [CrossRef]
33. Tian, Y.; Zhang, H.; Zhao, J.; Li, T.; Bie, B.X.; Luo, S.N.; Zhang, Z. High strain rate compression of epoxy based nanocomposites. *Compos. Part A Appl. Sci. Manuf.* **2016**, *90*, 62–70. [CrossRef]
34. Guo, Y.; Li, Y. Quasi-static/dynamic response of SiO_2-epoxy nanocomposites. *Mater. Sci. Eng. A* **2007**, *458*, 330–335. [CrossRef]
35. Ma, P.; Jiang, G.; Li, Y.; Zhong, W. The impact compression behaviors of silica nanoparticles—Epoxy composites. *J. Text. Sci. Technol.* **2015**, *01*, 1–11. [CrossRef]
36. Sekiguchi, Y. Effects of mixed micro and nano silica particles on the dynamic compressive performances of epoxy adhesive. *Appl. Adhes. Sci.* **2017**, *5*, 1–12. [CrossRef]
37. Fang, M.; Volotinen, T.T.; Kulkarni, S.K.; Belova, L.; Rao, K.V. Effect of embedding Fe3O4 nanoparticles in silica spheres on the optical transmission properties of three-dimensional magnetic photonic crystals. *J. Appl. Phys.* **2010**, *108*, 103501. [CrossRef]
38. Zotti, A.; Zuppolini, S.; Borriello, A.; Zarrelli, M. Thermal properties and fracture toughness of epoxy nanocomposites loaded with hyperbranched-polymers-based core/shell nanoparticles. *Nanomaterials* **2019**, *9*, 418. [CrossRef] [PubMed]
39. Jerabek, M.; Major, Z.; Lang, R.W. Uniaxial compression testing of polymeric materials. *Polym. Test.* **2010**, *29*, 302–309. [CrossRef]
40. Kolsky, H. An investigation of the mechanical properties of materials at very high rates of loading. *Proc. Phys. Soc. Sect. B* **1949**, *62*, 676–700. [CrossRef]
41. Gorham, D.A. The effect of specimen dimensions on high strain rate compression measurements of copper. *J. Phys. D Appl. Phys.* **1991**, *24*, 1489–1492. [CrossRef]
42. Gerlach, R.; Siviour, C.R.; Petrinic, N.; Wiegand, J. Experimental characterisation and constitutive modelling of RTM-6 resin under impact loading. *Polymer* **2008**, *49*, 2728–2737. [CrossRef]
43. Miao, Y.G.; Li, Y.L.; Liu, H.Y.; Deng, Q.; Shen, L.; Mai, Y.W.; Guo, Y.Z.; Suo, T.; Hu, H.T.; Xie, F.Q.; et al. Determination of dynamic elastic modulus of polymeric materials using vertical split Hopkinson pressure bar. *Int. J. Mech. Sci.* **2016**, *108–109*, 188–196. [CrossRef]
44. Sperandio, C.; Laachachi, A.; Ruch, D.; Poilâne, C.; Bourson, P.; Salvestrini, J.P.; Ahzi, S. Use of functionalized nanosilica to improve thermo-mechanical properties of epoxy adhesive joint bonding aluminium substrates. *J. Nanosci. Nanotechnol.* **2010**, *10*, 2844–2849. [CrossRef] [PubMed]
45. Morelle, X.P.; Chevalier, J.; Bailly, C.; Pardoen, T.; Lani, F. Mechanical characterization and modeling of the deformation and failure of the highly crosslinked RTM6 epoxy resin. *Mech. Time Depend. Mater.* **2017**, *21*, 419–454. [CrossRef]
46. Tamrakar, S.; Ganesh, R.; Sockalingam, S.; Haque, B.Z.; Gillespie, J.W. Strain rate-dependent large deformation inelastic behavior of an epoxy resin. *J. Compos. Mater.* **2020**, *54*, 71–87. [CrossRef]
47. Chen, L.P.; Yee, A.F.; Moskala, E.J. The molecular basis for the relationship between the secondary relaxation and mechanical properties of a series of polyester copolymer glasses. *Macromolecules* **1999**, *32*, 5944–5955. [CrossRef]
48. Rose, S.; Dizeux, A.; Narita, T.; Hourdet, D.; Marcellan, A. Time dependence of dissipative and recovery processes in nanohybrid hydrogels. *Macromolecules* **2013**, *46*, 4095–4104. [CrossRef]

49. Pan, Z.; Sun, B.; Shim, V.P.W.; Gu, B. Transient heat generation and thermo-mechanical response of epoxy resin under adiabatic impact compressions. *Int. J. Heat Mass Transf.* **2016**, *95*, 874–889. [CrossRef]
50. Gómez-del Río, T.; Rodríguez, J. Compression yielding of epoxy: Strain rate and temperature effect. *Mater. Des.* **2012**, *35*, 369–373. [CrossRef]

polymers

MDPI

Article

Facile and Green Process to Synthesize a Three-Dimensional Network Few-Layer Graphene/Carbon Nanotube Composite for Electromagnetic Interference Shielding

Yu-Hong Yeh [1], Kuei-Ting Hsu [2,*], Chia-Hung Huang [3,4] and Wei-Ren Liu [1,*]

1 Department of Chemical Engineering, R&D Center for Membrane Technology, Chung Yuan Christian University, 200 Chung Pei Road, Chungli District, Taoyuan City 320, Taiwan; ziv910543@gmail.com
2 Department of Chemical Engineering, Army Academy, No.750, Longdong Rd., Chung-Li 320, Taiwan
3 Department of Electrical Engineering, National University of Tainan, No.33, Sec. 2, Shulin St., West Central District, Tainan City 700, Taiwan; chiahung@mail.mirdc.org.tw
4 Metal Industries Research and Development Centre, Kaohsiung 701, Taiwan
* Correspondence: kt1768@gmail.com (K.-T.H.); wrliu1203@gmail.com (W.-R.L.); Tel.: +886-3-265-4140 (W.-R.L.)

Abstract: We propose an environmentally friendly liquid exfoliation approach and subsequent freeze-drying process for constructing a three-dimensional (3D) carbon-based network by using few-layer graphene (FLG) and carbon nanotubes (CNTs) for electromagnetic interference (EMI) shielding applications. Systematic characterizations—such as X-ray diffraction, scanning electron microscopy, and transmission electron microscopy—as well as Raman characterization and EMI shielding tests were performed. The results indicated that the as-synthesized 3D-FLG/CNT composite obtained through the freeze-drying process exhibited excellent electromagnetic interference shielding. The shielding effect of FLG could be improved from 15 to 22 dB by introducing CNTs. The CNTs inhibited restacking of FLG in the structure. We also compared two drying processes: oven drying and freeze-drying. The freeze-drying technique markedly improved the shielding effect of FLG/CNTs from 22 to 36 dB. The composition-optimized 3D-FLG/CNT composite could be a candidate material for use in EMI shielding.

Keywords: graphene; carbon nanotubes; electromagnetic interference shielding

Citation: Yeh, Y.-H.; Hsu, K.-T.; Huang, C.-H.; Liu, W.-R. Facile and Green Process to Synthesize a Three-Dimensional Network Few-Layer Graphene/Carbon Nanotube Composite for Electromagnetic Interference Shielding. *Polymers* **2022**, *14*, 1892. https://doi.org/10.3390/polym14091892

Academic Editors: Ting-Yu Liu and Yu-Wei Cheng

Received: 8 April 2022
Accepted: 29 April 2022
Published: 5 May 2022

1. Introduction

Electronic products have become increasingly widespread in daily life. Nevertheless, electronic devices emit electromagnetic waves that are invisible and harmful to the human body [1–8]. Being in an environment with many such devices for a long period may lead to a concern for health. Thus, the development of electromagnetic-wave-shielding materials to reduce or even eliminate electromagnetic waves is crucial [9–12]. In earlier days, the widely suggested shielding material was metal [13], which has a strong electromagnetic-wave-shielding effect (SE) due to its high electrical conductivity [14–17]. However, metal-based materials may undergo corrosion and oxidation. Their higher density also limits their application [18]. The most critical disadvantage is that the shielding offered by metal originates from its reflection efficiency [19,20]. The creation of secondary reflected electromagnetic waves does not solve the problem and can damage electronic devices and shorten their lifetime. Carbon-based materials are promising alternatives to metal due to their lightness, corrosion resistance, flexibility, and high conductivity [21–28]. Carbon-based materials such as few-layer graphene (FLG) and carbon nanotubes (CNTs) have been demonstrated to be potential electromagnetic-wave-shielding materials [29–50].

Liu et al. used chitosan to improve the microporous structure of the interface between graphene sheets, thereby increasing the conductivity of the modified graphene/iron pentacarbonyl porous membrane. The electromagnetic wave-SE was as high as 38 dB. The SE was 6.3 times higher than that of reduced graphene oxide/epoxy (6 dB). The conductivity of the chitosan-modified material (40.2 S/m) was 1340 times higher than that of the unmodified material (0.03 S/m) [36]. Fu et al. reported a highly flexible single-walled CNT (SWCNT)/graphene film. After folding 1000 times, the SWCNT/graphene film maintained excellent mechanical and electrical stability. A structure for multiple reflections was constructed between layers of SWCNT/graphene films to produce an excellent EMI SE of approximately 80 dB [37]. Bagotia et al. successfully synthesized a synergistic graphene/CNT hybrid nanostructure by connecting multiwalled CNTs (MWCNTs) between graphene layers; with the construction of a polycarbonate/vinyl methyl acrylate graphene MWCNT structure, this hybrid improved mechanical properties, electrical conductivity, and EMI shielding performance. The highest conductivity of this structure was 1.91×10^{-1} S/cm. The corresponding electromagnetic SE was 34 dB [38]. Zhang et al. developed foam comprising thermally reduced graphene oxide/CNTs (~0.65 g/cm^3); this foam had a microporous structure and exhibited excellent conductivity of 2.92 S/m and high EMI SE of 30.4 dB [39]. Zhu et al. used graphene and oxide CNTs (OCNTs) as precursors to construct a 3D conductive framework through simple and convenient filtration and calcination methods. For polydimethylsiloxane-coated r-OCNT/thermally-reduced graphene composite filter cake with a thickness of approximately 1 mm, a strong EMI SE of 67.3 dB was obtained in the frequency range of 8.2–12.4 GHz. After repeated bending 10,000 times, the SE of the composite material was still 62.7 dB [40].

In this paper, we propose a simple and facile jet cavitation process for synthesizing FLG and FLG/carbon nanotube (FLG/CNT) suspensions. One-dimensional CNTs were introduced into the skeleton of FLG to prevent restacking of FLG during drying processes. We compared two drying methods: oven drying and freeze-drying. The results indicated that freeze-dried samples had a highly porous structure and exhibited the highest EMI shielding performance of 40 dB.

2. Materials and Methods

2.1. Materials and Chemicals

Graphite was purchased from Knano Graphene Technology (Xiamen, China). MWCNTs were purchased from Jiangsu Cnano Technology (Zhenjiang, China). Paraffin wax was purchased from Sigma-Aldrich (Burlington, MA, USA).

2.2. Preparation of FLG/CNTs/Wax Composite

Scheme 1 displays the schema of the FLG, graphene nanotube (GNT) (FLG/CNT), FFLG (freeze-dried FLG), and FGNT (freeze-dried FLG/CNT) composite materials. The lower-left corner shows the FLG/CNT/wax composite. Graphite and carbon nanotubes were added to deionized water to attain a solid content of 5 wt.%. The liquid exfoliation process of graphite was implemented as follows. First, artificial graphite (purity > 99%, Knano Graphene Technology, Xiamen, China) and MWCNTs (purity > 99%, Jiangsu Cnano Technology, Zhenjiang, China) were used as feed materials for the delamination experiment. All materials were used as supplied without further purification. Deionized water was used for the preparation of all graphite suspensions and graphite/CNTs (5 wt.%). All delamination experiments were conducted in a low-temperature ultra-high-pressure continuous flow cell disrupter (JNBIO, JN 10C, Guangzhou, China). In this device, a graphite and graphite/CNT suspension was prepared through high pressure forcing the sample through a small orifice at high speed. Because of the shearing, hole, and impact, the graphite in the sample was crushed; the substances in the sample were dispersed and emulsified. FLG would be kept in suspension after liquid exfoliation at 14 °C–16 °C in a circulating water bath. Deionized water was used as a solvent. Graphene production was performed in two batch runs at 800 bar and two batch runs at 1200 bar. The as-synthesized FLG and the FLG/CNT suspension were dried through oven drying or freeze-drying processes. The condition for oven drying was 80 °C and the freeze-drying conditions were −55 °C under a pressure of 20 Pa. The final samples were oven-dried (FLG, GNT10) or freeze-dried (FFLG, FGNT10).

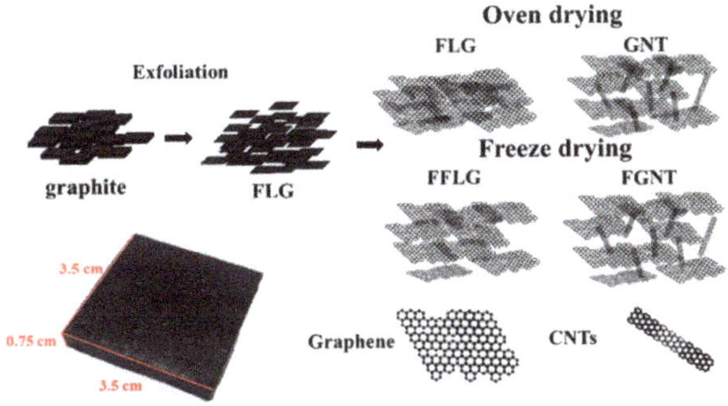

Scheme 1. Schema of few-layer graphene (FLG), graphene nanotube (GNT), freeze-dried FLG (FFLG), and freeze-dried GNT (FGNT) composite materials. Lower left corner shows the FLG/CNT/wax composite with dimensions of 3.5 cm × 3.5 cm × 0.75 cm.

2.3. *Preparation of EMI Shielding Testing Plates*

FLG or FLG/CNT powder was mixed with paraffin by 15 wt.% at 180 °C for 2 h. Subsequently, the mixture was poured into a mold (3.5 cm × 3.5 cm), and pressure of 17,000 pounds (1.5×10^6 N/m^2) was applied for 30 min. Finally, the bulk material was removed for electromagnetic-wave-shielding measurements.

2.4. *Characterizations*

The surface morphology and crystal structure of the materials were characterized using field emission scanning electron microscopy (SEM, JEOL/JSM-7600F, Tokyo, Japan), high-resolution scanning transmission electron microscopy (TEM, JEOL-JEM2100, Tokyo, Japan), and X-ray diffraction (XRD, Bruker D8, Billerica, MA, USA). Raman characterization was performed using a LabRAM ARAMIS Raman confocal microscope (HORIBA JobinYvon, Edison, NJ, USA) with 532-nm laser excitation. The lateral dimensions of FLG and CNTs were evaluated. The specific surface area was determined using the Brunauer–Emmett–Teller equation (Micromeritics Tristar 3000, Norcross, GA, USA). A four-point probe tester (KeithLink LRS4-T) was used for electrical performance analysis to measure the sheet resistance of FLG and CNTs and electrical conductivity of the nanocomposite. The sheet resistance and electronic conductivity were measured 10 times. Atomic force microscope (AFM) images were captured by a Bruker Dimension Icon. The samples for AFM were prepared by dropping the dispersion directly onto freshly cleaved mica wafer with an injector. The microwave scattering parameters (S11 and S21) of FLG and CNTs were measured using a vector network analyzer (Keysight E5071C) in the X-band frequency range of 8.2–12.4 GHz. The nanocomposites were sandwiched between the holder (22.86 mm × 10.16 mm) within the measurement device. The SE of total EMI shielding (SE_T), microwave absorption (SE_A), and microwave reflection (SE_R) were calculated using parameters S_{11} and S_{21} as follows: $R = |S_{11}|^2$, $T = |S_{21}|^2$ and $A = 1 - R - T$. The effective absorbance (A_{eff}) can be expressed as $A_{eff} = (1 - R - T)/(1 - R)$. The total EMI SE ($SE_T$) is the sum of the reflection from the material surface (SE_R), absorption of electromagnetic energy (SE_A), and multiple internal reflections (SE_M) of electromagnetic radiation. SE_T can thus be described as follows:

$$SE_T = SE_R + SE_A + SE_M$$

When $SE_T > 15$ dB, SE_M can be ignored. Therefore, SE_T is usually defined as follows:

$SE_T \approx SE_R + SE_A$

SE_R and SE_A can be calculated as follows:

$SE_R = -\log10(1 - R)$

$SE_A = -\log10(1 - A_{eff}) = -\log10[T/(1 - R)]$

3. Results and Discussion

Figure 1a,b shows the surface morphology of graphite and FLG, respectively. The lateral size of the graphite was 2–5 μm. The morphology of the graphite was clearly thick and irregular. Liquid exfoliation was performed to obtain FLG, which had a smaller lateral size and thickness of <10 nm. This indicated that the initial graphite could be effectively exfoliated into thin sheets through a jet cavitation device. A thickness line scan and an atomic force microscopy image of FLG are displayed in Figure 1c,d, respectively. The thickness of FLG was approximately 3 nm (~10 layers) [51,52].

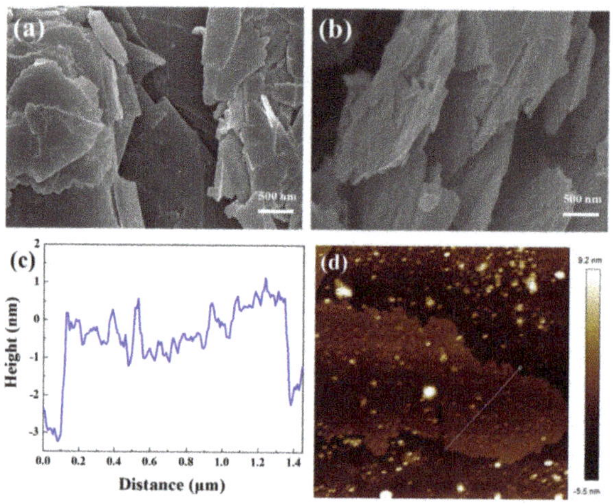

Figure 1. Scanning electron microscopy (SEM) images of (**a**) graphite and (**b**) FLG; (**c**) line scan and (**d**) atomic force microscopy image of FLG.

Figure 2a,b presents the SEM images of the FLG/CNT composites obtained through conventional drying, namely FLG and GNT10, respectively. Under optimized operating conditions, single-layer, double-layer, and multilayer chemically converted graphene may be differentiated based on their opacity in SEM images. CNTs were used to intercalate the graphene, and FLG combined with CNTs was coded as GNT10. After the CNTs had been added, it was observed that the CNTs were uniformly distributed on the surface of FLG. Figure 2c displays a TEM image of FLG, which exhibited the few-layer characteristics of graphene sheets. The TEM of GNT10 (Figure 2d) revealed that CNTs were intercalated between the graphene layers. This was attributed to the mechanical liquid-phase exfoliation method enabling the CNTs to be intercalated while expanding the few layers of graphene [53,54]. Thus, the CNTs were evenly distributed throughout FLG.

Figure 3a–d displays the SEM and TEM images of the FLG/CNT composites obtained through freeze-drying, namely FFLG and FGNT10, respectively. Conventional oven drying caused restacking of FLG (Figure 2a), resulting in a poor SE. After freeze-drying, however, surface morphology was smooth, with larger interlayer spacing, as shown in Figure 3a,c. The TEM image also indicates that the CNTs were distributed throughout FFLG, and that the layers were fewer (Figure 3b,d). The uniform dispersion of CNTs on FFLG, shown in Figure 3b,d, was due to the solvent being evaporated directly during the freeze-drying process, whereas only the solvent on the surface of FFLG was evaporated during conventional oven drying. Thus, the graphene layers were restacked, and the SE was lower.

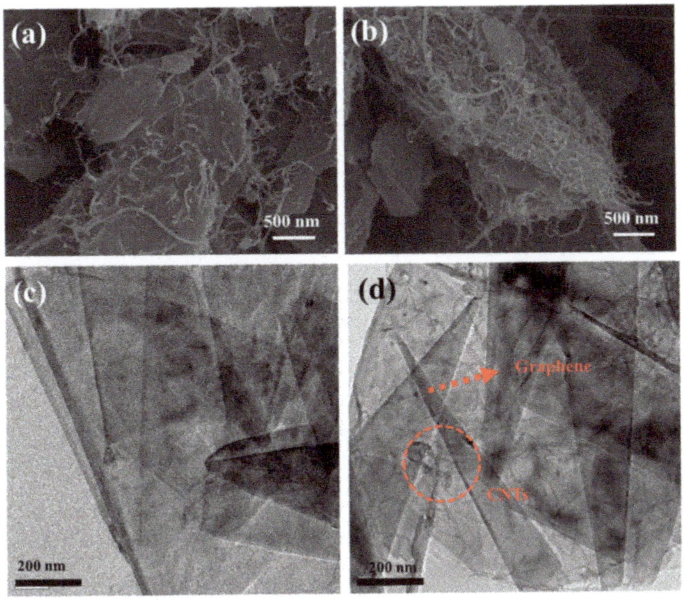

Figure 2. SEM images of (**a**) FLG and (**b**) GNT10; TEM images of (**c**) FLG and (**d**) GNT10.

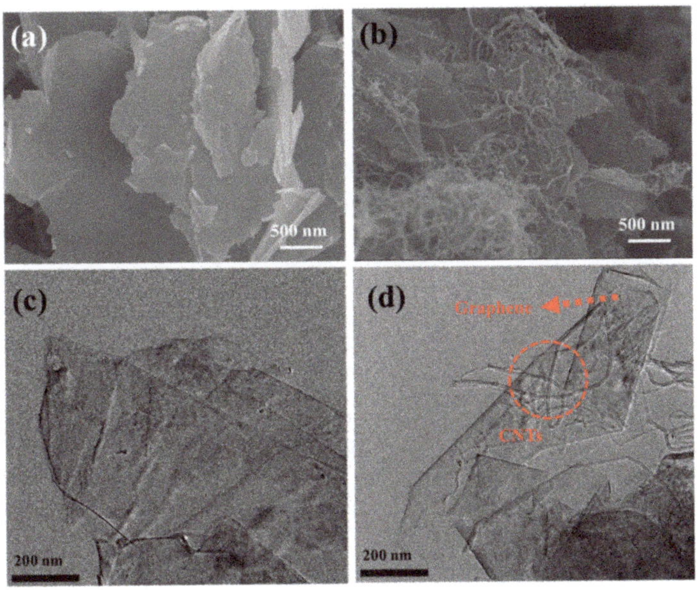

Figure 3. SEM images of (**a**) FFLG and (**b**) FGNT10; TEM images of (**c**) FFLG and (**d**) FGNT10.

Figure 4 displays a representative Raman spectrum of FLG (excited at 532 nm). The spectrum included a strong graphene band at 1594 cm^{-1}, corresponding to graphite. After exfoliation, the spectrum of FLG had a strong graphene band at 1578 cm^{-1}. The I_D/I_G ratio for FLG slightly increased from 0.131 to 0.133 after the jet cavitation process. The position of the peak corresponding to graphene was shifted nearly 20 cm^{-1} from 1594 to 1578 cm^{-1}. Graphite was thus successfully exfoliated into FLG [38]. Conversely, the graphene bands in the spectra of FFLG and FGNT10 were shifted from 1557 and 1566 cm^{-1}, respectively. The

I_D/I_G ratio for FGNT10 was 0.19. This was because a Raman deviation in the low-angle direction represents a decrease in the number of graphene layers.

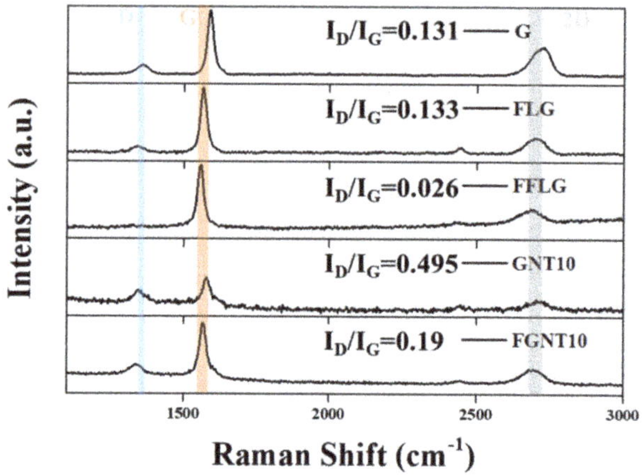

Figure 4. Raman spectra of graphite, FLG, FFLG, GNT10, and FGNT10.

The adsorption–desorption isotherms and corresponding pore size distributions of a series of FLG-based materials are presented in Figure 5. According to Figure 5a,c, the specific surface area of FFLG (30.6 m^2/g) was higher than that of FLG (29.7 m^2/g). As shown in Figure 5b, the distribution contained two peaks at approximately 4 and 50 nm. Figure 5d shows a narrow peak at approximately 3 nm and a broad peak at approximately 100 nm. The volume of the freeze-dried macropores was lower for the freeze-dried samples than for the oven-dried samples. The specific surface areas of GNT10 and FGNT10, shown in Figure 5e,g, were measured to be 49.8 and 51.0 m^2/g, respectively. The greater specific surface area may be due to the introduction of CNTs to prevent restacking of FLG. Figure 5f,h reveals no difference in the pore size distributions of GNT10 and FGNT10. To determine the percentages of micropores, mesopores, and macropores for these samples, we obtained the three corresponding pore distributions by using the BJH method, which are displayed in Figure 5i. Based on the pore size distributions of these FLG-based materials, these materials mainly contained macropores and mesopores. The specific surface area of FLG increased from 29.7 to 30.6 m^2/g (an increase of 1.54%) through freeze-drying and could be markedly enhanced to 49.8 m^2/g (an increase of 67.54%). The FGNT10 sample had 51% macropores, 49% mesopores, and 0% micropores. The composition of the material (FLG or FLG/CNTs) and the drying process (oven drying or freeze-drying) thus have critical effects on both the specific surface area and pore size distribution.

The EMI SE of FLG is shown in Figure 6a. FFLG and FGNT10 exhibited favorable absorption. Comparing GNT10 with FLG, the CNT-intercalated material exhibited higher absorption efficiency—22 versus 15 dB. Figure 6b shows the shielding benefit provided by reflection. The material we used, graphene, did not contribute considerably to the reflection benefit. Figure 6c illustrates the total SE, which is a combination of the results shown in Figure 6a,b. Figure 6d displays a histogram revealing the contributions made to EMI shielding in terms of absorption versus reflection for the graphite, FLG, FFLG, GNT10, and FGNT10 samples. FGNT10 provided an SE of nearly 36 dB. Moreover, the addition of CNTs to FLG to obtain GNT10 increased the SE from 15 to 22 dB. For few-layer graphene, most of the contribution to EMI shielding was from absorption. Graphite became restacked during oven drying; thus, the SE decreased from 17 to 15 dB. The EMI SE was improved to as high as 35 dB by the freeze-drying process.

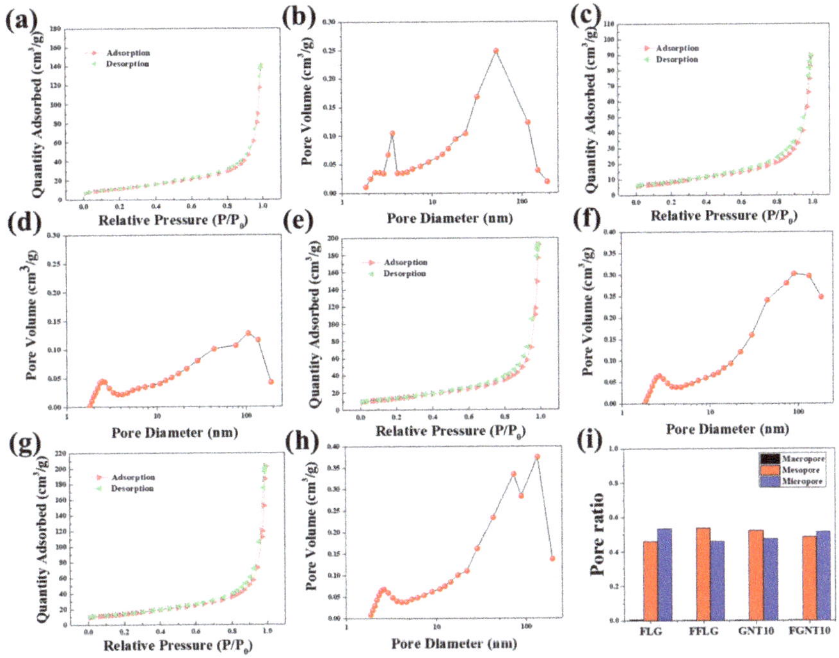

Figure 5. Adsorption–desorption isotherms (**a**,**c**,**e**,**g**), corresponding pore size distributions (**b**,**d**,**f**,**h**) and (i) micropores/mesopores/macropores distribution of graphite, FLG, FFLG, GNT10, and FGNT10.

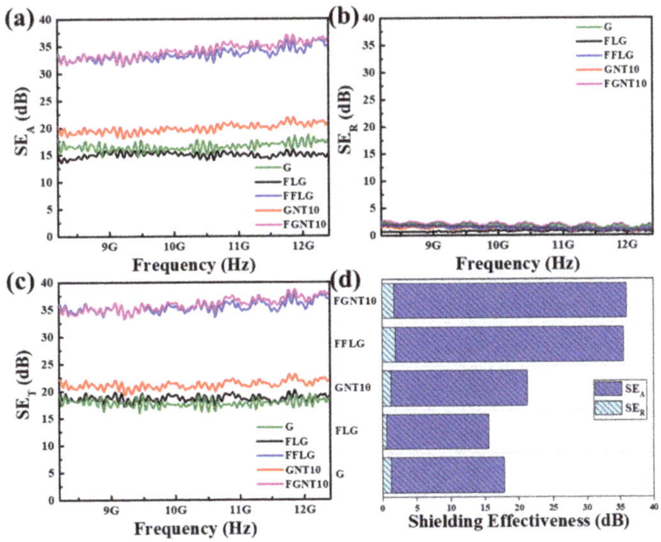

Figure 6. Electromagnetic shielding effect (SE) of G (graphene), FLG, FFLG, GNT10, and FGNT10 samples: (**a**) SE_A, (**b**) SE_R, and (**c**) S_{ET} in the X-band (8.2–12.4 GHz). (**d**) Electromagnetic-wave-shielding histogram of total SE.

A schema of electromagnetic wave shielding using GNT10 is presented in Figure 7a. Oven drying causes restacking of graphene layers during the drying process. Figure 7b displays a schema of the shielding achieved using FGNT10. EMI is attenuated through three major mechanisms: reflection, absorption, and multiple reflections. In cases where the shielding effect through absorption (i.e., absorption loss) is higher than 10 dB, most of the re-reflected waves are absorbed within the shield.

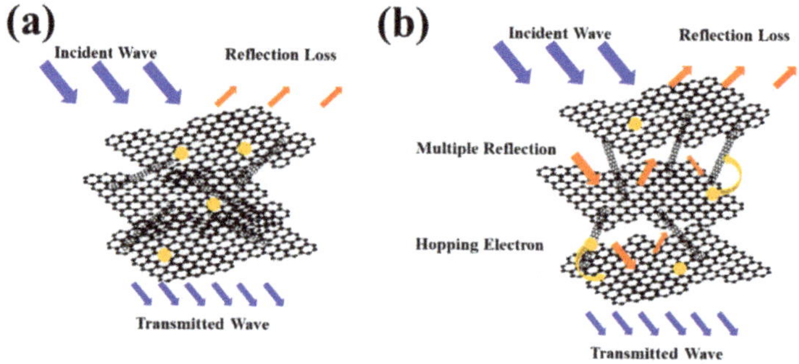

Figure 7. Schema of electromagnetic wave shielding by (**a**) GNT10 and (**b**) FGNT10.

Figure 8 shows the four-point probe measurement results for FLG, FFLG, GNT10, and FGNT10. CNT addition and freeze-drying improved the sheet resistance of material flakes. Furthermore, the addition of CNTs increased interlayer spacing and sheet resistance. Freeze drying did not cause the restacking of graphene layers; thus, the sheet resistance of the FGNT10 sample was 1.25×10^{-1} Ω/■.

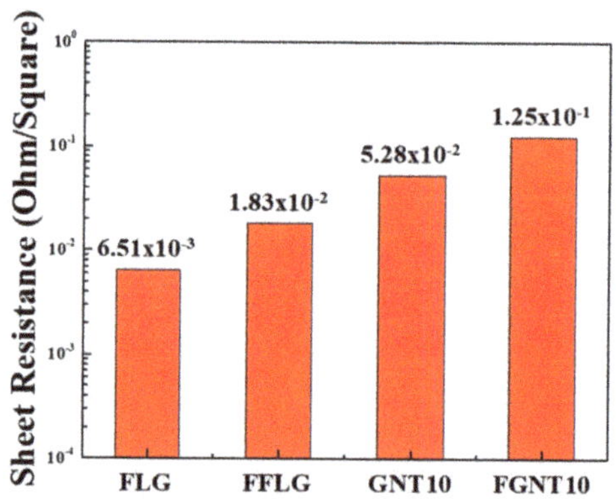

Figure 8. Sheet resistances of the FLG, FFLG, GNT10, and FGNT10 samples.

4. Conclusions

In summary, we proposed a liquid exfoliation method for preparing high-quality FLG. By introducing CNTs into FLG, the phenomenon of graphene restacking caused by oven drying can be avoided. The freeze-drying process increases the material's porosity and solves the problem of graphene restacking. The EMI shielding of FGNT10 was as high as 36 dB. By combining FLG, CNTs, and the freeze-drying approach, we obtained a sample, GNT10, with potential for EMI shielding in the range of 8–12 GHz.

Author Contributions: Y.-H.Y. wrote the paper; W.-R.L. and K.-T.H. conceived and designed the experiments; C.-H.H. analyzed the data. All authors have read and agreed to the published version of the manuscript.

Funding: This research was funded by Ministry of Science and Technology [MOST 109-2923-E-006-006, 110-2622-M-033-001, 109-2622-E-033-010, 110-2923-E-006-011, 110-3116-F-011-002, and 108-E-033-MY3].

Institutional Review Board Statement: Not applicable.

Informed Consent Statement: Not applicable.

Data Availability Statement: The data presented in this study are available on request from the corresponding author.

Conflicts of Interest: The authors declare no conflict of interest.

References

1. Tsydenova, O.; Bengtsson, M. Chemical hazards associated with treatment of waste electrical and electronic equipment. *Waste Manag.* **2011**, *31*, 45–58. [CrossRef] [PubMed]
2. Qin, F.; Brosseau, C. A review and analysis of microwave absorption in polymer composites filled with carbonaceous particles. *J. Appl. Phys.* **2012**, *111*, 061301. [CrossRef]
3. Che, R.C.; Peng, L.M.; Duan, X.F.; Chen, Q.; Liang, X.L. Microwave absorption enhancement and complex permittivity and permeability of Fe encapsulated within carbon nanotubes. *Appl. Phys. Lett.* **2004**, *16*, 401–405. [CrossRef]
4. Cao, W.Q.; Wang, X.X.; Yuan, J.; Wang, W.Z.; Cao, M.S. Temperature dependent microwave absorption of ultrathin graphene composites. *J. Mater. Chem. C* **2015**, *3*, 10017–10022. [CrossRef]
5. Chung, D.D. Electromagnetic interference shielding effectiveness of carbon materials. *Carbon* **2001**, *39*, 279–285. [CrossRef]
6. Radasky, W.A.; Baum, C.E.; Wik, M.W. Introduction to the special issue on high-power electromagnetics (HPEM) and intentional electromagnetic interference (IEMI). *IEEE Trans. Electromagn. Compat.* **2004**, *46*, 314–321. [CrossRef]
7. Chen, Z.; Xu, C.; Ren, W.; Ma, C.; Cheng, H.M. Lightweight and flexible graphene foam composites for high-performance electromagnetic interference shielding. *Adv. Mater.* **2013**, *25*, 1296–1300. [CrossRef]
8. MacGregor, D.G.; Slovic, P.; Morgan, M.G. Perception of risks from electromagnetic fields: A psychometric evaluation of a risk-communication approach. *Risk Anal.* **1994**, *14*, 815–828. [CrossRef]
9. Cheng, W.H.; Hung, W.C.; Lee, C.H.; Hwang, G.L.; Jou, W.S.; Wu, T.L. Low-cost and low-electromagnetic-interference packaging of optical transceiver modules. *J. Lightwave Technol.* **2004**, *22*, 2177. [CrossRef]
10. Wen, S.H.; Chung, D.D.L. Design and operation of a real-scale electromagnetic shielding evaluation system for reinforced composite construction materials. *Cem. Concr. Res.* **2018**, *34*, 329. [CrossRef]
11. Fauveaux, S.; Wojkiewicz, J.L.; Miane, J.L. Broadband electromagnetic shields using polyaniline composites. *Electromagnetics* **2003**, *23*, 617. [CrossRef]
12. Huang, C.J.; Chang, T.C. Studies on the electromagnetic interference shielding effectiveness of metallized PVAc-$AgNO_3$/PET conductive films. *J. Appl. Polym. Sci.* **2004**, *91*, 270. [CrossRef]
13. Azim, S.S.; Satheesh, A.; Ramu, K.K.; Ramu, S.; Venkatachari, G. Studies on graphite based conductive paint coatings. *Prog. Org. Coat.* **2006**, *55*, 1–4. [CrossRef]
14. Singh, A.P.; Mishra, M.; Sambyal, P.; Gupta, B.K.; Singh, B.P.; Chandra, A. Encapsulation of γFe_2O_3 decorated reduced graphene oxide in polyaniline core–shell tubes as an exceptional tracker for electromagnetic environmental pollution. *J. Mater. Chem. A* **2014**, *2*, 3581–3593. [CrossRef]
15. Zhang, X.J.; Wang, G.S.; Cao, W.Q.; Wei, Y.Z.; Liang, J.F.; Guo, L. Enhanced microwave absorption property of reduced graphene oxide (RGO)-$MnFe_2O_4$ nanocomposites and polyvinylidene fluoride. *ACS Appl. Mater. Interfaces* **2014**, *6*, 7471–7478. [CrossRef] [PubMed]
16. He, S.; Wang, G.S.; Lu, C.; Liu, J.; Wen, B.; Liu, H. Enhanced wave absorption of nanocomposites based on the synthesized complex symmetrical CuS nanostructure and poly (vinylidene fluoride). *J. Mater. Chem. A* **2013**, *1*, 4685–4692. [CrossRef]
17. Singh, B.; Choudhary, V.; Saini, P.; Pande, S.; Singh, V.; Mathur, R. Enhanced microwave shielding and mechanical properties of high loading MWCNT–epoxy composites. *J. Nanoparticle Res.* **2013**, *15*, 1554. [CrossRef]
18. Thomassin, J.M.; Jerome, C.; Pardoen, T.; Bailly, C.; Huynen, I.; Detrembleur, C. Polymer/carbon based composites as electromagnetic interference (EMI) shielding materials. *Mater. Sci. Eng. R Rep.* **2013**, *74*, 211–232. [CrossRef]
19. Choi, W.H.; Kim, C.G. Broadband microwave-absorbing honeycomb structure with novel design concept. *Compos. Part B Eng.* **2015**, *83*, 14. [CrossRef]
20. Lee, K.P.M.; Czajka, M.; Shanks, R.; Daver, F. Low-defect graphene–polyamide-6 composites and modeling the filler–matrix interface. *J. Appl. Polym. Sci.* **2020**, *137*, 48630. [CrossRef]
21. Yuan, Y.; Yin, W.; Yang, M.; Xu, F.; Zhao, X.; Li, J.; Peng, Q.; He, X.; Du, S.; Li, Y. Lightweight flexible and strong core-shell non-woven fabrics covered by reduced graphene oxide for high-performance electromagnetic interference shielding. *Carbon* **2018**, *130*, 59–68. [CrossRef]
22. Yan, D.X.; Pang, H.; Li, B.; Vajtai, R.; Xu, L.; Ren, P.G.; Wang, J.H.; Li, Z.M. Structured reduced graphene oxide/polymer composites for ultra-efficient electromagnetic interference shielding. *Adv. Funct. Mater.* **2015**, *25*, 559–566. [CrossRef]
23. Zhang, H.B.; Yan, Q.; Zheng, W.G.; He, Z.; Yu, Z.Z. Tough graphene-polymer microcellular foams for electromagnetic interference shielding. *ACS Appl. Mater. Interfaces* **2011**, *3*, 918–924. [CrossRef] [PubMed]
24. Alam, F.E.; Yu, J.; Shen, D.; Dai, W.; Li, H.; Zeng, X.; Yao, Y.; Du, S.; Jiang, N.; Lin, C.T. Highly conductive 3D segregated graphene architecture in polypropylene composite with efficient EMI shielding. *Polymers* **2017**, *9*, 662. [CrossRef]
25. Liu, X.; Chai, N.; Yu, Z.; Xu, H.; Li, X.; Liu, J.; Yin, X.; Riedel, R. Ultra-light, high flexible and efficient CNTs/Ti_3C_2-sodium alginate foam for electromagnetic absorption application. *J. Mater. Sci. Technol.* **2019**, *35*, 2859–2867. [CrossRef]

26. Wang, H.; Shao, Y.; Mei, S.; Lu, Y.; Zhang, M.; Sun, J.-K.; Matyjaszewski, K.; Antonietti, M.; Yuan, J. Polymer-derived heteroatom-doped porous carbon materials. *Chem. Rev.* **2020**, *120*, 9363–9419. [CrossRef]
27. Tuccitto, N.; Spitaleri, L.; Li-Destri, G.; Pappalardo, A.; Gulino, A.; Sfrazzetto, T.G. Supramolecular sensing of a chemical warfare agents simulant by functionalized carbon nanoparticles. *Molecules* **2020**, *25*, 5731. [CrossRef]
28. Jang, D.; Park, J.-E.; Kim, Y.-K. Evaluation of (CNT@CIP)-embedded magneto-resistive sensor based on carbon nanotube and carbonyl iron powder polymer composites. *Polymers* **2022**, *14*, 542. [CrossRef]
29. Song, P.; Liang, C.; Wang, L.; Qiu, H.; Gu, H.; Kong, J.; Gu, J. Obviously improved electromagnetic interference shielding performances for epoxy composites via constructing honeycomb structural reduced graphene oxide. *Compos. Sci. Technol.* **2019**, *181*, 107698. [CrossRef]
30. Alam, F.E.; Dai, W.; Yang, M.; Du, S.; Li, X.; Yu, J.; Jiang, N.; Lin, C.T. Potential applications of quantum dots in mapping sentinel lymph node and detection of micrometastases in breast carcinoma. *J. Breast Cancer* **2013**, *16*, 1–11. [CrossRef]
31. Li, X.H.; Li, X.; Liao, K.N.; Min, P.; Liu, T.; Dasari, A.; Yu, Z.Z. Thermally annealed anisotropic graphene aerogels and their electrically conductive epoxy composites with excellent electromagnetic interference shielding efficiencies. *ACS Appl. Mater. Interfaces* **2016**, *8*, 33230–33239. [CrossRef] [PubMed]
32. Zhang, Y.; Wang, L.; Zhang, J.; Song, P.; Xiao, Z.; Liang, C.; Qiu, H.; Kong, J.; Gu, J. Fabrication and investigation on the ultra-thin and flexible $Ti_3C_2T_x$/co-doped polyaniline electromagnetic interference shielding composite films. *Compos. Sci. Technol.* **2019**, *183*, 107833. [CrossRef]
33. Dai, W.; Ma, T.; Yan, Q.; Gao, J.; Tan, X.; Lv, L.; Hou, H.; Wei, Q.; Yu, J.; Wu, J.; et al. Metal-level thermally conductive yet soft graphene thermal interface materials. *ACS Nano* **2019**, *13*, 11561–11571. [CrossRef] [PubMed]
34. Nguyen, D.D.; Tai, N.H.; Chen, S.Y.; Chueh, Y.L. Controlled growth of carbon nanotube-graphene hybrid materials for flexible and transparent conductors and electron field emitters. *Nanoscale* **2012**, *4*, 632–638. [CrossRef] [PubMed]
35. Yu, D.; Dai, L. Self-assembled graphene/carbon nanotube hybrid films for supercapacitors. *J. Phys. Chem. Lett.* **2010**, *1*, 467–470. [CrossRef]
36. Liu, J.; Zhang, B.H.; Liu, Y.; Wang, Q.; Liu, Z.; Mai, Y.W.; Yu, Z.Z. Magnetic electrically conductive and lightweight graphene/iron pentacarbonyl porous films enhanced with chitosan for highly efficient broadband electromagnetic interference shielding. *Compos. Sci. Technol.* **2017**, *151*, 71–78. [CrossRef]
37. Fu, H.; Yang, Z.; Zhang, Y.; Zhu, M.; Jia, Y.; Chao, Z.; Hu, D.; Li, Q. SWCNT-modulated folding-resistant sandwich-structured graphene film for high-performance electromagnetic interference shielding. *Carbon* **2020**, *162*, 490–496. [CrossRef]
38. Bagotia, N.; Choudhary, V.; Sharma, D.K. Synergistic effect of graphene/multiwalled carbon nanotube hybrid fillers on mechanical, electrical and EMI shielding properties of polycarbonate/ethylene methyl acrylate nanocomposites. *Compos. Part B Eng.* **2019**, *159*, 378–388. [CrossRef]
39. Zhang, H.; Zhang, G.; Gao, Q.; Zong, M.; Wang, M.; Qin, J. Electrically electromagnetic interference shielding microcellular composite foams with 3D hierarchical graphene-carbon nanotube hybrids. *Compos. Part A* **2020**, *130*, 105773. [CrossRef]
40. Zhu, S.; Xing, C.; Wu, F.; Zuo, X.; Zhang, Y.; Yu, C.; Chen, M.; Li, W.; Li, Q.; Liu, L. Cake-like flexible carbon nanotubes/graphene composite prepared via a facile method for high-performance electromagnetic interference shielding. *Carbon* **2019**, *145*, 259–265. [CrossRef]
41. Liang, J.J.; Wang, Y.; Huang, Y.; Ma, Y.F.; Liu, Z.F.; Cai, J.M.; Zhang, C.D.; Gao, H.J.; Chen, Y.S. A highly efficient gas-phase route for the oxygen functionalization of carbon nanotubes based on nitric acid vapor. *Carbon* **2009**, *47*, 922–925. [CrossRef]
42. Yan, D.X.; Pang, H.; Xu, L.; Bao, Y.; Ren, P.G.; Lei, J.; Li, Z.M. Electromagnetic interference shielding of segregated polymer composite with an ultralow loading of in situ thermally reduced graphene oxide. *Nanotechnology* **2014**, *25*, 145705. [CrossRef] [PubMed]
43. Hsiao, S.T.; Ma, C.C.M.; Tien, H.W.; Liao, W.H.; Wang, Y.S.; Li, S.M.; Huang, Y.C. Using a non-covalent modification to prepare a high electromagnetic interference shielding performance graphene nanosheet/water-borne polyurethane composite. *Carbon* **2013**, *60*, 57–66. [CrossRef]
44. Yan, D.X.; Ren, P.G.; Pang, H.; Fu, Q.; Yang, M.B.; Li, Z.M. Efficient electromagnetic interference shielding of lightweight graphene/polystyrene composite. *J. Mater. Chem.* **2012**, *22*, 18772–18774. [CrossRef]
45. Song, W.L.; Cao, M.S.; Lu, M.M.; Bi, S.; Wang, C.Y.; Liu, J.; Yuan, J.; Fan, L.Z. Flexible graphene/polymer composite films in sandwich structures for effective electromagnetic interference shielding. *Carbon* **2014**, *66*, 67–76. [CrossRef]
46. Gedler, G.; Antunes, M.; Velasco, J.I.; Ozisik, R. Electromagnetic shielding effectiveness of polycarbonate/graphene nanocomposite foams processed in 2-steps with supercritical carbon dioxide. *Mater Lett.* **2015**, *160*, 41–44. [CrossRef]
47. Zhan, Y.; Oliviero, M.; Wang, J.; Sorrentino, A.; Buonocore, G.G.; Sorrentino, L.; Lavorgna, M.; Xia, H.; Iannace, S. Enhancing the EMI shielding of natural rubber-based supercritical CO_2 foams by exploiting their porous morphology and CNT segregated networks. *Nanoscale* **2019**, *11*, 1011–1020. [CrossRef]
48. Jia, L.C.; Yan, D.X.; Yang, Y.; Zhou, D.; Cui, C.H.; Bianco, E.; Lou, J.; Vajtai, R.; Li, B.; Ajayan, P.M.; et al. High strain tolerant EMI shielding using carbon nanotube network stabilized rubber composite. *Adv. Mater. Technol.* **2017**, *2*, 1700078. [CrossRef]
49. Abraham, J.; Arif, M.P.; Xavier, P.; Bose, S.; George, S.C.; Kalarikkal, N.; Thomas, S. Investigation into dielectric behaviour and electromagnetic interference shielding effectiveness of conducting styrene butadiene rubber composites containing ionic liquid modified MWCNT. *Polymer* **2017**, *112*, 102–115. [CrossRef]

50. Joseph, N.; Janardhanan, C.; Sebastian, M.T. Electromagnetic interference shielding properties of butyl rubber-single walled carbon nanotube composites. *Compos. Sci. Technol.* **2014**, *101*, 139–144. [CrossRef]
51. Giofrè, S.V.; Tiecco, M.; Celesti, C.; Patanè, S.; Triolo, C.; Gulino, A.; Spitaleri, L.; Scalese, S.; Scuderi, M.; Iannazzo, D. Eco-friendly 1,3-dipolar cycloaddition reactions on graphene quantum dots in natural deep eutectic solvent. *Nanomaterials* **2020**, *10*, 2549. [CrossRef] [PubMed]
52. Contino, A.; Maccarrone, G.; Fragalà, E.M.; Spitaleri, L.; Gulino, A. Conjugated gold–porphyrin monolayers assembled on inorganic surfaces. *Chem. Eur. J.* **2017**, *23*, 14937–14943. [CrossRef] [PubMed]
53. Wang, G.; Yang, J.; Park, J.; Gou, X.; Wang, B.; Liu, H.; Yao, J. Facile synthesis and characterization of graphene nanosheets. *J. Phys. Chem. C* **2008**, *112*, 8192–8195.
54. Tuccitto, N.; Riela, L.; Zammataro, A.; Spitaleri, L.; Li-Destri, G.; Sfuncia, G.; Nicotra, G.; Pappalardo, A.; Capizzi, G.; Sfrazzetto, T.G. Functionalized carbon nanoparticle-based sensors for chemical warfare agents. *ACS Appl. Nano Mater.* **2020**, *3*, 8182–8191. [CrossRef]

polymers

MDPI

Article

Novel Hybrid Nanomaterials Based on Poly-*N*-Phenylanthranilic Acid and Magnetic Nanoparticles with Enhanced Saturation Magnetization

Sveta Zhiraslanovna Ozkan [1,*], Aleksandr Ivanovich Kostev [1], Petr Aleksandrovich Chernavskii [1,2] and Galina Petrovna Karpacheva [1]

1 A.V. Topchiev Institute of Petrochemical Synthesis, Russian Academy of Sciences, 29 Leninsky Prospect, 119991 Moscow, Russia; kostev@ips.ac.ru (A.I.K.); chern5@inbox.ru (P.A.C.); gpk@ips.ac.ru (G.P.K.)
2 Department of Chemistry, Lomonosov Moscow State University, 1–3 Leninskie Gory, 119991 Moscow, Russia
* Correspondence: ozkan@ips.ac.ru

Abstract: A one-step preparation method for cobalt- and iron-containing nanomaterials based on poly-*N*-phenylanthranilic acid (P-N-PAA) and magnetic nanoparticles (MNP) was developed for the first time. To synthesize the MNP/P-N-PAA nanocomposites, the precursor is obtained by dissolving a Co (II) salt in a magnetic fluid based on Fe_3O_4/P-N-PAA with a core-shell structure. During IR heating of the precursor in an inert atmosphere at T = 700–800 °C, cobalt interacts with Fe_3O_4 reduction products, which results in the formation of a mixture of spherical Co-Fe, γ-Fe, β-Co and Fe_3C nanoparticles of various sizes in the ranges of $20 < d < 50$ nm and $120 < d < 400$ nm. The phase composition of the MNP/P-N-PAA nanocomposites depends significantly on the cobalt concentration. The reduction of metals occurs due to the hydrogen released during the dehydrogenation of phenylenamine units of the polymer chain. The introduction of 10–30 wt% cobalt in the composition of nanocomposites leads to a significant increase in the saturation magnetization of MNP/P-N-PAA (M_S = 81.58–149.67 emu/g) compared to neat Fe_3O_4/P-N-PAA (M_S = 18.41–27.58 emu/g). The squareness constant of the hysteresis loop is $\kappa_S = M_R/M_S$ = 0.040–0.209. The electrical conductivity of the MNP/P-N-PAA nanomaterials does not depend much on frequency and reaches 1.2×10^{-1} S/cm. In the argon flow at 1000 °C, the residue is 77–88%.

Keywords: magnetic nanomaterials; conjugated polymers; poly-*N*-phenylanthranilic acid; one-step synthesis; IR heating

Citation: Ozkan, S.Z.; Kostev, A.I.; Chernavskii, P.A.; Karpacheva, G.P. Novel Hybrid Nanomaterials Based on Poly-*N*-Phenylanthranilic Acid and Magnetic Nanoparticles with Enhanced Saturation Magnetization. *Polymers* **2022**, *14*, 2935. https://doi.org/10.3390/polym14142935

Academic Editors: Ting-Yu Liu and Yu-Wei Cheng

Received: 17 June 2022
Accepted: 16 July 2022
Published: 20 July 2022

1. Introduction

Nanocomposite magnetic materials based on polyconjugated polymers are new generation materials with physical and chemical properties required for modern technologies [1–5]. These magnetic nanocomposites can find potential application as hybrid electrocatalysts [6,7], as cathode materials for rechargeable batteries and fuel cells [8–10], as active materials in solar cells [11–13] and supercapacitors [14–18], for the remediation of water resources [19–24], ion-exchange materials [25–27], ion-specific electrodes [28–31], to produce sensors [32–36], as anticorrosive coatings [37–40], as electromagnetic radiation absorbing materials [41–50] and for medical applications (as antibacterial and antifungal agents, for controlled drug release) [38,51,52]. Moreover, high-magnetic-moment materials are basic components of devices used in electronic and optical fields. Wireless charging can be carried out by absorbing emitted from the power generators electromagnetic waves [53,54].

Progress in the creation of novel multifunctional magnetic nanomaterials is due to the development of new synthesis methods for nanostructures. The most common method of obtaining hybrid magnetic nanocomposites is oxidative polymerization of monomers (aniline [8,9,25,29,32,51,55,56], pyrrole [28,40,44,57] and thiophene [23]) in a reaction medium containing prefabricated magnetic nanoparticles (Fe_3O_4 [25,28,29,42,55,58], γ-Fe_2O_3 [32,43], α-Fe_2O_3 [37], Co_3O_4 [8,9,38,50] and $CoFe_2O_4$ [33,59]). One of the ways to prevent the metal nanoparticles aggregation is their stabilization in a polymer matrix during synthesis. Development of a complex synthetic strategy would make it possible to expand the range of magnetic nanoparticles based nanomaterials.

Hybrid magnetic nanomaterials with core-shell structure, where Fe_3O_4 nanoparticles form the core and poly-*N*-phenylanthranilic acid (P-N-PAA) is the shell, was obtained by a one-pot method [60]. The originality of the one-pot synthesis method lies in the fact that the Fe_3O_4/P-N-PAA preparation does not require prefabricated Fe_3O_4 nanoparticles. At the same time, the entire process of nanocomposites synthesis without intermediate stages of product extraction and purification is carried out in the same reaction vessel. The obtained nanomaterials are superparamagnetic. However, the saturation magnetization M_S does not exceed 27.58 emu/g.

Inverse suspension polymerization was applied to cover iron-magnetic materials with conjugated polymers. Prepared from inverse suspension-polymerized Fe_3O_4/polyaniline composites were subjected to calcination at 950 °C in the argon atmosphere to synthesize α-Fe (ferrite), Fe_3C (cementite) and $\alpha^{//}$-Fe_xN_y (ferric nitride) based high-magnetic-moment materials [53,54].

Earlier, we have proposed an original preparation method of magnetic nanomaterials under IR heating of polydiphenylamine (PDPA) and polyphenoxazine (PPOA) in the presence of metal salts [61–65]. It was shown that the phase state, as well as size and shape of the resulting magnetic particles depend on the polymer matrix nature. At the same time, an important role belongs to the polymer thermal stability and the presence of a sufficient amount of hydrogen in its structure capable of reducing metals during dehydrogenation. Under the same conditions of synthesis, IR heating of PDPA in the presence of $Co(OOCCH_3)_2 \cdot 4H_2O$ leads to the formation of α-Co and β-Co nanoparticles, their size ranging $2 < d < 8$ nm. In the case of PPOA, only β-Co nanoparticles are formed, their size ranging $4 < d < 14$ nm. IR heating of the precursor obtained by co-solution of PDPA or PPOA and salts of cobalt Co (II) and iron Fe (III) in an organic solvent leads to the formation only of bimetallic Co-Fe particles dispersed in a polymer matrix. For Co-Fe/PDPA and Co-Fe/PPOA nanomaterials, the saturation magnetization M_S is not higher than 20.43 and 27.28 emu/g, respectively (Table 1). Thus, magnetic nanomaterials with saturation magnetization exceeding M_S = 20–27 emu/g cannot be obtained by the proposed methods.

Table 1. Magnetic properties of nanomaterials.

Nanomaterials	T, °C	Co, wt%	Fe, wt%	Nano-Particles Size, nm	Me Phase Composition	H_C, Oe	M_S, emu/g	M_R, emu/g	M_R/M_S
Co/PDPA [61]	450	10	-	2–8	α-Co, β-Co	145	22.23	0.69	0.03
Co/PPOA [62]	500	10	-	4–14	β-Co	134	26.33	3.05	0.116
Co-Fe/PDPA [65]	600	5	10	8–30, 400–800	Co-Fe	5	20.43	0.06	0.003
Co-Fe/PPOA [63]	600	5	10	4–24, 400–1400	Co-Fe	55	27.28	0.7	0.025

The development of novel high-magnetic-moment materials seems to be an urgent task. The aim of the proposed work is to create nanocomposite magnetic materials with high saturation magnetization.

In this work, a one-step preparation method for cobalt- and iron-containing nanomaterials based on poly-*N*-phenylanthranilic acid (P-N-PAA) and magnetic nanoparticles (MNP) with enhanced saturation magnetization was developed for the first time. The nanocomposites were obtained by IR heating of precursors prepared by dissolving a Co (II) salt in a magnetic fluid based on Fe_3O_4/P-N-PAA with a core-shell structure. An analysis of the phase composition, morphology, magnetic, electrical and thermal properties of MNP/P-N-PAA nanomaterials depending on the conditions of synthesis was carried out.

2. Experimental Section

2.1. Materials

N-phenylanthranilic acid (diphenylamine-2-carboxylic acid) ($C_{13}H_{11}O_2N$) (analytical grade), chloroform (reagent grade), aqueous ammonia (reagent grade), sulfuric acid (reagent grade), DMF (Acros Organics, 99%), as well as iron (II) sulfate (Acros organics), iron (III) chloride (high purity grade) and $Co(OOCCH_3)_2 \cdot 4H_2O$ (pure grade), were used without further purification. Ammonium persulfate (analytical grade) was purified by recrystallization from distilled water. Aqueous solutions of reagents were prepared using distilled water.

2.2. Synthesis of Fe_3O_4/P-N-PAA

Fe_3O_4/P-N-PAA was synthesized by a method of one-pot synthesis in an interfacial process developed by the authors in [60]. To obtain the Fe_3O_4/P-N-PAA nanocomposite, firstly, Fe_3O_4 nanoparticles of the required concentration were synthesized via hydrolysis of a mixture of iron (II) and (III) salts with a molar ratio of 1:2 in an ammonium hydroxide solution at 55 °C. For that, 0.43 g of $FeSO_4 \cdot 7H_2O$ and 1.175 g of $FeCl_3 \cdot 6H_2O$ were dissolved in 20 mL of distilled water and heated to 55 °C, then 5 mL of NH_4OH were added. A N-PAA solution (0.1 mol/L, 1.0 g) in a mixture of chloroform (60 mL) and NH_4OH (5 mL) (volume ratio is 12:1) was added to the obtained aqueous alkaline suspension of Fe_3O_4 nanoparticles. The process was carried out at 55 °C under constant intensive stirring for 0.5 h. The suspension was cooled at room temperature under constant intensive stirring for 1 h. Then, for the in situ interfacial oxidative polymerization of N-PAA on the surface of Fe_3O_4, an aqueous solution (0.2 mol/L, 1.96 g) of ammonium persulfate (30 mL) was added to the Fe_3O_4/N-PAA suspension thermostated under constant stirring at 0 °C. Solutions of the organic and aqueous phases were mixed immediately without gradual dosing of reagents. The volume ratio of the organic and aqueous phases is 1:1 (V_{total} = 120 mL). The polymerization reaction continued for 3 h under constant intensive stirring at 0 °C. When the synthesis was completed, the reaction mixture was precipitated in a threefold excess of 1 M H_2SO_4. The obtained product was filtered off, washed repeatedly with distilled water until neutral reaction, and then vacuum-dried over KOH to constant weight. The Fe_3O_4/P-N-PAA yield is 0.741 g. The content of iron in the resulting Fe_3O_4/P-N-PAA is [Fe] = 16.4% (according to ICP-AES data).

The Fe_3O_4/P-N-PAA nanocomposite suspension in DMFA was prepared to obtain magnetic fluids. The stability of suspension was being observed for 6 months.

2.3. Synthesis of MNP/P-N-PAA

The following method was used to obtain the MNP/P-N-PAA nanocomposite [66]. To prepare a magnetic fluid, 0.2 g of Fe_3O_4/P-N-PAA were added to 15 mL of DMF in a 100 mL crystallization dish. Then, $Co(OOCCH_3)_2 \cdot 4H_2O$ of required concentration was dissolved in the resulting stable suspension. The Fe_3O_4/P-N-PAA concentration in the DMF solution was 2 wt%. In the initial Fe_3O_4/P-N-PAA nanocomposite, the content of iron is [Fe] = 16.4 and 38.5% (according to ICP-AES data). The content of cobalt is [Co] = 5–30 wt% relative to the weight of the neat nanocomposite. After removing the solvent (DMF) at T = 85 °C, the precursor consisting of Fe_3O_4/P-N-PAA and cobalt acetate salt was subjected to IR radiation using an automated IR heating unit [61] in an argon atmosphere at T = 700–800 °C for 2 min. The heating rate was 50 °C min^{-1}. The yield of MNP/P-N-PAA is 0.109 g (51.39%) at the cobalt content [Co] = 30 wt% relative to the Fe_3O_4/P-N-PAA weight. Figure 1 shows a flowchart with the step-by-step preparation of MNP/P-N-PAA nanomaterials.

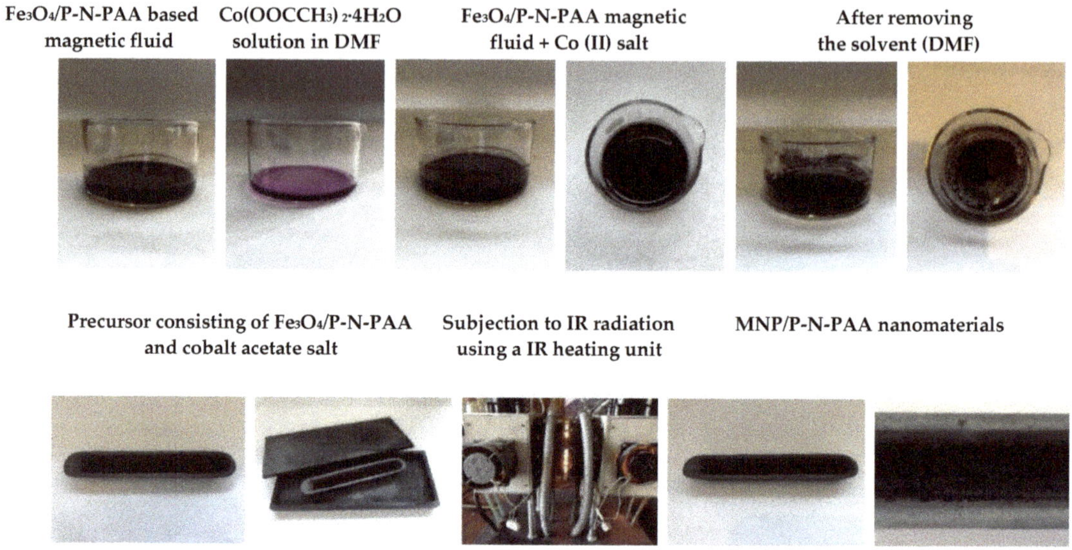

Figure 1. Flowchart with the step-by-step preparation of MNP/P-N-PAA nanomaterials.

2.4. Characterization

The IR heating unit, which is a laboratory quartz tube IR furnace [61], was used to synthesize the MNP/P-N-PAA nanocomposites. The halogen lamps were implemented as a radiation source. The rectangular graphite box with the samples was placed in a quartz reactor. The heating temperature was regulated via IR radiation intensity.

An inductively coupled plasma atomic emission spectroscopy method (ICP-AES) was used to measure quantitatively the metal content in the nanocomposites on a Shimadzu ICP emission spectrometer (ICPE-9000) (Kyoto, Japan).

A Difray-401 X-ray diffractometer (Scientific Instruments Joint Stock Company, Saint-Petersburg, Russia) with Bragg–Brentano focusing on CrK_α radiation, λ = 0.229 nm was used to perform in air X-ray diffraction study.

A Bruker IFS 66v FTIR spectrometer (Karlsruhe, Germany) was used to measure FTIR spectra in the range of 400–4000 cm^{-1}. The samples were prepared as KBr pressed pellets.

A Senterra II Raman spectrometer (Bruker, Karlsruhe, Germany) was used to record Raman spectra using a laser with the wavelength of 532 nm and the power of 0.25 mW, spectral resolution of 4 cm^{-1}.

A JEM-2100 transmission electron microscope (accelerating voltage of 200 kV) (JEOL, Akishima, Tokyo, Japan) and a Hitachi TM 3030 scanning electron microscope (Hitachi High-Technologies Corporation, Fukuoka, Japan) with magnification up to 30,000 and 30 nm resolution were used to perform an electron microscopic study. The size of nanoparticles is determined using the EsiVision software.

A vibration magnetometer was used to measure specific magnetization at room temperature depending on the magnetic field value [67].

An E7-20 precision LCR-meter (MC Meratest, Moscow, Russia) was used to measure the *ac* conductivity in the frequency range of 25.0 Hz–1.0 MHz. To measure the frequency dependence on the conductivity (σ_{ac}), samples as a tablet with a diameter of 6 mm and a thickness of 3–5 mm, pressed into a mold made of PTFE, were prepared. Brass electrodes were located on both sides of the mold. Tablet compression was performed by pressing the powder in the PTFE mold with a threaded connection. The design of the measuring cell was similar to the "swagelok cell", but without the spring. The torque force was 10 N·m.

A Mettler Toledo TGA/DSC1 (Giessen, Germany) was used to perform thermogravimetric analysis (TGA) in the dynamic mode in the range of 30–1000 °C in air and in the argon flow. The heating rate was 10 °C/min, and the argon flow velocity was 10 mL/min. The samples were analyzed in an Al_2O_3 crucible.

A Mettler Toledo DSC823^e calorimeter (Giessen, Germany) was used to perform differential scanning calorimetry (DSC). The heating rate was 10 °C/min in the nitrogen atmosphere, with the nitrogen flow rate of 70 mL/min.

3. Results and Discussion

3.1. Characterization of Nanomaterials

A one-step method was proposed to synthesize novel hybrid nanomaterials based on P-N-PAA and cobalt- and iron-containing magnetic nanoparticles (MNP). The choice of polymer was due to the fact that, unlike PDPA, the presence of carboxyl groups in the P-N-PAA structure promotes the formation of a nanomaterial with a core-shell structure, where Fe_3O_4 nanoparticles form the core, and P-N-PAA is the shell. The polymer shell effectively prevents the aggregation of nanoparticles, which makes it possible to use Fe_3O_4/P-N-PAA nanocomposites to obtain magnetic fluids suitable for dissolving metal salt. When the precursor obtained from Fe_3O_4/P-N-PAA and a Co (II) salt is IR heated in an inert atmosphere at T = 700–800 °C, metals are reduced due to the hydrogen released during dehydrogenation of phenylenamine units with the formation of a mixture of Co-Fe, γ-Fe, β-Co and Fe_3C magnetic nanoparticles. As for the core-shell structure, due to the reduction of Fe_3O_4, a Fe—OOC coordination bond formed via binding of the carboxylate-ion with iron in Fe_3O_4/P-N-PAA is broken. As a result, MNP/P-N-PAA nanocomposite materials are formed. They are cobalt- and iron-containing MNP of various compositions dispersed in a polyconjugated polymer matrix. Figure 2 shows a synthesis scheme of the MNP/P-N-PAA nanomaterials.

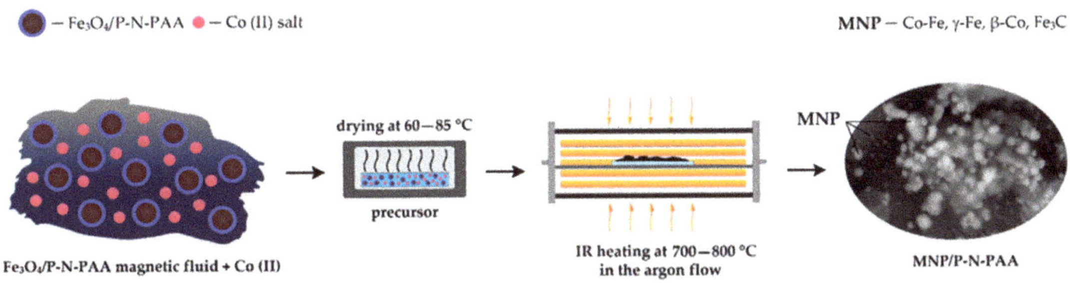

Figure 2. Scheme of the MNP/P-N-PAA nanomaterials synthesis.

The originality and distinctive feature of the proposed approach to the MNP/P-N-PAA synthesis is determined by the fact that the precursor is obtained via dissolving a Co (II) salt in a magnetic fluid based on the Fe_3O_4/P-N-PAA nanocomposite, which was synthesized by the authors and which has a core-shell structure [60]. In DMF, Fe_3O_4/P-N-PAA forms magnetic fluids combining the properties of a magnetic material and a liquid. That is, the presence of a polymer coating prevents the aggregation of Fe_3O_4 nanoparticles, ensuring the magnetic fluid stability over 6 months. An important role belongs to the hydrogen amount in the structure of polymer capable of reducing metals during dehydrogenation. During the synthesis of MNP/P-N-PAA from Fe_3O_4/P-N-PAA, cobalt interacts with Fe_3O_4 reduction products, which leads to the formation of cobalt- and iron-containing MNP of different compositions. Whereas, IR heating of the precursor obtained by co-solution of the polymer and salts of cobalt Co (II) and iron Fe (III) in an organic solvent leads to the formation only of bimetallic Co-Fe particles dispersed in a polymer matrix [61].

The novelty of the proposed approach is determined by the fact that the use of Fe_3O_4/P-N-PAA based magnetic fluid for dissolving the Co (II) salt makes it possible to expand the range of high-magnetic-moment nanoparticles in the nanomaterial composition. The developed one-step method for the formation of a nanocomposite material under IR heating helps to obtain Co-Fe, γ-Fe, β-Co and Fe_3C nanoparticles of various compositions directly during the synthesis of the nanocomposite without subjecting the polymer matrix to destruction. At the same time, applying incoherent IR radiation in a pulsed mode for the formation of the magnetic nanomaterial can significantly reduce energy costs.

As shown in Figure 3, IR heating of Fe_3O_4/P-N-PAA at 600 °C does not lead to any noticeable phase changes. XRD patterns of Fe_3O_4/P-N-PAA and Fe_3O_4/P-N-PAA, IR heated at 600 °C, demonstrate only the Fe_3O_4 phase, clearly identified by its reflection peaks in the range of scattering angles 2θ = 46.3°, 54.6°, 66.8°, 84.7°, 91.0° and 101.6° (CrK_α radiation) [60]. These diffraction peaks correspond to Miller indices (220), (311), (400), (422), (511) and (440) and characterize the simple cubic lattice of Fe_3O_4 (JCPDS 19-0629) [68]. At 800 °C, partial reduction of Fe_3O_4 with the formation of FeO, α-Fe, γ-Fe and Fe_4N nanoparticles takes place.

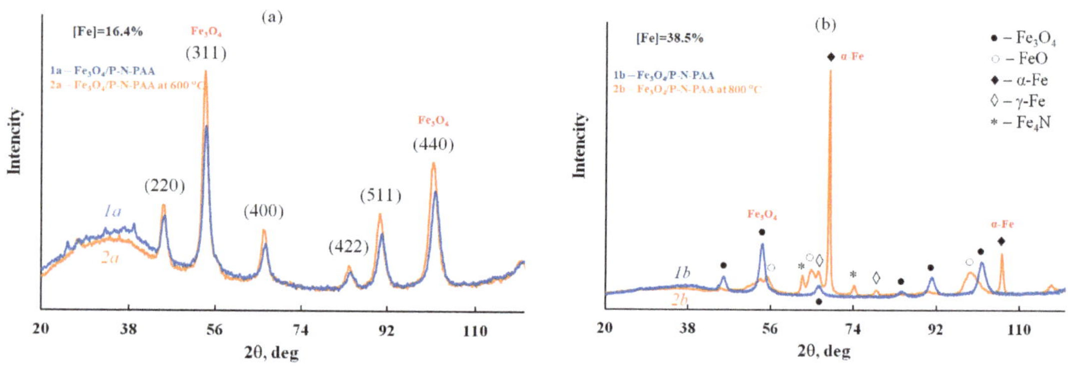

Figure 3. XRD patterns of Fe_3O_4/P-N-PAA (**1a**,**1b**) and Fe_3O_4/P-N-PAA, IR heated at 600 °C (**2a**) and 800 °C (**2b**).

IR heating of Fe_3O_4/P-N-PAA in the presence of Co (II) salt at 800 °C leads to the formation of a mixture of Co-Fe, γ-Fe, β-Co and Fe_3C magnetic nanoparticles, which was confirmed by X-ray phase analysis (Figure 4). In this case, the phase composition of the MNP/P-N-PAA nanocomposites depends on the cobalt concentration. In the FTIR spectrum of MNP/P-N-PAA, the absence of an intense absorption band at 572 cm^{-1}, corresponding to the stretching vibrations of the $\nu_{Fe\text{–}O}$ bond, is explained by Fe_3O_4 reduction (Figure 5).

According to elemental analysis data, when Fe_3O_4/P-N-PAA is IR heated in the presence of $Co(CH_3CO_2)_2{\cdot}4H_2O$, the dehydrogenation of phenylenamine units causes the decrease in the content of hydrogen from 1.9 to 0.1% (the C/H ratio increases from 12.5 to 70.7) (Table 2). The released hydrogen contributes to the reduction of metals. The C/N ratio changes insignificantly (from 8.6 to 9.0), which indicates that the polymer component degradation during IR heating is weakly expressed.

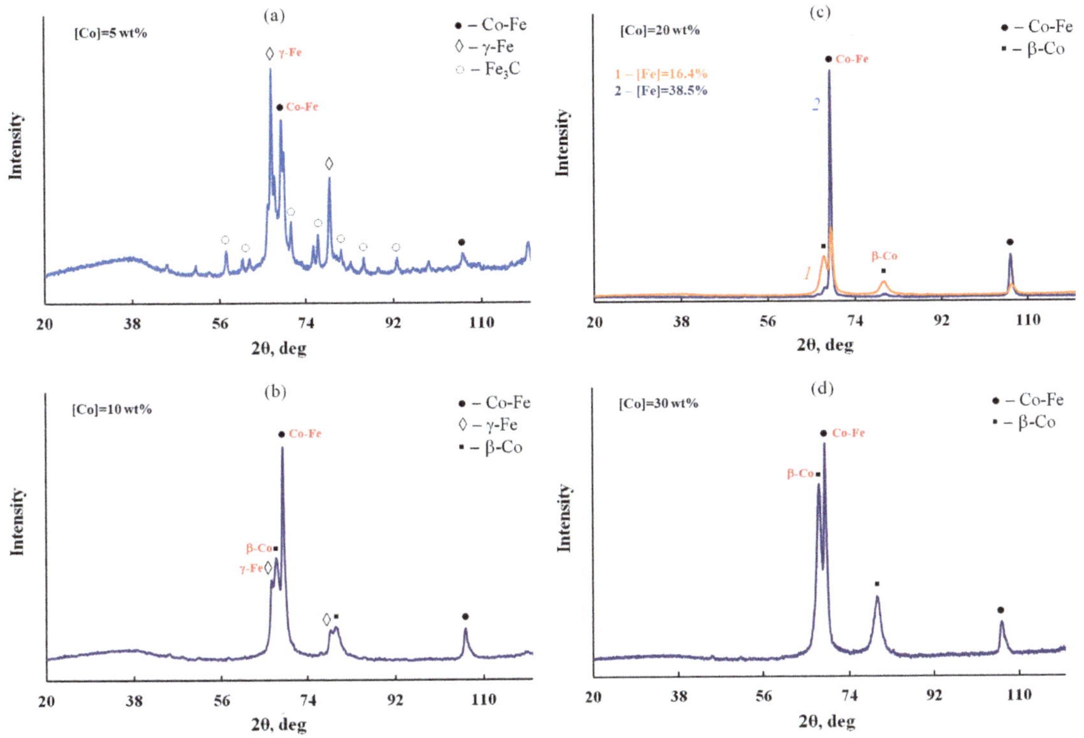

Figure 4. XRD patterns of MNP/P-N-PAA, obtained at 800 °C, [Co] = 5 (**a**), 10 (**b**), 20 (**c**) and 30 wt% (**d**).

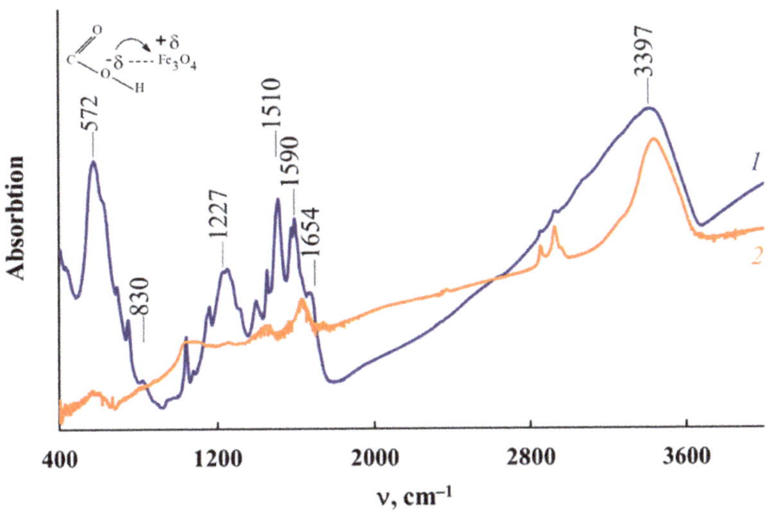

Figure 5. FTIR spectra of Fe_3O_4/P-N-PAA (*1*) and MNP/P-N-PAA (*2*).

Table 2. Inductively coupled plasma atomic emission spectroscopy (ICP-AES) data of nanocomposites and elemental analysis data of P-N-PAA component.

Materials	[Co] *, wt%	Co, %	Fe, %	C, %	N, %	H, %	O, %	C/N	C/H
P-N-PAA	-	-	-	60.7	8.2	5.8	25.3	7.7	10.5
Fe_3O_4/P-N-PAA	-	-	38.5	23.58	2.75	1.88	33.3	8.6	12.5
MNP/P-N-PAA	20	29.1	58.3	8.48	0.94	0.12	3.1	9.0	70.7

* [Co] wt% at the loading.

Figure 6 shows Raman spectra of P-N-PAA, neat Fe_3O_4/P-N-PAA and Fe_3O_4/P-N-PAA, IR heated at 600 and 800 °C, and MNP/P-N-PAA. As seen in Figure 6, as well as in the IR heated at 800 °C Fe_3O_4/P-N-PAA, there are two pronounced bands: a G band at 1596 cm^{-1} from sp^2 (aromatic) carbon atoms and a D band at 1350 cm^{-1} from sp^3 carbon atoms in the Raman spectrum of MNP/P-N-PAA. In the Raman spectra of both the neat Fe_3O_4/P-N-PAA and Fe_3O_4/P-N-PAA, IR heated at 600 °C, as well as in P-N-PAA, these bands are absent. The intensity ratio of these bands (I_D/I_G = 0.82), the high intensity and the width of the 2D band at 2800 cm^{-1} indicate that graphite-like structures are formed in the polymer component structure during IR heating of the precursor based on Fe_3O_4/P-N-PAA and a Co (II) salt in an inert atmosphere at 800 °C. XRD patterns of MNP/P-N-PAA nanocomposites demonstrate a wide halo in the range of scattering angles 2θ = 20–44°, which characterizes graphite-like structures (Figure 4). The broadening of all main absorption bands characterizing the chemical structure of the polymer component in the FTIR spectrum of MNP/P-N-PAA is also connected with the dehydrogenation of P-N-PAA phenylenamine units and the subsequent formation of graphite-like structures (Figure 5).

However, it should be noted that the presence of nitrogen and hydrogen atoms in the MNP/P-N-PAA structure (Table 2), as well as the absence of a sharp peak at diffraction angles 2θ = 39.36°, which characterizes carbon, on the XRD patterns of MNP/P-N-PAA (Figure 4) indicate an incomplete transformation of the polymer component to all-carbon structures. Apparently, during the reduction of Fe_3O_4 in the course of Fe_3O_4/P-N-PAA thermal transformations at high temperatures (700–800 °C), the partial formation of graphite-like structures occurs.

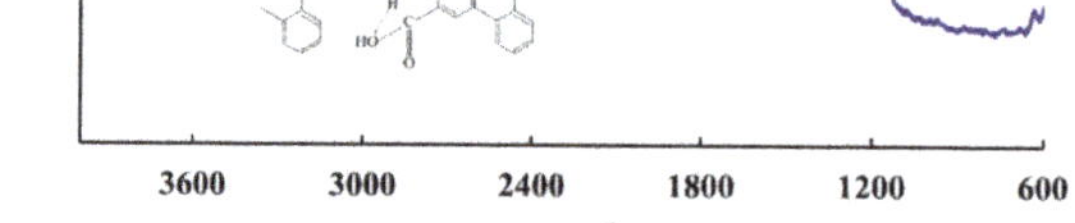

Figure 6. *Cont.*

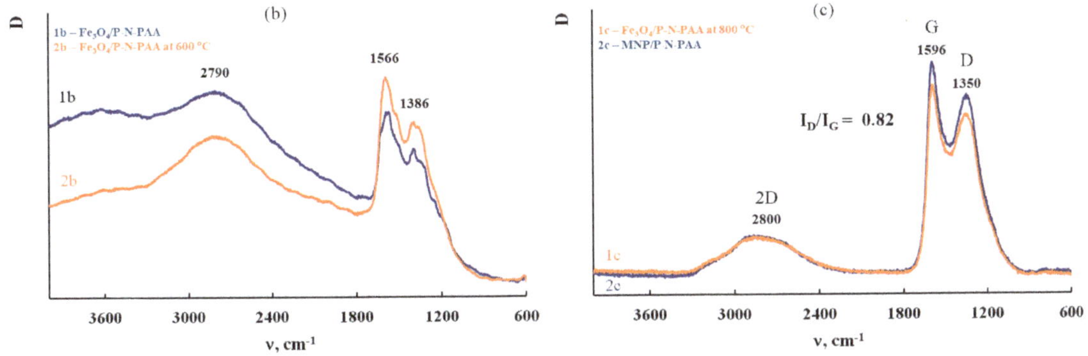

Figure 6. Raman spectra of P-N-PAA (**a**), Fe_3O_4/P-N-PAA (**1b**) and Fe_3O_4/P-N-PAA, IR heated at 600 (**2b**) and 800 °C (**1c**) and MNP/P-N-PAA (**2c**).

XRD patterns of all MNP/P-N-PAA nanocomposites show clearly identified reflection peaks of Co-Fe nanoparticles in the region of diffraction scattering angles $2\theta = 68.86°$, $106.36°$ (Figure 4). According to the database Miller indices, interplanar distances in the Co-Fe phase correspond to the Co-Fe solid solution. The formation of bimetallic Co-Fe nanoparticles is associated with the interaction of cobalt and Fe_3O_4 reduction products. The reflection peaks of β-Co nanoparticles with a face-centered cubic lattice can be identified at diffraction angles $2\theta = 67.52°$, $80.14°$. In the case of γ-Fe nanoparticles they are identified in the region of $2\theta = 66.56°$, $78.77°$. The formation of Fe_3C nanoparticles at [Co] = 5 wt% is associated with the interaction of Fe_3O_4 reduction products with the polymer matrix due to low concentration of cobalt.

According to TEM and SEM data, a bimodal nature of the MNP distribution is observed. A mixture of spherical nanoparticles sized $20 < d < 50$ nm and $120 < d < 400$ nm is formed (Figures 7 and 8). At the same time, according to XRD data for the MNP/P-N-PAA nanocomposite, the CSR size distribution curve is in the region of 3–55 nm with a peak at 13–14 nm (Figure 9). According to ICP-AES data, depending on the conditions of synthesis the content of Co = 8.6–38.0%, and Fe = 14.2–58.3% (Tables 2 and 3). According to elemental analysis data, the content of P-N-PAA component is 12.6–72.2%.

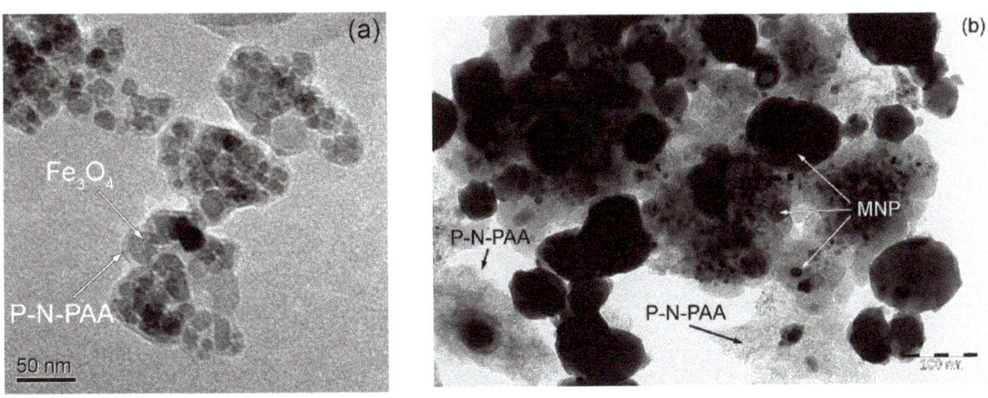

Figure 7. TEM images of Fe_3O_4/P-N-PAA (**a**) and MNP/P-N-PAA (**b**).

Figure 8. SEM images of Fe_3O_4/P-N-PAA (**a**,**b**) and MNP/P-N-PAA (**c**,**d**).

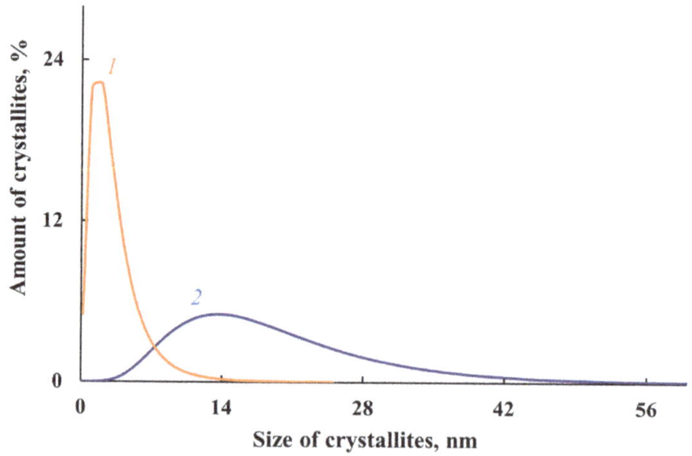

Figure 9. Crystallites size distribution in Fe_3O_4/P-N-PAA (*1*) and MNP/P-N-PAA (*2*).

Table 3. Magnetic properties of nanomaterials.

Nanomaterials	T, °C	[Co] *, wt%	[Fe] **, %	Co ***, %	Fe ***, %	**** MNP Phase Composition	H_C, Oe	M_S, emu/g	M_R, emu/g	M_R/M_S
Fe_3O_4/P-N-PAA	0	-	16.4	-	16.4	Fe_3O_4	0	18.41	0	0
	0	-	38.5	-	38.5	Fe_3O_4	0	27.58	0	0
Fe_3O_4/P-N-PAA	600	-	16.4	-	16.9	Fe_3O_4	0	17.02	0	0
	800	-	38.5	-	47.2	Fe_3O_4, FeO, α-Fe, γ-Fe, Fe_4N	25	12.41	0.28	0.022
MNP/P-N-PAA	800	5	16.4	8.6	19.2	Co-Fe, γ-Fe, Fe_3C	128	35.22	3.00	0.085
	800	10	16.4	13.6	17.4	Co-Fe, γ-Fe, β-Co	176	99.86	16.00	0.160
	800	20	16.4	28.3	14.2	Co-Fe, β-Co	170	81.58	12.80	0.156
	700	20	38.5	29.1	58.3	Co-Fe, β-Co	45	149.67	6.0	0.040
	800	30	16.4	38.0	17.7	Co-Fe, β-Co	200	95.70	20.00	0.209

* [Co] wt% at the loading. ** [Fe] content in the neat Fe_3O_4/P-N-PAA. *** According to ICP-AES data. **** MNP—Co-Fe, γ-Fe, β-Co and Fe_3C magnetic nanoparticles. H_C—coercive force, M_S—saturation magnetization and M_R—residual magnetization.

Energy dispersive X-ray spectroscopy (EDS) elemental mapping method was used to characterize the element distribution in the P-N-PAA based nanomaterials (Figure 10). Figure 11 demonstrates SEM-EDS mapping images of ferrum Fe, cobalt Co, carbon C, nitrogen N and oxygen O in Fe_3O_4/P-N-PAA and MNP/P-N-PAA. Table 4 shows the EDS analysis data of these nanocomposites. The EDS element mapping reveals a homogeneous distribution of Fe and Co elements. As seen in Table 4, the oxygen content in the MNP/P-N-PAA nanocomposite drops sharply due to the Fe_3O_4 reduction.

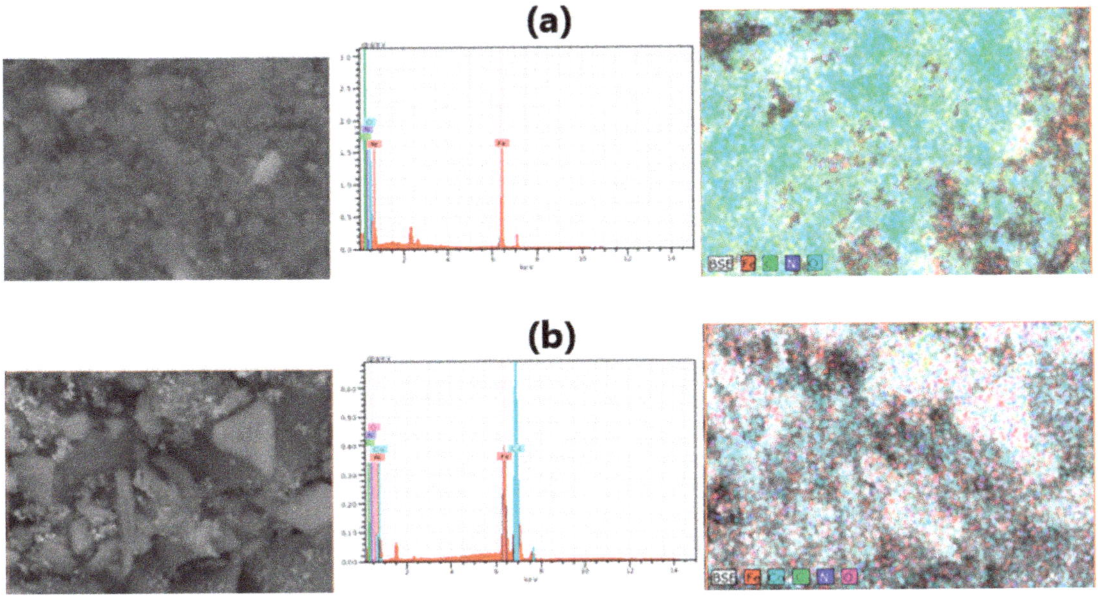

Figure 10. Representative SEM and EDX images of Fe_3O_4/P-N-PAA (**a**) and MNP/P-N-PAA (**b**).

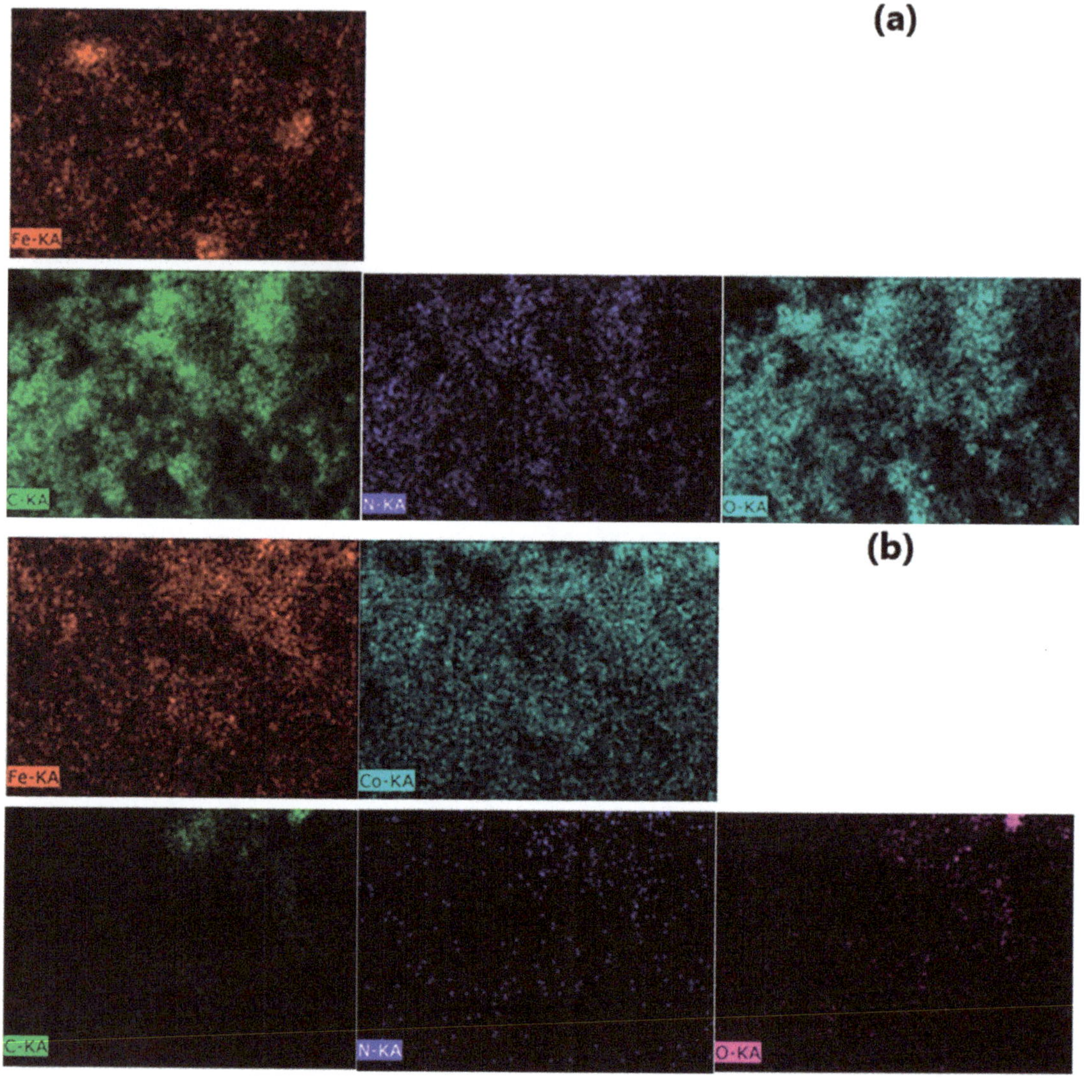

Figure 11. SEM-EDS mapping images of ferrum Fe, cobalt Co, carbon C, nitrogen N and oxygen O in Fe_3O_4/P-N-PAA (**a**) and MNP/P-N-PAA (**b**).

Table 4. EDS analysis data of nanomaterials.

Nanomaterials	MNP Phase Composition	* Polymer Component, %	Co		Fe		C		N		O	
			wt%	at%	wt%	at%	wt%	at%	wt%	at%	wt%	at%
Fe_3O_4/P-N-PAA	Fe_3O_4	28	-	-	12.99	3.43	49.45	60.76	8.84	9.32	28.72	26.49
MNP/P-N-PAA	Co-Fe, β-Co	16	72.44	55.26	19.63	15.80	7.05	26.39	0.19	0.62	0.68	1.92

* According to TGA data.

3.2. Magnetic Properties of Nanomaterials

The obtained MNP/P-N-PAA nanomaterials demonstrate a hysteresis character of remagnetization at room temperature. The dependency of magnetization on the value of the applied magnetic field is shown in Figures 12 and 13. The residual magnetization M_R of the MNP/P-N-PAA nanomaterials is 3–20 emu/g, the coercive force is H_C = 45–200 Oe (Table 3).

The squareness constant of the hysteresis loop is $\kappa_S = M_R/M_S$ = 0.040–0.209, which indicates a significant proportion of superparamagnetic nanoparticles. However, the content of superparamagnetic nanoparticles can be quantified only if the single-domain condition is satisfied. In this case, it is difficult to determine the quantity of superparamagnetic nanoparticles, since MNP are a mixture of spherical Co-Fe, γ-Fe, β-Co and Fe_3C nanoparticles of various sizes in the ranges of 20 < *d* < 50 nm and 120 < *d* < 400 nm.

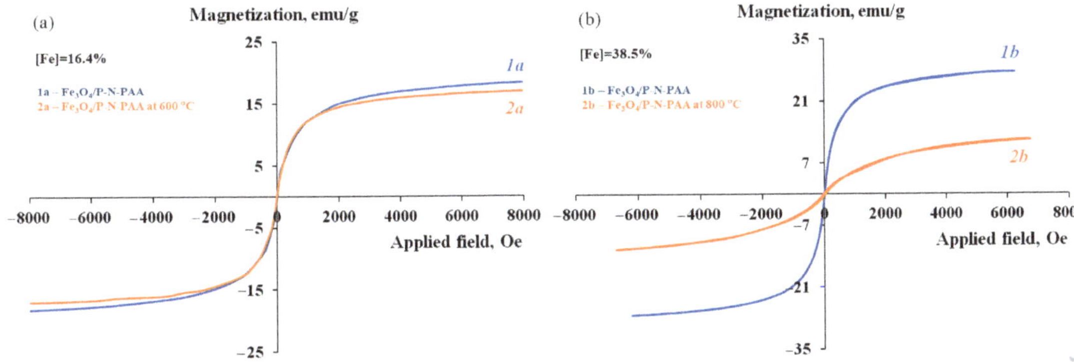

Figure 12. Magnetization of Fe_3O_4/P-N-PAA (**1a**,**1b**) and Fe_3O_4/P-N-PAA, IR heated at 600 °C (**2a**) and 800 °C (**2b**), as a function of applied magnetic field.

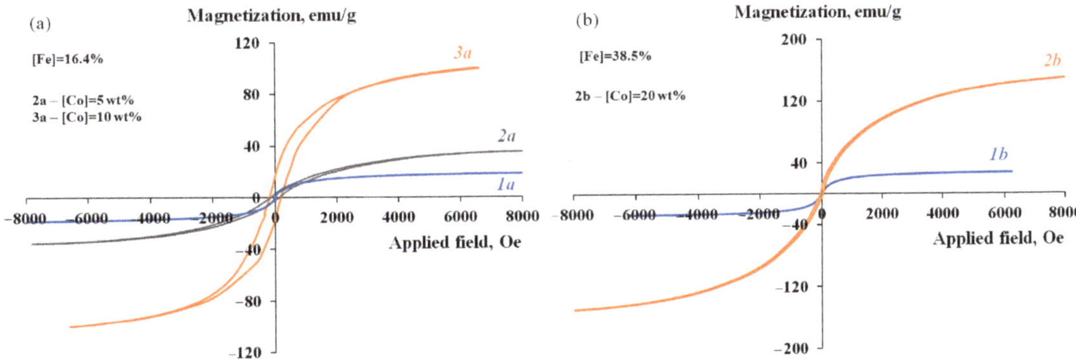

Figure 13. Magnetization of Fe_3O_4/P-N-PAA (**1a**,**1b**) and MNP/P-N-PAA, obtained at 800 °C, [Co] = 5 (**2a**), 10 (**3a**) and 20 wt% (**2b**), as a function of applied magnetic field.

As shown in Figure 12, IR heating of neat Fe_3O_4/P-N-PAA at 600 °C has little effect on its magnetic properties (M_S = 18.41–17.02 emu/g). At 800 °C, the saturation magnetization of Fe_3O_4/P-N-PAA decreases to 12.41 emu/g due to phase transformations of Fe_3O_4 into FeO, α-Fe, γ-Fe and Fe_4N.

It can be seen from Table 3, that the introduction of 10–30 wt% cobalt into nanocomposites leads to a significant increase in the saturation magnetization of MNP/P-N-PAA compared to the initial Fe_3O_4/P-N-PAA. The saturation magnetization of MNP/P-N-PAA nanomaterials grow with the increase in cobalt concentration and reaches M_S = 81.58–99.86 emu/g. An increase in the content of Fe_3O_4 nanoparticles in the initial Fe_3O_4/P-N-PAA nanocomposite ([Fe] = from 16.4 to 38.5%) leads to an even greater increase in the saturation magnetization of MNP/P-N-PAA—up to 149.67 emu/g, while in Fe_3O_4/P-N-PAA the value of M_S is 27.58 emu/g (Figure 13). Furthermore, as shown in Table 1, IR heating of the precursor obtained by co-solution of the polymer and salts of cobalt Co (II) and iron Fe (III) in an organic solvent leads to the formation of Co-Fe-based magnetic nanomaterials with saturation magnetization not exceeding M_S = 20–27 emu/g. Thus, the saturation magnetization of MNP/P-N-PAA grows with the increase in magnetic phase, but not linearly, since the MNP composition strongly depends on the Co and Fe content.

3.3. Electrical Properties of Nanomaterials

The frequency dependence on *ac* conductivity of the MNP/P-N-PAA nanocomposite obtained at 800 °C during 2 min at [Co] = 30 wt% at the loading compared to neat Fe_3O_4/P-N-PAA was studied (Figure 14). The metallic phase composition corresponds to the Co-Fe and β-Co phases (Figure 4d).

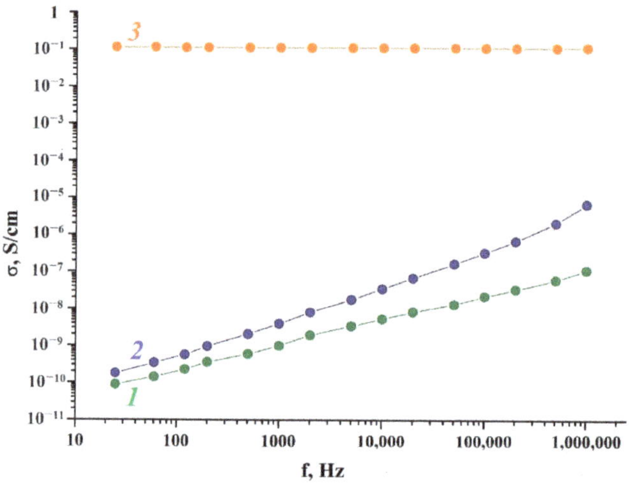

Figure 14. Dependence of conductivity on the frequency for P-N-PAA *(1)*, Fe_3O_4/P-N-PAA *(2)* and MNP/P-N-PAA, obtained at 800 °C *(3)*.

The dependence of conductivity (σ_{ac}) on the frequency is described by equation [69–71]:

$$\sigma_{ac} = \sigma_{dc} + A\omega^{n}$$

As can be seen from Table 5, neat P-N-PAA demonstrates a low conductivity value in the range of 25 Hz–1 MHz. The polymer electrical conductivity σ_{ac} increases linearly from 8.8 × 10^{-11} S/cm to 1.1 × 10^{-7} S/cm. The value of n = 0.75 indicates the hopping mechanism of conductivity ($0 \leq n \leq 1$), typical of most conductive polymers [69–72].

Table 5. The conductivity values of materials.

Materials	* σ_{ac}, S/cm		σ_{dc}, S/cm	n	*A*
P-N-PAA	8.8 × 10^{-11}	1.1 × 10^{-7}	2.8 × 10^{-12}	0.75	8.5 × 10^{-12}
Fe_3O_4/P-N-PAA	1.8 × 10^{-10}	6.7 × 10^{-6}	1.3 × 10^{-10}	0.99	1.3 × 10^{-12}
MNP/P-N-PAA	1.1 × 10^{-1}	1.2 × 10^{-1}	1.1 × 10^{-1}	0.98	6.5 × 10^{-10}

* σ—The *ac* conductivity at 25 Hz and 1 MHz.

At low frequencies, the Fe_3O_4/P-N-PAA nanocomposite is characterized by weak dependence of electrical conductivity on frequency. As frequency grows, the Fe_3O_4/P-N-PAA electrical conductivity increases gradually by four orders of magnitude to 6.7×10^{-6} S/cm.

The electrical conductivity of the MNP/P-N-PAA nanomaterial is significantly higher than the conductivity of Fe_3O_4/P-N-PAA and does not depend much on frequency (1.1×10^{-1} S/cm–1.2×10^{-1} S/cm). The formation of an extended conjugated system during heat treatment of Fe_3O_4/P-N-PAA, as well as the presence of MNP of various compositions in the polymer matrix causes an increase in the degree of percolation and leads to a rise in electrical conductivity of MNP/P-N-PAA. Apparently, the formation of large conducting regions leads to exceeding the percolation threshold.

As can be seen in Table 5, as well as for neat Fe_3O_4/P-N-PAA, for MNP/P-N-PAA the exponential parameter n lies in the range of $0 \leq n \leq 1$, which is typical for systems with a hopping mechanism of charge transfer. The effect of tunneling in the nanocomposites is minimal. The *dc* conductivity plays an important role at low frequencies, whereas the *ac* conductivity of nanocomposites shows an increase with the growth in current frequency.

3.4. Thermal Properties of Nanomaterials

Thermal stability of the MNP/P-N-PAA nanocomposites prepared at 800 °C for 2 min at [Co] = 5 and 30 wt% at the loading was studied by TGA and DSC methods. Figure 15 shows the dependence of temperature on the decrease in the weight of MNP/P-N-PAA compared to neat Fe_3O_4/P-N-PAA at heating up to 1000 °C in the argon flow and in air.

The MNP/P-N-PAA nanocomposites are characterized by high thermal stability that exceeds considerably the thermal stability of neat Fe_3O_4/P-N-PAA. Weight loss at low temperatures on the TGA thermograms is associated with the removal of moisture (Figure 15). The DSC thermograms of nanomaterials show an endothermic peak at 108 °C (Figure 16). When re-heated, this peak is absent, which confirms the moisture removal. As can be seen, in air, after removing moisture, the same shape of weight loss curves until 500 °C is observed for all materials due to the thermooxidative degradation of polymer component.

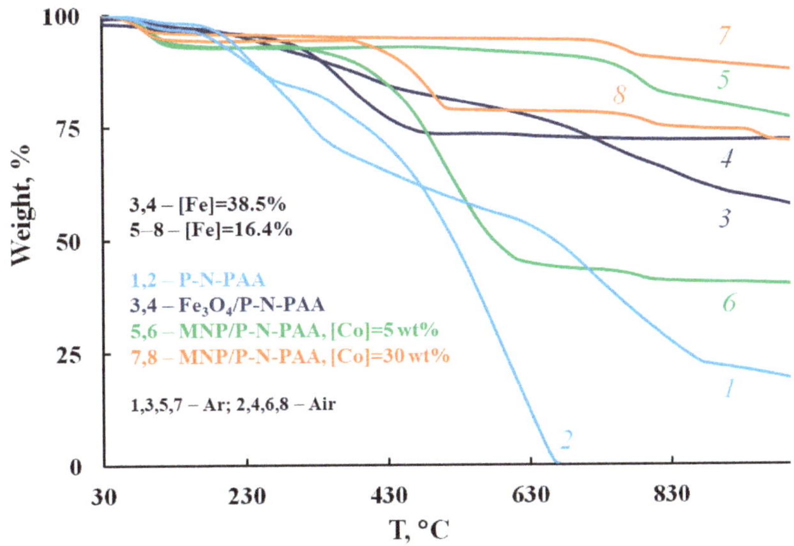

Figure 15. TGA thermograms of P-N-PAA *(1, 2)*, Fe_3O_4/P-N-PAA *(3, 4)* and MNP/P-N-PAA, obtained at 800 °C, [Co] = 5 *(5, 6)* and 30 wt% *(7, 8)*, in the argon flow *(1, 3, 5, 7)* and in air *(2, 4, 6, 8)*.

In an inert medium in MNP/P-N-PAA, a gradual weight loss is observed, and depending on the concentration of cobalt at 1000 °C, the residue is 77–88% (Table 6). At the same time, in Fe_3O_4/P-N-PAA, the residue accounts for 58%. In air, the lower thermal stability of MNP/P-N-PAA obtained at [Co] = 5 wt% is associated with the phase state of the MNP nanoparticles, as well as with a Fe_3O_4 content in neat Fe_3O_4/P-N-PAA ([Fe] = 16.4 wt% according to ICP-AES data) (Table 6). As can be seen on the TGA thermograms of MNP/P-N-PAA nanocomposites prepared at [Co] = 5 and 30 wt%, in air the content of the polymer component is 48 and 16%, respectively. The processes of thermooxidative degradation of MNP/P-N-PAA begin at 330 390 °C. The DTG curve of MNP/P-N-PAA shows removal of water and thermal decomposition of polymer component. The degradation of polymer component occurs within the range of 320–620 °C, with the maximum at 509 °C (Figure 17). Phase transformations of MNP nanoparticles occur in the range of 720–810 °C, with the maximum at 790 °C.

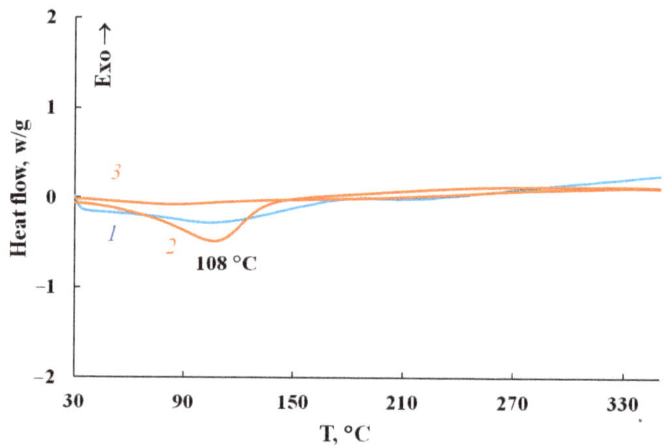

Figure 16. DSC thermograms of Fe_3O_4/P-N-PAA (*1*) and MNP/P-N-PAA (*2, 3*) at heating in the nitrogen flow (*1, 2*—first heating, *3*—second heating).

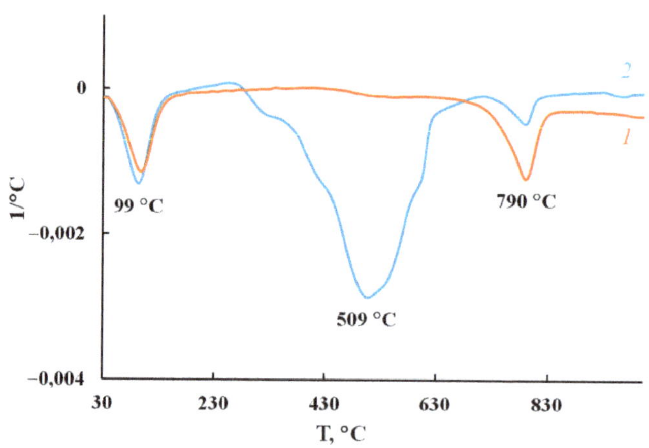

Figure 17. DTG curves of MNP/P-N-PAA, obtained at [Co] = 5 wt%, in the argon flow (*1*) and in air (*2*).

Table 6. Thermal properties of nanomaterials.

Materials	[Co] *, wt%	[Fe] **, %	Co ***, %	Fe ***, %	**** MNP Phase Composition	^ $T_{5\%}$, °C	^* $T_{20\%}$, °C	^^ $T_{50\%}$, °C	^^^ Residue, %
P-N-PAA	-	-	-	-	-	185/205	357/299	523/663	0/20
Fe_3O_4/P-N-PAA	-	38.5	-	38.5	Fe_3O_4	258/230	405/557	>1000/>1000	72/58
MNP/P-N-PAA	5	16.4	8.6	19.2	Co-Fe, γ-Fe, Fe_3C	102/111	459/>1000	579/>1000	40/77
	30	16.4	38.0	17.7	Co-Fe, β-Co	108/371	507/>1000	>1000/>1000	71/88

* [Co] wt% at the loading. ** [Fe] content in the neat Fe_3O_4/P-N-PAA. *** According to ICP-AES data. **** MNP—Co-Fe, γ-Fe, β-Co, and Fe_3C magnetic nanoparticles. ^ $T_{5\%}$, ^* $T_{20\%}$, ^^ $T_{50\%}$—5, 20 and 50 % weight losses (air/argon). ^^^ residue at 1000 °C (air/argon).

4. Conclusions

The proposed one-step method offers the possibility of obtaining cobalt- and iron-containing nanomaterials based on the Fe_3O_4/P-N-PAA nanocomposite synthesized by the authors. Under IR heating of a precursor consisting of Fe_3O_4/P-N-PAA and a Co (II) salt, cobalt interacts with Fe_3O_4 reduction products to form a mixture of spherical Co-Fe, γ-Fe, β-Co and Fe_3C nanoparticles with sizes of $20 < d < 50$ nm and $120 < d < 400$ nm. The phase composition of MNP/P-N-PAA nanocomposites depends on the cobalt concentration. The formation of the nanoparticles occurs directly during the nanocomposites synthesis under IR heating conditions, which makes it possible to expand the magnetic nanoparticles range. The originality of the proposed approach to MNP/P-N-PAA synthesis is determined by the fact that the preparation of a precursor by dissolving a Co (II) salt in a magnetic fluid based on Fe_3O_4/P-N-PAA with a core-shell structure leads to the formation of cobalt- and iron-containing nanomaterials with enhanced saturation magnetization. The introduction of 10–30 wt% cobalt into the composition of nanocomposites leads to a significant increase in the saturation magnetization of MNP/P-N-PAA (M_S = 81.58–149.67 emu/g) compared to neat Fe_3O_4/P-N-PAA (M_S = 18.41–27.58 emu/g). The hysteresis loop squareness constant $\kappa_S = M_R/M_S = 0.040$–$0.209$ proves a significant share of superparamagnetic nanoparticles in MNP/P-N-PAA. The electrical conductivity of the MNP/P-N-PAA nanomaterials (1.1×10^{-1} S/cm–1.2×10^{-1} S/cm) is significantly higher than the conductivity of Fe_3O_4/P-N-PAA (6.7×10^{-6} S/cm) and does not depend much on frequency in the range of 25 Hz–1 MHz. The MNP/P-N-PAA nanocomposites are characterized by high thermal stability. In an inert atmosphere at 1000 °C, the residue is up to 88%, whereas the Fe_3O_4/P-N-PAA residue is 58%. The obtained magnetic nanomaterials can be used for modern technologies as materials that absorb electromagnetic radiation, to create wireless fast charging power source chargers, supercapacitors, electrochemical current sources, energy converters, contrast materials for magnetic resonance imaging, electromagnetic screens, in magnetic information recording systems, in high-temperature processes as protective coatings for construction materials, etc.

Author Contributions: Conceptualization, investigation and writing—original draft, S.Z.O.; supervision, writing—review & editing, G.P.K.; investigation, A.I.K. and P.A.C. All authors have read and agreed to the published version of the manuscript.

Funding: This research received no external funding.

Institutional Review Board Statement: Not applicable.

Informed Consent Statement: Not applicable.

Acknowledgments: This work was carried out within the State Program of TIPS RAS. This work was performed using the equipment of the Shared Research Center "Analytical center of deep oil processing and petrochemistry of TIPS RAS".

Conflicts of Interest: The authors declare no conflict of interest.

References

1. Godovsky, D.Y. Device applications of polymer-nanocomposites. *Adv. Polym. Sci.* **2000**, *153*, 163–205. [CrossRef]
2. Karpacheva, G.P. Hybrid magnetic nanocomposites including polyconjugated polymers. *Polym. Sci. Ser. C* **2016**, *58*, 131–146. [CrossRef]
3. Idumah, C.I. Novel trends in conductive polymeric nanocomposites, and bionanocomposites. *Synth. Met.* **2021**, *273*, 116674. [CrossRef]
4. Ravindra, S.; Behzad, Y.; Kitirote, S.; Harikaranahalli, W.; Shivaraju, P. Potentiality of polymer nanocomposites for sustainable environmental applications: A review of recent advances. *Polymer* **2021**, *233*, 124184. [CrossRef]
5. Muñoz-Bonilla, A.; Sánchez-Marcos, J.; Herrasti, P. Magnetic nanoparticles-based conducting polymer nanocomposites. In *Conducting Polymer Hybrids*; Kumar, V., Kalia, S., Swart, H.C., Eds.; Springer Series on Polymer and Composite Materials; Chapter 2; Springer International Publishing AG: Cham, Switzerland, 2017; pp. 45–80, ISSN1 2364-1878, ISSN2 2364-1886 (electronic). ISBN1 978-3-319-46456-5, ISBN2 978-3-319-46458-9 (eBook). [CrossRef]
6. Al Betar, A.-R.F. Enhanced electrocatalytic water oxidation using cobalt-based polyaniline hybrid assembly. *Synth. Met.* **2021**, *275*, 116738. [CrossRef]
7. Chen, X.; Chen, Y.; Luo, X.; Guo, H.; Wang, N.; Su, D.; Zhang, C.; Liu, T.; Wang, G.; Cui, L. Polyaniline engineering defect-induced nitrogen doped carbon-supported Co_3O_4 hybrid composite as a high-efficiency electrocatalyst for oxygen evolution reaction. *Appl. Surf. Sci.* **2020**, *526*, 146626. [CrossRef]
8. Shambharkar, B.H.; Umare, S.S. Production and characterization of polyaniline/Co_3O_4 nanocomposite as a cathode of Zn–polyaniline battery. *Mater. Sci. Eng. B* **2010**, *175*, 120–128. [CrossRef]
9. Talooki, E.F.; Ghorbani, M.; Rahimnejad, M.; Lashkenari, M.S. Investigating the effects of in-situ fabrication of a binder-free Co_3O_4-polyaniline cathode towards enhanced oxygen reduction reaction and power generation of microbial fuel cells. *Synth. Met.* **2019**, *258*, 116225. [CrossRef]
10. Huang, W.-Y.; Chang, M.-Y.; Wang, Y.-Z.; Huang, Y.-C.; Ho, K.-S.; Hsieh, T.-H.; Kuo, Y.-C. Polyaniline based Pt-electrocatalyst for a proton exchanged membrane fuel cell. *Polymers* **2020**, *12*, 617. [CrossRef]
11. Sadegh, F.; Modarresi-Alam, A.R.; Noroozifar, M.; Kerman, K. A facile and green synthesis of superparamagnetic Fe_3O_4@PANI nanocomposite with a core–shell structure to increase of triplet state population and efficiency of the solar cells. *J. Env. Chem. Eng.* **2021**, *9*, 104942. [CrossRef]
12. Wang, K.; Yi, C.; Liu, C.; Hu, X.; Chuang, S.; Gong, X. Effects of magnetic nanoparticles and external magnetostatic field on the bulk heterojunction polymer solar cells. *Sci. Rep.* **2015**, *5*, 9265. [CrossRef] [PubMed]
13. Meng, L.; Watson, B.W., II; Qin, Y. Hybrid conjugated polymer/magnetic nanoparticle composite nanofibers through cooperative non-covalent interactions. *Nanoscale Adv.* **2020**, *2*, 2462–2470. [CrossRef]
14. Lv, H.; Pan, Q.; Song, Y.; Liu, X.-X.; Liu, T. A review on nano-/microstructured materials constructed by electrochemical technologies for supercapacitors. *Nano-Micro Lett.* **2020**, *12*, 118. [CrossRef] [PubMed]
15. Das, T.; Verma, B. Synthesis of polymer composite based on polyaniline-acetylene black-copper ferrite for supercapacitor electrodes. *Polymer* **2019**, *168*, 61–69. [CrossRef]
16. Kumar, M.P.; Lathika, L.M.; Mohanachandran, A.P.; Rakhi, R.B. A high-performance flexible supercapacitor anode based on polyaniline/Fe_3O_4 composite@carbon cloth. *ChemistrySelect* **2018**, *3*, 3234–3240. [CrossRef]
17. Alsulami, Q.A.; Alharbi, L.M.; Keshk, S.M.A.S. Synthesis of a graphene oxide/$ZnFe_2O_4$/polyaniline nanocomposite and its structural and electrochemical characterization for supercapacitor application. *Int. J. Energy Res.* **2021**, *46*, 2438–2445. [CrossRef]
18. Rajkumar, S.; Elanthamilan, E.; Merlin, J.P.; Sathiyan, A. Enhanced electrochemical behaviour of $FeCo_2O_4$/PANI electrode material for supercapacitors. *J. Alloys Compd.* **2021**, *874*, 159876. [CrossRef]
19. Govan, J. Recent advances in magnetic nanoparticles and nanocomposites for the remediation of water resources. *Magnetochemistry* **2020**, *6*, 49. [CrossRef]
20. Hlongwane, G.N.; Sekoai, P.T.; Meyyappan, M.; Moothi, K. Simultaneous removal of pollutants from water using nanoparticles: A shift from single pollutant control to multiple pollutant control. *Sci. Total Environ.* **2019**, *656*, 808–833. [CrossRef]
21. Tahoon, M.A.; Siddeeg, S.M.; Salem Alsaiari, N.S.; Mnif, W.; Ben Rebah, F. Effective heavy metals removal from water using nanomaterials: A review. *Processes* **2020**, *8*, 645. [CrossRef]
22. Rahdar, A.; Rahdar, S.; Labuto, G. Environmentally friendly synthesis of Fe_2O_3@SiO_2 nanocomposite: Characterization and application as an adsorbent to aniline removal from aqueous solution. *Environ. Sci. Pollut. Res.* **2020**, *27*, 9181–9191. [CrossRef] [PubMed]
23. Kamel, A.H.; Hassan, A.A.; Amr, A.E.-G.E.; El-Shalakany, H.H.; Al-Omar, A.M. Synthesis and characterization of $CuFe_2O_4$ nanoparticles modified with polythiophene: Applications to mercuric ions removal. *Nanomaterials* **2020**, *10*, 586. [CrossRef] [PubMed]
24. Momina, K.A. Study of different polymer nanocomposites and their pollutant removal efficiency: Review. *Polymer* **2021**, *217*, 123453. [CrossRef]
25. Rezvani, M.; Asgharinezhad, A.A.; Ebrahimzadeh, H.; Shekari, N. A polyaniline-magnetite nanocomposite as an anion exchange sorbent for solid-phase extraction of chromium (VI) ions. *Microchim. Acta* **2014**, *181*, 1887–1895. [CrossRef]
26. Lei, C.; Wang, C.; Chen, W.; He, M.; Huang, B. Polyaniline@magnetic chitosan nanomaterials for highly efficient simultaneous adsorption and in-situ chemical reduction of hexavalent chromium: Removal efficacy and mechanisms. *Sci. Total Environ.* **2020**, *733*, 139316. [CrossRef] [PubMed]

27. Khan, M.I.; Almesfer, M.K.; Elkhaleefa, A.; Shigidi, I.; Shamim, M.Z.; Ali, I.H.; Rehan, M. Conductive polymers and their nanocomposites as adsorbents in environmental applications. *Polymers* **2021**, *13*, 3810. [CrossRef]
28. Bhaumik, M.; Leswifi, T.Y.; Maity, A.; Shrinivasu, V.V.; Onyango, M.S. Removal of fluoride from aqueous solution by polypyrrole/Fe_3O_4 magnetic nanocomposite. *J. Hasardous Mater.* **2011**, *186*, 150–159. [CrossRef]
29. Jokar, M.; Foroutani, R.; Safaralizadeh, M.H.; Farhadi, K. Synthesis and characterization of polyaniline/Fe_3O_4 magnetic nanocomposite as practical approach for fluoride removal process. *Ann. Res. Rev. Biol.* **2014**, *4*, 3262–3273. [CrossRef]
30. Bober, P.; Minisy, I.M.; Acharya, U.; Pfleger, J.; Babayan, V.; Kazantseva, N.; Hodan, J.; Stejskal, J. Conducting polymer composite aerogel with magnetic properties for organic dye removal. *Synth. Met.* **2020**, *260*, 116266. [CrossRef]
31. Zare, E.N.; Motahari, A.; Sillanpää, M. Nanoadsorbents based on conducting polymer nanocomposites with main focus on polyaniline and its derivatives for removal of heavy metal ions/dyes: A review. *Environ. Res.* **2018**, *162*, 173–195. [CrossRef]
32. Sen, T.; Shimpi, N.G.; Mishra, S.; Sharma, R. Polyaniline/γ-Fe_2O_3 nanocomposite for room temperature LPG sensing. *Sens. Actuators B* **2014**, *190*, 120–126. [CrossRef]
33. Alharthy, R.D.; Saleh, A. A novel trace-level ammonia gas sensing based on flexible PAni-$CoFe_2O_4$ nanocomposite film at room temperature. *Polymers* **2021**, *13*, 3077. [CrossRef] [PubMed]
34. Batool, R.; Akhtar, M.A.; Hayat, A.; Han, D.; Niu, L.; Ahmad, M.A.; Nawaz, M.H. A nanocomposite prepared from magnetite nanoparticles, polyaniline and carboxy-modified graphene oxide for non-enzymatic sensing of glucose. *Microchim. Acta* **2019**, *186*, 267. [CrossRef] [PubMed]
35. Bandgar, D.K.; Navale, S.T.; Naushad, M.; Mane, R.S.; Stadler, F.J.; Patil, V.B. Ultra-sensitive polyaniline–iron oxide nanocomposite room temperature flexible ammonia sensor. *RSC Adv.* **2015**, *5*, 68964–68971. [CrossRef]
36. Malook, K.; Khan, H.; Shah, M. Ammonia sensing behavior of polypyrrole-bimetallic oxide composites. *Polym. Composite.* **2020**, *41*, 2610–2615. [CrossRef]
37. Umare, S.S.; Shambharkar, B.H. Synthesis, characterization, and corrosion inhibition study of polyaniline-α-Fe_2O_3 nanocomposite. *J. Appl. Polym. Sci.* **2013**, *127*, 3349–3355. [CrossRef]
38. Kumar, H.; Luthra, M.; Punia, M.; Singh, R.M. Co_3O_4/PANI nanocomposites as a photocatalytic, antibacterial and anticorrosive agent: Experimental and theoretical approach. *Colloid Interface Sci. Commun.* **2021**, *45*, 100512. [CrossRef]
39. Kumar, H.; Boora, A.; Yadav, A.; Rajni; Rahul. Polyaniline-metal oxide-nano-composite as a nano-electronics, optoelectronics, heat resistance, and anticorrosive material. *Results Chem.* **2020**, *2*, 100046. [CrossRef]
40. Zadeh, M.K.; Yeganeh, M.; Shoushtari, M.T.; Esmaeilkhanian, A. Corrosion performance of polypyrrole-coated metals: A review of perspectives and recent advances. *Synth. Met.* **2021**, *274*, 116723. [CrossRef]
41. Zeng, X.J.; Cheng, X.Y.; Yu, R.H.; Stucky, G.D. Electromagnetic microwave absorption theory and recent achievements in microwave absorbers. *Carbon* **2020**, *168*, 606–623. [CrossRef]
42. Luo, X.; Li, H.; Deng, D.; Zheng, L.; Wu, Y.; Luo, W.; Zhang, M.; Gong, R. Preparation and excellent electromagnetic absorption properties of dendritic structured Fe_3O_4@PANI composites. *J. Alloys Compd.* **2022**, *891*, 161922. [CrossRef]
43. Mallikarjuna, N.N.; Manohar, S.K.; Kulkarni, P.V.; Venkataraman, A.; Aminabhavi, T.M. Novel high dielectric constant nanocomposites of polyaniline dispersed with γ-Fe_2O_3 nanoparticles. *J. Appl. Polym. Sci.* **2005**, *97*, 1868–1874. [CrossRef]
44. Janem, N.; Azizi, Z.S.; Tehranchi, M.M. Microwave absorption and magnetic properties of thin-film Fe_3O_4@polypyrrole nanocomposites: The synthesis method effect. *Synth. Met.* **2021**, *282*, 116948. [CrossRef]
45. Maruthi, N.; Faisal, M.; Raghavendra, N. Conducting polymer based composites as efficient EMI shielding materials: A comprehensive review and future prospects. *Synth. Met.* **2021**, *272*, 116664. [CrossRef]
46. Xu, F.; Ma, L.; Huo, Q.; Gan, M.; Tang, J. Microwave absorbing properties and structural design of microwave absorbers based on polyaniline and polyaniline/magnetite nanocomposite. *J. Magn. Magn. Mater.* **2015**, *374*, 311–316. [CrossRef]
47. Zhang, B.; Du, Y.; Zhang, P.; Zhao, H.; Kang, L.; Han, X.; Xu, P. Microwave absorption enhancement of Fe_3O_4/polyaniline core/shell hybrid microspheres with controlled shell thickness. *J. Appl. Polym. Sci.* **2013**, *130*, 1909–1916. [CrossRef]
48. Wu, Z.C.; Tan, D.G.; Tian, K.; Hu, W.; Wang, J.J.; Su, M.X.; Li, L. Facile preparation of core-shell Fe_3O_4@polypyrrole composites with superior electromagnetic wave absorption properties. *J. Phys. Chem. C* **2017**, *121*, 15784–15792. [CrossRef]
49. Tran, N.; Lee, M.Y.; Jeong, W.H.; Phan, T.L.; Tuan, N.Q.; Lee, B.W. Thickness independent microwave absorption performance of La-doped $BaFe_{12}O_{19}$ and polyaniline composites. *J. Magn. Magn. Mater.* **2021**, *538*, 168299. [CrossRef]
50. Jia, H.; Xing, H.; Ji, X.; Gao, S. Synergistic effect of hexagonal flake Co_3O_4@PANI core–shell composites with excellent microwave-absorbing properties. *J. Mater. Sci. Mater. Electron.* **2019**, *30*, 3386–3395. [CrossRef]
51. Manzoor, S.; Yasmin, G.; Raza, N.; Fernandez, J.; Atiq, R.; Chohan, S.; Iqbal, A.; Manzoor, S.; Malik, B.; Winter, F.; et al. Synthesis of polyaniline coated magnesium and cobalt oxide nanoparticles through eco-friendly approach and their application as antifungal agents. *Polymers* **2021**, *13*, 2669. [CrossRef]
52. Yang, F.; Zhang, X.; Song, L.; Cui, H.; Myers, J.; Tingting, B.; Zhou, Y.; Chen, Z.; Ning, G. Controlled drug release and hydrolysis mechanism of polymer–magnetic nanoparticle composite. *ACS Appl. Mater. Interfaces* **2015**, *7*, 9410–9419. [CrossRef] [PubMed]
53. Hsieh, T.-H.; Ho, L.-C.; Wang, Y.-Z.; Ho, K.-S.; Tsai, C.-H.; Hung, L.-F. New inverse emulsion-polymerized iron/polyaniline composites for permanent, highly magnetic iron compounds via calcination. *Polymers* **2021**, *13*, 3240. [CrossRef] [PubMed]
54. Wang, Y.-Z.; Cheng, Y.-W.; Ho, L.-C.; Huang, W.-Y.; Ho, K.-S.; Syu, Y.-T. Superparamagnetic, high magnetic α-Fe & α''-$Fe_{16}N_2$ mixture prepared from inverse suspension-polymerized Fe_3O_4@polyaniline composite. *Polymers* **2021**, *13*, 2380. [CrossRef] [PubMed]

55. de Araújo, A.C.V.; de Oliveira, R.J.; Alves Junior, S.; Rodrigues, A.R.; Machado, F.L.A.; Cabral, F.A.O.; de Azevedo, W.M. Synthesis, characterization and magnetic properties of polyaniline-magnetite nanocomposites. *Synth. Met.* **2010**, *160*, 685–690. [CrossRef]
56. Aphesteguy, J.C.; Jacobo, S.E. Synthesis of a soluble polyaniline–ferrite composite: Magnetic and electric properties. *J. Mater. Sci.* **2007**, *42*, 7062–7068. [CrossRef]
57. Sapurina, I.; Bubulinca, C.; Trchova, M.; Prokes, J.; Stejskal, J. Conducting polypyrrole and polypyrrole/manganese dioxide composites prepared with a solid sacrificial oxidant of pyrrole. *Synth. Met.* **2021**, *278*, 116807. [CrossRef]
58. Khan, A.; Aldwayan, A.S.; Alhoshan, M.; Alsalhi, M. Synthesis by in situ chemical oxidative polymerization and characterization of polyaniline/iron oxide nanoparticle composite. *Polym. Int.* **2010**, *59*, 1690–1694. [CrossRef]
59. Tanriverdi, E.E.; Uzumcu, A.T.; Kavas, H.; Demir, A.; Baykal, A. Conductivity study of polyaniline-cobalt ferrite (PANI-$CoFe_2O_4$) nanocomposite. *Nano-Micro Lett.* **2011**, *3*, 99–107. [CrossRef]
60. Karpacheva, G.P.; Ozkan, S.Z.; Eremeev, I.S.; Bondarenko, G.N.; Dzidziguri, E.L.; Chernavskii, P.A. Synthesis of hybrid magnetic nanomaterial based on polydiphenylamine-2-carboxylic acid and Fe_3O_4 in the interfacial process. *Eur. Chem. Bull.* **2014**, *3*, 1001–1007. [CrossRef]
61. Ozkan, S.Z.; Dzidziguri, E.L.; Chernavskii, P.A.; Karpacheva, G.P.; Efimov, M.N.; Bondarenko, G.N. Metal-polymer nanocomposites based on polydiphenylamine and cobalt nanoparticles. *Nanotechnol. Russ.* **2013**, *8*, 452–460. [CrossRef]
62. Karpacheva, G.P.; Ozkan, S.Z.; Dzidziguri, E.L.; Chernavskii, P.A.; Eremeev, I.S.; Efimov, M.N.; Ivantsov, M.I.; Bondarenko, G.N. Hybrid metal-polymer nanocomposites based on polyphenoxazine and cobalt nanoparticles. *Eur. Chem. Bull.* **2015**, *4*, 135–141. [CrossRef]
63. Ozkan, S.Z.; Karpacheva, G.P.; Dzidziguri, E.L.; Chernavskii, P.A.; Bondarenko, G.N.; Efimov, M.N.; Pankina, G.V. One step synthesis of hybrid magnetic material based on polyphenoxazine and bimetallic Co-Fe nanoparticles. *Polym. Bull.* **2017**, *74*, 3043–3060. [CrossRef]
64. Ozkan, S.Z.; Karpacheva, G.P.; Dzidziguri, E.L.; Efimov, M.N.; Bondarenko, G.N.; Shandryuk, G.A.; Chernavskii, P.A.; Pankina, G.V. Iron-containing magnetic nanocomposites based on polyphenoxazine. *J. Polym. Res.* **2019**, *26*, 176. [CrossRef]
65. Ozkan, S.Z.; Karpacheva, G.P.; Efimov, M.N.; Vasilev, A.A.; Muratov, D.G.; Petrov, V.A.; Chernavskii, P.A.; Pankina, G.V. One-step synthesis, characterization and properties of novel hybrid electromagnetic nanomaterials based on polydiphenylamine and Co-Fe particles in the absence and presence of single-walled carbon nanotubes. *RSC Adv.* **2021**, *11*, 24772–24786. [CrossRef] [PubMed]
66. Ozkan, S.Z.; Kostev, A.I.; Karpacheva, G.P. RU Patent for the Invention of "Nanocomposite Magnetic Material Based on a Polyconjugated Polymer and a Mixture of Magnetic Nanoparticles, and a Method for Production Thereof". Russian Patent Application NO. 2768158 C1, 23 March 2022.
67. Chernavskii, P.A.; Pankina, G.V.; Lunin, V.V. Magnetometric methods of investigation of supported catalysts. *Russ. Chem. Rev.* **2011**, *80*, 579–604. [CrossRef]
68. Soloveva, A.Y.; Ioni, Y.V.; Gubin, S.P. Synthesis of Fe_3O_4 nanoparticles on the surface of graphene. *Mendeleev Commun.* **2016**, *26*, 38–39. [CrossRef]
69. Jonscher, A.K. The universal dielectric response. *Nature* **1977**, *267*, 673–679. [CrossRef]
70. Dyre, J.C.; Schrøder, T.B. Universality of ac conduction in disordered solids. *Rev. Mod. Phys.* **2000**, *72*, 873–892. [CrossRef]
71. Xie, P.; Li, Y.; Hou, Q.; Sui, K.; Liu, C.; Fu, X.; Zhang, J.; Murugadoss, V.; Fan, J.; Wang, Y.; et al. Tunneling-induced negative permittivity in Ni/MnO nanocomposites by a bio-gel derived strategy. *J. Mater. Chem. C* **2020**, *8*, 3029–3039. [CrossRef]
72. Yoshimoto, S.; Ohashi, F.; Kameyama, T. Simple preparation of sulfate anion-doped polyaniline-clay nanocomposites by an environmentally friendly mechanochemical synthesis route. *Macromol. Rapid Commun.* **2004**, *25*, 1687–1691. [CrossRef]

polymers MDPI

Article

A Comparative Study on X-ray Shielding and Mechanical Properties of Natural Rubber Latex Nanocomposites Containing Bi_2O_3 or $BaSO_4$: Experimental and Numerical Determination

Arkarapol Thumwong [1], Manchusa Chinnawet [2], Preawpraw Intarasena [2], Chanis Rattanapongs [2], Shinji Tokonami [3], Tetsuo Ishikawa [4] and Kiadtisak Saenboonruang [2,5,6,7,*]

1. Department of Materials Science, Faculty of Science, Kasetsart University, Bangkok 10900, Thailand
2. Department of Applied Radiation and Isotopes, Faculty of Science, Kasetsart University, Bangkok 10900, Thailand
3. Institute of Radiation Emergency Medicine, Hirosaki University, Aomori 0368564, Japan
4. Department of Radiation Physics and Chemistry, Fukushima Medical University, Hikarigaoka 9601295, Japan
5. Kasetsart University Research and Development Institute (KURDI), Kasetsart University, Bangkok 10900, Thailand
6. Specialized Center of Rubber and Polymer Materials in Agriculture and Industry (RPM), Faculty of Science, Kasetsart University, Bangkok 10900, Thailand
7. Special Research Unit of Radiation Technology for Advanced Materials (RTAM), Faculty of Science, Kasetsart University, Bangkok 10900, Thailand

* Correspondence: kiadtisak.s@ku.th; Tel.: +66-2-562-5555 (ext. 646219)

Citation: Thumwong, A.; Chinnawet, M.; Intarasena, P.; Rattanapongs, C.; Tokonami, S.; Ishikawa, T.; Saenboonruang, K. A Comparative Study on X-ray Shielding and Mechanical Properties of Natural Rubber Latex Nanocomposites Containing Bi_2O_3 or $BaSO_4$: Experimental and Numerical Determination. *Polymers* **2022**, *14*, 3654. https://doi.org/10.3390/polym14173654

Academic Editors: Ting-Yu Liu and Yu-Wei Cheng

Received: 5 August 2022
Accepted: 1 September 2022
Published: 2 September 2022

Abstract: This work experimentally determined the X-ray shielding and morphological, density, and tensile properties of sulfur-vulcanized natural rubber latex (SVNRL) nanocomposites containing varying content of nano-Bi_2O_3 or nano-$BaSO_4$ from 0 to 200 phr in 100 phr increments, with modified procedures in sample preparation to overcome the insufficient strength of the samples found in other reports. The experimental X-ray shielding results, which were numerically verified using a web-based software package (XCOM), indicated that the overall X-ray attenuation abilities of the SVNRL nanocomposites generally increased with increasing filler content, with the 0.25-mm-thick SVNRL films containing 200 phr of the filler providing the highest overall X-ray shielding properties, as evidenced by the highest values of lead equivalence (Pb_{eq}) of 0.0371 mmPb and 0.0326 mmPb in Bi_2O_3/SVNRL nanocomposites, and 0.0326 mmPb and 0.0257 mmPb in $BaSO_4$/SVNRL nanocomposites, for 60 kV and 100 kV X-rays, respectively. The results also revealed that the addition of either filler increased the tensile modulus at 300% elongation (M300) and density but decreased the tensile strength and the elongation at break of the Bi_2O_3/SVNRL and $BaSO_4$/SVNRL nanocomposites. In addition, the modified procedures introduced in this work enabled the developed nanocomposites to acquire sufficient mechanical and X-ray shielding properties for potential use as medical X-ray protective gloves, with the recommended content of Bi_2O_3 and $BaSO_4$ being in the range of 95–140 phr and 105–120 phr, respectively (in accordance with the requirements outlined in ASTM D3578-19 and the value of Pb_{eq} being greater than 0.02 mmPb). Consequently, based on the overall outcomes of this work, the developed Bi_2O_3/SVNRL and $BaSO_4$/SVNRL nanocomposites show great potential for effective application in medical X-ray protective gloves, while the modified procedures could possibly be adopted for large-scale production.

Keywords: natural rubber latex; Bi_2O_3; $BaSO_4$; X-ray shielding; gloves; mechanical properties

1. Introduction

High-energy electromagnetic (EM) waves, especially X-rays and gamma rays, are currently utilized in various applications, including the quantification of elements, compounds, and radionuclides contained in commercial products, plants, and foods [1–3]; medical and industrial imaging [4,5]; cancer diagnostics and therapy [6,7], and quality control for medical and industrial products [8]. Despite their wide utilization and increasing demand, the adverse effects of excessive exposure to various types of radiation could seriously harm both users and the public, for whom the symptoms may vary from skin reddening, nausea, vomiting, diarrhea, hair loss, and cancer to even death, depending on several factors, such as the exposure dose and duration, the receiver's age and sex, as well as the levels of sensitivity and responses of the exposed organs to radiation [9–11]. Consequently, to reduce or prevent the risk of such adverse effects, a radiation safety principle, namely "As Low As Reasonably Achievable", or ALARA, must be strictly followed in all facilities, involving the management of the radiation exposure working time and distance, as well as the utilization of appropriate and effective radiation-shielding equipment [12].

Several types of products have been developed and used specifically for X-ray shielding, for which the selection of materials for the equipment has depended on factors such as the nature of usage, specific requirements for intended applications, and the available budget to acquire the equipment. Particularly for applications requiring excellence in flexibility, strength, and elongation at break of the products, as well as the ability to accommodate high content of radiation-protective fillers, natural rubber (NR) composites have emerged as one of the most promising candidates to serve such purposes and needs [13–15]. In addition to these properties, NR composites offer other favorable properties, such as being natural products with biodegradability, which is consistent with today's increasing demand for environmentally friendly materials [16,17]. Examples of NR composites used in X-ray protection include those containing Bi_2O_3 [18,19], WO_3 [20], $BaSO_4$ [21,22], Pb [23], and PbO [24] as radiation-protective fillers, for which the fillers can considerably enhance the X-ray attenuation ability of a composite with respect to pristine NR. Among the aforementioned fillers, Bi_2O_3 and $BaSO_4$ are the two most common lead-free fillers that are suitable to be used as X-ray attenuators due to their economical accessibility, the high atomic numbers (Z) of Bi and Ba (Z = 83 and Z = 56, respectively), and the high densities (ρ) of Bi_2O_3 and $BaSO_4$ (ρ = 8.9 g/cm^3 and ρ = 4.5 g/cm^3, respectively), which result in substantially enhanced interaction probabilities between the materials and incident X-rays [25]. Furthermore, Bi_2O_3/NR and $BaSO_4$/NR composites, exhibiting comparable or even greater X-ray attenuation abilities than those containing Pb and Pb compounds [26], are considered less hazardous for production and use, as evidenced by the higher permissible exposure limits and thresholds for Bi_2O_3 and $BaSO_4$ than those for Pb [27].

Due to possible exposure to both primary and secondary X-rays by medical personnel working directly with or close to radioactive sources, the use of appropriate X-ray protective gloves is crucial for their safety. Recently, formulations and processes for the production of X-ray protective gloves have been developed based on sulfur-vulcanized and gamma-vulcanized natural rubber latex (SVNRL and GVNRL, respectively) composites containing nano-Bi_2O_3 [28]. The results indicated that while both curing systems offered improved X-ray shielding properties to the samples after the addition of nano-Bi_2O_3, only the GVNRL composites had sufficient tensile properties to surpass the requirements for medical examination gloves, according to ASTM D3578-19 [28,29], while the SVNRL composites suffered substantial drops in their strength after the addition of high filler content. This unsatisfactory result was due to the nano-Bi_2O_3 molecules obstructing the functionality of the main activators and accelerators during the process of vulcanization, which prevented the complete curing of the samples and, hence, reduced the overall strength of the samples. Consequently, despite the beneficial evidence of nanoparticles used for X-ray protection, the inferior mechanical strength of the SVNRL gloves (a common method to prepare latex gloves commercially [30]) has prevented the implementation of such a procedure in actual large-scale production. Hence, a new method or procedure for sample preparation must be developed to make the production of X-ray protective gloves possible using commonly available technology and equipment.

Hence, the current work aimed to develop novel procedures for the production of SVNRL composites containing either nano-Bi_2O_3 or nano-$BaSO_4$, with the filler content varying from 0 to 100 and 200 phr, for potential use as medical X-ray protective gloves (the maximum filler content was 200 phr, based on our previous work that indicated the recommended filler content of 90–170 phr for GVNRL composites [28]). The properties of the nanocomposites investigated in this work were: X-ray shielding (based on the linear attenuation coefficient (μ), the mass attenuation coefficient (μ_m), the half value layer (HVL), and the lead equivalence (Pb_{eq})) and morphological, physical (density), and mechanical (tensile modulus at 300% elongation (M300), tensile strength, and elongation at break). Furthermore, to verify the reliability and correctness of the experimental results for X-ray shielding measurements, the obtained results were compared with those numerically computed using a web-based software package (XCOM [31]), to determine the recommended filler content for the attenuation of 60 kV and 100 kV X-rays, and subsequently compared to the requirements outlined in ASTM D3578-19 and the value of $Pb_{eq} > 0.02$ mmPb for medical X-ray-protective gloves. The outcomes of this work do not only present new data on SVNRL nanocomposites for X-ray attenuation but also offer improved procedures for sample preparation that would be beneficial and suitable for actual production at larger scales.

2. Experimental Section

2.1. Materials and Chemicals

High-ammonia natural rubber latex (HA-NRL) samples, with total solid and dry rubber content of 61.0% (ISO 124: 2014) and 60.3% (ISO 126: 2005), respectively, were supplied by the Office of Rubber Authority of Thailand (RAOT), Bangkok, Thailand. Names, contents, and the roles of chemicals used for the sample preparation are shown in Table 1. Nano-Bi_2O_3 and nano-$BaSO_4$ were obtained from Shanghai Ruizheng Chemical Technology Co., Ltd. (Shanghai, China), distilled water was supplied by the Faculty of Science, Kasetsart University (Bangkok, Thailand), and other chemicals were supplied by the RAOT (Bangkok, Thailand). The images of nano-Bi_2O_3 and nano-$BaSO_4$, taken using a scanning electron microscope (SEM; Quanta 450 FEI: JSM-6610LV, Eindhoven, the Netherlands), are shown in Figure 1, indicating that the average particle sizes of nano-Bi_2O_3 and nano-$BaSO_4$ were 234.9 nm and 287.6 nm, respectively, as determined using the ImageJ software version 1.50i. It should be noted that in order to improve the compatibility between the added chemicals and the NRL matrix, all chemicals used in this work (except KOH and Teric 16A16) were prepared using a stainless-steel ball mill by diluting each pure chemical with vultamol, bentonite, and distilled water for 72 h (final weight content of the chemical: vultamol: bentonite: distilled water was 50:1:1:48). It should be noted that the nanoparticles of Bi_2O_3 and $BaSO_4$ were selected for this investigation due to their superior radiation-shielding and mechanical properties in the nanocomposites in comparison with those containing microparticles at the same filler content found in previous reports [32,33].

Table 1. Material formulations of SVNRL nanocomposites and their chemical names, content, and roles.

Chemical	Content (phr)	Role
50% *w*/*w* nano-Bi_2O_3 or nano-$BaSO_4$	0, 100, and 200	X-ray protective filler
10% *w*/*w* potassium hydroxide (KOH)	0.2	Stabilizer
10% *w*/*w* Teric 16A16	0.02	Stabilizer
50% *w*/*w* sulfur (S)	0.8	Crosslinking agent
50% *w*/*w* zinc diethyl dithiocarbamate (ZDEC)	0.4	Accelerator
50% *w*/*w* zinc-2-mercaptobenzthiazole (ZMBT)	0.4	Accelerator
50% *w*/*w* titanium dioxide	1.0	Pigment
50% *w*/*w* wingstay-L	1.0	Antioxidant
50% *w*/*w* zinc oxide (ZnO)	1.0	Activator
Distilled water (H_2O)	170.5	Solvent

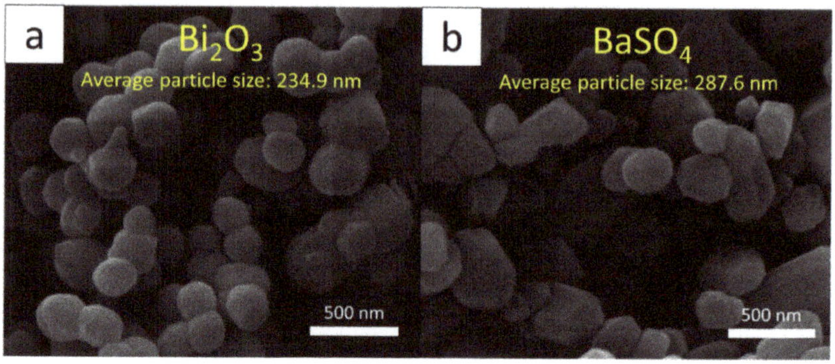

Figure 1. SEM images of (**a**) nano-Bi_2O_3 and (**b**) nano-$BaSO_4$ (60,000× magnification).

2.2. Preparation of SVNRL Mixture

NRL was mechanically stirred using an automatic top stirrer (Eurostar 60 digital, IKA, Bangkok, Thailand) at a rotation speed of 300 rpm for 60 min. Then, all chemicals listed in Table 1 (except nano-Bi_2O_3 and nano-$BaSO_4$) were consecutively added to the stirred NRL (from top to bottom order), with a 2 min interval between each chemical, and the stirring was continued for another 60 min. Then, the NRL mixture was stored in a closed container at room temperature for 72 h before the addition of the nano-Bi_2O_3 or nano-$BaSO_4$ to the NRL mixture. The mixture was continuously stirred for another 60 min and kept in a closed container for further use. It should be noted that this step (adding nano-Bi_2O_3/nano-$BaSO_4$ after the pre-vulcanization process of 72 h) was different from our previous work [28]. This procedure was modified to reduce the effects of nano-Bi_2O_3/nano-$BaSO_4$ on obstructing the functionality of the main activators and accelerators during vulcanization, which helped the SVNRL mixture to achieve a higher degree of curing that could potentially improve the overall mechanical properties of the nanocomposites [28].

2.3. Preparation of Nano-Bi_2O_3/SVNRL and Nano-$BaSO_4$/SVNRL Gloves

The procedure to prepare nano-Bi_2O_3/SVNRL and nano-$BaSO_4$/SVNRL gloves followed the steps outlined in our previous work [28]. In summary, after thorough washing, the ceramic molds were oven-dried at 70 °C for 40 min, dipped in a 35% coagulant consisting of $Ca(NO_3)_2$, Teric 16A16, 50% $CaCO_3$, and distilled water (RAOT, Bangkok, Thailand) with the final weight content of 35.0:0.1:5.0:59.9, respectively, for 5 sec, and oven-dried again at 70 °C for 2 min. Then, the dried molds were dipped in the nano-Bi_2O_3/SVNRL or nano-$BaSO_4$/SVNRL mixture for 40 sec, carefully flicked and rotated (at least 3 times), and oven-dried at 70 °C for 5 min. The molds were dipped in 70 °C distilled water for 5 min to rinse off all remaining chemicals and dried again at 100 °C for 40 min. The nano-Bi_2O_3/SVNRL or nano-$BaSO_4$/SVNRL gloves were peeled off the molds and processed using chlorination to remove any powder that remained on the surface of the gloves [28]. Figure 2 shows images of the prepared nano-Bi_2O_3/SVNRL and nano-$BaSO_4$/SVNRL samples containing 200 phr of the fillers, which clearly indicate smooth and uniform surfaces, while the colors of the nano-Bi_2O_3/SVNRL and nano-$BaSO_4$/SVNRL samples were yellow and white, respectively (the same as the colors of the nano-Bi_2O_3 and nano-$BaSO_4$ particles).

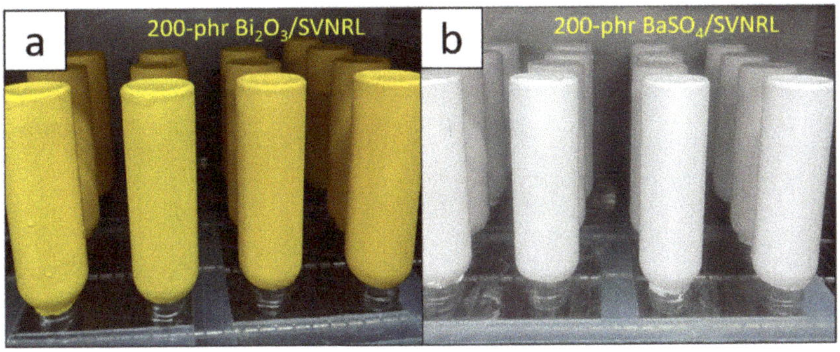

Figure 2. Images showing physical appearance and color of (**a**) nano-Bi_2O_3/SVNRL and (**b**) nano-$BaSO_4$/SVNRL composites containing 200 phr of their respective fillers.

2.4. Characterization

2.4.1. X-ray Shielding Properties

The X-ray shielding properties of the Bi_2O_3/SVNRL and $BaSO_4$/SVNRL nanocomposites were investigated at the Secondary Standard Dosimetry Laboratory (SSDL), the Office of Atoms for Peace (OAP), Bangkok, Thailand. The X-ray shielding parameters of interest were the X-ray transmission ratio (I/I_0), the linear attenuation coefficient (μ), the mass attenuation coefficient (μ_m), the half value layer (HVL), and the lead equivalence (Pb_{eq}), with their relationships shown in Equations (1)–(4):

$$I = I_0 e^{-\mu x} \tag{1}$$

$$\mu_m = \frac{\mu}{\rho} \tag{2}$$

$$HVL = \frac{\ln(2)}{\mu} \tag{3}$$

$$Pb_{eq} = \frac{\mu x}{\mu_{Pb}} \tag{4}$$

where I_0 is the intensity of incident X-rays, I is the intensity of transmitted X-rays, x is the thickness, μ is the linear attenuation coefficient, μ_m is the mass attenuation coefficient, ρ is the density, HVL is the half value layer, Pb_{eq} is the lead equivalence, and μ_{Pb} is the linear attenuation coefficient of a pure lead sheet. For Equation (4), the values of μ_{Pb} were 63.06 cm^{-1} and 25.99 cm^{-1} for the X-ray energies of 45 keV and 80 keV, respectively, numerically determined using the XCOM software (National Institute of Standards and Technology, Gaithersburg, MD, USA). These were the average values of incident X-rays emitted from an X-ray tube with supplied voltages of 60 and 100 kV, respectively, in our setup. The X-rays were collimated using a 1 mm pinhole to achieve a narrow-beam setup and pointed directly to the center of the 0.25-mm-thick SVNRL nanocomposites. The transmitted X-rays were detected and counted using a free air ionization chamber (Korea Research Institute of Standards and Science, KRISS; Daejon, Korea) that was powered by a high-voltage power unit (Keithley 247, Cleveland, OH, USA) and connected to an electrometer (Keithley 6517B, Cleveland, OH, USA). The X-rays used in this work were controlled by an X-ray system (YXLON MGC41, Hudson, NY, USA) and the energies were selected based on ISO 4037-1:2019. The schematic setup for the X-ray shielding measurement is shown in Figure 3 [25].

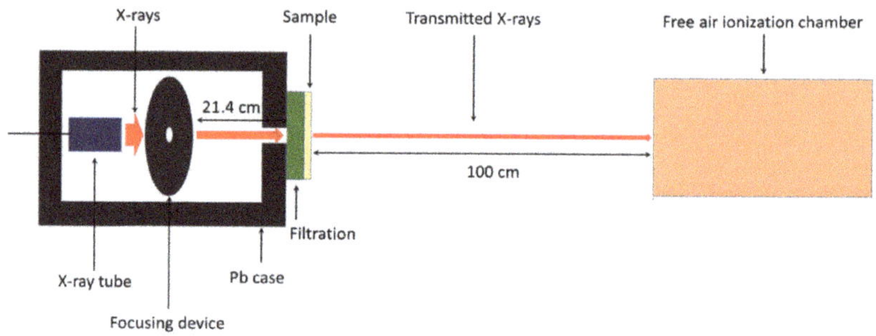

Figure 3. Schematic setup for X-ray shielding measurement.

To verify the correctness and reliability of the experimental results, the numerical determination based on the XCOM software was conducted at X-ray energies of 45 keV and 80 keV and the results were compared with those obtained experimentally [34]. Notably, since XCOM provided only the value of μ_m, the density (ρ) for each formulation, which was used for the calculation of μ, HVL, and Pb_{eq}, was theoretically determined using Equation (5):

$$\rho = \frac{C_{NR} + C_F}{\frac{C_{NR}}{\rho_{NR}} + \frac{C_F}{\rho_F}} \quad (5)$$

where ρ_{NR} (ρ_F) is the density of NR (radiation-protective filler) and C_{NR} (C_F) is the content of the NR (radiation-protective filler).

2.4.2. Morphology and Density Measurement

The morphology, the dispersion of nano-Bi_2O_3 and $BaSO_4$ particles, and the dispersion of Bi and Ba elements in the SVNRL composites were determined using scanning electron microscopy (SEM) with energy-dispersive X-ray spectroscopy (EDX; Quanta 450 FEI: JSM-6610LV, Eindhoven, The Netherlands). All samples were coated with gold using a sputter coater (Quorum SC7620: Mini Sputter Coater/Glow Discharge System, Laughton, UK) prior to the SEM-EDX images being taken.

The density of each sample was determined using a densitometer (MH-300A, Shanghai, China) with a precision of 0.001 g/cm^3. The determination was carried out based on the Archimedes principle [35].

2.4.3. Mechanical Properties

The tensile properties, consisting of tensile modulus at 300% elongation (M300), tensile strength, and elongation at break, were determined using a universal testing machine (TM Tech, TM-G5K, Bangkok, Thailand) according to ASTM D412-06 standard testing. The tensile testing speed used for all samples was 500 mm/min.

2.5. Determination of Recommended Filler Content for Medical X-ray Protective Gloves

The determination of the recommended filler content for the production of medical X-ray protective gloves based on Bi_2O_3/SVNRL and $BaSO_4$/SVNRL nanocomposites was conducted by comparing the X-ray shielding properties of the 0.25-mm-thick samples, with a minimum required Pb_{eq} value of 0.02 mmPb, as well as their tensile properties, with those outlined in ASTM D3578-19, which states that, for medical examination gloves, the tensile strength and the elongation at break must be higher than 14 MPa and 650%, respectively [29]. Then, ranges of filler content that provided sufficient X-ray shielding and tensile properties in accordance with the above requirements could be selected and recommended for actual use.

3. Results and Discussion

3.1. Density

The densities of the pristine SVNRL, nano-Bi_2O_3/SVNRL, and nano-$BaSO_4$/SVNRL composites are shown in Table 2. The results revealed that the densities of the samples increased with increasing filler content, with nano-Bi_2O_3/SVNRL having slightly higher densities than nano-$BaSO_4$/SVNRL with the same filler content. This was due to the much higher densities of Bi_2O_3 and $BaSO_4$ compared to the pristine SVNRL (ρ_{NR} = 0.93 g/cm^3, $\rho_{Bi_2O_3}$ = 8.9 g/cm^3, and ρ_{BaSO_4} = 4.5 g/cm^3), leading to enhanced overall densities of the composites [36]. Notably, these density results were later used for the calculation of μ_m from I/I_0 and μ (Equations (1) and (2)).

Table 2. Densities of pristine SVNRL, nano-Bi_2O_3/SVNRL, and nano-$BaSO_4$/SVNRL composites containing filler content of 0, 100 phr, and 200 phr.

Sample	Filler Content (phr)	Density (g/cm^3)
Pristine SVNRL	0	0.93 ± 0.01
Nano-Bi_2O_3/SVNRL	100	1.67 ± 0.02
	200	1.95 ± 0.01
Nano-$BaSO_4$/SVNRL	100	1.44 ± 0.01
	200	1.71 ± 0.02

3.2. X-ray Shielding Properties

Table 3 shows the values of μ, μ_m, HVL, and Pb_{eq} of the pristine SVNRL, nano-Bi_2O_3/SVNRL, and nano-$BaSO_4$/SVNRL composites, at the X-ray supplied voltages of 60 kV and 100 kV. The results indicated that the overall X-ray shielding abilities of the SVNRL nanocomposites increased with increasing filler content, as seen by the highest values of μ, μ_m, and Pb_{eq}, and the lowest values of HVL, observed in samples containing 200 phr of the fillers. Furthermore, Table 3 shows that the ability to attenuate X-rays of the nanocomposites was reduced at higher X-ray energies, as evidenced by the lower values of μ, μ_m, and Pb_{eq} and the higher values of HVL observed in the samples tested using 100 kV X-rays.

Table 3. Linear attenuation coefficients (μ), mass attenuation coefficients (μ_m), half value layer (HVL), and lead equivalence (Pb_{eq}) of pristine SVNRL, nano-Bi_2O_3/SVNRL, and nano-$BaSO_4$/SVNRL composites, at X-ray supplied voltages of 60 kV and 100 kV.

Properties	X-Ray Supplied Voltage	Pristine SVNRL	Bi_2O_3/SVNRL		$BaSO_4$/SVNRL	
			100 phr	200 phr	100 phr	200 phr
μ (cm^{-1})	60 kV	0.24 ± 0.01	6.19 ± 0.20	9.36 ± 0.38	5.23 ± 0.12	8.23 ± 0.13
	100 kV	0.18 ± 0.01	2.17 ± 0.07	3.38 ± 0.13	1.75 ± 0.01	2.67 ± 0.06
μ_m (cm^2/g)	60 kV	0.26 ± 0.01	3.71 ± 0.12	4.80 ± 0.19	3.63 ± 0.08	4.81 ± 0.07
	100 kV	0.19 ± 0.01	1.30 ± 0.04	1.74 ± 0.07	1.21 ± 0.01	1.56 ± 0.04
HVL (cm)	60 kV	2.87 ± 0.01	0.11 ± 0.01	0.07 ± 0.01	0.13 ± 0.01	0.08 ± 0.01
	100 kV	3.84 ± 0.09	0.32 ± 0.01	0.20 ± 0.01	0.40 ± 0.01	0.26 ± 0.01
Pb_{eq} (mm Pb)	60 kV	0.0010 ± 0.0001	0.0245 ± 0.0008	0.0371 ± 0.0015	0.0207 ± 0.0005	0.0326 ± 0.0005
	100 kV	0.0017 ± 0.0001	0.0209 ± 0.0006	0.0326 ± 0.0012	0.0168 ± 0.0001	0.0257 ± 0.0006

The positive relationship between the filler content and X-ray shielding properties was due to the relatively larger Z values of Bi and Ba compared to the C and H in NR, as well as the higher densities of Bi_2O_3 and $BaSO_4$ compared to NR. These characteristics greatly enhanced the interaction probabilities between incident X-rays and the materials through the dominant and effective photoelectric absorption, which is related to the photoelectric cross-section (σ_{pe}) and Z, as shown in Equation (6):

$$\sigma_{pe} \propto \frac{Z^n}{(h\nu)^3} \quad (6)$$

where h is Planck's constant and ν is the frequency of the incident X-rays that is directly proportional to the energy, via Equation (7):

$$E = h\nu \quad (7)$$

As depicted in Equation (6), the addition of Bi_2O_3 and $BaSO_4$ in the SVNRL matrix led to higher numbers of heavy elements (Bi and Ba) available in the composites, resulting in larger σ_{pe} values and, hence, a better ability to attenuate incident X-rays [37]. The increases in the numbers of Bi and Ba elements in the SVNRL composites containing 200 phr of the fillers could be illustrated by considering elemental mapping obtained using the SEM-EDX images (Figure 4), with Figure 4b,d showing the elemental distributions for samples with 200 phr filler and revealing higher concentrations of Bi and Ba atoms, respectively, than in Figure 4a,c, which represent the elemental distributions for samples with 100 phr filler, respectively.

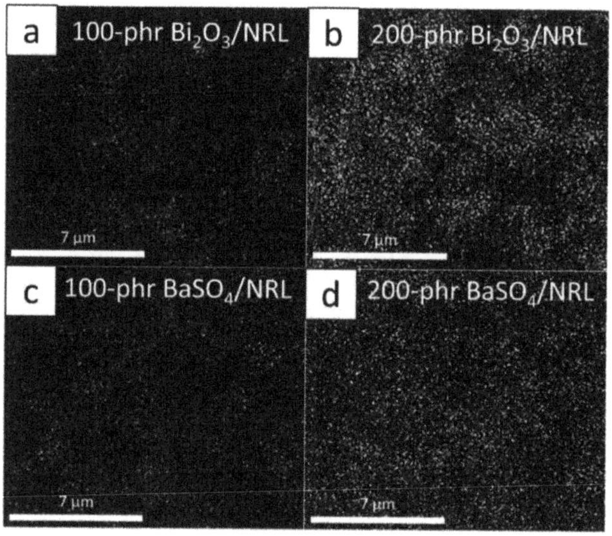

Figure 4. Dispersion of (**a**,**b**) Bi and (**c**,**d**) Ba elements in (**a**) 100 phr Bi_2O_3/SVNRL, (**b**) 200 phr Bi_2O_3/SVNRL, (**c**) 100 phr $BaSO_4$/SVNRL, and (**d**) 200 phr $BaSO_4$/SVNRL composites. The images were taken using SEM-EDX with 10,000× magnification.

Another interesting result shown in Table 3 was that nano-Bi_2O_3/SVNRL had slightly higher X-ray shielding properties than the nano-$BaSO_4$/SVNRL composites at both supplied voltages. This behavior could be explained by comparing the values of μ_m for Bi_2O_3 and $BaSO_4$, obtained from XCOM at various X-ray energies (Figure 5), which indicated that the μ_m values for both Bi_2O_3 and $BaSO_4$ were similar at the 45 keV X-rays (representing the average X-ray energy of those emitted from the X-ray tube with a supplied voltage of 60 kV), while Bi_2O_3 clearly had a higher μ_m than that of $BaSO_4$ at the 80 keV X-rays (representing the average X-ray energy of those emitted from the X-ray tube with a supplied voltage of 100 kV), leading to the more pronounced enhancement in X-ray attenuation ability in nano-Bi_2O_3/SVNRL composites. Notably, although both Bi_2O_3 and $BaSO_4$ had similar μ_m values at the 45 keV X-rays, the densities of nano-Bi_2O_3/SVNRL were greater compared to those of nano-$BaSO_4$/SVNRL (Table 2), leading to greater amplification of the overall X-ray shielding properties in nano-Bi_2O_3/SVNRL (determined at the same filler content). This phenomenon could also be mathematically explained using Equation (2), which implies a direct relationship between μ and ρ. Notably, the sharp increases in μ_m at particular X-ray energies in Figure 5 (such as 37.4 keV and 90.5 keV) were due to the K-absorption (K-edge) and L-absorption (L-edge) of Ba and Bi (the X-ray energies that are slightly above the binding energy of the electron shell of the atoms), for which the σ_{pe} or the interaction probabilities between incident X-rays and the compounds abruptly increased at these energies [38].

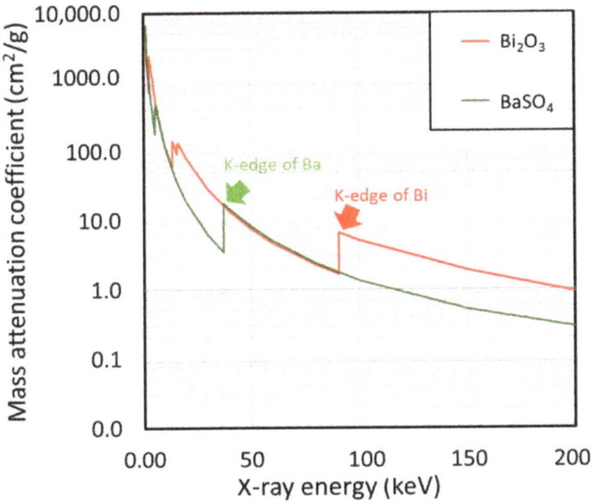

Figure 5. Mass attenuation coefficients (μ_m) of Bi_2O_3 and $BaSO_4$ at varying X-ray energies (1–200 keV), determined using XCOM.

In addition, Table 3 suggests that the X-ray shielding properties of pristine SVNRL, nano-Bi_2O_3/SVNRL, and nano-$BaSO_4$/SVNRL composites at the 60 kV X-rays were greater than those at the 100 kV X-rays. This behavior could be explained using Equation (6), which implies that σ_{pe} is inversely proportional to v^3 or E^3, resulting in less interaction probabilities with incident X-rays at higher energies [39]. The dependence of σ_{pe} could also be observed in Figure 5, which reveals overall decreases in the μ_m values of Bi_2O_3 and $BaSO_4$ with increasing X-ray energies.

To verify the correctness and reliability of the experimental results, the μ_m values of pristine SVNRL, nano-Bi_2O_3/SVNRL, and nano-$BaSO_4$/SVNRL composites at filler content of 100 phr and 200 phr were compared with those numerically determined using XCOM (Figure 6a). The comparison indicated strong agreement between the μ_m values obtained experimentally and numerically, with the percentage of difference being less than 2% for samples containing 0 and 100 phr of the fillers and being less than 7% for samples containing 200 phr of the fillers. The discrepancies between the two results could have been due to several factors, such as the fact that the experimental X-ray energies emitted from the X-ray tube were actually in spectra, with the average energies being around 45 keV and 80 keV (rather than discrete energies, as in the case of XCOM), which could cause deviations in the X-ray shielding measurements [40,41]. Nonetheless, the small percentages of difference (less than 7%) implied that the experimental results were reliable and the μ_m values obtained from XCOM could be further used for the prediction of μ, HVL, and Pb_{eq} values for all filler content values in the range 0–200 phr.

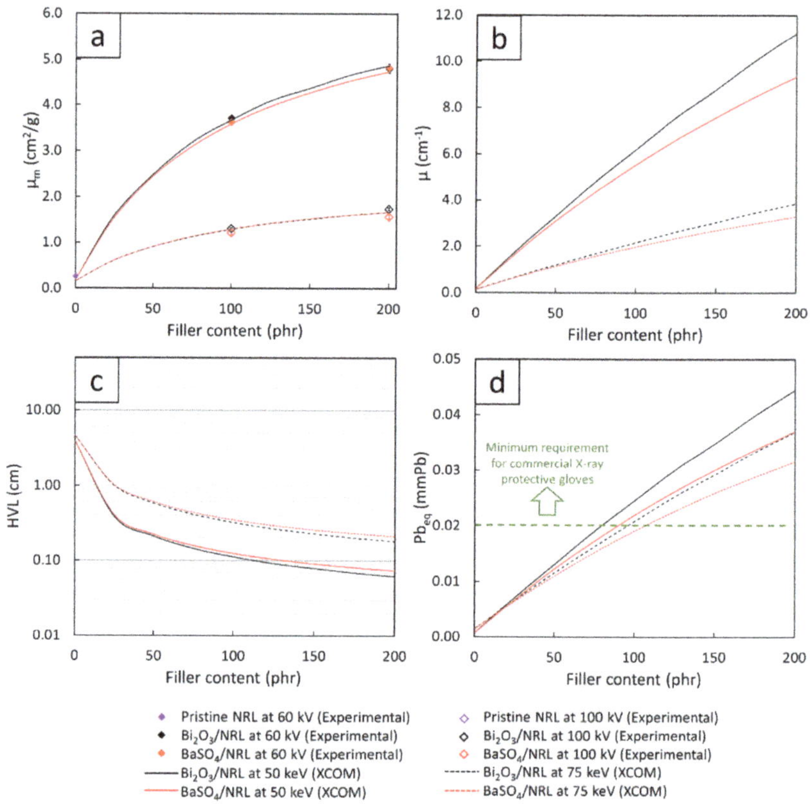

Figure 6. (**a**) Mass attenuation coefficients (μ_m), (**b**) linear attenuation coefficients (μ), (**c**) half value layer (HVL), and (**d**) lead equivalence (Pb_{eq}) of Bi_2O_3/SVNRL and $BaSO_4$/SVNRL composites containing varying filler content from 0 to 200 phr. The dotted line in (**d**) represents the minimum requirement for commercial X-ray protective gloves.

Figure 6b,d, which show the numerical values of μ, HVL, and Pb_{eq}, determined using XCOM of the Bi_2O_3/SVNRL and $BaSO_4$/SVNRL composites with varying filler content from 0 to 200 phr, confirm the dependence of the X-ray shielding properties of the samples on the filler type and content, as well as the X-ray energy, shown in Table 3. For Figure 6d, the results implied that Bi_2O_3/SVNRL composites required less filler content to meet the minimum requirement of Pb_{eq} being greater than 0.02 mmPb compared to those from the $BaSO_4$/SVNRL composites (determined at the same X-ray energy). Again, these behaviors were observed due to the higher values for μ_m (Figure 5) and ρ of Bi_2O_3 than those of $BaSO_4$, which made the former a better X-ray attenuator than the latter [29,42].

3.3. Mechanical Properties

Figure 7 shows the tensile properties, including tensile modulus at 300% elongation, tensile strength, and elongation at break, of the nano-Bi_2O_3/SVNRL and nano-$BaSO_4$/SVNRL composites. The results indicated that increases in the filler content led to an increase in the tensile modulus but decreases in the tensile strength and elongation at break. The increase in tensile modulus after the addition of the fillers to SVNRL could have been due to the high rigidity of the nano-Bi_2O_3 and nano-$BaSO_4$ particles, which enhanced the overall rigidity and, subsequently, the tensile modulus of the nanocomposites [43,44]. On the other hand, the addition of the nano-Bi_2O_3 and nano-$BaSO_4$ particles resulted in reductions in the tensile strength and elongation at break, probably due to the poor interfacial compatibility between the fillers and the NRL matrix (rubber–filler interactions), which led to visible voids inside the matrix [45]. Another factor that could have contributed to the decreases in the properties was the increase in filler–filler interactions at higher filler content, which resulted in higher particle agglomeration and worse particle dispersion in samples with 200 phr filler content (Figure 8c,e) than in samples with 0 and 100 phr filler content (Figure 8a,b,d) [46].

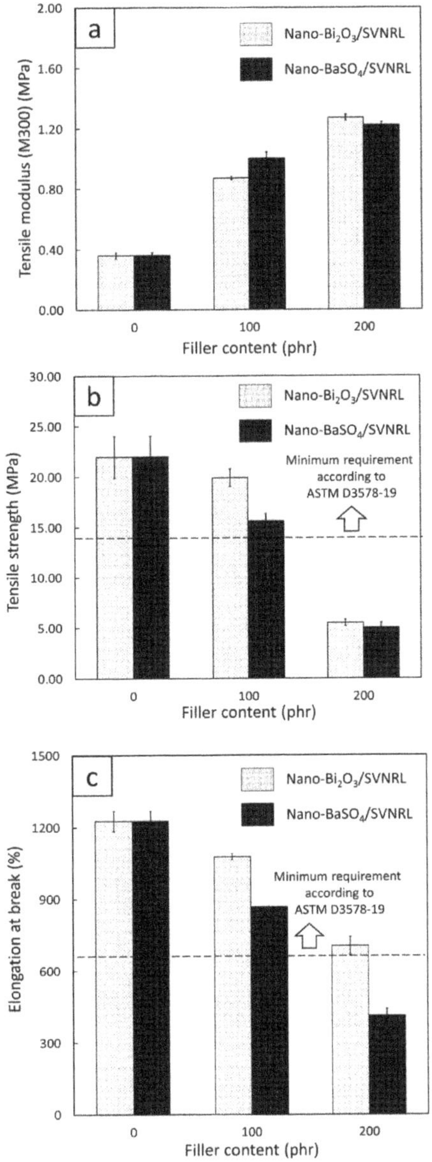

Figure 7. (**a**) Tensile modulus at 300% elongation (M300), (**b**) tensile strength, and (**c**) elongation at break of nano-Bi_2O_3/SVNRL and nano-$BaSO_4$/SVNRL composites. The dotted lines in (**b**) and (**c**) represent the minimum requirements for tensile strength and elongation at break, respectively, for medical examination gloves according to ASTM D3578-19.

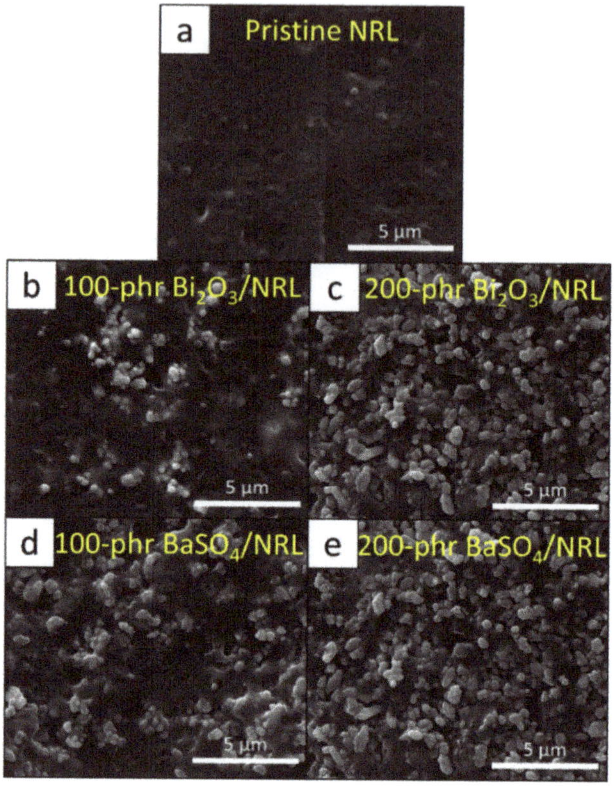

Figure 8. SEM images showing morphology and particle dispersion for (**a**) pristine SVNRL, (**b**) 200 phr Bi_2O_3/SVNRL, (**c**) 200 phr Bi_2O_3/SVNRL, (**d**) 100 phr $BaSO_4$/SVNRL, and (**e**) 200 phr $BaSO_4$/SVNRL composites (10,000× magnification).

Figure 7 also reveals that the nano-Bi_2O_3/SVNRL composites had higher tensile strength and elongation at break than the nano-$BaSO_4$/SVNRL composites, determined at the same filler content. This was mainly due to the higher density of nano-Bi_2O_3 particles than nano-$BaSO_4$ particles; hence, when both fillers were added to the samples at the same weight content, less volume of nano-Bi_2O_3 would be actually added to the composites, resulting in fewer voids and less particle agglomeration created in the nano-Bi_2O_3/SVNRL composites. Nonetheless, Figure 7b,c imply that both SVNRL nanocomposites containing less than 100 phr filler had higher experimental values of tensile strength and elongation at break than those outlined in ASTM D3578-19 for medical examination gloves (greater than 14 MPa and 650%, respectively, represented as horizontal dotted lines in this figure). Notably, these mechanical results could be further considered along with the results from the X-ray shielding measurement to determine the recommended filler content that allowed the nanocomposites to satisfy all the requirements for medical X-ray protective gloves.

As mentioned in the experimental section above, the current work modified the procedure for sample preparation by postponing the addition of nano-Bi_2O_3 or nano-$BaSO_4$ until after the completion of rubber vulcanization (72 h after sulfur was added to the SVNRL mixture). The effects of this improved procedure on the tensile strengths of the samples are shown in Table 4, which indicates that the current tensile strengths of the nano-Bi_2O_3/SVNRL composites were higher than those in a previous work for all nano-Bi_2O_3 contents investigated [28], especially for the 100 phr content, which showed an almost 3-fold increase in the values. This improvement in tensile strength could have been due to the postponed addition of nano-Bi_2O_3 reducing the obstruction effects of the filler on the functionality of the main activators and accelerators, allowing higher degrees of vulcanization to occur prior to the addition of nano-Bi_2O_3, which consequently improved the overall strengths of the samples [26]. This outcome would be crucial for the actual production of medical X-ray protective gloves based on SVNRL as the achieved tensile strengths were greater than the strength requirement (ASTM D3578-19), which was unobtainable in the previous report.

Table 4. Mean (± standard deviation) tensile strength of nano-Bi_2O_3/SVNRL composites in current work compared to previous work [28].

Bi_2O_3 Content (phr)	Tensile Strength (MPa)	
	Current Work	Previous Work [28]
100	19.90 ± 0.87	7.23 ± 1.49
200	6.85 ± 0.75	5.47 ± 0.30

3.4. Determination of Recommended Filler Content

To determine the recommended filler content for the actual production of medical X-ray protective gloves based on the requirements outlined in ASTM D3578-19 (tensile strength > 14 MPa) and ensuring a value of $Pb_{eq} > 0.02$ mmPb, the relationships between the experimental tensile strength and Pb_{eq} of the nano-Bi_2O_3/SVNRL and nano-$BaSO_4$/SVNRL composites were plotted, as shown in Figure 9, with interpolation between data points. While most of the formulations investigated in this work did not simultaneously satisfy both the X-ray shielding and mechanical requirements, the samples containing approximately 95–140 phr of nano-Bi_2O_3 and 105–120 phr of nano-$BaSO_4$ offered sufficient characteristics to satisfy the requirements; thus, these filler ranges could be regarded as the recommended filler content levels. Notably, the nano-Bi_2O_3/SVNRL composites had larger ranges of recommended filler content than the nano-$BaSO_4$/SVNRL composites, which could have been due to the greater levels of X-ray attenuation ability and overall mechanical strength found in the nano-Bi_2O_3/SVNRL composites. In addition, these findings confirmed the useability of Bi_2O_3 and $BaSO_4$ as effective fillers for radiation protection, which were also found in other shielding materials such as glasses and concrete [47–50].

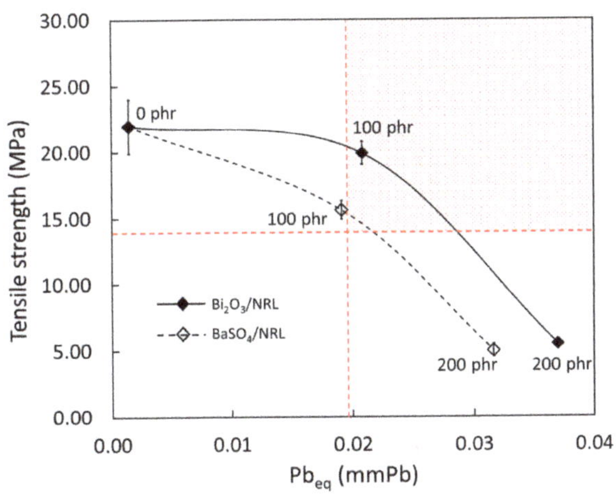

Figure 9. Relationship between tensile strength and Pb_{eq} (from XCOM) of nano-Bi_2O_3/SVNRL (solid line) and nano-$BaSO_4$/SVNRL (dotted line) composites. The shaded area in the top-right corner indicates the conditions for which samples satisfy both tensile strength and Pb_{eq} requirements for medical X-ray protective gloves (ASTM D3578-19).

4. Conclusions

This work developed medical X-ray protective gloves based on nano-Bi_2O_3/SVNRL and nano-$BaSO_4$/SVNRL composites, with varying filler content of 0 to 200 phr in 100 phr increments. The results suggested that the increases in filler content increased the values of μ, μ_m, HVL, Pb_{eq}, density, and tensile modulus at 300% elongation but decreased the tensile strength and elongation at break of the nanocomposites. The experimental results of X-ray shielding measurement were also numerically verified using XCOM, which indicated strong agreement between the two methods (less than 7% difference), implying the reliability and correctness of the results. Furthermore, after considering the X-ray shielding and mechanical properties of both composites, nano-Bi_2O_3/SVNRL with filler content of 95–140 phr and nano-$BaSO_4$/SVNRL with filler content of 105–120 phr satisfied the minimum requirements of Pb > 0.02 mmPb and tensile strength > 14 MPa outlined in the commercial X-ray protective gloves standard and ASTM D3578-19, respectively. In addition, the modified sample preparation procedures introduced in this work resulted in improved tensile properties of the SVNRL composites (not obtainable in the previous work), potentially making the method suitable for implementation in actual large-scale production.

Author Contributions: Conceptualization, K.S.; formal analysis, A.T., M.C., P.I., C.R., S.T., T.I. and K.S.; funding acquisition, K.S.; investigation, A.T., M.C., P.I., C.R., S.T., T.I. and K.S.; methodology, A.T., M.C., P.I., C.R., S.T., T.I. and K.S.; supervision, K.S.; validation, A.T., M.C., P.I., C.R., S.T., T.I. and K.S.; visualization, K.S.; writing—original draft K.S.; writing—review and editing, A.T., M.C., P.I., C.R., S.T., T.I. and K.S. All authors have read and agreed to the published version of the manuscript.

Funding: This research was financially supported by the Office of the Ministry of Higher Education, Science, Research and Innovation; the Thailand Science Research and Innovation through the Kasetsart University Reinventing University Program 2021, and the Kasetsart University Research and Development Institute (KURDI), Bangkok, Thailand, grant number FF(KU)25.64.

Acknowledgments: The Kasetsart University Research and Development Institute (KURDI) and the Specialized Center of Rubber and Polymer Materials in Agriculture and Industry (RPM) provided publication support.

Conflicts of Interest: The authors declare no conflict of interest. The funders had no role in the design of the study; in the collection, analyses, or interpretation of data; in the writing of the manuscript, or in the decision to publish the results.

References

1. Saenboonruang, K.; Phonchanthuek, E.; Pransandee, K. Soil-to-plant transfer factors of natural radionuclides (^{226}Ra and ^{40}K) in selected Thai medicinal plants. *J. Environ. Radioact.* **2018**, *184–185*, 1–5. [CrossRef] [PubMed]
2. Timothy Oyedotun, T.D. X-ray fluorescence (XRF) in the investigation of the composition of earth materials: A review and an overview. *Geol. Ecol. Landsc.* **2018**, *2*, 148–154. [CrossRef]
3. Kumar, A.; Lingfa, P. Sodium bentonite and kaolin clays: Comparative study on their FT-IR, XRF, and XRD. *Mater. Today Proc.* **2020**, *22*, 737–742. [CrossRef]
4. Afshar, P.; Heidarian, S.; Naderkhani, F.; Oikonomou, A.; Plataniotis, K.N.; Mohammadi, A. COVID-CAPS: A capsule network-based framework for identification of COVID-19 cases from X-ray images. *Pattern Recognit. Lett.* **2020**, *138*, 638–643. [CrossRef] [PubMed]
5. Withers, P.J.; Bouman, C.; Carmignato, S.; Cnudde, V.; Grimaldi, D.; Hagen, C.K.; Maire, E.; Manley, M.; Plessis, A.D.; Stock, S.R. X-ray computed tomography. *Nat. Rev. Methods Primers* **2021**, *1*, 18. [CrossRef]
6. Lu, L.; Sun, M.; Lu, Q.; Wu, T.; Huang, B. High energy X-ray radiation sensitive scintillating materials for medical imaging, cancer diagnosis and therapy. *Nano Energy* **2021**, *79*, 105437. [CrossRef]
7. Zheng, L.; Zhu, R.; Chen, L.; Fu, Q.; Li, J.; Chen, C.; Song, J.; Yang, H. X-ray sensitive high-Z metal nanocrystals for cancer imaging and therapy. *Nano Res.* **2021**, *14*, 3744–3755. [CrossRef]
8. De Prados, M.; Fulladosa, E.; Gou, P.; Munoz, I.; Garcia-Perez, J.V.; Benedito, J. Non-destructive determination of fat content in green hams using ultrasound and X-rays. *Meat Sci.* **2015**, *104*, 37–43. [CrossRef]
9. Bartley, K.; Metayer, C.; Selvin, S.; Ducore, J.; Buffler, P. Diagnostic X-rays and risk of childhood leukaemia. *Int. J. Epidemiol.* **2010**, *39*, 1628–1637. [CrossRef]
10. Claus, E.B.; Calvocoressi, L.; Bondy, M.L.; Schildkraut, J.M.; Wiemels, J.L.; Wrensch, M. Dental X-rays and risk of meningioma. *Cancer* **2012**, *118*, 4530–4537. [CrossRef]

11. Memon, A.; Godward, S.; Williams, D.; Siddique, I.; Al-Saleh, K. Dental X-rays and the risk of thyroid cancer: A case-control study. *Acta Oncol.* **2010**, *49*, 447–453. [CrossRef] [PubMed]
12. Gislason-Lee, A.J. Patient X-ray exposure and ALARA in the neonatal intensive care unit: Global patterns. *Pediatr. Neonatol.* **2021**, *62*, 3–10. [CrossRef] [PubMed]
13. Tzounis, L.; Debnath, S.; Rooj, S.; Fischer, D.; Mader, E.; Das, A.; Stamm, M.; Heinrich, G. High performance natural rubber composites with a hierarchical reinforcement structure of carbon nanotube modified natural fibers. *Mater. Des.* **2014**, *58*, 1–11. [CrossRef]
14. Kaewsakul, W.; Sahakaro, K.; Dierkes, W.K.; Noordermeer, J.W.M. Optimization of mixing conditions for silica-reinforced natural rubber tire tread compounds. *Rubb. Chem. Technol.* **2012**, *85*, 277–294. [CrossRef]
15. Prasertsri, S.; Rattanasom, N. Mechanical and damping properties of silica/natural rubber composites prepared from latex system. *Polym. Test.* **2011**, *30*, 515–526. [CrossRef]
16. Sareena, C.; Sreejith, M.P.; Ramesan, M.T.; Purushothaman, E. Biodegradation behaviour of natural rubber composites reinforced with natural resource fillers – monitoring by soil burial test. *J. Reinf. Plast. Compos.* **2013**, *33*, 412–429. [CrossRef]
17. Abraham, E.; Elbi, P.A.; Deepa, B.; Jyotishkumar, P.; Pothen, L.A.; Narine, S.S.; Thomas, S. X-ray diffraction and biodegradation analysis of green composites of natural rubber/nanocellulose. *Polym. Degrad. Stab.* **2012**, *97*, 2378–2387. [CrossRef]
18. Thumwong, A.; Jitsuna, D.; Saenboonruang, K. Comparative X-ray shielding properties of single-layered and multi-layered Bi_2O_3/NR composites: Simulation and numerical Studies. *Polymers* **2022**, *14*, 1788. [CrossRef]
19. Liao, Y.C.; Xu, D.G.; Zhang, P.C. Preparation and characterization of Bi_2O_3/XNBR flexible films for attenuating gamma rays. *Nucl. Sci. Tech.* **2018**, *29*, 99. [CrossRef]
20. Al-Ghamdi, H.; Hemily, H.M.; Saleh, I.H.; Ghataas, Z.F.; Abdel-Halim, A.A.; Sayyed, M.I.; Yasmin, S.; Almuqrin, A.H.; Elsafi, M. Impact of WO_3-nanoparticles on silicone rubber for radiation protection efficiency. *Materials* **2022**, *15*, 5706. [CrossRef]
21. Plangpleng, N.; Charoenphun, P.; Polpanich, D.; Sakulkaew, K.; Buasuwan, N.; Onjun, O.; Chuamsaamarkkee, K. Flexible gamma ray shielding based on natural Rubber/$BaSO_4$ nanocomposites. *Radiat. Phys. Chem.* **2022**, *199*, 110311. [CrossRef]
22. Sanpo, N.; Tharajak, J.; Kanthang, P. Determination of radiation attenuation coefficients of rubber containing barite. *Appl. Mech. Mater.* **2017**, *866*, 204–207.
23. El-Khatib, A.M.; Doma, A.S.; Abu-Rayan, A.E.; Aly, N.S.; Abbas, M.I. Novel composites made of natural and waste rubber loaded with lead nanoparticles for gamma radiation shielding. *Surf. Innov.* **2022**, *10*, 312–328. [CrossRef]
24. Kalkornsuranee, E.; Intom, S.; Lehman, N.; Johns, J.; Kothan, S.; Sengloyluan, K.; Chaiphaksa, W.; Kaewkhao, J. Mechanical and gamma radiation shielding properties of natural rubber composites: Effects of bismuth oxide (Bi_2O_3) and lead oxide (PbO). *Mater. Res. Innov.* **2022**, *26*, 8–15. [CrossRef]
25. Poltabtim, W.; Wimolmala, E.; Markpin, T.; Sombatsompop, N.; Rosarpitak, V.; Saenboonruang, K. X-ray shielding, mechanical, physical, and water absorption properties of wood/PVC composites containing bismuth oxide. *Polymers* **2021**, *13*, 2212. [CrossRef]
26. Konruang, S.; Srisook, A.; Buaphet, P.; Tayeh, F.; Naebpetch, W. Preparation of lead-free X-ray shielding materials based on natural rubber/barium sulfate composites. *ASEAN J. Sci. Technol. Rep.* **2022**, *25*, 59–66. [CrossRef]
27. Kuempel, E.D.; Geraci, C.L.; Schulte, P.A. Risk assessment approaches and research needs for nanomaterials: An examination of data and information from current studies. In *Nanotechnology—Toxicological Issues and Environmental Safety and Environmental Safety*; NATO Science for Peace and Security Series; Springer: Dordrecht, The Netherlands, 2007.
28. Thumwong, A.; Wimolmala, E.; Markpin, T.; Sombatsompop, N.; Saenboonruang, K. Enhanced X-ray shielding properties of NRL gloves with nano-Bi_2O_3 and their mechanical properties under aging conditions. *Radiat. Phys. Chem.* **2021**, *186*, 109530. [CrossRef]
29. Ramirez Cadavid, D.A.; Layman, R.R.; Nishino, T.; Slutzky, J.L.; Li, Z.; Cornish, K. Guayule natural rubber latex and Bi_2O_3 films for X-ray attenuating medical gloves. *Materials* **2022**, *15*, 1184. [CrossRef]
30. Sheikh, S.H.; Yin, X.; Ansarifar, A.; Yendall, K. The potential of kaolin as a reinforcing filler for rubber composites with new sulfur cure systems. *J. Reinf. Plast. Compos.* **2017**, *36*, 1132–1145. [CrossRef]
31. Poltabtim, W.; Toyen, D.; Saenboonruang, K. Theoretical determination of high-energy photon attenuation and recommended protective filler contents for flexible and enhanced dimensionally stable wood/NR and NR composites. *Polymers* **2021**, *13*, 869. [CrossRef]
32. Mesbahi, A.; Ghiasi, H. Shielding properties of the ordinary concrete loaded with micro- and nano-particles against neutron and gamma radiations. *Radiat. Phys. Chem.* **2018**, *136*, 27–31. [CrossRef] [PubMed]
33. Shik, N.A.; Gholamzadeh, L. X-ray shielding performance of the EPVC composites with micro- or nanoparticles of WO_3, PbO or Bi_2O_3. *Appl. Radiat. Isot.* **2018**, *139*, 61–65. [CrossRef] [PubMed]
34. Saenboonruang, K.; Poltabtim, W.; Thumwong, A.; Pianpanit, T.; Rattanapongs, C. Rare-earth oxides as alternative high-energy photon protective fillers in HDPE composites: Theoretical aspects. *Polymers* **2021**, *13*, 1930. [CrossRef]
35. Kires, M. Archimedes' principle in action. *Phys. Educ.* **2007**, *42*, 484. [CrossRef]
36. Pianpanit, T.; Saenboonruang, K. High-energy photon attenuation properties of lead-free and self-healing poly (vinyl alcohol) (PVA) hydrogels: Numerical determination and simulation. *Gels* **2022**, *8*, 197. [CrossRef] [PubMed]
37. Yu, H.; Chen, X.; Zhou, Y.; Chen, D.; Zhang, L. Impact of photoelectric effect on X-ray density logging and its correction. *Appl. Radiat. Isot.* **2020**, *156*, 108785. [CrossRef]

38. Atak, H.; Shikhaliev, P.M. Photon counting x-ray imaging with *K*-edge filtered x-rays: A simulation study. *Med. Phys.* **2016**, *43*, 1385–1400. [CrossRef]
39. Noor Azman, N.Z.; Siddiqui, S.A.; Low, I.M. Characterisation of micro-sized and nano-sized tungsten oxide-epoxy composites for radiation shielding of diagnostic X-rays. *Mater. Sci. Eng. C* **2013**, *33*, 4952–4957. [CrossRef]
40. Poludniowski, G.G.; Evans, P.M. Calculation of x-ray spectra emerging from an x-ray tube. Part I. Electron penetration characteristics in x-ray targets. *Med. Phys.* **2007**, *34*, 2164–2174. [CrossRef]
41. Poludniowski, G.G. Calculation of x-ray spectra emerging from an x-ray tube. Part II. X-ray production and filtration in X-ray targets. *Med. Phys.* **2007**, *34*, 2175–2186. [CrossRef]
42. Abdolahzadeh, T.; Morshedian, J.; Ahmadi, S. Preparation and characterization of nano $WO_3/Bi_2O_3/GO$ and $BaSO_4/GO$ dispersed HDPE composites for X-ray shielding application. *Polyolefins J.* **2022**, *9*, 73–83.
43. Amestoy, H.; Diego, P.; Meaurio, E.; Muñoz, J.; Sarasua, J.-R. Crystallization behavior and mechanical properties of poly(ε-caprolactone) reinforced with barium sulfate submicron particles. *Materials* **2021**, *14*, 2368. [CrossRef] [PubMed]
44. Gaballah, M.; Issa, S.A.M.; Saddeek, Y.B.; Elsaman, R.; Susoy, G.; Erguzel, T.T.; Alharbi, T.; Tekin, H.O. Mechanical and nuclear radiation shielding properties of different boro-tellurite glasses: A comprehensive investigation on large Bi_2O_3 concentration. *Phys. Scr.* **2020**, *95*, 085701. [CrossRef]
45. Leblanc, J.L. Rubber–filler interactions and rheological properties in filled compounds. *Prog. Polym. Sci.* **2002**, *27*, 627–687. [CrossRef]
46. Brechet, Y.; Cavaille, J.Y.; Chabert, E.; Chazeau, L.; Dendievel, R.; Flandin, L.; Gauthier, C. Polymer based nanocomposites: Effect of filler-filler and filler-matrix interactions. *Adv. Eng. Mater.* **2001**, *3*, 571–577. [CrossRef]
47. Kilic, G.; El Agawany, F.I.; Ilik, B.O.; Mahmoud, K.A.; Ilik, E.; Rammah, Y.S. Ta_2O_5 reinforced Bi_2O_3–TeO_2–ZnO glasses: Fabrication, physical, structural characterization, and radiation shielding efficacy. *Opt. Mater.* **2021**, *112*, 110757. [CrossRef]
48. Hegazy, H.H.; Al-Buriahi, M.S.; Alresheedi, F.; El-Agawany, F.I.; Sriwunkum, C.; Neffati, R.; Rammah, Y.S. Nuclear shielding properties of B_2O_3–Bi_2O_3–SrO glasses modified with Nd_2O_3: Theoretical and simulation studies. *Ceram. Int.* **2021**, *47*, 2772–2780. [CrossRef]
49. Rammah, Y.S.; El-Agawany, F.I.; Gamal, A.; Olarinoye, I.O.; Ahmed, E.M.; Abouhaswa, A.S. Responsibility of Bi_2O_3 Content in Photon, Alpha, Proton, Fast and Thermal Neutron Shielding Capacity and Elastic Moduli of $ZnO/B_2O_3/Bi_2O_3$ Glasses. *J. Inorg. Organomet. Polym.* **2021**, *31*, 3505–3524. [CrossRef]
50. Tasnim, A.; Sahadath, M.H.; Khan, M.N.I. Development of high-density radiation shielding materials containing $BaSO_4$ and investigation of the gamma-ray attenuation properties. *Radiat. Phys. Chem.* **2021**, *189*, 109772. [CrossRef]

polymers

MDPI

Article

New Hybrid Nanocomposites with Catalytic Properties Obtained by In Situ Preparation of Gold Nanoparticles on Poly (Ionic Liquid)/Poly (4-Vinylpyridine) Nanofibers

Oscar Ramírez [1], Matías Leal [2], Ximena Briones [3], Marcela Urzúa [4], Sebastián Bonardd [5,6], Cesar Saldías [1] and Angel Leiva [1,*]

1 Departamento de Físico Química, Facultad de Química y de Farmacia, Pontificia Universidad Católica de Chile, Santiago 7820436, Chile
2 Centro de Bioinformática y Biología Integrativa, Facultad de Ciencias de la Vida, Universidad Andrés Bello, Santiago 8370035, Chile
3 Departamento de Química Orgánica y Fisicoquímica, Facultad de Ciencias Químicas y Farmacéuticas, Universidad de Chile, Santiago 8380544, Chile
4 Departamento de Química, Facultad de Ciencias, Universidad de Chile, Santiago 7800003, Chile
5 Departamento de Química Orgánica, Universidad de La Laguna, Avda. Astrofísico Francisco Sánchez 3, La Laguna, 38206 Tenerife, Spain
6 Instituto de Bio-Orgánica Antonio González, Universidad de La Laguna, Avda. Astrofísico Francisco Sánchez 2, La Laguna, 38206 Tenerife, Spain
* Correspondence: aleivac@uc.cl

Citation: Ramírez, O.; Leal, M.; Briones, X.; Urzúa, M.; Bonardd, S.; Saldías, C.; Leiva, A. New Hybrid Nanocomposites with Catalytic Properties Obtained by In Situ Preparation of Gold Nanoparticles on Poly (Ionic Liquid)/Poly (4-Vinylpyridine) Nanofibers. *Polymers* **2022**, *14*, 3782. https://doi.org/10.3390/polym14183782

Academic Editors: Ting-Yu Liu and Yu-Wei Cheng

Received: 18 August 2022
Accepted: 7 September 2022
Published: 9 September 2022

Abstract: In this work, we report the obtaining of new hybrid nanocomposites with catalytic activity formed by nanofibers of polymer blends and gold nanoparticles. The nanofibers were obtained by electrospinning blends of a poly (ionic liquid) (PIL) and its precursor polymer, poly (4-vinyl pyridine) (P4VPy). The characteristics of the nanofibers obtained proved to be dependent on the proportion of polymer in the blends. The nanofibers obtained were used to synthesize, in situ, gold nanoparticles on their surface by two-step procedure. Firstly, the adsorption of precursor ions on the nanofibers and then their reduction with sodium borohydride to generate gold nanoparticles. The results indicated a significant improvement in the performance of PIL-containing nanofibers over pure P4VPy NFs during ion adsorption, reaching a 20% increase in the amount of adsorbed ions and a 6-fold increase in the respective adsorption constant. The catalytic performance of the obtained hybrid systems in the reduction reaction of 4-nitrophenol to 4-aminophenol was studied. Higher catalytic conversions were obtained using the hybrid nanofibers containing PIL and gold nanoparticles achieving a maximum conversion rate of 98%. Remarkably, the highest value of kinetic constant was obtained for the nanofibers with the highest PIL content.

Keywords: poly ionic liquids; nanofibers; electrospinning; gold nanoparticles; hybrid nanocomposites

1. Introduction

The rapid development of nanoscience and nanotechnology in recent decades has led to significant advances in the applications of nanomaterials in many technological fields. Thus, the special physical and chemical properties of nanomaterials have been widely explored to develop or improve applications aimed at solving important current global challenges, such as environmental remediation, sustainable energy, and climate change. Indeed, a great deal of scientific activity has been developed to obtain nanomaterials able to contribute, for example, to capture and efficiently use carbon dioxide from the air, degrading toxic pollutants in water and contributing to process of energy conversion and storage [1–3]. Thus, the use of catalytic materials can help to improve the efficiency of diverse processes, their economy, and even contribute to reduce the generation of waste and pollutants. In this context, the exploration and development of advanced materials to be used in nanocatalysis represents an important challenge in the short-medium term, in order to contribute to sustainable chemistry [4].

Thanks to the important achievements of nanoscience and nanotechnology, it is now possible to prepare with great precision many catalytic materials at the nanometer scale, achieving a significant improvement in their performance thanks to the so-called "nanoeffects". These nanoeffects, although not fully understood to date, are attributed to structural, electronic and quantum size effects [5]. Nanocatalysts can be formed by composites, compounds, alloys, or elemental solids and can also be obtained in different formats (e.g., powder, thin films, fibers). Among the nanomaterials with catalytic properties, nanoparticles (NPs) of noble metals stand out, since, in addition to providing high surface areas favoring the interaction between the reactants and the catalyst surface, they display their unique chemical, electronic, and optical properties through the well-known plasmonic effect [6].

Notwithstanding the above, the use of nanomaterials as catalysts also presents some difficulties. For example, a considerable decrease in the activity of catalytic NPs has been observed after a few cycles of use, which has been attributed to aggregation and surface passivation phenomena. Indeed, when NPs agglomerate, they tend to lose their nanometer size and transform into bulk materials losing their special properties. Another problem really challenging is the difficulty to recover the catalytic nanoparticles from the reaction medium for further use. To solve these problems a suitable strategy is to support the nanoparticles on structures of different type and nature [7], achieving that the supports not only facilitate the recovery and reuse of the nanocatalyst, but also prevent the aggregation of the NPs during successive reaction cycles or under continuous flow conditions. Interestingly, with the advancement of knowledge it has been established that in many cases the role played by the supports in practice may be rather more complex and even not yet clearly understood. Some supports can perform, directly or indirectly, different functions during a catalytic reaction, providing specific defect sites. These zones would allow to anchor and even stabilize metal nanoparticles favoring activity and transfer of electrons [8].

In general, it is desirable that the ideal supports are chemically inert, have a large surface area and uniform size, and retain the nanocatalyst efficiently [9]. Among the most used supports for nanocatalyst immobilization are inorganic fibers, glass mats, and different polymers [10]. An alternative that has shown great promise for supporting and stabilizing nanoparticles, preserving their special properties over time and after repeated use in different chemical environments, is the use of polymeric nanofibers (NFs) [11,12]. One of the advantages offered by polymeric NFs is that these materials can be obtained on a macroscopic scale in the form of mat-like structures. These mats are characterized by high flexibility and elasticity, larges specific surface area, and show high degree of porosity, of small pore sizes and high pore interconnectivity. In addition, it has been demonstrated that electrospun nanofiber films can be modified and grafted with different functionalities, providing them with a high potential to be used for various purposes. Thanks to these special properties, polymeric nanofiber mats can be advantageously used in a variety of emerging environmental applications, such as air filtration, heavy metal ion adsorption, self-cleaning applications, and water purification processes. Consequently, considerable effort has been made to obtain hybrid catalytic nanomaterials of the Polymer Nanofibers containing metal Nanoparticles, in which the special characteristics of both nanocatalysts and nanofibers converge [12,13].

Inspired by the above, we have prepared a hybrid nanomaterial consisting of polymeric nanofibers and gold nanoparticles, and we studied its performance as a heterogeneous catalyst using the well-known reduction reaction of 4-nitrophenol (4NP) as a model. Noticeably, in this simple reduction reaction with sodium boron hydride catalyzed by gold nanoparticles, 4-NP, a compound considered a priority pollutant by the USEPA, is effectively reduced to form the considerably less toxic compound 4-aminephenol (4-

AP) under mild reaction conditions, and the reaction can be easily monitored by UV-vis spectroscopy [14,15].

The polymeric nanofibers were obtained by electrospinning mixtures of the poly ionic liquid, Poly penthyl 4-vinylpyridine bis trifluoromethane sulfonamide (PP4VPy-NTf_2^-) and poly 4-vinyl pyridine PV4Py, while the gold nanoparticles were synthesized in situ on the nanofibers by reduction in ionic precursors adsorbed on them. The polymeric ionic liquids PILs, also known as poly (ionic liquids) are a special type of polyelectrolytes that contain some of the chemical groups of ILs, cations and anions. PILs keep some of the special properties of Ils, such as, negligible vapor pressure, ionic conductivity, thermal stability, tuneable solution properties, nonflammability, and high chemical and electrochemical stabilities, as well as good compatibility. Many of the special properties that characterize ILs are mainly attributable to their ionic nature and are, therefore, also present in PILs. A particularly interesting property of PILs is their high capacity to interact and their compatibility with different materials, both organic and inorganic, hybrids, e.g., metal-organic and biomacromolecules [16,17]. Based on the above, PILs are expected to present a good performance as support for catalytic Au nanoparticles, mainly due to the potential generation of favorable interactions between NFs and NPs, providing high stability to the system along with other special properties such as flexibility, processability and mechanical durability, all highly desirable for supported catalysts. Therefore, the main novelty and also the main aim of this work lies in the inclusion of a PIL in the nanofibers and the study of its effect in the in situ gold nanoparticles and in the catalytic performance of the final hybrid nanomaterial. To the best of our knowledge, to date, information on the use of PIL-containing NFs as supports for catalytic NPs is at least scarce, so we seek to contribute to knowledge in this interesting area.

2. Materials and Methods

2.1. Materials

Poly (4-vinylpyridine) (P4VPy) $\overline{M}_n$ = 160,000 g mol^{-1}, potassium tetrachloroaurate (III) ($KAuCl_4$, 99.9%), bis(trifluoromethane)sulfonamide lithium salt ($LiNTf_2$, 99.9%) and pentyl bromide (99.9%) were purchased from Sigma–Aldrich. Ethyl acetate, methanol, *N*, *N*-dimethyl formamide (DMF), dichloromethane (DCM) and sodium borohydride ($NaBH_4$, 98.0%) were purchased by Merck. Milli-Q water (18.2 MΩ/cm) was used in all required experiments. All solvents used were of analytical grade and used without further purification.

2.2. Synthesis of Poly(Ionic Liquid) PP4VPy-NTf_2^-

Poly (pentyl 4-vinylpyridine bis trifluoromethane sulphonamide) (PP4VPy-NTf_2^-) was synthesized by a two-step process previously reported [18]. Firstly, the quaternization of pyridine rings was carried out by using pentyl bromide as alkylating agent. Then, in a 50 mL round-bottom flask, the counterion exchange was carried out by dissolving 1 g of quaternized PP4VPy-Br^- in 20 mL of Milli-Q water and adding a $LiNTf_2$ aqueous solution containing 6 g of salt in 12 mL of water. After stirring for 30 min at room temperature, the white precipitated formed was filtered, washed several times with water and finally lyophilized.

2.3. Nanofibers Preparation

Polymer nanofibers were obtained by electrospinning of PP4VPy-NTf_2^- and P4VPy blend solutions in a DMF/DCM (1:1 *v*/*v*) solvent mixture. The solutions were prepared in 100/0, 90/10, 80/20, 70/30, and 50/50 ratios of P4VPy/PP4VPy-NTf_2^-, maintaining a total polymer concentration of 20% *w*/*v*, and were stirred at room temperature during 24 h before electrospinning. A Linari Engineering electrospinning system was used, the operating parameters were a voltage of 15 kV, a tip-collector distance between 10–20 cm and a flow rate of 1.5 mL/h.

2.4. FTIR and 1H-NMR Analysis

FTIR analysis of samples was performed using a VECTOR 22 FTIR (Bruker, Germany). FTIR spectra were recorded between 600–4000 cm^{-1} after the accumulation of 20 scans with a resolution of 2 cm^{-1}. ^{1}H-NMR and ^{19}F-NMR measurements were carried out at room temperature using a BRUKER AVANCE III HD-400 spectrometer (400 MHz). Samples were prepared by dissolving 50 mg of material in DMSO-d6 (0.5 mL).

2.5. Conductimetric Measurements

The electrical conductivity of polymer solutions was performed using a digital conductivity meter CX-701 (Agrotools meettechniek) with a glass electrode and a cell constant of 1 cm^{-1}. The conductivity of the electrode was first calibrated using a KCl solution with a conductivity (σ) of 1.41 mS/cm.

2.6. Thermogravimetric Analysis

TGA measurements were performed in a Mettler TGA/SDTA 851^{e} calorimetric system and processed using the STAR^{e} program interface. Measurements were carried out in alumina pans by heating the sample between 25 and 800 °C at 20 °C/min under a nitrogen flow of 20 mL/min.

2.7. Scanning Electron Microscopy FESEM

The visualization of obtained nanofibers was carried out using field emission scanning electron microscopy (FE-SEM) on a FEI QUANTA FEG 250 microscope equipped with an Oxford X-MAX50 energy dispersive X-ray spectroscopy (EDX) analyzer. Nanofibers were collected on silicon wafers and coated with an ultrathin layer of gold. Electron micrographs were recorded in high-vacuum mode under an acceleration voltage of 10.0 kV. The analysis of the morphology, average-diameter size, and diameter size distribution was performed using the Image-J image processing program.

2.8. X-ray Photoelectron Spectroscopy

X-ray photoelectron spectroscopy (XPS) spectra were collected by a SPECS equipped with an PHOIBOS 150 analyzer, 1D-DLD detector and a focus 500 monochromatic excitation source (X-ray Al Kα hν = 1486.71 eV) using a Flood gun to compensate charge effects.

2.9. $AuCl_4^-$ Adsorption Experiments on PP4VPy-NTf_2^-/P4VPy Nanofibers

Measurements were performed using 10 mL of a $KAuCl_4$ 40 ppm stock solution and 10 mg of nanofiber mat. Adsorption kinetics were carried out by using UV-visible spectroscopy, and following the intensity decrease in the band centered at 305 nm, ascribed to the $AuCl_4^-$ specie. Aliquots from the supernatant were analyzed at different time intervals until reaching an equilibrium-state. The gold concentration was determined by UV-visible spectroscopy using a calibration curve. The adsorption capacity, q_t (mg/g) was calculated using the following equation:

$$q_t = \left(\frac{C_0 - C_t}{m}\right)V_0 \quad (1)$$

where C_0 and C_t correspond to the initial gold concentration and the concentration measured at time t. V_0 is the initial volume of solution and m is the mass of adsorbent used in the adsorption.

2.10. Synthesis of Gold Nanoparticles Supported on PP4VPy-NTf_2^-/P4VPy Nanofibers

The preparation of PP4VPy-NTf_2^-/P4VPy nanofibers containing gold nanoparticles was achieved following a two-step consecutive process. Firstly, a nanofiber mat (10 mg) was immersed in 10 mL of a 40 ppm $KAuCl_4$ solution and kept with gentle stirring for 1 h. The complete adsorption was confirmed by UV-visible spectroscopy corroborating the disappearance of the band at 305 nm. Once the adsorption was concluded, the nanofiber mat was washed with abundant water for 1 h and dried under vacuum for 24 h at 35 °C. In a second step, the reduction in the adsorbed gold ions was carried out using $NaBH_4$ as reducing agent. Nanofiber mats were immersed into 10 mL of Milli-Q water under constant stirring for 10 min. Then, 0.5 mL of a $NaBH_4$ solution (25 mg/mL) was added, visualizing immediately a color change in the material, going from pale-yellow to purple, accusing the formation of gold nanoparticles. After 30 min of reaction, nanofibers mats were removed, washed with abundant water, and dried under vacuum for 24 h at 35 °C.

2.11. Catalytic Studies Based on the 4-Nitrophenol (4NP) Reduction

The catalytic activity of nanocomposites was tested using, as a model reaction, the well-known catalyzed reduction in 4NP, using a methodology previously reported by our group [19]. Briefly, 5 mL of a stock solution of 4NP (60 µM) and 1 mg of catalyst were placed into a glass vial equipped with a magnetic stirrer. Subsequently, 350 µL of a 20 mg/mL $NaBH_4$ solution was added to start the reaction. The evolution of the reaction was studied by UV-Vis spectroscopy by following the disappearance of the band centered at 400 nm, assigned to the p-nitrophenolate specie. The data were analyzed by adjusting the results to a pseudo-first order kinetics.

3. Results and Discussion

For the poly ionic liquid PP4VPy-NTf_2^- synthesis, initially, P4VPy was reacted with an excess of pentyl bromide (alkylating agent) to achieve almost complete quaternization of the pyridinyl groups. A quaternization degree of 99% was obtained by FTIR and ^{1}H-NMR analysis, as reported in a previous work [18]. Then, the exchange of bromide for NTf_2^- anions, was performed by direct dissolution of the $LiNTf_2$ salt on a solution of the polyelectrolyte PP4VPy-Br [20]. Finally, purification of PP4VPy-NTf_2^- PIL was performed by consecutive dialysis, filtration and lyophilization. The successful synthesis of PP4VPy-NTf_2^- was corroborated by FTIR, ^{1}H-NMR and ^{19}F-NMR product characterization. A scheme of synthesis route and the corresponding spectra are shown in Figure 1.

The FTIR spectrum (Figure 1B) shows the typical bands expected for the quaternized $PP4VPy^+$ backbone, accompanied by new ones ascribed to vibrational modes of NTf_2^- counterions. The quaternization process was verified by the appearance of an intense band at 1647 cm^{-1} corresponding to the vibrational modes of the quaternized pyridine rings, and the absence of the signal attributed to the vibrations of the uncharged pyridine rings which typically appears at 1600 cm^{-1}. Moreover, the increased intensity of bands between 2800–3000 cm^{-1}, would denote a higher aliphatic content in the polymer due to the incorporation of pentyl moieties. On the other hand, the absorption bands at 1350, 1250, 1205, 1130, and 1050 cm^{-1}, are ascribed to NTf_2^- counterions. Figure 1C shows the ^{1}H-NMR spectrum of PP4VPyNTf_2^-, the most relevant signals are, at 8.74 and 7.38 ppm corresponding to protons of the quaternized pyridine rings, and signals at 1.81, 1.34, and 0.91 ppm protons of pentyl chains attached to the nitrogen atom of pyridine groups. Notably, the analysis of ^{1}H-NMR spectrum indicated a quantitative polymer quaternization. Similarly, the ^{19}F-NMR spectrum analysis allowed corroborate the presence of NTf_2^- counterions (specifically, $-CF_3$ signal at 78.65 ppm) in the PIL structure.

Polyelectrolytes are among the most difficult polymers to be electrospun since this type of polymer has special properties in solution due to the presence of charges along their chains [21]. To minimize this difficulty, nanofibers were obtained by electrospinning employing solutions prepared from mixtures of PP4VPy-NTf_2^- and P4VPy in different proportions, using DMF/DCM (1:1 v/v) as solvent. Figure 2 shows SEM images of the obtained nanofibers obtained from the P4VPy and P4VPy/PP4VPyNTf_2^- blends.

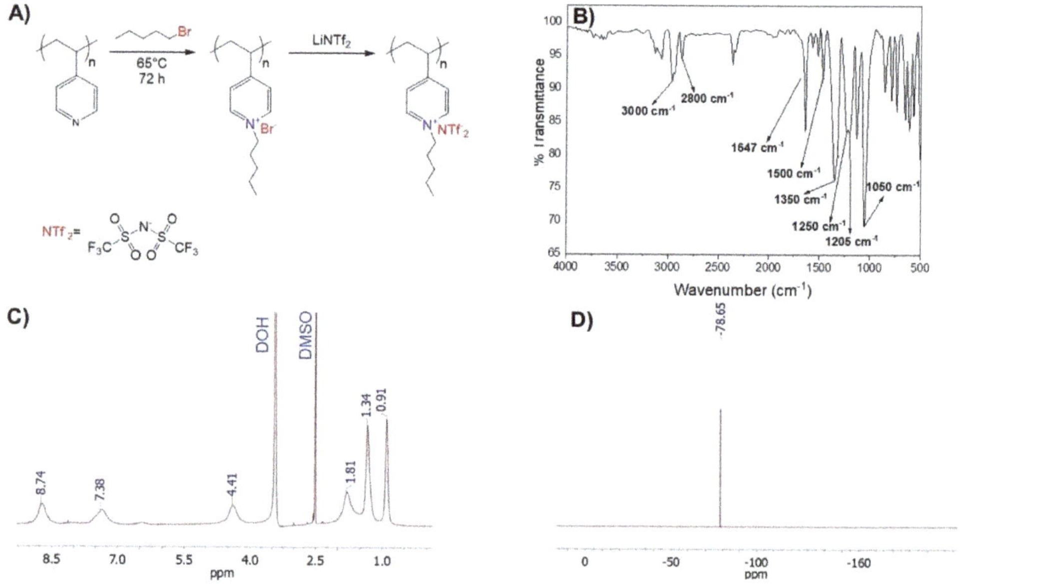

Figure 1. (**A**) Scheme of synthesis $PP4VPyNTf_2^-$ poly(ionic liquid) synthesis; (**B**) FT-IR spectrum and (**C**) ^1H- and (**D**) ^{19}F- NMR spectra of the poly(ionic liquid).

Images revealed the obtainment of uniform and bead-free nanofibers using pure P4VPy and their blends with compositions up to 30% w/w of $PP4VPyNTf_2^-$. In contrast, the blend containing 50% w/w of PIL produced a material consisting of a high content of beads connected to each other by rather irregular fibers. A significant detriment of the mechanical properties of fibrous materials accompanied by a drastic effect on their surface properties due to the presence of beads have been reported [22]. Notably, by increasing the amount of PIL in blends, mats with higher fiber density and lower fiber diameter values were obtained, Figure 3A. In addition, the analysis of the size distributions shows that the diameter of the nanofibers becomes narrower as the PIL content in the electrospun solution increases. These results are ascribed to the increase in conductivity of the solutions as the PIL content increases (insert Figure 3A), giving rise to an increase in the repulsive forces in the solution jet during the electrospinning process, phenomenon that significantly reduce the diameter of the fibers [23]. On the other hand, a proper relation between the PIL content of solutions and obtained nanofibers was determined by means of EDX analysis, with a rising trend for the values of relative content of fluorine regarding carbon (% F/% C), as is shown in Figure 3B.

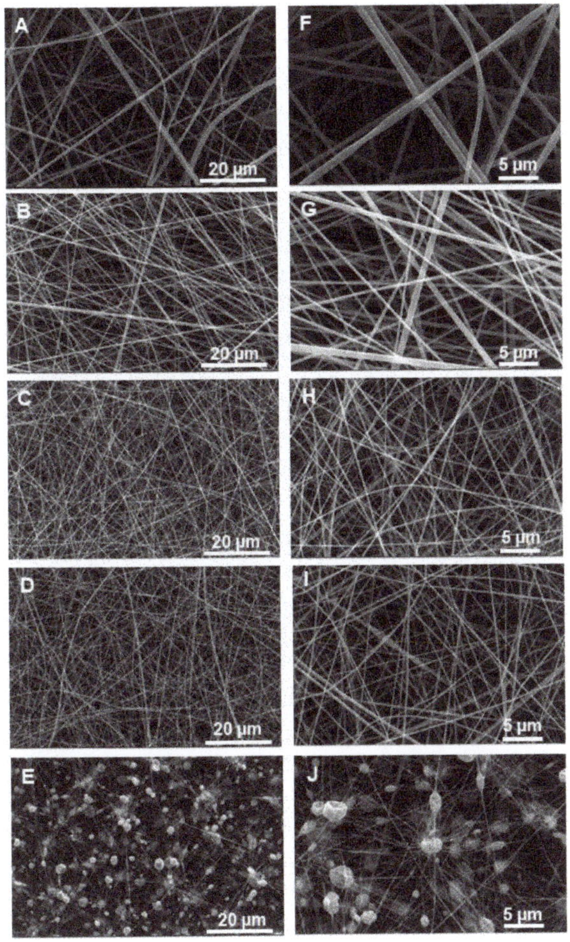

Figure 2. FE-SEM micrographs of pristine P4VPy nanofibers (**A**) and P4Vpy/PP4VpyNTf$_2^-$ nanofibers with PIL compositions of 10 (**B**), 20 (**C**), 30 (**D**), and 50% *w*/*w* (**E**). (**F–J**) the same with higher resolution.

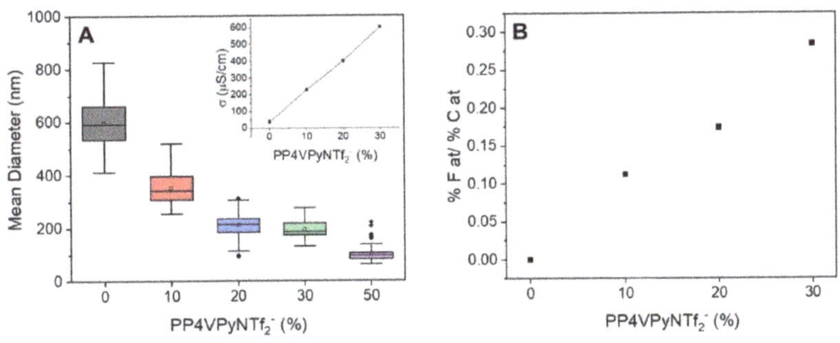

Figure 3. (**A**) Diameters Distribution of P4VPy and P4VPy/PP4VPyNTf$_2^-$ nanofibers and corresponding solution conductivities (insert). (**B**) EDX analysis of the relative content of fluorine respect to carbon atoms, % F at/% C at.

The nanofibers obtained with different amounts of PIL were characterized by FTIR spectroscopy and thermogravimetric analysis (TGA). Figure 4A shows the FTIR spectra of bare P4VPy NFs and of all P4VPy/PP4VPyNTf$_2$$^-$ blend NFs. The characteristic signal centered at 1600 cm^{-1} (signal 2 in spectra) which correspond to vibrations of the non-quaternized pyridine rings, are observed in all nanofibers. Additionally, new signals ascribed to PIL are detected. As in the PIL spectrum (Figure 1B), the FTIR spectra of the P4VPy/PP4VPyNTf$_2$$^-$ NFs present a signal attributed to the quaternized pyridine rings, labelled as 1, and signals corresponding to the NTf$_2$$^-$ counterions, from 3 to 7. As expected, these signals showed an intensity increase in those blends having higher PIL content.

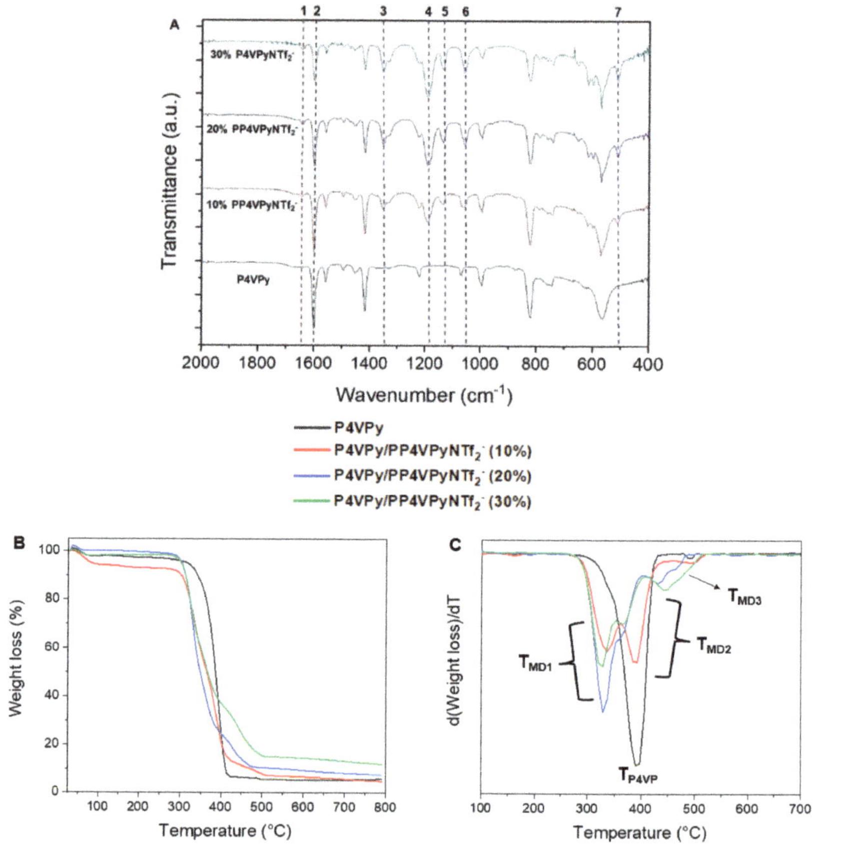

Figure 4. (**A**) FT-IR spectra of the P4VPy and P4VPy/PP4VPyNTf$_2$$^-$ nanofibers, (**B**) thermal decomposition profiles of P4VPy and P4VPy/PP4VPNTf$_2$$^-$ nanofibers, and (**C**) curves of the corresponding weight loss derivatives with temperature.

Figure 4B,C show the thermal degradation profiles and their corresponding differential curves of the nanofibers of P4VPy and P4VPy/PP4VPyNTf$_2$$^-$ blends. From these results, relevant information such as onset degradation temperatures (T_{onset}), maximum decomposition rate temperatures (T_{MAX}), and percentage of residues were obtained. The above-mentioned values are summarized in Table 1.

Table 1. Maximum weight loss rate T_{max}, onset of thermal degradation T_{onset} and percentage of final residue, of P4VP NFs, and NFs with different PIL contents.

Sample	T_{max} ST1	T_{max} ST2	T_{max} ST3	T_{onset}	Residue (%)
P4VPy	392.3	/	/	382.3	5.9
P4VPy/PP4VPyNTf$_2^-$ (10%)	334.8	388.7	465.4	326.3	7.4
P4VPy/PP4VPyNTf$_2^-$ (20%)	329.0	367.7	427.3	310.6	12.3
P4VPy/PP4VPyNTf$_2^-$ (30%)	323.8	367.7	443.6	306.9	13.8

All samples showed an initial weight loss at around 100 °C, which is attributed to the removal of water molecules. All nanofibers except P4VPy exhibited a multistage degradation profile consisting of three stages. Reported works indicate that the first stage could be related to the degradation of the quaternized pyridinyl moieties since, apparently, the quaternization process induces a selective labilization of the pyridine structure [24]. On the other hand, the second stage would correspond to the decomposition of pentyl chains. Interestingly, the temperatures of this stage in the blends containing 20 and 30% PIL are lower than that of blend containing 10% of PIL. The loss weight of third stage, would be related to the volatilization of fragments coming from NTf$_2^-$ counterions [25]. This is consistent with the intensity increase in DTGA curves as the PIL content increases. Although the incorporation of PIL tends to reduce the onset temperature of NFs decomposition, all nanofiber mats showed good thermal stability by degrading above 300 °C, allowing them to be considered as suitable materials for a wide spectrum of applications.

Once the characterization of the nanofibers was completed, their capacity to adsorb Au (III), in the form of the AuCl$_4^-$ anions, was studied, in order to subsequently obtain gold nanoparticles "in situ" as has been reported in previous works [26]. The kinetic of AuCl$_4^-$ adsorption onto P4VPy NFs, and NFs with different PIL contents was analyzed by plotting the adsorbed capacity (q) onto the P4VPy/PP4VPyNTf$_2^-$ nanofibers over time, Figure 5.

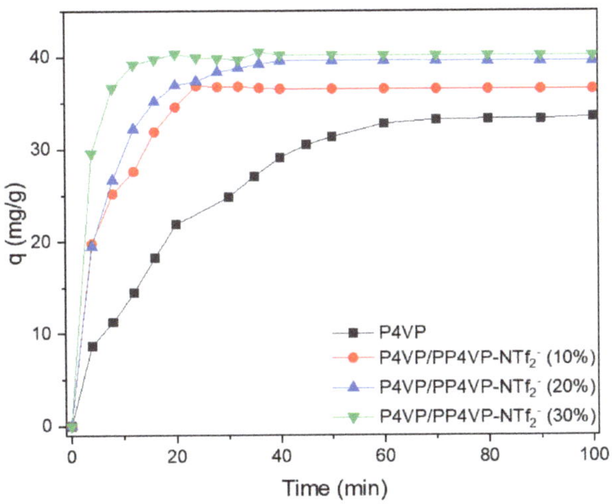

Figure 5. Effect of contact time on the adsorption capacity of P4VPy and P4VPy/PP4VPyNTf$_2^-$ nanofibers.

From the above Figure, it can be seen that all nanofibers containing PIL showed a faster adsorption rate than P4VP NFs in the early stage of the process. In this sense, while P4VPy NFs required around 70 min to reach a plateau value, all PIL-based nanofibers achieved the above situation in remarkably less time. Additionally, an increase in the adsorption capacity was observed as the PIL content in nanofibers rises. Therefore, the inclusion of PIL proved to be an effective strategy to improve the $AuCl_4^-$ adsorption process both kinetically and in terms of capacity.

To better understand the adsorption mechanism governing the coordination of gold ions, the experimental data shown in Figure 6 were fitted using the well-known linear expression of a pseudo-first order kinetic model (Equation (2)).

$$Ln(q_e - q_t) = Ln(q_e) - k_1 t \tag{2}$$

In this equation, t corresponds to the contact time between nanofibers and the $AuCl_4^-$ solution, q_t and q_e are the adsorption capacity at time t and at the equilibrium state (mg/g), respectively, while k_1 (min^{-1}) is the pseudo first-order kinetic constant. The main kinetic parameters, such as k_1 and q_e, are summarized in Table 2.

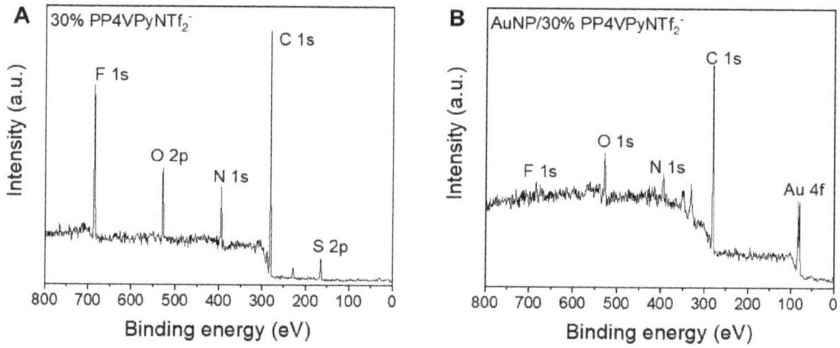

Figure 6. XPS survey spectra of P4VPy/PP4VPyNTf$_2^-$ nanofibers (**A**) before and (**B**) after the adsorption of gold ions.

Table 2. Adsorption kinetics parameters for Au (III) adsorption onto nanofibers.

Sample		Pseudo-First Order	
	q_e	k_1 (min^{-1})	R^2
P4VP	32.22	0.047	0.977
P4VPy/PP4VPyNTf$_2^-$ (10%)	32.58	0.049	0.973
P4VPy/PP4VPyNTf$_2^-$ (20%)	34.23	0.127	0.996
P4VPy/PP4VPyNTf$_2^-$ (30%)	37.42	0.281	0.993

Note that kinetic constant and q_e values increase as the PIL content in the nanofibrous material increases. Notably, the sample containing 30% PIL, in addition to being able to adsorb a higher amount of gold ions, exhibited a kinetic constant around six times higher than the one measured for P4VPy nanofibers. A plausible explanation to the previous results could be related to an increase in the charged species along the surface of the nanofibers containing the PP4VPyNTf$_2^-$ PIL, generating a faster adsorption of the ionic species due to electrostatic interactions. It is even reasonable to speculate the existence of an anion exchange process in which the NTf_2^- entities of the PIL are replaced by $AuCl_4^-$ species as new counterions. To verify this hypothesis, XPS measurements of PP4VPyNTf$_2^-$ NFs before and after $AuCl_4^-$ adsorption were performed, and the changes in the elemental composition of the nanofiber surface were analyzed. Figure 6 shows the survey XPS spectra belonging to PP4VPyNTf$_2^-$ (30%).

The XPS spectra show the main signals corresponding to the atoms that are part of the P4VPy and PP4VPyNTf$_2^-$. As expected, the appearance of the Au 4f signal is observed after the adsorption of gold ions occurs, corresponding to the Au $4f_{7/2}$ and $4f_{5/2}$ doublet at 83.1 eV and 87.0 eV, respectively. In addition, it is interesting to note the intensity decrease ascribed to the F 1s and S 2p signals, which would evidence a substitution of the NTf_2^- counterion upon adsorption of complex gold species on the surface of the fibers, suggesting an ion exchange mechanism.

To achieve a deeper understanding of the adsorption mechanism, a more detailed analysis of the main atoms present in the nanofibers XPS spectrum was performed with a particular interest in the region corresponding to nitrogen 1s. This is based on reports that have previously suggested that the N atom of the pyridinic ring has significant interactions with gold atoms [27]. Figure 7 shows that the nitrogen region exhibits multiple signals that can be deconvoluted into signals with binding energies of 400.1, 397.2, and 396.6 eV, being assigned to nitrogen atoms present in quaternized pyridine rings (N^+). In addition, it is interesting to note the intensity decrease ascribed to non-quaternized pyridine rings (N) and the counterion specie (N^-), respectively. After gold ion adsorption, the nitrogen region undergoes remarkable changes, although the deconvolution of this region again indicated the presence of three different nitrogen species, N^+, N, and N^-, in this case the signals appear 398.4, 397.0, 396.5 eV, respectively. Thus, the most noticeable changes in the region of the XPS spectrum of nitrogen are, the shift of the N^+ signal to lower binding energies, and a decrease in the intensity of the N^- signal. Consequently, these results would indicate the involvement of N^+ during the adsorption process, along with the loss of some anionic species during the process. Therefore, according to the previous analysis it is possible to hypothesize that the adsorption would take place by an ionic exchange of NTf_2^- by $AuCl_4^-$ species.

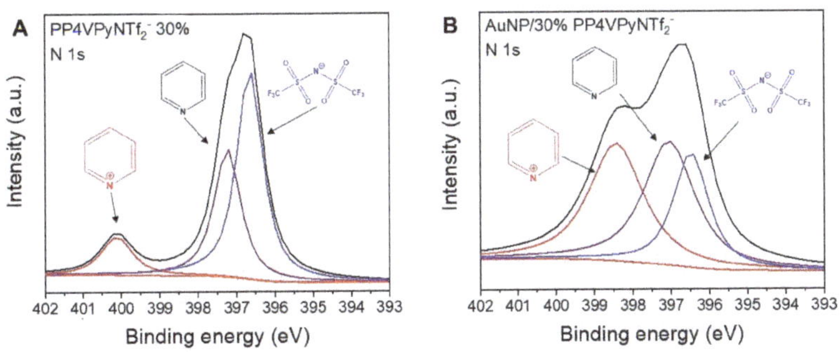

Figure 7. XPS deconvoluted peaks in the Nitrogen region of P4VPy/PP4VPyNTf$_2^-$ (30%) nanofibers (**A**) before and (**B**) after the adsorption of gold ions.

Once the $AuCl_4^-$ adsorption capacity of the NFs was determined, the nanofibers containing adsorbed $AuCl_4^-$ were subjected to a reduction process for the in situ generation of gold nanoparticles. This experiment was carried out adsorbing 1% *w*/*w* of Au^{+3} considering the total mass of the nanofiber film, in order to ensure the same amount of metal in all prepared nanocomposites.

After demonstrating the coordination capacity of these materials, the obtained nanofibers containing adsorbed $AuCl_4^-$ species were tested in the elaboration of nanocomposites by reducing the gold precursor into metal nanoparticles. The experiments consisted of the treatment of nanofiber mat samples containing 1% *w*/*w* gold ions with a sodium borohydride solution, as described in the experimental section.

To observe the nanoparticles obtained by SEM, the organic fraction of the nanocomposites obtained was removed by calcining the nanofibers containing the nanoparticles by controlled heating up to 900 °C and depositing the residue obtained on a copper grid. The obtained images of the nanoparticles and the analysis of the corresponding size distribution are shown in Figure 8. The size of nanoparticles synthesized in situ on pure P4VPy nanofibers was approximately 43 nm with a broad size distribution, while interestingly, when the amount of PIL in NFs increases, the nanoparticle size decreases to a minimum average size of 36 nm in the case of fibers containing 30% PP4VPy-NTf_2^-. Notably, along with the decrease in nanoparticle size, the nanoparticles also exhibited a narrower particle size distribution as the PIL content increased.

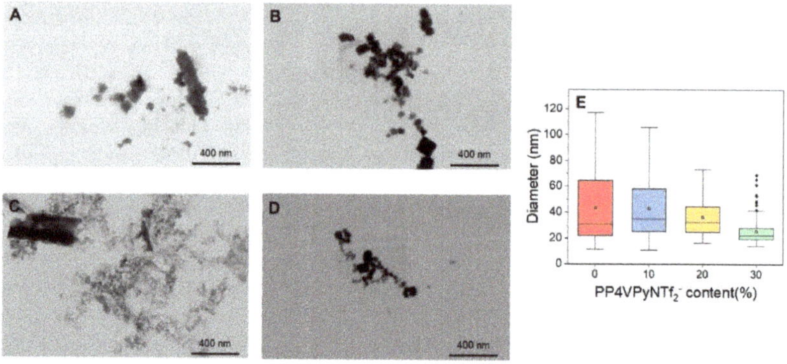

Figure 8. FE-SEM images of synthesized gold nanoparticles in (**A**) P4VPy, and P4VPy/PP4VPyNTf$_2^-$ with (**B**) 10%, (**C**) 20%, (**D**) 30% of PIL, and size distribution of nanoparticles diameter (**E**).

Finally, as detailed in experimental section, the performance of nanofibers with different PIL contents and containing gold nanoparticles as heterogeneous catalysts was evaluated. The recognized model catalytic reaction to test the catalytic activity of metal nanoparticles, reduction in 4-nitrophenol to 4-aminophenol by sodium borohydride, was used. This reduction reaction can be easily followed by UV-visible spectroscopy, due to the characteristic light absorption presented by 4-nitrophenolate species at 400 nm [28].

The kinetic data were determined using a pseudo-first order kinetic model, as Equation (3) indicates.

$$Ln[4NP]_t = Ln[4NP]_0 - k_{app}t \quad (3)$$

where $[4NP]_0$ and $[4NP]_t$ correspond to 4-NP concentration at the beginning of the reaction and at t minutes after this was started, respectively, t is the reaction time, and k_{app} is the pseudo first-order kinetic constant of the reaction. Using the Lambert–Beer relation, Equation (3) can be converted to Equation (4).

$$LnAbs_t = LnAbs_0 - k_{app}t \quad (4)$$

where Abs_t and Abs_0 correspond to 4-*NP* absorbance at the beginning of the reaction and at t minutes after this started, respectively.

The results obtained, conversion profiles, pseudo first order models, and k_{app} values, both for pure P4VPy nanofibers and with different poly liquid ionic contents, containing in all cases gold NPs, are shown in Figure 9.

In all cases, a rapid increase in the conversion percentage was initially observed, which slowed down as time progressed, until the highest conversion values were reached after 140 min of reaction. Interestingly, the highest value of 4 NP to 4 AP conversion (96%) was obtained with the nanofibers with the highest PIL content.

Additionally, although the values of the kinetic constants are of the same order, an evident increase in the magnitude of k_{app}, corresponding to 10.5%, was observed for the reaction catalyzed with the nanofibers with 30% w/w of PIL and gold NPs, with respect to those with only P4VPy and gold NPs.

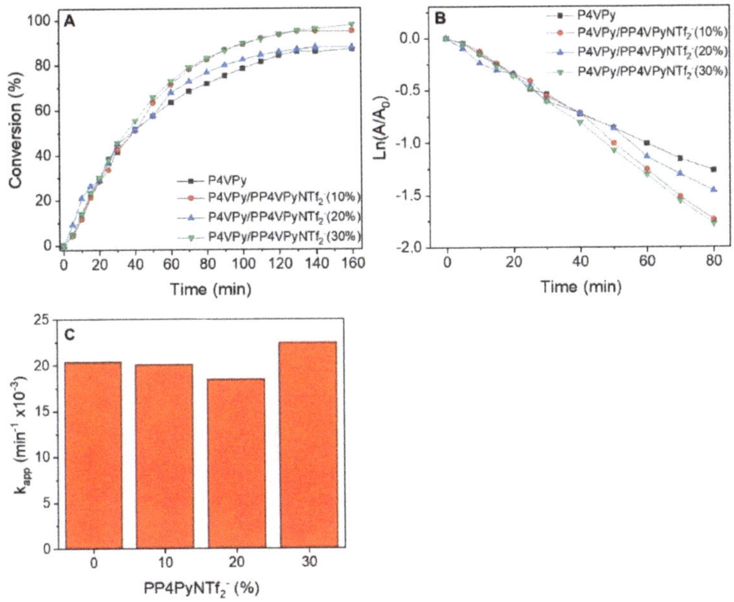

Figure 9. (**A**) Conversion profiles, (**B**) pseudo-first order model, and (**C**) apparent kinetic constant of the catalysts for the reduction in 4-nitrophenol into 4-aminophenol as a model reaction.

4. Conclusions

PIL/PV4Py blends nanofibers with different compositions were obtained by electrospinning technique from solutions in the binary solvent DMF/DCM. Importantly, as the PIL content in the solution increased, the diameter of the NFs obtained decreased, while the amount per unit area of the same ones incremented. The process of adsorption of Au III ions and their subsequent in situ reduction, allowed to obtain Au NPs supported on PIL/P4VPy nanofibers. The presence of abundant charged groups on the polymer NFs due to the inclusion of PIL favored the adsorption process of $AuCl_4^-$, through an exchange phenomenon with the counterion NTf_2^-. Hybrid systems, thus obtained, (NFs-Gold NPs) showed catalytic activity in the reduction reaction of 4-nitrophenol to 4-aminophenol using sodium borohydride. In terms of catalytic activity, higher conversion values were obtained with the nanofibers containing PIL and gold nanoparticles compared to those constituted of P4VPy and gold nanoparticles. Remarkably, the highest magnitude of the kinetic constant was obtained for the nanofibers with higher PIL content.

Accordingly, the inclusion and content of PILs in NFs play a fundamental role in the catalytic efficiency and performance of the Au NPs-PIL/P4VPy NFs hybrid nanomaterial. Undoubtedly, this allows envisioning potential technological applications in the field of heterogeneous catalysis for this type of materials.

Author Contributions: Conceptualization, A.L. and S.B.; methodology, M.L., X.B., O.R., M.U. and C.S.; validation, A.L., C.S., M.U. and S.B.; formal analysis, O.R.; investigation, A.L., O.R., M.L. and X.B.; resources, A.L., O.R., S.B. and M.U.; data curation, X.B., M.L. and O.R.; writing—original draft preparation, A.L., O.R., S.B. and C.S.; project administration, A.L.; funding acquisition, A.L. All authors have read and agreed to the published version of the manuscript.

Funding: This research was funded by FONDECYT grant number 1211124 and FONDAP grant number 15110019.

Institutional Review Board Statement: Not applicable.

Data Availability Statement: Not applicable.

Acknowledgments: O.R. thanks to ANID fellowship 21191002. X.B. thanks to FONDECYT grant 11190546.

Conflicts of Interest: The authors declare no conflict of interest.

References

1. Deng, Y.; Li, J.; Miao, Y.; Izikowitz, D. A comparative review of performance of nanomaterials for Direct Air Capture. *Energy Rep.* **2021**, *7*, 3506–3516. [CrossRef]
2. Paramsothy, M. Alleviating climate change and pollution with nanomaterials. *Nanomaterials* **2020**, *10*, 358. [CrossRef] [PubMed]
3. Zhang, Q.; Uchaker, E.; Candelaria, S.L.; Cao, G. Nanomaterials for energy conversion and storage. *Chem. Soc. Rev.* **2013**, *42*, 3127–3171. [CrossRef] [PubMed]
4. Prinsen, P.; Luque, R. Introduction to nanocatalysts. In *Nanoparticle Design and Characterization for Catalytic Applications in Sustainable Chemistry*; Royal Society of Chemistry: London, UK, 2019.
5. Gawande, M.B.; Goswami, A.; Asefa, T.; Guo, H.; Biradar, A.V.; Peng, D.-L.; Zboril, R.; Varma, R.S. Core–shell nanoparticles: Synthesis and applications in catalysis and electrocatalysis. *Chem. Soc. Rev.* **2015**, *44*, 7540–7590. [CrossRef] [PubMed]
6. Rodrigues, T.S.; da Silva, A.G.; Camargo, P.H. Nanocatalysis by noble metal nanoparticles: Controlled synthesis for the optimization and understanding of activities. *J. Mater. Chem. A* **2019**, *7*, 5857–5874. [CrossRef]
7. Ndolomingo, M.J.; Bingwa, N.; Meijboom, R. Review of supported metal nanoparticles: Synthesis methodologies, advantages and application as catalysts. *J. Mater. Sci.* **2020**, *55*, 6195–6241. [CrossRef]
8. Sankar, M.; He, Q.; Engel, R.V.; Sainna, M.A.; Logsdail, A.J.; Roldan, A.; Willock, D.J.; Agarwal, N.; Kiely, C.J.; Hutchings, G.J. Role of the support in gold-containing nanoparticles as heterogeneous catalysts. *Chem. Rev.* **2020**, *120*, 3890–3938. [CrossRef]
9. Fernandes, A.B.; Pavliuk, M.V.; Paun, C.; Carvalho, A.C.; Nomura, C.S.; Lewin, E.; Lindblad, R.; Camargo, P.H.; Sa, J.; Bastos, E.L. Recoverable and Reusable Polymer Microbead-Supported Metal Nanocatalysts for Redox Chemical Transformations. *ACS Appl. Nano Mater.* **2020**, *3*, 1722–1730. [CrossRef]
10. Singh, S.; Mahalingam, H.; Singh, P.K. Polymer-supported titanium dioxide photocatalysts for environmental remediation: A review. *Appl. Catal. A Gen.* **2013**, *462*, 178–195. [CrossRef]
11. Du, Y.; Wei, S.; Tang, M.; Ye, M.; Tao, H.; Qi, C.; Shao, L. Palladium nanoparticles stabilized by chitosan/PAAS nanofibers: A highly stable catalyst for Heck reaction. *Appl. Organomet. Chem.* **2020**, *34*, e5619. [CrossRef]
12. Cordoba, A.; Saldias, C.; Urzúa, M.; Montalti, M.; Guernelli, M.; Focarete, M.L.; Leiva, A. On the Versatile Role of Electrospun Polymer Nanofibers as Photocatalytic Hybrid Materials Applied to Contaminated Water Remediation: A Brief Review. *Nanomaterials* **2022**, *12*, 756. [CrossRef] [PubMed]
13. Zhang, W.; He, Z.; Han, Y.; Jiang, Q.; Zhan, C.; Zhang, K.; Li, Z.; Zhang, R. Structural design and environmental applications of electrospun nanofibers. *Compos. Part A Appl. Sci. Manuf.* **2020**, *137*, 106009. [CrossRef] [PubMed]
14. Qin, L.; Yi, H.; Zeng, G.; Lai, C.; Huang, D.; Xu, P.; Fu, Y.; He, J.; Li, B.; Zhang, C. Hierarchical porous carbon material restricted Au catalyst for highly catalytic reduction of nitroaromatics. *J. Hazard. Mater.* **2019**, *380*, 120864. [CrossRef] [PubMed]
15. Fu, Y.; Huang, T.; Jia, B.; Zhu, J.; Wang, X. Reduction of nitrophenols to aminophenols under concerted catalysis by Au/g-C3N4 contact system. *Appl. Catal. B Environ.* **2017**, *202*, 430–437. [CrossRef]
16. Qian, W.; Texter, J.; Yan, F. Frontiers in poly (ionic liquid) s: Syntheses and applications. *Chem. Soc. Rev.* **2017**, *46*, 1124–1159. [CrossRef]
17. Zhang, S.-Y.; Zhuang, Q.; Zhang, M.; Wang, H.; Gao, Z.; Sun, J.-K.; Yuan, J. Poly (ionic liquid) composites. *Chem. Soc. Rev.* **2020**, *49*, 1726–1755. [CrossRef]
18. Briones, X.; Tapia, R.A.; Campodónico, P.R.; Urzúa, M.; Leiva, Á.; Contreras, R.; González-Navarrete, J. Synthesis and characterization of poly (ionic liquid) derivatives of N-alkyl quaternized poly (4-vinylpyridine). *React. Funct. Polym.* **2018**, *124*, 64–71. [CrossRef]
19. Bonardd, S.; Saldías, C.; Ramírez, O.; Radić, D.; Recio, F.J.; Urzúa, M.; Leiva, A. A novel environmentally friendly method in solid phase for in situ synthesis of chitosan-gold bionanocomposites with catalytic applications. *Carbohydr. Polym.* **2019**, *207*, 533–541. [CrossRef]
20. Bittner, B.; Wrobel, R.J.; Milchert, E. Physical properties of pyridinium ionic liquids. *J. Chem. Thermodyn.* **2012**, *55*, 159–165. [CrossRef]
21. Josef, E.; Guterman, R. Designing Solutions for Electrospinning of Poly (ionic liquid) s. *Macromolecules* **2019**, *52*, 5223–5230. [CrossRef]
22. Tarus, B.; Fadel, N.; Al-Oufy, A.; El-Messiry, M. Effect of polymer concentration on the morphology and mechanical characteristics of electrospun cellulose acetate and poly (vinyl chloride) nanofiber mats. *Alex. Eng. J.* **2016**, *55*, 2975–2984. [CrossRef]
23. Angammana, C.J.; Jayaram, S.H. Analysis of the Effects of Solution Conductivity on Electrospinning Process and Fiber Morphology. *IEEE Trans. Ind. Appl.* **2011**, *47*, 1109–1117. [CrossRef]
24. Toral, M.I.; González-Navarrete, J.; Leiva, A.; Ríos, H.E.; Urzúa, M.D. Chromium retention properties of N-alkyl quaternized poly (4-vinylpyridine). *Eur. Polym. J.* **2009**, *45*, 730–737. [CrossRef]
25. Ferdeghini, C.; Guazzelli, L.; Pomelli, C.S.; Ciccioli, A.; Brunetti, B.; Mezzetta, A.; Ciprioti, S.V. Synthesis, thermal behavior and kinetic study of N-morpholinium dicationic ionic liquids by thermogravimetry. *J. Mol. Liq.* **2021**, *332*, 115662. [CrossRef]

26. Ramirez, O.; Bonardd, S.; Saldías, C.; Radic, D.; Leiva, A. Biobased chitosan nanocomposite films containing gold nanoparticles: Obtainment, characterization, and catalytic activity assessment. *ACS Appl. Mater. Interfaces* **2017**, *9*, 16561–16570. [CrossRef]
27. Ryoo, H.; Kim, K.; Shin, K.S. Adsorption and aggregation of gold nanoparticles onto poly (4-vinylpyridine) film revealed by Raman spectroscopy. *Vib. Spectrosc.* **2010**, *53*, 158–162. [CrossRef]
28. Iben Ayad, A.; Luart, D.; Ould Dris, A.; Guénin, E. Kinetic analysis of 4-nitrophenol reduction by "water-soluble" palladium nanoparticles. *Nanomaterials* **2020**, *10*, 1169. [CrossRef]

polymers

MDPI

Article

Nonwoven Mats Based on Segmented Biopolyurethanes Filled with MWCNT Prepared by Solution Blow Spinning

Pablo Ramos [1], Tamara Calvo-Correas [2], Arantxa Eceiza [2] and Javier González-Benito [1,*]

1 Dept. Materials Science and Engineering, Universidad Carlos III de Madrid, IQMAAB, Av. Universidad 30, 28911 Madrid, Spain
2 Materials + Technologies' Group, Faculty of Engineering of Gipuzkoa, Department of Chemical and Environmental Engineering, University of the Basque Country (UPV/EHU), Pza. Europa 1, 20018 Donostia-San Sebastian, Spain
* Correspondence: javid@ing.uc3m.es; Tel.: +34-916248870

Abstract: To prepare nonwoven mats constituted by submicrometric fibers of thermally responsive biopolyurethanes (TPU) modified with multiwalled carbon nanotubes (MWCNT), solution blow spinning (SBS) was used. The TPU was the product of synthesis using poly(butylene sebacate)diol, PBSD, ethyl ester L-lysine diisocyanate (LDI), and 1,3-propanediol (PD) (PBSe:LDI:PD) as reactants. TPU was modified by adding different amounts of MWCNT (0, 0.5, 1, 2, and 3 wt.%). The effect of the presence and amount of MWCNT on the morphology and structure of the materials was studied using field-emission scanning electron microscopy (FESEM) and Fourier-transform infrared spectroscopy (FTIR), respectively, while their influence on the thermal and electric behaviors was studied using differential scanning calorimetry (DSC) and capacitance measurements, respectively. The addition of MWCNT by SBS induced morphological changes in the fibrous materials, affecting the relative amount and size of submicrometric fibers and, therefore, the porosity. As the MWCNT content increased, the diameter of the fibers increased and their relative amount with respect to all morphological microfeatures increased, leading to a more compact microstructure with lower porosity. The highly porous fibrous morphology of TPU-based materials achieved by SBS allowed turning a hydrophilic material to a highly hydrophobic one. Percolation of MWCNT was attained between 2 and 3 wt.%, affecting not only the electric properties of the materials but also their thermal behavior.

Keywords: polyurethanes; solution blow spinning; dielectrics; carbon nanotubes

Citation: Ramos, P.; Calvo-Correas, T.; Eceiza, A.; González-Benito, J. Nonwoven Mats Based on Segmented Biopolyurethanes Filled with MWCNT Prepared by Solution Blow Spinning. *Polymers* **2022**, *14*, 4175. https://doi.org/10.3390/polym14194175

Academic Editors: Ting-Yu Liu and Yu-Wei Cheng

Received: 11 September 2022
Accepted: 1 October 2022
Published: 5 October 2022

1. Introduction

Research on materials with low Young's modulus and especial electric properties is recently receiving great attention because they are highly promising in the field of stretch sensors [1]. When these materials are stretched, at least one electrical property changes, which can then be measured with respect to the strain of the material. Although there are several possible applications, those focused on monitoring different parts of the human body are currently more than relevant [2]. In addition to having the expected electrical performance, these materials must be easily adapted to the surface of the body; thus, they should be soft and lightweight. Accordingly, dielectric elastomer sensors (DESs) seem to be the best candidates.

On the other hand, the design and preparation of elastic materials with relatively high dielectric constants to fabricate capacitors with intricate shapes or elastic devices for energy harvesting are other interesting issues to consider. For all these applications, room-temperature elastomers seem to be the best choice and, among them, thermoplastics are recommended because they can usually be more easily processed.

Among the most promising elastomers, thermoplastic polyurethanes (TPU) are receiving special attention because they can fulfill the requirements of elasticity, ease of processing, and price [3]. For example, Zahid et al. studied nanocomposites based on thermoplastic polyurethane and reduced graphene oxide to increase electromagnetic interference [4]. Ning et al. prepared blends based on carboxylated nitrile rubber and thermoplastic polyurethane with macroscopic homogeneous high-performance dielectric properties [5]. Ke et al. reported on a strategy to tune the dielectric constant and electric loss of thermoplastic polyurethane composites by using hybrids of carbon nanostructures and graphene nanoplatelets [6]. Furthermore, in the last decades, interest in designing polyurethanes derived from renewable sources has increased because of economic, environmental, and social concerns [7–9], since TPUs, in general, have many fields of applications, such as in coatings, adhesives, thermoplastic elastomers, and composites. TPUs are generally constituted by a soft segment (SS; high-molecular-weight polyester or polyether macrodiol) and a hard segment (HS; diisocyanate and low-molecular-weight diol or diamine chain extender). From a micro- or nanostructural point of view, the most important characteristic in these kinds of systems is probably the phase separation occurring because of incompatibility between hard and soft segments. In principle, the extension of phase separation, as well as its associated morphology, can greatly influence the final properties of the TPU. Furthermore, this phase separation depends on copolymer composition, length, hydrogen bonding, and crystallization extent among others [10–15]; in turn, these characteristics may be influenced by the processing method used, addition of fillers, and temperature. Therefore, understanding the relationship among structure, morphology, processing, and properties is essential when properly designing materials to tune properties focused on specific applications.

In most cases, polyurethanes are dielectrics with low permittivity and little interest in terms of electrical applications. However, PTUs conveniently modified with conductive particles, such as carbon nanotubes (CNT), may lead to elastomers with special electric properties. These kinds of systems are especially interesting since previous reports indicated that polyether- and polyester-based polyurethanes allow quite uniformly dispersing even nonfunctionalized carbon nanotubes [16–18]. Due to their large aspect ratio, high conductivity, high elastic modulus, excellent mechanical properties, and chemical stability, carbon nanotubes offer many advantages when they are used as fillers in polymer composites [19]. For instance, due to their high aspect ratio, they can lead to low electrical percolation thresholds, and they can act as effective nano-reinforcements in materials with improved or tailored mechanical when used at low concentrations [20–23]. In the case of using thermoplastic polyurethanes as polymer matrices, Barick et al. [23] showed that the conductivity increases with increasing nanotube loading, and the tensile strength and modulus improved in comparison with the neat TPU. Moreover, with the possibility of using elastomeric TPU at room temperature, more applications are available. For example, Shin et al. fabricated a type of conductive sensor based on polyurethane filled with CNTs which could strain up to 300% but with a low gauge factor, ranging from 0.34 to 1.07 [24].

There exist different methods to fabricate CNT/TPU composites, e.g., in situ polymerization of monomers in the presence of CNTs [25], melt mixing process [26], casting CNT/polymer dispersions to prepare films [27], ball mixing and compression molding [28], direct mixing of polymer melt/solution with the CNT powders [29–32], and mixing of CNT and polymer in solvent and, after its evaporation, processing of composite pellets in a screw extruder [3].

From a technological point of view, it would be interesting to prepare this kind of materials using processes that allow applying them in situ and, therefore, with even the ability of restoration. Additionally, incorporation of a high amount of CNTs into polymers increases the stiffness of the resulting composites and decreases their stretchability [33], which can be inhibitory when highly flexible and soft materials are needed. Therefore, methods to prepare these materials aimed at reducing this stiffening effect are necessary for certain applications. Although some approaches have been tried, such as making perforations using a mechanical punching system [34], they do not ensure in situ material preparation.

Solution blow spinning (SBS) [35–38] is a relatively recent method that, due to its characteristics, might satisfy all the abovementioned requirements. Furthermore, SBS allows producing materials with morphologies highly affected by the processing conditions, being possible to obtain materials ranging from nonwoven mats (constituted by small fibers with submicrometric diameters) to continuous smooth films. Using this method, a polymer solution (or even a suspension of particles in a polymer solution) is ejected from a concentric nozzle to a certain collector. During the time of flight of the solution the solvent evaporates, with the particles typically being uniformly trapped within the polymer [39–41]. Therefore, conveniently choosing the SBS processing conditions enables fabricating highly porous nanocomposites composed of submicrometric fibers, which should favor stretchability. Preparation by SBS of polyurethane composites incorporating multiwalled carbon nanotubes was already reported by Kuk et al. [42]. However, they only confirmed the possibility of obtaining good dispersion of MWNT and gave some data about mechanical properties and thermal stability. Nevertheless, other SBS conditions and TPUs with different chemical structures should be investigated. Furthermore, they did not go deeper into the characterization, e.g., in terms of structure. Lastly, at least simple electrical tests should be conducted if the potential use, e.g., in sensors and capacitors, is expected for these materials.

Among the available TPUs, thermally activated shape memory polymers offer new challenges in terms of preparing multifunctional materials. If they are filled with conductive nanoparticles such as nanotubes, in addition to tailoring electrical properties, their ability to control shape by temperature should enable new functions for more specific applications. Recently, Calvo-Correas et al. synthesized and characterized microphase separated segmented thermoplastic polyurethanes with different mechanical properties and a wide range of transition temperatures, presenting quite controllable thermally activated shape memory [43]. They are based on a reaction involving macrodiol poly(butylene sebacate)diol arising from castor oil (PBSe), ethyl ester L-lysine diisocyanate (LDI), and 1,3-propanediol as a chain extender (PD).

In the present work, nonwoven mats made of a thermally responsive thermoplastic biopolyurethane modified with multiwalled carbon nanotubes (MWCNTs) were prepared by solution blow spinning (SBS). The effect of the presence and the concentration of MWCNT on the morphology and structure of these materials were studied in order to understand their influence on the thermal, wettability, and electric behavior in terms of capacitance.

2. Materials and Methods

2.1. Materials

As a biobased thermoplastic polyurethane (TPU) polymer matrix, synthesized by the GMT group of the University of the Basque Country, UPV/EHU was used (M_n = 100,000 g·mol^{-1} and polydispersity index, Γ = 2.5) [43]. This polyurethane is based on the reaction mixture formed by poly(butylene sebacate)diol derived from castor oil (PBSe; hydroxyl index of 32.01 mg KOH·g^{-1}, and a number-average molecular weight of 3505 g·mol^{-1}) [15], ethyl ester L-lysine diisocyanate (LDI; supplied by CHEMOS GmBH), and corn sugar-based 1,3-propanediol (PD; supplied by Quimidroga S.A.), in a molar ratio PBSe:LDI:PD of 1:5:4, corresponding to a 29 wt.% LDI/PD segment content (calculated as weight percentage of LDI and PD with respect to total biopolyurethane weight) and 75 wt.% biobased carbon content. On the other hand, multiwalled carbon nanotubes (MWCNTs) supplied by Sigma-Aldrich were used (more than 95 wt.% purity, 97.9% carbon, diameters ranging from 6 to 9 nm, density of 2.1 g·cm^{-3}, and length of 5 μm); the specification sheet of the product does not indicate any functionalization. Tetrahydrofuran (THF; 99.9 wt.%) and acetone (99.9 wt.%) from Sigma-Aldrich were mixed in a proportion of 3:2 v/v to prepare the solvent to be used in the solution blow spinning process.

2.2. Material Preparation

Before SBS processing, it is necessary to prepare a polymer solution with some characteristics (viscosity, boiling point, surface tension, etc.) favoring fiber formation under the effect of the ejecting gas [44]. Therefore, a preliminary study about TPU solubility and SBS conditions was carried out to ensure the production of fibers.

Although THF was demonstrated to be one of the best solvents of polyurethanes, it was decided to mix it with the maximum amount possible of acetone, a volatile solvent, to decrease as much as possible the harmfulness of the final solvent. Thus, 40% (v/v) acetone was the limit composition found at room temperature (24 °C) to ensure dissolution of TPU when the composition of the polymer solution was 10% (w/v; dissolving 1 g of polymer in 10 mL of solvent). To prepare the systems to be blow-spun, 1 g of TPU was first dissolved in 6 mL of THF by magnetic stirring, and then, depending on the sample to be prepared, either 4 mL of acetone or 4 mL of a suspension of MWCNT in acetone was added. When preparing suspensions in the TPU solutions, an initial suspension of MWCNT in 4 mL of just acetone was prepared, before subjecting it to a sonication process in an ultrasonic bath for 30 min (this step was carried out to help a previous disaggregation of nanotubes), and then this suspension was added to the TPU solution to achieve the final suspension to be blow-spun. The suspensions were prepared in such a way that, at the end of the process, after the solvent evaporation, materials with compositions 0.0%, 0.5%, 1.0%, 2.0%, and 3.0% by weight of MWCNT were obtained. All preparations were made at 24 °C. The prepared materials were designated as x% MWCNT, where x is the wt.% of MWCNT.

The solution blow spinning was carried out using a machine designed and made in our laboratory [45] inspired by one patented by Medeiros et al. [35,36]. The SBS machine was composed by a nozzle connected to an air compressor with a pressure regulator and a plastic syringe coupled to an automatic pump (NE 1000 X, New Era System, Inc., Farmingdale, NY, USA). The nozzle was made from a cylinder of aluminum perforated along the long axis (diameter 1 mm) to introduce a glass capillary (inner and outer diameters of 0.5 and 0.7 mm, respectively) positioned along the central position with the help of a silicon stopper and protruding 2 mm from the nozzle exit. The aluminum cylinder was also perpendicularly perforated to introduce air at 0.5 bars. With the help of the pump, 10 mL of the mixtures (solution and suspensions) were injected with the syringe at 0.25 mL·min^{-1} in the nozzle to subsequently be ejected at 24 °C via the action of the pressurized air, before finally being deposited on a rotating cylindrical collector wrapped with aluminum foil located at a 15 cm working distance. Figure 1 shows representative images of the materials deposited on the collector.

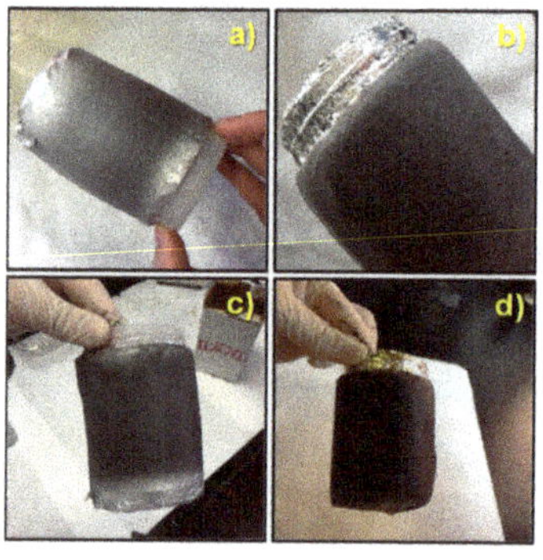

Figure 1. Images of solution blow-spun PBSe:LDI:PD polyurethane-based nanocomposites on the collector: (**a**) 0.5%; (**b**) 1.0%; (**c**) 2.0%; (**d**) 3.0% by weight of MWCNT.

2.3. *Characterization Techniques*

The morphology of the solution blow-spun samples was studied using field-emission scanning electron microscopy (FESEM) with a TENEO field-emission scanning electron microscope (FEI). The acceleration voltage was set at 1.5 or 2 kV, and the images were obtained from the signal arising from (a) secondary electrons detected using an Everhart Thornley detector (ETD), and (b) backscattered electrons collected using a T1 detector operating in the composite mode A + B. In order to improve the conductivity of the samples, they were carbon-coated by vapor deposition using a Leica EM ACE200 coater. The morphology of the samples and the porosity were studied by image analysis using the free software Image J (NIH, Bethesda, MD, USA). Furthermore, porosity was also determined in terms of the volume of air, V_{air}, occluded within the samples as the ratio between volume of air respect to the total volume of the sample, V_s.

$$V_{air} = V_s - V_B = V_s - \frac{\omega_S}{\delta_B}, \tag{1}$$

$$Porosity = P = \frac{V_{air}}{V_s} = 1 - \frac{\omega_S}{V_s \cdot \delta_B}, \tag{2}$$

where V_s is the volume occupied by the sample, and V_B is the bulk volume or the real volume occupied by the material. On the other hand, ω_s and δ_B are the sample weight and bulk density, respectively. Samples were cut with circular shape, and V_s was determined as

$$V_s = A_s \cdot d = \left(\frac{1}{4}\pi D^2\right) \cdot d, \tag{3}$$

where A_s, d, and D are the surface area, thickness, and diameter of the sample, respectively. Thicknesses of the tested specimens were measured with a micrometer Digimatic micrometer (Mitutoyo Corporation, Kawasaki, Japan) with an accuracy of ± 1 µm and diameters with a conventional caliper. Moreover, specimens were weighted out using a balance with an accuracy of 0.1 mg. Lastly, bulk density was determined using a pycnometer (Micromeritics AccuPyc 1330, Neurtek, Eibar, Spain) obtaining a value of 1.1575 ± 0.0013 ($g \cdot cm^{-3}$). For each sample, at least three specimens were used to measure size parameters and densities to finally give results in terms of the corresponding averages.

The influence of the SBS processing and presence of MWCNT on the TPU structure was studied using Fourier-transform infrared spectroscopy (FTIR) with an FTIR Spectrum GX spectrometer (Perkin-Elmer, Waltham, MA, USA). Spectra were obtained in the transmission mode with the samples directly deposited by SBS over IR transparent KBr discs and averaging interferograms obtained from 20 scans. Spectra were recorded in the range 400 to 4000 cm^{-1} with a resolution of 4 cm^{-1}.

Thermal behavior was studied using differential scanning calorimetry (DSC) with a Mettler Toledo 822e. Around 5 mg of sample was weighed out in 50 µL aluminum pans that were subsequently sealed and perforated. Heating and cooling scans were carried out under nitrogen atmosphere. The thermal program used consisted of (i) a first heating scan from -100 to 120 °C at 20 $°C \cdot min^{-1}$ to study the thermal behavior of samples as they were directly obtained from the SBS process, (ii) isothermal treatment at 120 °C for 5 min to ensure full erasing of processing history, and (iii) cooling scan from 120 to -100 °C at 20 $°C \cdot min^{-1}$ to study, during cooling, the effect of the presence of MWCNT. Thermal transitions considered were (a) glass transitions, characterized by the glass transition temperature, T_g (determined from the inflexion point when a heat capacity change of the sample takes place), and (b) melting and crystallization characterized by melting temperature, T_m (extracted from the peak associated to the endothermic transition observed in the DSC trace obtained during the heating scan) and the crystallization temperature, T_c (obtained from the peak associated to the exothermic transition observed in the DSC trace obtained during the cooling scan). Lastly, the crystallinity fraction was determined from the following expression:

$$X_c = \frac{\Delta H_m}{(1 - \phi_{MWCNT})\Delta H_t}, \tag{4}$$

where ΔH_m is the enthalpy of fusion obtained from the area of the endothermic peak obtained arising from the heating scan. ΔH_t is the enthalpy of fusion when a 100% crystalline polymer is considered, defined as $\Delta H_t = \Delta H_{PBSe} \cdot \phi_{PBSe}$, where ΔH_{PBSe} is the enthalpy of fusion for the macrodiol, ΔH_{PBSe}= 142 (J·g^{-1}) [43], and ϕ_{PBSe} is the weight fraction of the macrodiol in the polymer under consideration, ϕ_{PBSe} = 0.71.

Wettability behavior was studied by contact angle measurements using the sessile drop method with an OCA-15 KRÜSS GmbH tensiometer. Distilled and deionized water was used as testing liquid taking per sample the average contact angle from at least 10 measurements of drops of 3 μL at room temperature, 1 s after their deposition over the samples surface at 1 μL/s.

Lastly, the dielectric measurements were carried out at room temperature with a simple digital multimeter Electro DH Mod.: 60.131. Ensuring good contact, circular specimens of 16 mm were located in between two circular-plane aluminum electrodes (diameter 12 mm) of a home-designed measurement cell connected to the multimeter [1]. Pressure exerted to the specimens in the cell was always the same thanks to the use of a torque wrench (1.18 N·m applied torque). The dielectric constant, ε, of the materials was calculated using the simple expression used for a capacitor of parallels plates:

$$\varepsilon = \frac{C \cdot d}{A \cdot \varepsilon_0}, \tag{5}$$

where A is the surface area of the circular plates, ε_o = 8.85 × 10^{-12} F·m^{-1} is the permittivity of vacuum, C is the capacitance, and d is the thickness of the specimens.

3. Results and Discussion

SEM images at different magnifications from ×100 to ×5000 were used to carry out a morphological study. In Figure 2, as an example, SEM images of the SBS neat TPU sample, without MWCNT, are presented.

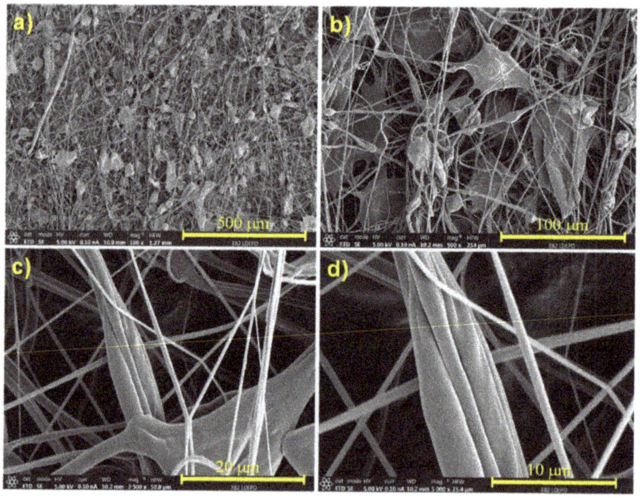

Figure 2. SEM images of the SBS neat TPU, sample without MWCNT, at different magnifications: (**a**) 100×; (**b**) 500×; (**c**) 2500×; (**d**) 5000×.

Regardless of the concentration of MWCNT, the remaining materials prepared in this work showed similar morphologies (Figure 3). The materials mainly constituted fibers of diameters ~1 μm, sometimes interconnected through plane and/or nearly spherical regions where higher material accumulation existed.

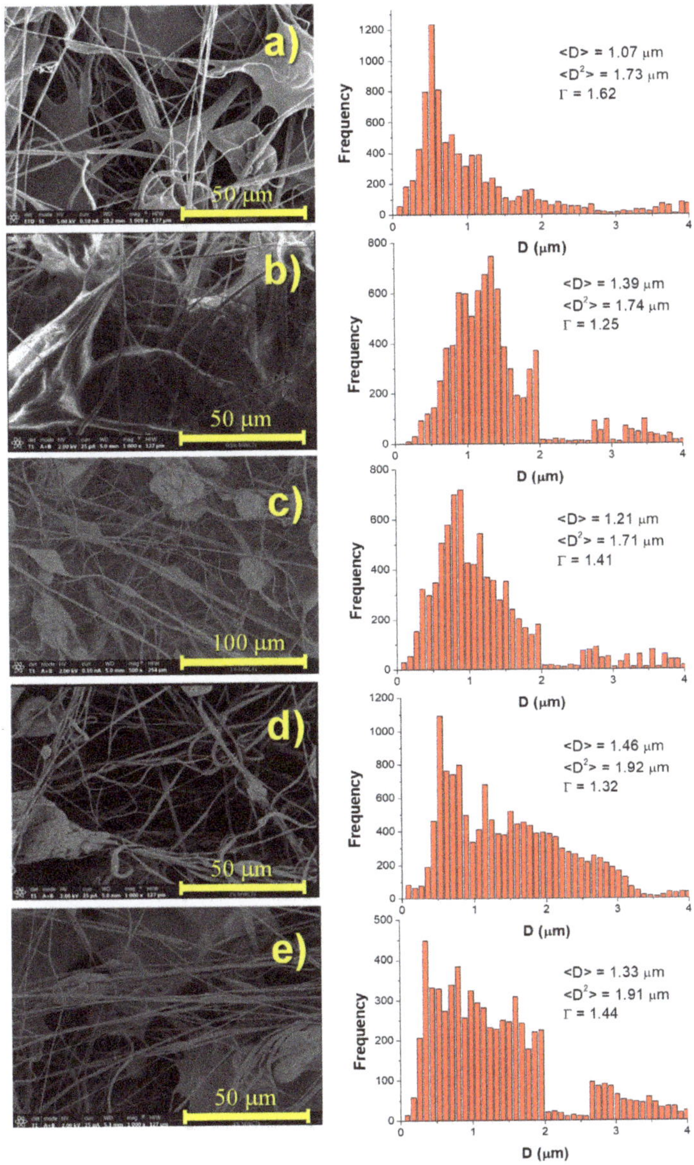

Figure 3. SEM images, diameter distributions, values of average diameter ($<D>$), second moment of the distribution ($<D_2>$), and dispersion of diameters (Γ) of the materials under study: (**a**) 0.0%; (**b**) 0.5%; (**c**) 1.0%; (**d**) 2.0%; (**e**) 3.0% MWCNT.

In particular, these regions of interconnected fibers constituted fibers that were probably aggregated during the process of fabrication. In addition to isolated fibers and lumps, bundles of fibers could be observed (Figure 2d). When comparing materials with different concentrations of MWCNT (Figure 3), it can be seen that, as the concentration of MWCNT increased, the proportion of fibers with respect to the remaining microstructural features (beads, lumps, etc.) also increased.

Using the software Image J, deeper morphological analysis in terms of diameters of fibers, D, and porosity, P, was carried out. In Figure 3, SEM images and diameter distributions for every material prepared in this work are presented together with the corresponding values of the average diameter ($<D>$; first moment of the distributions, Equation (6)), the second moment of the distribution ($<D_2>$; Equation (7)), and the dispersion of diameters (Γ; given by the quotient between the second and first moments of the distributions, Equation (8)).

$$< D > = \frac{\sum_{i=1}^{n} D_i \cdot f_i}{\sum_{i=1}^{n} f_i}, \tag{6}$$

$$< D^2 > = \frac{\sum_{i=1}^{n} D_i^2 \cdot f_i}{\sum_{i=1}^{n} D_i \cdot f_i}, \tag{7}$$

$$\Gamma = \frac{< D^2 >}{< D >}, \tag{8}$$

where D_i is the diameter of a particular fiber i, and fi is the fraction of fibers with a particular diameter value D_i.

As can be seen in Figure 3, when adding carbon nanotubes, there was a general tendency of an increase in the size of the regions with material accumulation (in the form of lumps) and a slight increase in the average diameter of the fibers mainly combined with a broadening of the distribution of diameters. Therefore, the presence of MWCNT slightly influenced the morphology, potentially having an important effect on final performance of the materials.

Changes in the porosity of the materials are among the most important consequences of variations in morphology. The size and relative amount of micro- and submicro-constituents will influence morphology and, therefore, specific properties (properties per unit of mass) and surface properties. Values of porosity were obtained as a mean value arising from at least three measurements on three specimens. In the particular case of neat TPU, the porosity obtained using Equation (2) was $P = 49\% \pm 3\%$, which is quite similar to that obtained using the Image J software ($44\% \pm 1\%$). This result clearly evidences that morphological analysis through the use of Image J led to information quite close to that obtained from more conventional methods. According to Equation (2), the porosities estimated for the remaining materials were as follows: P (0.5% MWCNT) = $52\% \pm 8\%$, P (1.0% MWCNT) = $45\% \pm 8\%$, P (2.0% MWCNT) = $45\% \pm 9\%$, and P (3.0% MWCNT) = $41\% \pm 10\%$. Therefore, despite the error of the final values, data indicate a tendency of decreased porosity with the relative increase in MWCNT. It seems, therefore, that there is a relationship between morphological features and porosity. The results obtained suggest that an increase in the proportion of fibers must be the main factor affecting the porosity. More homogeneous microfeatures, such as fibers, may allow better coupling, leading to more compact microstructures with lower porosity.

On the other hand, a very important issue to consider, directly affected by the morphology of the materials, is the wettability behavior because of the different environments where the materials under study might work. In order to evaluate the wettability behavior, contact angle measurements were carried out using, as testing liquid, distilled and deionized water. Figure 4 shows representative droplets of water over the surfaces of the materials under study together with the average values of the contact angle obtained. In every case, the materials showed highly hydrophobic behavior with values in the range 126°–129° without any dependence on MWCNT content. Here, it is important to highlight that the water contact angle clearly changed with respect to that obtained for the bulk neat material TPU prepared by hot pressing ($78.4° \pm 0.4°$) [43]. Furthermore, the wettability of carbon nanotubes has been reported to be intermediate since water contact angles are around 90° [46,47]. Therefore, a clearly hydrophilic material (TPU) becomes highly hydrophobic simply by using solution blow spinning as the processing method. It is, therefore, clear that the wettability behavior of these materials is highly affected by the morphology of the specimen tested, in turn induced by the processing method used (SBS).

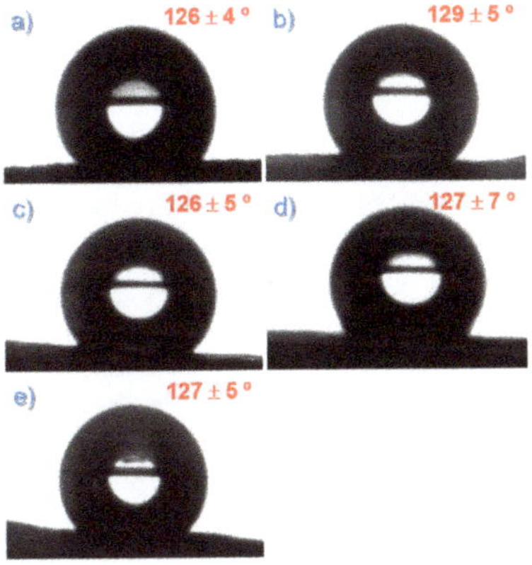

Figure 4. Images of water droplets over the surface of the prepared materials together with the average values of the contact angles obtained: (**a**) 0.0%; (**b**) 0.4%; (c) 1.0%; (**d**) 2.0%; (**e**) 3.0% of MWCNT.

Two models are commonly used to describe the wettability behavior of rough materials: Wenzel [48] and Cassie–Baxter [49]. In the first, Wenzel defined the relationship between roughness and wettability considering that an increase of surface roughness enhances wettability caused by the chemistry of the surface. Following Wenzel's model and for the materials under consideration, if the surface is hydrophilic, it would become even more hydrophilic when surface roughness increases. However, our results point out just the opposite. Therefore, the Cassie–Baxter model seems to be more adequate to explain the wettability mechanism of the SBS materials prepared in this work. This approach considers that a higher fraction of air available to the water drop leads to a higher contact angle [49]. Following this reasoning, one would expect that the contact angle changes as a function of porosity. However, the differences in porosity between the materials considered were lower than 22% and possibly even lower when considering the error. Therefore, large changes in contact angle were not expected (Figure 4).

Possible structural changes induced by SBS and the presence of MWCNT were evaluated through the analysis of FTIR spectra. In Figure 5, FTIR spectra of all the prepared materials are presented. To facilitate visualization of the main absorption bands, two regions were chosen, and baseline corrections were applied (Figure 5a,b).

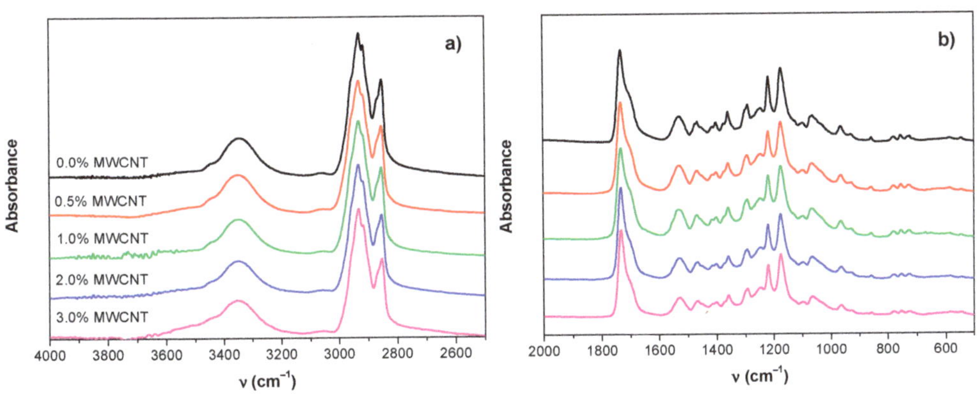

Figure 5. FTIR spectra of all the prepared samples: (**a**) high-energy region; (**b**) low-energy region.

As can be seen, there were no significant differences between the spectra obtained for the blow-spun neat TPU polymer and the spectrum obtained for the hot press neat TPU [43]. Furthermore, as can be seen in Figure 5, all spectra showed peaks located in the same positions, indicating that the incorporation of MWCNT almost did not lead to any structural change. In particular, the typical absorption bands assigned for the same polymer (TPU) in the form of the film obtained by hot pressing [43] were identified. For example, the broad band appearing between 3200 and 3400 cm^{-1} was composed of overlapped absorption bands associated with the N–H stretching vibration (free and hydrogen bonded) of urethane groups. Moreover, peaks at 2929 cm^{-1} and 2854 cm^{-1} with shoulders at 2955 cm^{-1} and 2866 cm^{-1} could be assigned to the C–H stretching vibrations of $-CH_2$ and $-CH_3$, respectively. On the other hand, the most prominent peak observed at 1732 cm^{-1} with a shoulder at about 1703 cm^{-1} corresponded to the stretching vibrational mode of the free and hydrogen bonded –C=O carbonyl of the urethane linkage (–HN–COO–) [43,50] and the ester group of the PBSe, respectively. Considering that MWCNTs were not modified by oxidation (they were used as received), special adhesion between them and TPU through specific interactions (mainly hydrogen bonds) was not expected. Therefore, any changes in the contribution of the hydrogen-bonded C=O group were mainly due to changes in the degree of phase separation [51] induced by the presence of MWCNT. Hydrogen bonding was expected between the NH groups as the proton donors and the oxygens as proton acceptors in the carbonyls of the LDI/PD segment and in the esters of the macrodiol. The hydrogen bonding index (HBI) can be defined as the ratio of absorption bands A_{1703}/A_{1732}, where A_i represents the area or absorbance of the band at a wavenumber i determined using a Gaussian curve fitting method (Figure 6). Lastly, the degree of phase separation (DPS) was obtained using the following equation [52]:

$$DPS = \frac{HBI}{HBI + 1} \times 100. \quad (9)$$

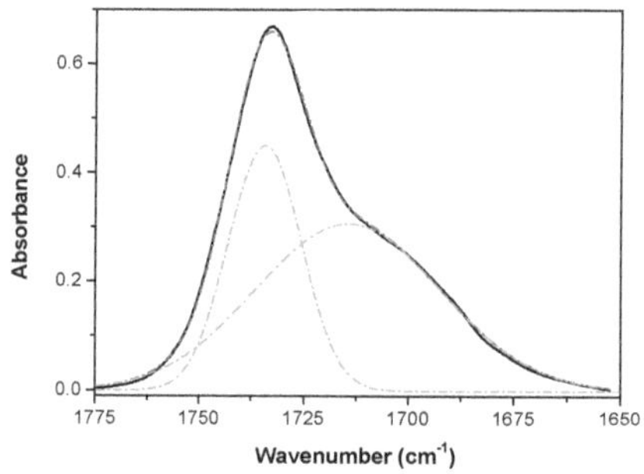

Figure 6. Example of Gaussian deconvolution to obtain the HBI as the ratio of absorption bands A_{1703}/A_{1732} (TPU with 2 wt.% MWCNT).

The values obtained for the DPS of the different materials prepared were as follows: 64% (0.0% MWCNT), 64% (0.5% MWCNT), 67% (1.0% MWCNT), 63% (2.0% MWCNT), and 63% (3.0% MWCNT). These results indicate that the addition of MWCNT scarcely affected the phase separation. There was a slight increase in phase separation upon adding a small number of nanotubes up to 1%; then, at higher MWCNT content (2% and 3%), the degree of phase separation was recovered. This result may be interpreted considering that, without reaching MWCNT percolation, when the concentration of nanotubes was enough, specific interactions between nanotubes and polymer chain associated with the PBSe could slightly increase the inter- and intramolecular interactions by hydrogen bonds between the LDI/PD segments. However, when percolation was reached between 1% and 3% MWCNT, as reported for other polymers filled with nanotubes [1], less of the nanotube surface was available for the interactions with polymer.

Analyzing the absorbance ratios, using the most intense band (C=O band) where some variations are expected, it is possible to extract more information about possible interactions between MWCNT and the polymer. For instance, taking the band at 2930 cm^{-1} assigned to the antisymmetric stretching mode of the methylene group $-CH_2$, the following evolution of its absorbance ratio with respect to the carbonyl group can be observed: A_{2930}/A_{1730}, 0.44 (0.0% MWCNT); 0.46 (0.5% MWCNT); 0.46 (1.0% MWCNT); 0.33 (2.0% MWCNT); 0.30 (3.0% MWCNT). Thus, when adding a small number of nanotubes (0.5%, 1.0% MWCNT), the ratio A_{2930}/A_{1730} increased slightly compared to the SBS neat polymer but decreased in the samples with the highest content of carbon nanotubes (2.0% and 3.0% MWCNT). This fact could be explained again considering a direct relationship between the available MWCNT surface and specific polymer–nanotube interactions. When the content of MWCNT is low enough, good dispersion with isolated nanotubes will be within the polymer favoring interactions between the surface of the nanotubes and the methylene groups of the polymer chains. However, by increasing the concentration of MWCNT (2.0% and 3.0% MWCNT), interconnections between the nanotubes will occur, favoring aggregate formation, thus reducing specific interactions with the polymer chains, in accordance with the analysis of the evolution of absorption band of hydrogen bonded carbonyl groups. It is also necessary to highlight that, considering the proportion of MWCNT with respect to the polymer, the contribution in the FTIR spectra of the bands associated with C–H stretching modes of the carbon nanotubes should be negligible respect to those arising from the polymer.

In Figure 7, thermograms corresponding to the first heating and subsequent cooling DSC scans are shown. Throughout the heating (Figure 7a), regardless of the concentration of carbon nanotubes, only a clear endothermic peak, T_m, (61 °C) with a shoulder, $T_{shoulder}$, at a lower temperature (45 °C) could be observed. They were associated with the melting temperature of an ordered microdomain mainly formed by PBSe. In the case of semicrystalline polymers, it is usually stated that the melting process depends on the distribution of crystallite sizes, the presence of different crystal forms, domain sizes, and different degrees of order in the crystalline structure [1,53,54]. Therefore, crystalline heterogeneity might be the main reason why the shoulder appeared.

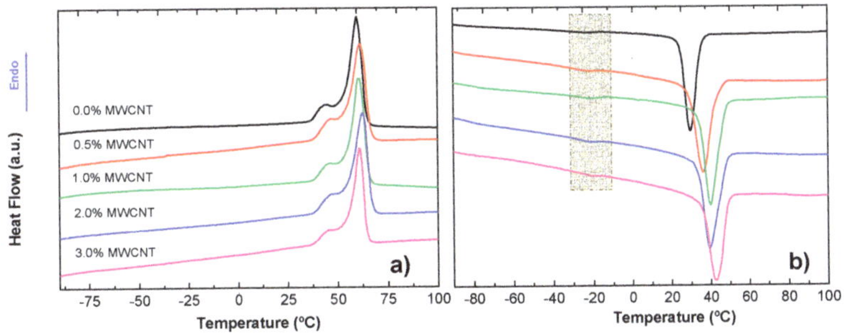

Figure 7. Thermograms corresponding to the (**a**) first heating and (**b**) subsequent cooling DSC scans.

Furthermore, for temperatures of both the peak and the shoulder (Table 1), there was a slight shift to higher temperatures when MWCNTs were added to the TPU, suggesting that the presence of nanotubes favored the crystallization, probably due to an enhanced preferential chain orientation when phase separation was favored, as already observed for other PU systems filled with modified MWCNTs [55]. In fact, when the concentration of MWCNTs was increased to 3%, the fraction of crystallinity, X_c, increased by 10% (Table 1). On the other hand, the T_g values reported for the same polymer, but prepared by hot pressing, which appeared at −40 °C (assigned to the PBSe-based domain) and 15 °C (assigned to the LDI/PD-based domain) [43], cannot be clearly distinguished in the thermograms of Figure 7a. The main reason for difficulties in the observation of T_g is the high crystallinity degree of the materials (Table 1), whose effect may be, in addition, enhanced by the preferential orientation of the polymer chains when the materials are prepared in the form of fibers due to the SBS process. However, it seems that, when materials are prepared by SBS, less crystallinity is achieved in comparison with similar materials prepared by hot pressing, at least when they are slowly cooled [43].

Table 1. Thermal parameters obtained from the first heating and cooling DSC scans of the prepared materials.

%MWCNT	T_m (°C)	$T_{shoulder}$ (°C)	X_c	T_c (°C)
0.0	60.1	44.4	0.76	29.6
0.5	61.5	46.7	0.80	35.9
1.0	60.8	46.0	0.82	39.9
2.0	62.8	46.7	0.79	39.6
3.0	61.1	46.4	0.88	42.6

In the DSC traces of the cooling scans (Figure 7b), the exothermic crystallization peak at higher temperatures can be clearly observed as the amount of MWCNT increased (Table 1). This result indicates the existence of a nucleating effect caused by the presence of carbon nanotubes. On the other hand, at least one T_g (−18 °C) can be observed (selected area in Figure 7b). This result may be explained by considering that, after erasing the processing history (SBS), the extra chain order given by the SBS process disappeared, allowing better visualization of the glass transition temperature. Additionally, the crystallization degree was at least 10% lower because the chosen cooling rate must have been relatively fast for this type of system.

Lastly, in order to better visualize the effect of the presence of MWCNT on the electrical behavior, a plot of the permittivity (calculated from Equation (5)) as a function of the carbon nanotube concentration is shown in Figure 8. It can be observed that the permittivity slightly increased at low concentrations of MWCNT up to about 1 wt.%, followed by an abrupt increase at 2 wt.%. This behavior indicates the existence of a clear transition (electrical percolation) in the system from dielectric to conductive material in the range 2–3 wt.%. In fact, values of percolation fraction from 2 to 3 wt.% are commonly found in the literature for many thermoplastic polymers filled with MWCNT [37] and even for TPU filled with MWCNTs [56,57]. Designing materials aimed at controlling the point at which this transition occurs is more than essential, since it would be possible to prepare materials with a wide range of electrical properties ranging from materials having relatively low dielectric constant to materials with a certain conductivity that might even be even tuned with temperature (piroelectricity).

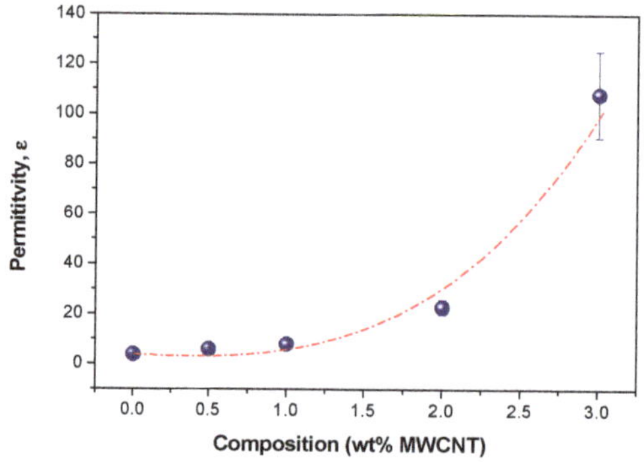

Figure 8. Permittivity (calculated from Equation (5)) as a function of MWCNT content.

4. Conclusions

Solution blow spinning was satisfactorily used to prepare highly porous fibrous materials based on thermally responsive biopolyurethanes modified with multiwalled carbon nanotubes (MWCNTs). When adding carbon nanotubes there was a change in the morphology mainly observed as an increase in the relative number of submicrometric fibers as the amount of MWCNT increased, which was finally reflected by a decrease in porosity.

The Cassie–Baxter approach explained the wettability behavior of the SBS TPU-based materials prepared in this work where hydrophilic material became highly hydrophobic simply by using solution blow spinning as the processing method.

The addition of MWCNT scarcely affected the phase separation in the TPU; however, there was a slight increase in phase separation upon adding a small amount of carbon nanotubes up to 1 wt.%; then, at higher MWCNT content of 2 and 3 wt.%, the degree of phase separation was recovered.

When percolation was reached between 1 and 3 wt.% MWCNT, there were changes not only in the electric properties but also in the thermal behavior. At high concentrations of MWCNT, less of the carbon nanotube surface was available for the polymer chains to interact. When the content of MWCNT was low enough, they were uniformly dispersed within the polymer, favoring interactions between the surface of the nanotubes and the methylene groups of the polymer chains. However, upon increasing the concentration of MWCNTs (2 and 3 wt.%), interconnections between the carbon nanotubes occurred, favoring aggregate formation and, therefore, reducing specific interactions with the polymer chains.

It was demonstrated that SBS is a convenient method to prepare TPU-based materials with tailored properties with potential applications as non-wettable DESs. Here, it is important to highlight that, using SBS, these materials can be prepared in situ to be perfectly adapted to almost any surface, regardless of its shape.

Author Contributions: Conceptualization, J.G.-B.; methodology, J.G.-B. and A.E.; software, P.R. and J.G.-B.; validation, P.R., T.C.-C., A.E. and J.G.-B.; formal analysis, P.R. and J.G.-B.; investigation, J.G.-B. and A.E.; resources, J.G.-B. and A.E.; data curation, P.R. and J.G.-B.; writing—original draft preparation, J.G.-B.; writing—review and editing, J.G.-B. and A.E.; supervision, J.G.-B.; funding acquisition, J.G.-B. and A.E. All authors read and agreed to the published version of the manuscript.

Funding: This research was funded by the Fondos de Investigación de Fco. Javier González Benito, política de reinversión de costes generales, Universidad Carlos III de Madrid [2012/00130/004], the Acción Estratégica en Acción Estratégica en Materiales nanocompuestos multifuncionales, Universidad Carlos III de Madrid [2011/00287/003], and the Project PID2020-112713RB-C22–C21 supported by AEI [Ministerio de Ciencia e Innovación of Spain], the University of the Basque Country (UPV/EHU) and (GIU18/216 Research Group).

Institutional Review Board Statement: Not applicable.

Data Availability Statement: The data of this study are available upon request.

Acknowledgments: The authors greatly appreciate the advice, suggestions, and comments with regard to the manuscript given by María Ángeles Corcuera Maeso (University of the Basque Country).

Conflicts of Interest: The authors declare no conflict of interest.

References

1. González-Benito, J.; Olmos, D.; Martínez-Tarifa, J.M.; González-Gaitano, G.; Sánchez, F.A. PVDF/$BaTiO_3$/carbon nanotubes ternary nanocomposites prepared by ball milling: Piezo and dielectric responses. *J. Appl. Polym. Sci.* **2019**, *136*, 47788. [CrossRef]
2. Huang, B.; Li, M.; Mei, T.; McCoul, D.; Qin, S.; Zhao, Z.; Zhao, J. Wearable stretch sensors for motion measurement of the wrist joint based on dielectric elastomers. *Sensors* **2017**, *17*, 2708. [CrossRef]
3. Lepak-Kuc, S.; Podsiadły, B.; Skalski, A.; Janczak, D.; Jakubowska, M.; Lekawa-Raus, A. Highly Conductive Carbon Nanotube-Thermoplastic Polyurethane Nanocomposite for Smart Clothing Applications and Beyond. *Nanomaterials* **2019**, *9*, 1287. [CrossRef] [PubMed]
4. Zahid, M.; Nawab, Y.; Gulzar, N.; Rehan, Z.A.; Shakir, M.F.; Afzal, A.; Abdul Rashid, I.; Tariq, A. Fabrication of reduced graphene oxide (RGO) and nanocomposite with thermoplastic polyurethane (TPU) for EMI shielding application. *J. Mater. Sci. Mater. Electron.* **2020**, *31*, 967–974. [CrossRef]

5. Ning, N.; Qin, H.; Wang, M.; Sun, H.; Tian, M.; Zhang, L. Improved dielectric and actuated performance of thermoplastic polyurethane by blending with XNBR as macromolecular dielectrics. *Polymer* **2019**, *179*, 121646. [CrossRef]
6. Ke, K.; McMaster, M.; Christopherson, W.; Singer, K.D.; Manas-Zloczower, L. Effects of branched carbon nanotubes and graphene nanoplatelets on dielectric properties of thermoplastic polyurethane at different temperatures. *Compos. Part B Eng.* **2019**, *166*, 673–680. [CrossRef]
7. Lligadas, G.; Ronda, J.C.; Galiá, M.; Cádiz, V. Plant oils as platform chemicals for polyurethane synthesis: Current state-of-the-art. *Biomacromolecules* **2010**, *11*, 2825–2835. [CrossRef]
8. Hojabri, L.; Kong, X.; Narine, S.S. Fatty Acid-Derived diisocyanate and biobased polyurethane produced from vegetable oil: Synthesis, polymerization, and characterization. *Biomacromolecules* **2009**, *10*, 884–891. [CrossRef]
9. Datta, J.; Głowińska, E. Chemical modifications of natural oils and examples of their usage for polyurethane synthesis. *J. Elastomers Plast.* **2014**, *46*, 33–42. [CrossRef]
10. Corcuera, M.A.; Rueda, L.; Fernandez D'Arlas, B.; Arbelaiz, A.; Marieta, C.; Mondragon, I.; Eceiza, A. Microstructure and properties of polyurethanes derived from castor oil. *Polym. Degrad. Stab.* **2010**, *95*, 2175–2184. [CrossRef]
11. Rueda-Larraz, L.; d'Arlas, B.F.; Tercjak, A.; Ribes, A.; Mondragon, I.; Eceiza, A. Synthesis and microstructure-mechanical property relationships of segmented polyurethanes based on a PCL-PTHF-PCL block copolymer as soft segment. *Eur. Polym. J.* **2009**, *45*, 2096–2109. [CrossRef]
12. Wang, C.S.; Kenney, D.J. Effect of Hard Segments on Morpholog, y and Properties of Thermoplastic Polyurethanes. *J. Elastomers Plast.* **1995**, *27*, 182–199. [CrossRef]
13. Do Kim, H.; Lee, T.J.; Huh, J.H.; Lee, D.J. Preparation and properties of segmented thermoplastic polyurethane elastomers with two different soft segments. *J. Appl. Polym. Sci.* **1999**, *73*, 345–352. [CrossRef]
14. Tsai, Y.M.; Yu, T.L.; Tseng, Y.H. Physical properties of crosslinked polyurethane. *Polym. Int.* **1998**, *47*, 445–450. [CrossRef]
15. Saralegi, A.; Rueda, L.; Fernández-D'Arlas, B.; Mondragon, I.; Eceiza, A.; Corcuera, M.A. Thermoplastic polyurethanes from renewable resources: Effect of soft segment chemical structure and molecular weight on morphology and final properties. *Polym. Int.* **2013**, *62*, 106–115. [CrossRef]
16. Sen, R.; Zhao, B.; Perea, D.; Itkis, M.E.; Hu, H.; Love, J.; Bekyarova, E.; Haddon, R.C. Preparation of single-walled carbon nanotube reinforced polystyrene and polyurethane nanofibers and membranes by electrospinning. *Nano Lett.* **2004**, *4*, 459–464. [CrossRef]
17. Kimmel, D.; Slobodiam, P.; Petrás, D.; Zatloukal, M.; Olejnik, R.; Sáha, P. Polyurethane/multiwalled carbon nanotube nanowebs prepared by an electrospieeing process. *J. Appl. Polym. Sci.* **2009**, *111*, 2711–2714. [CrossRef]
18. Meng, J.; Kong, H.; Han, Z.; Wang, C.; Zhu, G.; Xie, S.; Xu, H. Enhancement of nanofibrous scaffold of multiwalled carbon nanotubes/polyurethane composite to the fibroblasts growth and biosynthesis. *J. Biomed. Mater. Res. Part A* **2009**, *88*, 105–116. [CrossRef]
19. Baughman, R.H.; Zakhidov, A.A.; De Heer, W.A. Carbon nanotubes—The route toward applications. *Science* **2002**, *297*, 787–792. [CrossRef]
20. Ajayan, P.M.; Schadler, L.S.; Giannaris, C.; Rubio, A. Single-walled carbon nanotube-polymer composites: Strength and weakness. *Adv. Mater.* **2000**, *12*, 750–753. [CrossRef]
21. Tran, L.; Kim, J. A Comparative Study of the Thermoplastic Polyurethane/Carbon Nanotube and Natural Rubber/Carbon Nanotube Composites According to Their Mechanical and Electrical Properties. *Fibers Polym.* **2018**, *19*, 1948–1955. [CrossRef]
22. Natarajan, T.S.; Eshwaran, S.B.; Stöckelhuber, K.W.; Wießner, S.; Pötschke, P.; Heinrich, G.; Das, A. Strong Strain Sensing Performance of Natural Rubber Nanocomposites. *ACS Appl. Mater. Interfaces* **2017**, *9*, 4860–4872. [CrossRef] [PubMed]
23. Barick, A.K.; Tripathy, D.K. Preparation, characterization and properties of acid functionalized multi-walled carbon nanotube reinforced thermoplastic polyurethane nanocomposites. *Mater. Sci. Eng. B Solid State Mater. Adv. Technol.* **2011**, *176*, 1435–1447. [CrossRef]
24. Shin, M.K.; Oh, J.; Lima, M.; Kozlov, M.E.; Kim, S.J.; Baughman, R.H. Elastomeric conductive composites based on carbon nanotube forests. *Adv. Mater.* **2010**, *22*, 2663–2667. [CrossRef]
25. Xia, H.; Song, M. Preparation and characterization of polyurethane-carbon nanotube composites. *Soft Matter* **2005**, *1*, 386–394. [CrossRef]
26. Pötschke, P.; Häußler, L.; Pegel, S.; Steinberger, R.; Scholz, G. Thermoplastic polyurethane filled with carbon nanotubes for electrical dissipative and conductive applications. *KGK Rubberpoint* **2007**, *60*, 432–437.
27. Koerner, H.; Price, G.; Pearce, N.A.; Alexander, M.; Vaia, R.A. Remotely actuated polymer nanocomposites—Stress-recovery of carbon-nanotube-filled thermoplastic elastomers. *Nat. Mater.* **2004**, *3*, 115–120. [CrossRef]
28. Sun, W.J.; Xu, L.; Jia, L.C.; Zhou, C.G.; Xiang, Y.; Yin, R.H.; Yan, D.X.; Tang, J.H.; Li, Z.M. Highly conductive and stretchable carbon nanotube/thermoplastic polyurethane composite for wearable heater. *Compos. Sci. Technol.* **2019**, *181*, 107695. [CrossRef]
29. Bilotti, E.; Zhang, R.; Deng, H.; Baxendale, M.; Peijs, T. Fabrication and property prediction of conductive and strain sensing TPU/CNT nanocomposite fibres. *J. Mater. Chem.* **2010**, *20*, 9449–9455. [CrossRef]
30. Christ, J.F.; Aliheidari, N.; Pötschke, P.; Ameli, A. Bidirectional and stretchable piezoresistive sensors enabled by multimaterial 3D printing of carbon nanotube/thermoplastic polyurethane nanocomposites. *Polymers* **2019**, *11*, 11. [CrossRef]
31. Zhang, R.; Deng, H.; Valenca, R.; Jin, J.; Fu, Q.; Bilotti, E.; Peijs, T. Carbon nanotube polymer coatings for textile yarns with good strain sensing capability. *Sens. Actuators A Phys.* **2012**, *179*, 83–91. [CrossRef]

32. Zhang, R.; Dowden, A.; Deng, H.; Baxendale, M.; Peijs, T. Conductive network formation in the melt of carbon nanotube/thermoplastic polyurethane composite. *Compos. Sci. Technol.* **2009**, *69*, 1499–1504. [CrossRef]
33. Rangari, V.K.; Yousuf, M.; Jeelani, S.; Pulikkathara, M.X.; Khabashesku, V.N. Alignment of carbon nanotubes and reinforcing effects in nylon-6 polymer composite fibers. *Nanotechnology* **2008**, *19*, 245703. [CrossRef] [PubMed]
34. Sekitani, T.; Noguchi, Y.; Hata, K.; Fukushima, T.; Aida, T.; Someya, T. A rubberlike stretchable active matrix using elastic conductors. *Science* **2008**, *321*, 1468–1472. [CrossRef]
35. Medeiros, E.S.; Glenn, G.M.; Klamczynski, A.P.; Orts, W.J.; Mattoso, L.H.C. Solution Blow Spinning. U.S. Patent 8,641,960 B1, 4 February 2014.
36. Medeiros, E.S.; Glenn, G.M.; Klamczynski, A.P.; Orts, W.J.; Mattoso, L.H.C. Solution blow spinning: A new method to produce micro- and nanofibers from polymer solutions. *J. Appl. Polym. Sci.* **2009**, *113*, 2322–2330. [CrossRef]
37. González-Benito, J.; Torres, D.; Ballesteros, C.; Ruiz, V.M.; Teno, J. PVDF based nanocomposites produced by solution blow spinning, structure and morphology induced by the presence of MWCNT and their consequences on some properties. *Colloid Polym. Sci.* **2019**, *297*, 1105–1118. [CrossRef]
38. Teno, J.; González-Gaitano, G.; González-Benito, J. Poly (ethylene-co-vinyl acetate) films prepared by solution blow spinning: Surface characterization and its relation with *E. coli* adhesion. *Polym. Test.* **2017**, *60*, 140–148. [CrossRef]
39. González-Benito, J.; Teno, J.; González-Gaitano, G.; Xu, S.; Chiang, M.Y. PVDF/TiO_2 nanocomposites prepared by solution blow spinning: Surface properties and their relation with *S. Mutans* adhesion. *Polym. Test.* **2017**, *58*, 21–30. [CrossRef]
40. Teno, J.; González-Gaitano, G.; González-Benito, J. Nanofibrous polysulfone/TiO_2 nanocomposites: Surface properties and their relation with E. coli adhesion. *J. Polym. Sci. Part B Polym. Phys.* **2017**, *55*, 1575–1584. [CrossRef]
41. Lorente, M.A.; González-Gaitano, G.; González-Benito, J. Preparation, Properties and Water Dissolution Behavior of Polyethylene Oxide Mats Prepared by Solution Blow Spinning. *Polymers* **2022**, *14*, 1299. [CrossRef]
42. Kuk, E.; Ha, Y.M.; Yu, J.; Im, I.T.; Kim, Y.; Jung, Y.C. Robust and Flexible Polyurethane Composite Nanofibers Incorporating Multi-Walled Carbon Nanotubes Produced by Solution Blow Spinning. Macromol. *Mater. Eng.* **2016**, *301*, 364–370.
43. Calvo-Correas, T.; Santamaria-Echart, A.; Saralegi, A.; Martin, L.; Valea, A.; Corcuera, M.A.; Eceiza, A. Thermally-responsive biopolyurethanes from a biobased diisocyanate. *Eur. Polym. J.* **2015**, *70*, 173–185. [CrossRef]
44. Dias, G.C.; Cellet, T.S.P.; Santos, M.C.; Sanches, A.O.; Malmonge, L.F. PVDF nanofibers obtained by solution blow spinning with use of a commercial airbrush. *J. Polym. Res.* **2019**, *26*, 87. [CrossRef]
45. Ruiz, V.M.; Sirera, R.; Martínez, J.M.; González-Benito, J. Solution blow spun graded dielectrics based on poly(vinylidene fluoride)/multi-walled carbon nanotubes nanocomposites. *Eur. Polym. J.* **2020**, *122*, 109397. [CrossRef]
46. Standoa, G.; Łukawskib, D.; Lisieckic, F.; Janasa, D. Intrinsic hydrophilic character of carbon nanotube networks. *App. Surf. Sci.* **2019**, *463*, 227–233. [CrossRef]
47. Zhang, L.; Wang, J.; Fuentes, C.A.; Zhang, D.; Van Vuure, A.W.; Seo, J.W.; Seveno, D. Wettability of carbon nanotube fibers. *Carbon* **2017**, *122*, 128–140. [CrossRef]
48. Wenzel, R.N. Resistance of solid surfaces to wetting by water. *Ind. Eng. Chem.* **1936**, *28*, 988–994. [CrossRef]
49. Cassie, A.B.D.; Baxter, S. Wettability of porous surfaces. *Trans. Faraday Soc.* **1944**, *40*, 546–551. [CrossRef]
50. Ugarte, L.; Fernández-D'Arlas, B.; Valea, A.; Gonzaĺez, M.L.; Corcuera, M.A.; Eceiza, A. Morphology-properties relationship in high-renewable content polyurethanes. *Polym. Eng. Sci.* **2014**, *54*, 2282–2291. [CrossRef]
51. Seymour, R.W.; Estes, G.M.; Cooper, S.L. Infrared Studies of Segmented Polyurethan Elastomers. I. Hydrogen Bonding. *Macromolecules* **1970**, *3*, 579–583. [CrossRef]
52. Chen, T.K.; Tien, Y.I.; Wei, K.H. Synthesis and characterization of novel segmented polyurethane/clay nanocomposites. *Polymer* **2000**, *41*, 1345–1353. [CrossRef]
53. Olmos, D.; Domínguez, C.; Castrillo, P.D.; Gonzalez-Benito, J. Crystallization and final morphology of HDPE: Effect of the high energy ball milling and the presence of TiO_2 nanoparticles. *Polymer* **2009**, *50*, 1732–1742. [CrossRef]
54. Sánchez, F.A.; Redondo, M.; González-Benito, J. Influence of $BaTiO_3$ Submicrometric Particles on the Structure, Morphology, and Crystallization Behavior of Poly(vinylidene fluoride). *J. Appl. Polym. Sci.* **2015**, *132*, 41497. [CrossRef]
55. Fernández-d'Arlas, B.; Khan, U.; Rueda, L.; Coleman, J.N.; Mondragon, I.; Corcuera, M.A.; Eceiza, A. Influence of hard segment content and nature on polyurethane/multiwalled carbon nanotube composites. *Compos. Sci. Technol.* **2011**, *71*, 1030–1038. [CrossRef]
56. Pouladzadeh, F.; Katbab, A.A.; Haghighipour, N.; Kashi, E. Carbon nanotube loaded electrospun scaffolds based on thermoplastic urethane (TPU) with enhanced proliferation and neural differentiation of rat mesenchymal stem cells: The role of state of electrical conductivity. *Eur. Polym. J.* **2018**, *105*, 286–296. [CrossRef]
57. Teymouri, M.; Kokabi, M.; Alamdarnejad, G. Conductive shape-memory polyurethane/multiwall carbon nanotube nanocomposite aerogels. *J. Appl. Polym. Sci.* **2020**, *137*, 48602. [CrossRef]

polymers

Review

Graphene/Polymer Nanocomposites: Preparation, Mechanical Properties, and Application

Se Jung Lee †, Seo Jeong Yoon † and In-Yup Jeon *

Nanoscale Sciences and Technology Institute, Department of Chemical Engineering, Wonkwang University, 460 Iksandae-ro, Iksan 54538, Jeonbuk, Korea
* Correspondence: iyjeon79@wku.ac.kr
† These authors contributed equally to this work.

Abstract: Although polymers are very important and vastly used materials, their physical properties are limited. Therefore, they are reinforced with fillers to relieve diverse restrictions and expand their application areas. The exceptional properties of graphene make it an interesting material with huge potential for application in various industries and devices. The interfacial interaction between graphene and the polymer matrix improved the uniform graphene dispersion in the polymer matrix, enhancing the general nanocomposite performance. Therefore, graphene functionalization is essential to enhance the interfacial interaction, maintain excellent properties, and obstruct graphene agglomeration. Many studies have reported that graphene/polymer nanocomposites have exceptional properties that enable diverse applications. The use of graphene/polymer nanocomposites is expected to increase sustainably and to transform from a basic to an advanced material to offer optimum solutions to industry and consumers.

Keywords: graphene; reinforcement; nanocomposite; mechanical properties; application

Citation: Lee, S.J.; Yoon, S.J.; Jeon, I.-Y. Graphene/Polymer Nanocomposites: Preparation, Mechanical Properties, and Application. *Polymers* **2022**, *14*, 4733. https://doi.org/10.3390/polym14214733

Academic Editors: Ting-Yu Liu and Yu-Wei Cheng

Received: 13 October 2022
Accepted: 3 November 2022
Published: 4 November 2022

1. Introduction

Polymers are important materials in modern society as they are relatively inexpensive and easy to process compared to other materials (e.g., metals). However, their inadequate physical properties restrict their application in diverse areas. Therefore, polymers are usually enhanced with fillers of diverse sizes to relieve various restrictions and expand their application areas [1–3]. Nanoscale fillers, which refer to the size of the dispersed phase being less than 100 nm, have at least one characteristic length scale in the order of nanometers and vary essentially from isotropic to highly anisotropic sheet- or needle-like morphologies [3–5]. These were divided into three major types according to the nanoscale filler dimensions. First, two dimensions (2D) are on the nanometer scale, while the second is larger, forming an elongated one-dimensional (1D) structure that includes nanofibers or nanotubes (e.g., carbon nanofibers and nanotubes [6] or halloysite nanotubes [7]). Nanoscale fillers are iso-dimensional low-aspect-ratio nanoparticles (e.g., spherical silica [8], semiconductor nanoclusters [9], and quantum dots [10]). Third, two-dimensional (2D) nanoscale fillers (e.g., layered silicate [11], graphene [5,12], or MXene [13,14]) have the form of sheets that are one to a few nanometers and hundreds to thousands of nanometers thick.

Graphene has proven to be the most powerful material in the world, and has increasingly attracted attention as a promising candidate to substitute the position of carbon-based materials in the enhancement of the mechanical, electrical and thermal properties of polymers. Graphene is a one-atom-thick carbon layer arranged in a honeycomb structure. Additionally, several reports have introduced graphene into polymers for graphene/polymer nanocomposites [15–17], as graphene enhances the performance of polymers, including the mechanical strength, electrical conductivities, thermal stabilities, electro-chemical activity, and impart gas barrier properties [18–23]. However, the use of graphene has proven challenging due to the complicated preparation process, low solubility, and agglomeration by van der Waals interactions [24]. To resolve this, graphene-like materials with similar structures have been prepared from graphite or other carbon sources to maintain the several advantages of graphene while possessing oxygen-containing functional groups. Therefore, a method of developing new polymer/graphene nanocomposites and improving their physical properties is attracting attention.

2. Graphene

2.1. Preparation

Several studies have reported a variety of graphene synthesis techniques for pragmatic applications (Table 1) [25]. Each contributes to the diverse properties of the final material and has many possibilities for mass production. Thus, diverse graphene preparation processes were evaluated and compared with graphene quality (or purity) and the process (scalability, cost, and yield). Each method has its own characteristics; therefore, the choice of production method should be executed each time, according to the graphene applications.

Table 1. Advantages and disadvantages of the graphene preparation method.

Methods	Advantage	Disadvantage	Ref.
Mechanical Exfoliation	High-quality Simplest process	Small production scale	[26]
Chemical Vapor Deposition	High-quality Large-area graphene	Complicated process High energy demand	[27]
Chemical Oxidation (Hummers method)	Fast reaction Fewer defect	High contamination and degradation	[25]
Liquid-Phase Exfoliation	Mass production Upscaling production	Poor solubility Eco-friendly	[25] [28] [29]
Electrochemical Exfoliation	Single step Eco-friendly	Expensive	[25] [30]
Mechanochemical Reaction	Mass production High-quality Edge-selectively	High energy consumption	[31] [32]

2.1.1. Mechanical Exfoliation

Mechanical exfoliation is the simplest process for preparing standalone graphene [26]. In this technique, graphite is repeatedly exfoliated utilizing tape and then transferred to a substrate. As a result, this can yield the highest quality, but is only used for lab scale and prototyping, as it is impossible to scale-up the process.

2.1.2. Chemical Vapor Deposition (CVD)

CVD is the most beneficial process for fabricating high-quality single-layer graphene for use in various devices [27]. Graphene with a large area can be fabricated by exposing diverse hydrocarbon precursors to a metal at high temperatures. The unique mechanism for graphene formation based on the metal substrate begins with the growth of carbon atoms nucleated on the metal after hydrocarbon decomposition, and then the carbon nuclei develop gradually into large domains. Graphene transfer from a metal to other substrates is very difficult due to its chemical inertness, which causes defects and wrinkles in the final graphene. Additionally, thermal fluctuations affect graphene stability. Therefore, despite its complicated process and high energy demands, CVD is the most beneficial process for large-area graphene.

2.1.3. Chemical Oxidation

Graphite oxide (graphene oxide or GO) was created via the Staudenmaier and Hummers methods. For example, graphite was oxidized with strong oxidative reagents (e.g., $KClO_3$, $KMnO_4$, and $NaNO_3$) and concentrated H_2SO_4, HNO_3, or their mixture [33]. Among these, the Hummers method is the most popular, wherein $KMnO_4$ and $NaNO_3$ are used as oxidizing agents for the oxidized graphite in the presence of concentrated H_2SO_4 (Figure 1). Generally, the safer and more scalable Hummers method is used to generate GO. GO contains several oxygen-containing functional groups (e.g., −OH, −COOH, −O−, and −C=O). Graphite is transformed into GO, which accompanies the exfoliation of the graphite layer. GO is readily dispersive in water and organic solvents and has poor electrical insulation and thermal instability properties. Therefore, to recover electrical and thermal properties, the reduction reaction for GO, termed as reduced graphene oxide (rGO), is required. Unfortunately, the complete reduction of GO to graphene has not occurred yet.

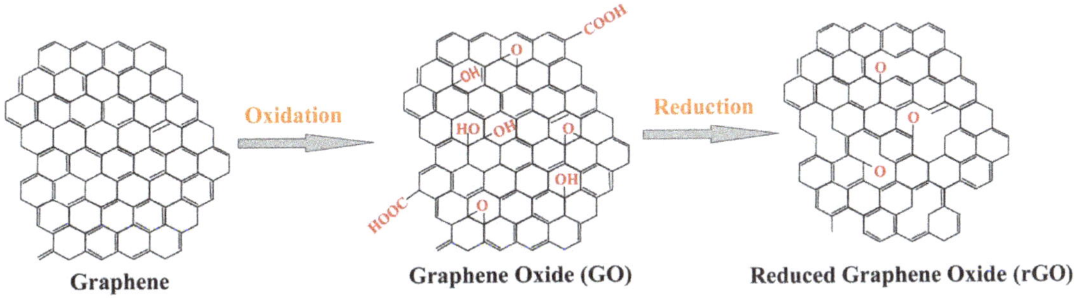

Figure 1. Conversion of graphene into GO and rGO. Reprinted from Ref. [34].

2.1.4. Liquid-Phase Exfoliation

To obtain graphene, liquid-phase exfoliation necessitates three methods: (1) dispersion in a solvent or surfactant, (2) graphite exfoliation, and (3) purification through the separation of graphene and solvent removal [28,29]. The sonication time is a crucial factor as longer sonication time results in a lower graphene layer. Post sonication, the graphite and thicker graphene were removed by ultracentrifugation. Thus, the characteristics of graphene (e.g., yield and number of layers) can be controlled by varying the amount of graphite, sonication time, and centrifugation speed. Owing to poor contact between the graphene sheets, the electrical properties of the produced graphene were similar to those of GO. Additionally, due to very low graphene solubility, the use of a large amount of solvent increases the cost and is ecologically unfriendly.

2.1.5. Electrochemical Exfoliation

Electrochemical exfoliation is generated by the anodic oxidation of the graphite-based electrode. Graphene with a few layers to the anode electrode is created, but its yield is very low and its properties are similar to those of GO, which comprises several oxygen-containing functional groups [30]. This technique offers a single step with an easy operation process that requires a production time over a period of minutes per hour compared with other methods that take a very long time (a few days) for preparation and purification. Additionally, the use of liquid electrolytes or aqueous surfactants makes the process eco-friendly, and that of $LiClO_4$ as an electrolyte can prepare GO-like materials, preventing dangerous and toxic chemical materials from producing GO.

2.1.6. Mechanochemical Reaction

To prepare large-quantity, high-quality, edge-selectively functionalized graphitic nanoplatelets (EFGnPs), a mechanochemical reaction was activated by ball milling (Figure 2) [31]. High-speed spinning metal balls (kinetic energy) are utilized to break the graphitic C−C bond that generates the chemical reaction of broken edges (C−X bond formation, X = heteroatom functional groups) and the physical delamination of graphitic layers into GnPs (graphene nanoplatelets, graphitic nanoplatelets, or GNP(s)). The X groups acted as physical wedges, preventing the GnP restacking. The graphitic C−C bond break induced by the kinetic energy of high-speed metal balls generates active carbon species, which have sufficient reactivity to combine with the appropriate reactants, which can be in diverse phases including vapors (CO_2 [31], N_2 [35], F_2 [36], Cl_2 [35], NF_3 [37], and propylene [38]), liquids (SO_3 [39], Br_2 [35], heptane [40], and styrene [41]), solids (I_2 [35], red P [42], S_8 [43], Te [44], and Sb [45]), or their mixtures (CO_2/SO_3 [39] and SO_3/I_2 [46]). EFGnPs are oxygen-containing and comprise the desired functional groups or atoms at the edges. The oxygen-containing functional groups at the edges led to good EFGnP solubility in diverse solvents, allowing improved processability.

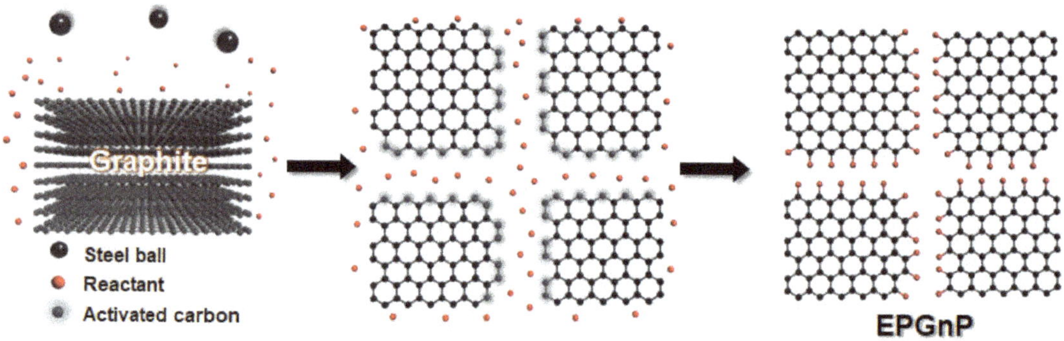

Figure 2. Conceptual diagram for preparation of EFGnPs by mechanochemical reaction.

2.2. Analysis Techniques

The unique properties of graphene have been verified by diverse analytical techniques. The atomic structure of graphene was verified by transmission electron microscopy (TEM) (Figure 3a), which is helpful for analyzing diverse structures [28]. Scanning tunneling microscopy (STM) offers information regarding graphene morphology and electronic properties in three dimensions (Figure 3b) [47]. Atomic force microscopy (AFM) was utilized to verify the number of layers as the graphene layer is 0.4 nm thick [48]. The specific surface area of graphene was estimated using the Brunauer–Emmett–Teller (BET) method with N_2 adsorption–desorption isotherms [49]. X-ray diffraction (XRD) verified the exfoliation and intercalation into graphene layers and graphene formation (Figure 3c) [50]. X-ray photoelectron spectroscopy (XPS) was used to verify the chemical structure (e.g., chemical modification or functionalization) (Figure 3d) [50]. Raman spectroscopy was utilized to verify the number of layers and functionalization of graphene or graphene-like materials [25] which can recognize monolayer graphene, wherein the intensity of the 2D band is at least twice that of the G band (Figure 3e). Additionally, the 2D band of the two-layer graphene moved to a higher wavenumber, and the intensity of the 2D band was lower than that of the G band. Increasing graphene layers broadened the 2D band and developed a shoulder. UV–vis spectroscopy verified graphene dispersion in diverse solvents using Beer's law and the linear relationship between absorbance and solution concentration (Figure 3f) [50].

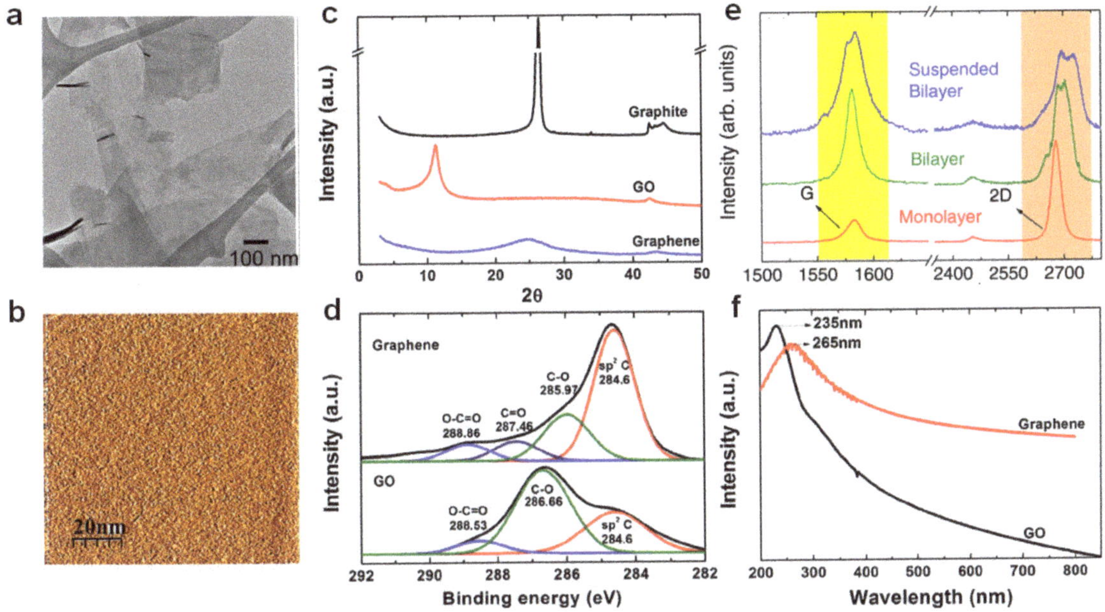

Figure 3. (**a**) Flakes of few-layer graphene on a holey carbon TEM grid. Adapted with permission from Ref. [51]. (**b**) STM images of graphene. Reprinted with permission from Ref. [47]. Copyright 2012 Hong et al. (**c**) XRD patterns of graphite, graphene oxide and graphene. Adapted with permission from Ref. [50]. Copyright 2014 Johra and Lee. (**d**) XPS spectra of GO and graphene [50]. (**e**) Raman spectra of monolayer, bilayer, and suspended bilayer graphene. Adapted with permission from Ref. [52]. Copyright 2011 AIP Publishing. (**f**) UV–vis spectra of GO and graphene. Adapted with permission from Ref. [50]. Copyright 2014 Elsevier.

2.3. Properties

The tremendous attention on graphene has resulted from its exceptional properties, which makes it an attractive material with enormous potential for applications in diverse industries and devices [32]. Its unique physical, mechanical, and electrical properties (e.g., specific surface area (2630 m^2/g), intrinsic mobility (200,000 $cm^2/V{\cdot}s$), tensile strength (130 GPa), Young's modulus (~1.0 TPa), thermal conductivity (~5000 $W/m{\cdot}K$), optical transmittance (~97.7%), and electrical conductivity (~10^6 S/cm)) have drawn lots of attention [48,53–55]. However, these characteristics depend on the graphene purity and thickness, which are major factors that determine the properties of graphene-based materials.

2.4. Graphene Functionalization

The intrinsic graphene properties are reduced dramatically as GO includes oxygen-containing functional groups and diverse defects on the basal plane and at the edges, and aggregation, which is derived from the high specific surface area and the strong van der Waals force between graphene sheets [56–58]. Proper GO functionalization inhibits aggregation, maintains excellent properties, and assigns new characteristics. Generally, GO functionalization can be divided into covalent and noncovalent methods (Figure 4) [59].

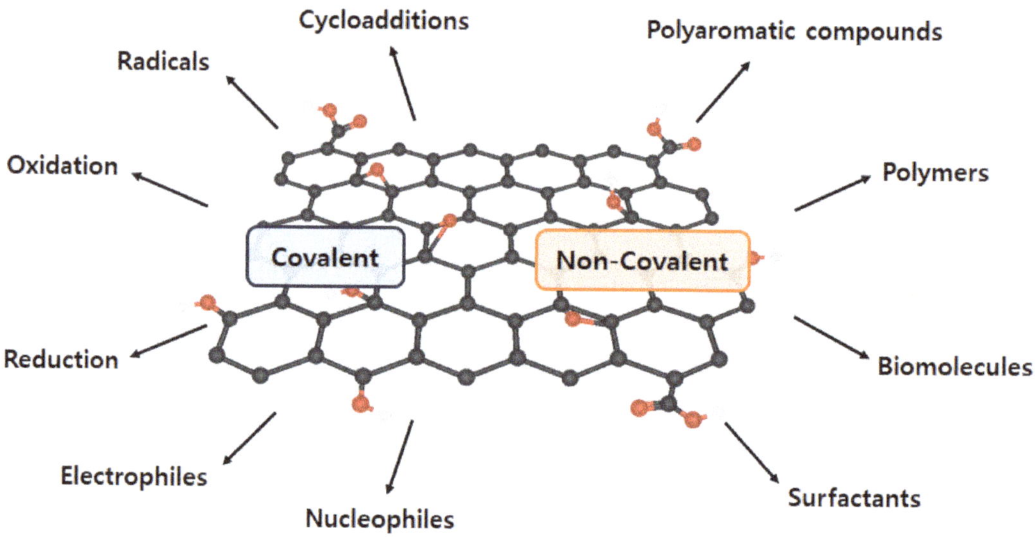

Figure 4. Various covalent and non-covalent graphene functionalization.

2.4.1. Covalent Functionalization

Covalent functionalization introduces new functional groups to enhance the performance of graphene and graphene-like nanomaterials. GO comprises a variety of oxygen-containing functional groups (e.g., −COOH, −OH, −C=O, and −O−) that can be applied to general chemical reactions (e.g., isocyanation, carboxylic acylation, epoxy ring opening, diazotization, and addition) [60,61]. Additionally, other functionalizations, including carbon skeleton, diazotization, Diels–Alder reaction, and click chemistry reaction, can be conducted using the C=C bond [62–64]. Therefore, a double-bond addition reaction with C=C is performed to form a new C−C single bond, which is linked to the benzene derivative with a reactive functional group by a sigma bond. However, covalent graphene functionalization changes the hybridization from sp^2 to sp^3, resulting in π-conjugation bond damage.

2.4.2. Non-Covalent Functionalization

Non-covalent functionalization can be generated through supramolecular chemistry including π–π interactions, hydrophobic forces, hydrogen bonding, van der Waals forces, ionic bonding, and electrostatic effects [65,66]. The greatest advantage of non-covalent functionalization is the maintenance of the bulk structure and specific properties of graphene (or GO) and to increase GO dispersibility and stability. However, its weakness is that other substances (e.g., solvents and surfactants) are included.

3. Preparation of Graphene/Polymer Nanocomposites

3.1. Solution Method

The solution method is the most vastly used technique to prepare polymer nanocomposites on the laboratory scale due to its high utility through the use of diverse solvents and graphene along with a fast and simple process [67–69]. The polymer was dissolved in an appropriate solvent. The two substances were then blended using simple mixing, shear mixing, or ultrasonication for graphene dispersal in the polymer matrix. It is crucial to ensure solvent removal for nanocomposite manufacturing and homogeneous dispersion during the process. This is because the premier properties are significantly affected by the remaining solvent as they can plasticize the nanocomposites and exist at the surface. In the solvent evaporation process, it is important to increase the solubility in the solvent due to the re-aggregation of graphene [32]. Additionally, very low solubility of some polymers in common solvents and the use of large solvent quantities of solvents are some of the limitations of this method.

3.2. Melting Method

The melting method is an industrial-friendly process to predominantly produce thermoplastic-based nanocomposites owing to its rapid and economic process [2,70]. This significantly affects polymer disentanglement in the molten state. Therefore, the polymer chains move freely and mix thoroughly during the molten state. The polymer nanocomposites prepared by melt mixing exhibited satisfactory dispersion of the fillers. However, the manufacturing process for the mixing temperature should be careful because the polymer may degrade at a high temperature [25]. Additionally, high shear forces are required for efficient mixing that leads to the formation of folds/wrinkles or even breaking of the nanoplatelets, decreasing the effective modulus. After melt mixing, additional steps including hot pressing and injection molding were necessary, which are affected by the dispersion, structure, and orientation of the fillers.

3.3. In-Situ Polymerization

In situ polymerization enables the filler to be grafted onto the polymer, which improves the compatibility and interface between the constituents [71–74]. The monomer and graphene were dispersed in an appropriate solvent and ultrasonicated for homogeneous dispersion. An initiator was then incorporated to the mixture for polymerization. However, the polymerization process itself generally increases the viscosity of the mixture, making further processing difficult, which eventually limits the loading fraction. In some cases, this method requires the use of a solvent and requires additional refinement steps to remove the solvent [32]. Its advantages include ease of manipulation, simplicity, scalability, low cost, and low environmental anxiety.

3.4. Electrochemical Reaction

Electrospinning is used for preparing fine nanofibers with average diameters ranging from nanometers to micrometers [2,75]. Electrospun fibers have attractive characteristics, including a high surface/volume ratio, which induces a low density, high pore volume, and exceptional mechanical strength. Additionally, process factors including the type, molecular weight, viscosity, solvent, applied voltage, needle-to-collector distance, and flow rate play a critical role in obtaining the desired properties [76,77].

Electrodeposition is a simple and fast method for producing nanocomposites via electrochemical reactions [2]. It involves the electro-polymerization of a graphene/polymer nanocomposite from the monomer, doping matter (if needed), and GO. Electrodeposition occurs at a specific potential and stops when a pertinent amount of charge has passed. Therefore, the nanocomposite encompassed the electrode surface.

4. Mechanical Properties

Carbon-based polymer nanocomposites have been used to improve the mechanical performance of pure polymers by increasing the interaction between polymers and carbon materials. Among them, graphene/polymer nanocomposites exhibit significantly improved mechanical properties compared to those of pure polymers [32,78,79]. The merits of graphene compared to other fillers enable smooth changes in the properties of nanocomposites at very low percolation thresholds due to the very high aspect ratio of graphene. GO, which comprises oxygen-containing functional groups, is hydrophilic and can interact well with polymers. Therefore, it can improve the mechanical parameters (e.g., Young's modulus, fracture toughness, fracture time, thermal stability, electrical conductivity, and gas barrier properties) [2,80,81]. The incorporation of graphene within the matrix increases the crosslinking of the polymer chains, which leads to an increase in the mechanical properties owing to the exceptional mechanical properties of graphene (tensile strength of ~130 GPa and Young's modulus of ~1 TPa) [82,83]. However, due to the filler, the movement of the polymer chain is inhibited, so the tension toughness and strain of nanocomposites may tend to decrease. Therefore, many papers have reported that the mechanical properties of graphene/polymer nanocomposites are considerably enhanced compared to those of pure polymers (Table 2). The table was assembled according to the type of graphene and polymer.

Table 2. Mechanical properties of diverse graphene/polymer nanocomposites.

Graphene	Matrix	Process	Filler Loading	Tensile Strength (MPa)	Young's Modulus (MPa)	Strain (%)	Ref.
TrGO	UHMWPE	Solution process	1 wt.%	3100	106,000		[84]
GNP	LLDPE	Solution process	5 wt.%	25.3	189.4		[40]
MGO	LLDPE	Solution process	3 wt.%	19.9			[85]
GNP	HDPE	Melt mixing	23 wt.%	34.84		7	[86]
TRG	HDPE	Polymerization	5.2 wt.%	12.9	624.4	9.7	[87]
GNP	PP	Melt mixing	10 wt.%		1963.2	18.20	[88]
GNP	PP	Melt mixing	5.5 wt.%		1900	7.00	[89]
GNP	PP	Melt mixing	5 wt.%	38		6.99	[90]
GNs	PP	Melt mixing	3 wt.%	61.57	2314.61	19.09	[91]
fGO	PP	Melt mixing	1 wt.%	38.7	562		[92]
Graphene	PP	Melt mixing	0.1 wt.%	33	1250	1150	[93]
Graphene	PP	Melt mixing	1 wt.%	37	1760	130	[93]
GNP	PP	Solution process	5 wt.%	55.85	7239	9.07	[38]
GNP	Epoxy	Solution process	10 vol%	1	26		[94]
Graphene	Epoxy	Melt mixing	0.5 wt.%	23.01	8000		[95]
GNP	Epoxy	Solution process	6 wt.%	53	3400	2	[96]
GNP	Epoxy	Melt mixing	6 wt.%	35.5		1.49	[97]
GNP	Epoxy	Melt mixing	4 wt.%	75.8		4.55	[98]
GNP	Epoxy	Solution process	0.3 wt.%	72.4	1990	8.2	[99]
fGr	Epoxy	Solution process	0.1 wt.%	83.43			[100]
rGO	PVA	Wet spinning	2 wt.%	867	15,900		[101]
GO	PVA	Solution process	0.3 wt.%	63			[102]
GO	PVA	Solution process	3 wt.%	110		36	[103]
rGO	PVA	Solution process	0.02 wt.%	45.6	162		[104]
rGO	PVA	Solution process	3.5 wt.%	29	520	22	[105]
GS	PVA	Solution process	1 wt.%	67.7		139	[106]
rGO	PVA	Solution process	0.7 wt.%	154	4900	5.1	[107]
GNs	PVA	Solution process	1.8 vol%	42	1040		[108]
GO	PVA	Solution process	0.7 wt.%	87.6	3450		[109]
GO	PVA	Solution process	2 wt.%	37.8	67.3	294	[110]

Table 2. *Cont.*

Graphene	Matrix	Process	Filler Loading	Tensile Strength (MPa)	Young's Modulus (MPa)	Strain (%)	Ref.
Graphene	PVA	Electrospinning	6 wt.%	19.2	638	113	[111]
GO	PVA	Polymerization	0.04 wt.%	50.8	2123	208	[112]
GNP	PVA	Electrospinning	1 wt.%	11		130	[113]
rGO	PVA	Electrospinning	2 wt.%	5.51	85.67		[114]
GNs	PU	Polymerization	2 wt.%	36.3		535	[72]
fGNP	PU	Polymerization	1.5 wt.%	23.4		6.7	[115]
fGO	PU	Solution process	0.4 wt.%	19.6		1035.3	[116]
fGO	PU	Electrospinning	1 wt.%	8.9	41.4	515.6	[117]
Graphene	PU	Solution process	3 wt.%	22.9	2.7	474	[118]
fGS	PU	Solution process	1 wt.%	11.9		448	[119]
fGNs	PU	Solution process	2 wt.%	20.2		138	[120]
TrGO	PU	Polymerization	2 wt.%	10.6	35.1	715	[121]
GSs	PU	Solution process	15 wt.%	25	1200	220	[122]
MrGO	PU	Solution process	0.608 wt.%	34.30		186.24	[123]
fGS	PU	Solution process	1 wt.%	40		590	[124]
GNP	PS	Solution process	5 wt.%	11.54	809.4		[41]
GSs	PS	Polymerization	0.9 wt.%	41.42	2280		[125]
GO	PS	Solution process	2 wt.%	43.5	3580	1.3	[126]
FLG	PS	Solution process	0.9 wt.%	13.98		11.3	[127]
FLG	PS	Solution process	0.7 wt.%	16.03		17	[128]
fGNs	PS	Solution process	1 wt.%	78.2		2.38	[129]
GO	PS	Solution process	1.02 wt.%	13.60		1.185	[130]
MLG	PVC	Polymerization	0.3 wt.%		50.5		[131]
GNP	PVC	Solution process	2.5 wt.%	24	11.42	33.5	[132]
mGO	PVC	Solution process	5 wt.%	37.87	1694.23	2.61	[133]
rGO	PVC	Melt mixing	1 wt.%	16.16	46.9	362	[134]
rGO	PVC	Solution process	0.2 wt.%	35	1700	5	[135]
GO	PA6	Polymerization	0.1 wt.%	123	722	269	[71]
fGO	PA6	Solution process	1 wt.%		1900		[136]
fGr	PA6	Polymerization	0.2 wt.%	68.4		100	[137]
GO	PA6	Polymerization	0.65 wt.%	64.9			[138]
fGr	PA6	Polymerization	0.1 wt.%	500		20.7	[139]

4.1. Graphene/Polyethylene Nanocomposites

In the case of ultrahigh-molecular-weight polyethylene (UHMWPE)/thermally reduced graphene oxide (TrGO), the tensile strength and Young's modulus improved with an increase in the TrGO content due to the exceptional inherent TrGO strength and stiffness [84]. In the case of 1 wt.% TrGO, the tensile strength of the pure UHMWPE increased from 1.1 to 3.1 GPa (182% increase) and the Young's modulus improved by over three times. The drawing ratio is also a factor that improves the mechanical properties of the composite. When the composite film is not drawn, the filler is randomly distributed within the PE matrix. As the drawing ratio increases, the filler cluster expands so that the TrGO flakes are aligned along the drawing direction, and the size of the cluster decreases, which means that the flakes are exfoliated. The exfoliated flakes increase the specific area in contact with the PE matrix molecule and the TrGO fillers and improve load transfer from PE to TrGO. Additionally, the significant increase in the mechanical properties at low TrGO content was ascribed predominantly to the uniform TrGO dispersion and very high specific surface areas that help transfer the load from the polymer matrix to the strong fillers.

High-density polyethylene (HDPE)/graphene nanocomposites were prepared directly from HDPE and graphene using a twin-screw extruder. HDPE stiffness increased significantly with increasing graphene content, with a maximum Young's modulus (187% increase) [86]. Therefore, the tensile strength increased to 34%, whereas the elongation at break reduced to 91%. The improvement in tensile strength without decreasing the elongation is due to the exceptional mechanical strength of graphene, which is a uniform dispersion in the matrix and strong interfacial interactions [140].

Heptene-functionalized graphitic nanoplatelets (HGN) produced by ball milling were incorporated as reinforcing fillers for LLDPE (linear low-density polyethylene) [40]. The tensile strength and Young's modulus of the HGN/LLDPE nanocomposites were higher than those of pure LLDPE (Figure 5a,b). The HGN/LLDPE_5 (5 wt.% HGN content) displayed the best mechanical properties among the pure LLDPE and HGN/LLDPE nanocomposites. Therefore, the tensile strength, yield strength, Young's modulus, and tensile toughness of the HGN/LLDPE_5 nanocomposites were enhanced to 43.8, 39.4, 39.5, and 125.8%, respectively, compared with those of pure LLDPE. Comparing the fracture surfaces, pure LLDPE appeared smooth and plain with wave-shaped morphology, but HGN/LLDPE _5 displayed rough and canyon-like structures (Figure 5c,d), indicating that the specimen was ripped steadily by an accumulated stress that transferred efficiently from HGNs to LLDPE due to good dispersion of the HGNs into LLDPE.

4.2. *Graphene/Polypropylene Nanocomposites*

Polypropylene (PP)/reduced graphene oxide (PP/rGO_X) nanocomposites with low rGO concentrations (X = rGO content, 0.05–1.0 wt.%) were created via twin-screw extrusion [141]. The tensile strength and Young's modulus of PP/rGO_1 increased by 11.15 and 41.45%, respectively, compared with those of pure PP. The enhanced mechanical properties were induced by the exceptional rGO dispersion due to the combination of shear and extensional flows during extrusion and good compatibility between rGO and PP.

PP/graphene (polypropylene coated graphene) nanocomposites with exfoliated PP-coated graphene, which can prevent the restacking of graphene sheets during melt blending, demonstrated exceptional interaction and effective load transfer between the polymer and graphene [93]. In the case of 0.1 wt.% graphene, the yield strength, tensile strength and Young's modulus were enhanced to 30 MPa (by 36%), 33 MPa (by 38%), and 1.25 GPa (by 23%), respectively, compared with pure PP. Moreover, the elongation at break remained almost unchanged, indicating that the nanocomposite toughness was not reduced. In the case of 1.0 wt.% graphene content, the yield strength (38 MPa), tensile strength (37 MPa), and Young's modulus (1.76 GPa) increased by 75%, 54%, and 74%, respectively, compared with pure PP. This is ascribed to the uniform graphene dispersion and effective load transfer from the polymer to graphene through interfacial interactions.

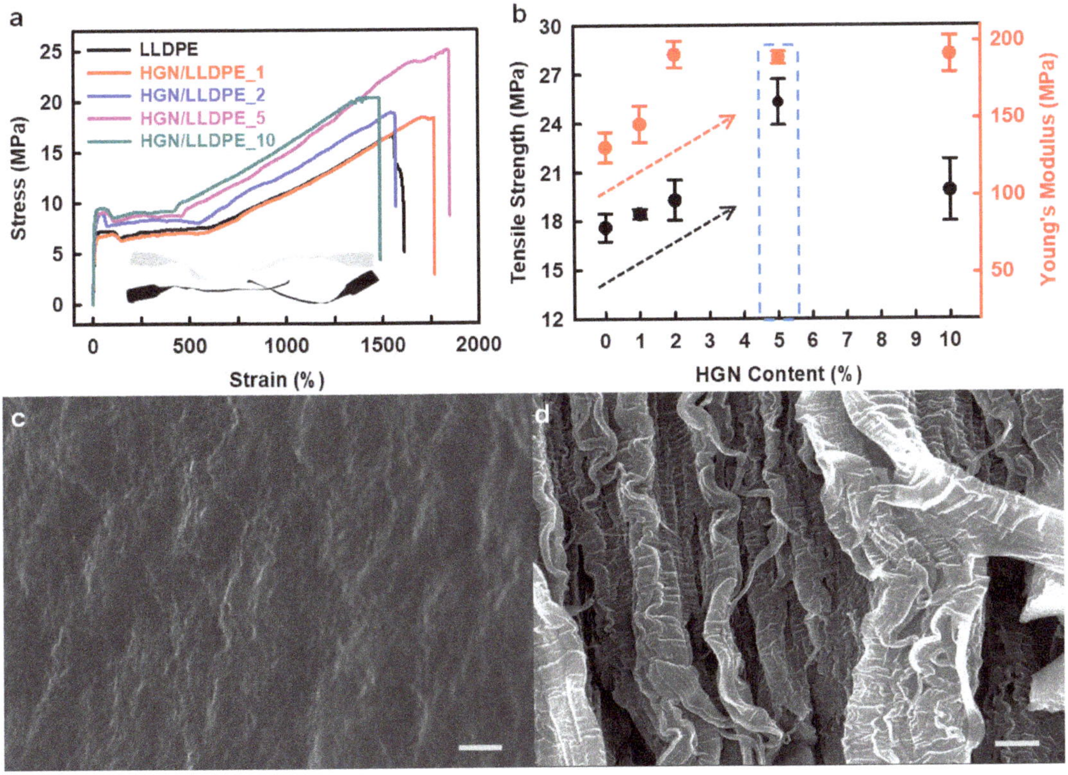

Figure 5. (**a**) stress–strain (S–S) curves of the pure LLDPE and HGN/LLDPE nanocomposites. (**b**) Tensile strengths and Young's modules of the pure LLDPE and HGN/LLDPE nanocomposites according to HGN content. SEM images of the fractured surface: (**c**) pure LLDPE, and (**d**) HGN/LLDPE_5. Scale bars are 10 μm. Reprinted with permission from Ref. [40]. Copyright 2020 Elsevier.

Edge-propylene graphitic nanoplatelets (EPGnP/PP)/PP_X nanocomposites were prepared utilizing a solution method [38]. EPGnP/PP_5 (EPGnP content of 5 wt.%) displayed that tensile strength, yield strength, and Young's modulus increased to 90.5, 90.8, and 249.5%, respectively, compared to pure PP (Figure 6a,b). This is due to the molecular-level dispersion of the EPGnPs into the PP matrix, and the strong interfacial interaction between EPGnP and PP through the EPGnP propylene functional groups. It verified the uniform dispersion of EPGnPs into the PP matrix using HR-TEM (Figure 6c,d). However, as the content of EPGnP increased, the tensile toughness of the EPGnP/PP_X nanocomposite decreased because the strain of the EPGnP/PP_X nanocomposite decreased gradually because the addition of multi-dimensional fillers limits the movement of the matrix. Although the filler is uniformly dispersed within the matrix, there is little elongation, and EPGnP behaves like a physical cross-linking point within the PP matrix.

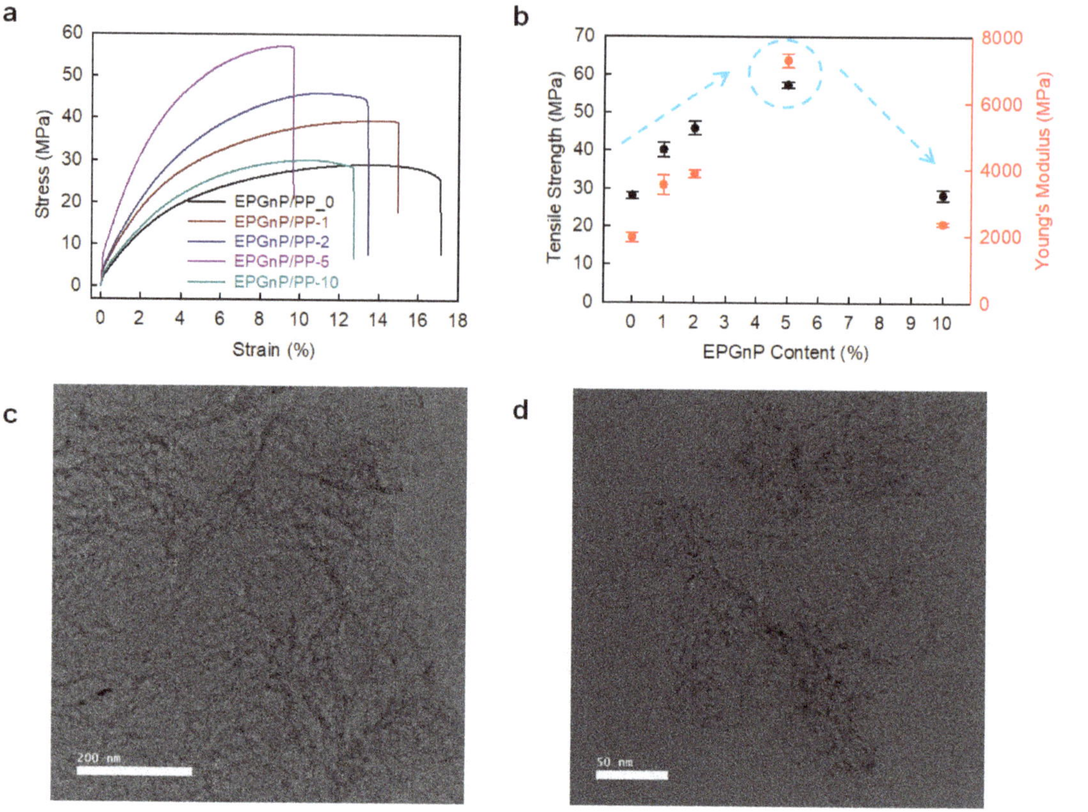

Figure 6. (**a**) S–S curves and (**b**) tensile strength and Young's modules of EPGnP/PP_X nanocomposites. HR-TEM images of EPGnP/PP_5 nanocomposites: (**c**) low-magnification; (**d**) high-magnification. Reprinted with permission from Ref. [38]. Copyright 2021 Elsevier.

4.3. Graphene/Epoxy Nanocomposites

The epoxy/GNP_X nanocomposites manufactured by melt mixing demonstrated an increase in the GNP content, tensile strength, and Young's modulus of the nanocomposite [94]. In epoxy/GNP_10 (GNP content:10 vol.%), the tensile strength and Young's modulus increased to 1.0 ± 0.08 MPa (66.7) and 26 ± 1.7 MPa (1344%), respectively, compared with pure epoxy. The tensile strength was fixed at 5 vol.% GNP. The tensile strength increased to 1.4 ± 0.07 MPa at 5 vol.% GNP and then was decreased to 1.0 ± 0.08 MPa at 10 vol.%, indicating an increase of 133.3% at 5 vol.% and a decrease of 28.6% at 10 vol.%. The increased tensile strength at low graphene content is attributed to the exceptional graphene dispersion and powerful mechanical interaction between graphene and the polymer.

Diverse epoxy nanocomposites with pristine- and Triton-graphene (graphene functionalized non-covalently with Triton-X100) have been prepared [100]. The corresponding stress of the nanocomposite was higher than that of pure epoxy for the same strain, indicating that the nanocomposite modulus was improved by the incorporation of pristine- and Triton-graphene. At the breaking point on the S–S curves, the strength and elongation at break of the nanocomposite with Triton-graphene were obviously higher than those of pure epoxy and pristine graphene-filled epoxy nanocomposite. The tensile strength of pristine graphene was almost unchanged compared to that of pure epoxy, whereas that of the Triton-graphene (0.1 wt.% loading) was greatly enhanced by 57% (from 52.98 ± 5.82 to 83.43 ± 5.90 MPa), although the elastic modulus still demonstrated a similar increase to that of the pristine graphene nanocomposites. As depicted in the SEM images of the Triton-graphene/epoxy nanocomposite fracture surface, well-dispersed graphene and a relatively good graphene/polymer interface effectively improved the load transfer efficiency with the polymer and thus increased the tensile strength.

4.4. Graphene/Poly(Vinyl Alcohol) Nanocomposites

PVA/GO-X (X = GO content) nanocomposites were prepared utilizing the solution method [110]. Compared with pure PVA, a maximum increase of approximately 50% in the tensile strength was observed for the nanocomposite (X = 0.3 and 2.0). Additionally, its elongation at break or failure strain increased to 13–22%, indicating that the exceptional GO dispersion at a lower content led to a significant increase in the mechanical properties. However, the elastic modulus of the PVA/GO-2 was larger than PVA/GO-0.3 nanocomposite due to the higher GO content, which possesses a high elastic modulus.

PVA/GO-X (X = GO content) nanocomposites were prepared via in situ polymerization [112]. The tensile strength and Young's modulus of the PVA/GO-X nanocomposites improved with increasing GO content at low content. Therefore, the tensile strength and Young's modulus of PVA/GO-0.04 were enhanced from 50.8 MPa and 2123 MPa compared with pure PVA (42.3 MPa and 1477 MPa). However, elongation behavior decreased slightly from 215% to 211.5% due to the strong H-bonding interaction between the GO surface and the PVA matrix, that prevents and reduces the movement of polymer chains. The enhanced mechanical properties at low GO contents are related to the combination of GO with excellent tensile strength and Young's modulus, homogeneous GO dispersion, and strong hydrogen bonding between the oxygen-containing functional groups of GO and PVA chains.

4.5. Graphene/Polyurethane Nanocomposites

PU-GX (X = graphene content) nanocomposites were prepared with waterborne biodegradable PU and graphene at diverse graphene contents (0, 1, 3, and 5 wt.%) [118]. Graphene enhanced the mechanical properties of PU-GX nanocomposites. The tensile strengths of PU, PU-G1, PU-G3, and PU-G5 nanocomposites were approximately 35.3, 23.3, 22.9, and 11.4 MPa, respectively. Additionally, the elongation of PU-G5 (317%) decreased compared to that of pure PU (640%) as the incorporation of graphene to pure PU induces the nanocomposites to be less elastic.

PU/graphene nanofibers were obtained by electrospinning a PU/graphene solution comprising GO, f-GO (PCL-functionalized GO), and r-GO at diverse concentrations (0, 0.1, 0.5, and 1.0 wt.%) [117]. The modulus and breaking stress of the PU/graphene nanocomposite nanofibers were higher than those of pure PU. The breaking stress of the nanocomposite nanofiber at 0.1 wt.% f-GO was 9.3 MPa, which is 1.2 times higher than pure PU nanofibers (7.8 MPa). The nanofiber modulus increased with increasing graphene content. The Young's moduli of the nanofibers with 1 wt.% f-GO displayed 41.4 MPa, whereas that of pure PU nanofiber webs was 31.5 MPa. Thus, f-GO was very well dispersed in the polymer matrix compared with GO and r-GO and functioned for excellent load transfer.

Hyperbranched aromatic polyamide-functionalized graphene sheets (GS-HBA) were prepared with 3,5-diaminobenzoic acid (DABA) as a monomer [122]. GS-HBA demonstrated good dispersion in thermoplastic polyurethane (TPU) and strong adhesion with TPU via hydrogen bonding, which efficiently enhanced the load transfer from TPU to the graphene sheets.

As the GS-HBA content increased, the stress gradually increased, and the strain steadily decreased. The S–S curve of the 15 wt.% GS-HBA nanocomposite was similar to pure TPU. The highest Young's modulus of the nanocomposite with 15 wt.% GS-HBA was 1.2 GPa, that is one order of magnitude higher than pure TPU (0.09 GPa). The yield tensile strengths of nanocomposites with 15 wt.% GS-HBA (25 MPa) increased about seven-fold compared with pure TPU (3.4 MPa), that is comparable to UHMWPE (about 22 MPa). Initially, the ultimate tensile strength increased from 18 to 37 MPa, and then decreased slightly to approximately 30 MPa when the GS-HBA content increased to over 2.5 wt.%. GS-HBA may advance strain hardening and strain-induced crystallization of soft segments (SS) owing to hydrogen bonding. As the GS-HBA content increased, the strain at the break of the nanocomposites decreased.

4.6. Graphene/Polystyrene Nanocomposites

PS/FLG-X (X = FLG content) nanocomposites were prepared by combining graphene produced via liquid-phase exfoliation with polystyrene [127]. The Young's moduli of the nanocomposites with FLG from 0.1 to 0.9 wt.% increased from 55.4 to 59.8% compared to pure PS (0.202 GPa) and achieved this at the peak value of the PS/FLG-0.9 composite. This is because FLG impedes the chain mobility of the host polymer. An increase in the ultimate tensile strength was observed based on the graphene value. The smallest value is at PS/FLG-0.1 and the highest is at 13.98 MPa of the PS/FLG-0.9 nanocomposite; the overall improvement in UTS was 74.75%. The polymer chain interacted with graphene and formed a bond, increasing the tensile strength of the nanocomposite.

Edge-styrene graphitic nanoplatelets (StGnPs) were prepared directly using a mechanochemical reaction with graphite and styrene, and StGnP/PS-X (X = StGnP content) nanocomposites were prepared utilizing solution methods [41]. The mechanical properties of the StGnP/PS nanocomposite depended on the StGnP content. The StGnP/PS-5 nanocomposite exhibited the best tensile strength (11.54 MPa), Young's modulus (809.4 MPa), yield strength (3.67 MPa), and tensile toughness (0.89 MPa) (Figure 7). The considerably enhanced mechanical properties of the StGnP/PS nanocomposites compared with those of pure PS can be ascribed to the stress being transferred comfortably from PS to StGnPs through the molecular-level StGnP dispersion into PS and strong interfacial interaction between the StGnPs and PS.

4.7. Graphene/Poly(Vinyl Chloride) Nanocomposites

Multi-layer graphene (MLG) PVC nanocomposites (MLG/PVC-X, X = MLG content) were prepared by in situ polymerization and melt mixing [131]. The yield strength and Young's modulus of the MLG/PVC nanocomposites strongly depended on the MLG content. The yield strength and Young's modulus first increased and then decreased with MLG incorporation. The yield strength and Young's modulus of the MLG/PVC-0.3 nanocomposites increased by 11 and 8%, respectively, compared to those of pure PVC. The exceptional dispersion of MLG and strong interactions led to the efficient transfer of the load and interruption of the motion of PVC chains, resulting in the excellent mechanical properties of the nanocomposites.

PVC nanocomposites embedded with graphene nanoplatelets (GNP/PVC) were prepared utilizing a solution method [132]. The GNP/PVC nanocomposites exhibited a far higher elastic modulus and strength than pure PVC. For the GNP/PVC nanocomposites (2.5 wt.% GNP), elastic modulus and ultimate tensile strength were 1.42 GPa (108.77%) and 24.0 MPa (100%), respectively, compared to pure PVC (5.47 GPa and 12.0 MPa). The exceptional mechanical properties of the GNP/PVC nanocomposites were attributed to the strong interaction between GNP and PVC and high GNP modulus.

4.8. Graphene/Polyamide Nanocomposites

GO/nylon 6 was prepared via a solution-mixing method [136]. The incorporated graphene induced a noteworthy enhancement in the mechanical strength and Young's modulus. At 1.0 wt.% graphene, the best yield strength (91 MPa) and Young's modulus (1.90 GPa) increased to 18 and 28%, respectively, compared to pure Nylon 6.

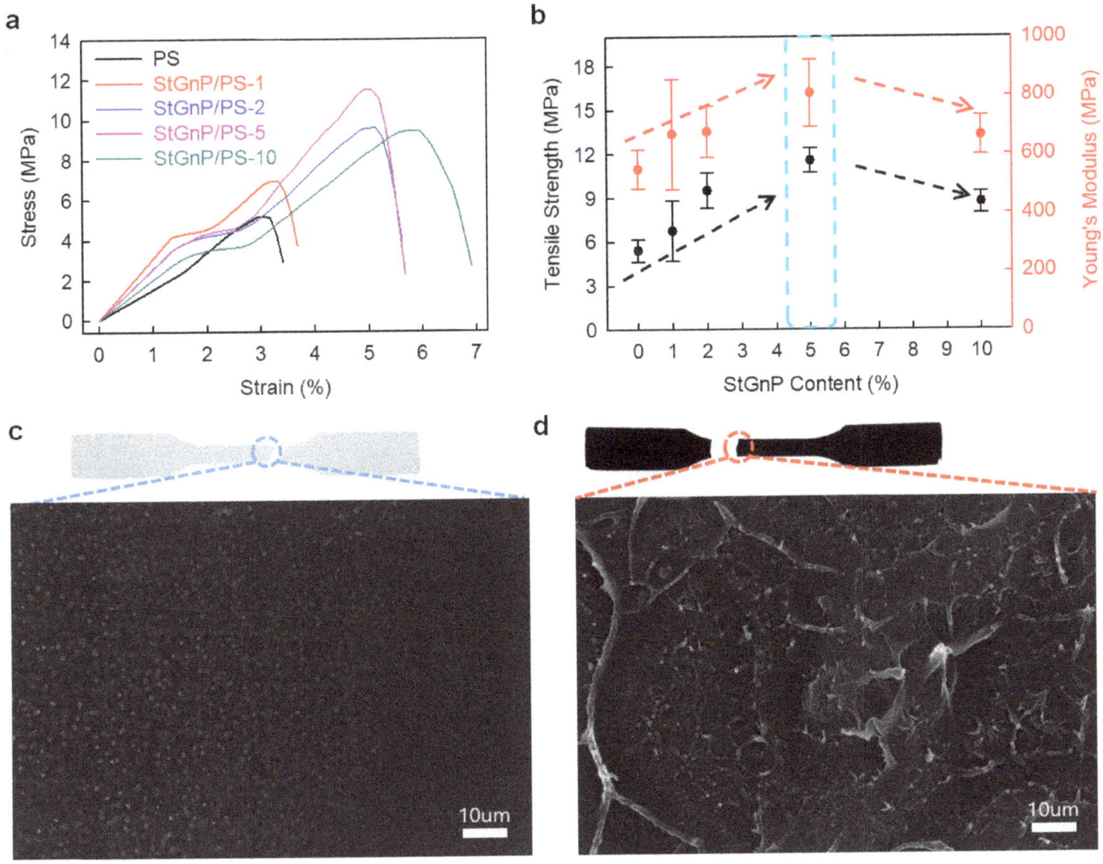

Figure 7. (**a**) S–S curves of pure PS and StGnP/PS nanocomposites. (**b**) Tensile strength and Young's modulus of pure PS and StGnP/PS nanocomposites. SEM images of the fractured surfaces after the tensile test: (**c**) pure PS and (**d**) StGnP/PS-5 nanocomposite. Reprinted with permission from Ref. [41]. Copyright 2021 Kang and Kim.

Nylon 6/sulfonated graphene nanocomposites (NSG-X, X = SG content) were prepared via in situ ring-opening polymerization of ε-caprolactam and sulfonated graphene (SG) with polar sulfonic acid [137]. Compared with pure nylon 6, NSG-0.2 demonstrated an improvement in tensile, impact, and especially bending strength, due to the buildup effect of SG with high specific surface areas. The bending and impact strengths increased by 32.2 and 5.7%, respectively, compared to those of pure nylon 6. However, in the case of NSG-3.0 with the highest SG content, the mechanical properties decreased. The low mechanical properties are due to the low molecular weight of grafting nylon 6 chains on the SG sheets. It also decreased with the elongation at break of the NSG composites with increasing SG content, due to the high density of grafting and the inability molecular "slippage". Therefore, mechanical properties may be maintained and improved by using an appropriate SG content.

4.9. Analysis Methods of Properties for Graphene/Polymer Nanocomposites

The mechanical properties and thermal stability of the graphene/polymer nanocomposite are influenced by the internal microstructure and crystallinity. Thermal properties of polymers can be confirmed with various measuring equipment (DSC, TGA, DMA etc.) and the T_m, T_g, T_c of nanocomposites can be found from the measurement data [142]. In particular, glass-transition temperature, T_g, is one of the important factors in polymer treatment and production applications, and it depends on the molecular weight of the polymer chain and the spatial structure of the polymer [143].

Differential scanning calorimetry (DSC) is used to study the thermal behavior of the nanocomposites and to follow their melting and crystallization behavior [89]. Liang et al., [109] used DSC to compare T_g of pure PVA with the T_g of graphene/PVA nanocomposites. As a result, it can be observed that T_g of the graphene/PVA nanocomposite increased compared to pure PVA. The increase in T_g indicates that the polymer chain was limited by the H-bonding interaction. Load transfer is largely dependent on the interfacial interaction between the filler and the wrapped polymer matrix [144]. In addition, the molecular-level dispersion of graphene sheets in the PVA matrix and the large aspect ratio of graphene are also advantageous for stress transfer through graphene/PVA interfaces. Therefore, significant improvement in tensile stress and modulus of the nanocomposite was greatly improved. The crystallization temperature (T_c) may also be obtained through DSC measurement. As the content of GNP in the PP/GNP nanocomposites increases, T_c shifted to a higher temperature, indicating that the presence of GNP as a nucleating agent in nanocomposites promotes the crystallization of PP [145,146]. The addition of GNP provides more surfaces in the nucleation process that facilitates the crystallization process [89].

The thermal stability and composition of the nanocomposites are investigated using thermogravimetric analysis (TGA). Additionally, TGA was used to confirm filler loading and evaluate the effect of filler loading on the thermal stability in nanocomposites. The degradation performance was improved as filler loading increased, due to the hindered effect of the filler upon the diffusion of oxygen and volatile products throughout the nanocomposite material. That is, it can be seen that the thermal stability of the polymer was improved by adding graphene [89].

Tensile test and dynamic mechanical analysis (DMA) have been conducted to evaluate the enhancement properties of nanofillers [147]. According to the storage modulus measured by DMA, it can be seen that the values of the IL-G/PU nanocomposites are higher than the value of the pure PU. The data of tan δ measured by DMA showed that as the ILG content increases, the T_g of the composites gradually transfers to higher temperatures [123]. The high specific surface area of the IL-G sheet provides an excellent contact area with the PU matrix, obtaining a strong interfacial attraction through π-π and/or cation-π as well as electrostatic interactions and van der Waals forces. Additionally, the presence of tertiary amine on the IL-G sheets may interact with the carbonyl group in the PU chain through hydrogen bonding [148,149]. The synergistic effects improve T_g of the PU nanocomposite by constraining the mobility of PU chains segment [123].

5. Applications

An optimized graphene content in a polymer matrix is typically used for structural reinforcement areas [150–153]. Owing to the potential multifunctional properties of polymers with graphene, graphene/polymer nanocomposites have been utilized in several areas. The noteworthy enhancement of the mechanical performance of polymers suggests the use of graphene in diverse applications to achieve a combination of high strength and light weight. Additionally, it has demonstrated a potential to enhance the safety, reliability, and cost-effectiveness of graphene/polymer nanocomposites, which can increase the range of material options in the aerospace, automotive, marine, sports, biomedical, and energy industries (Figure 8).

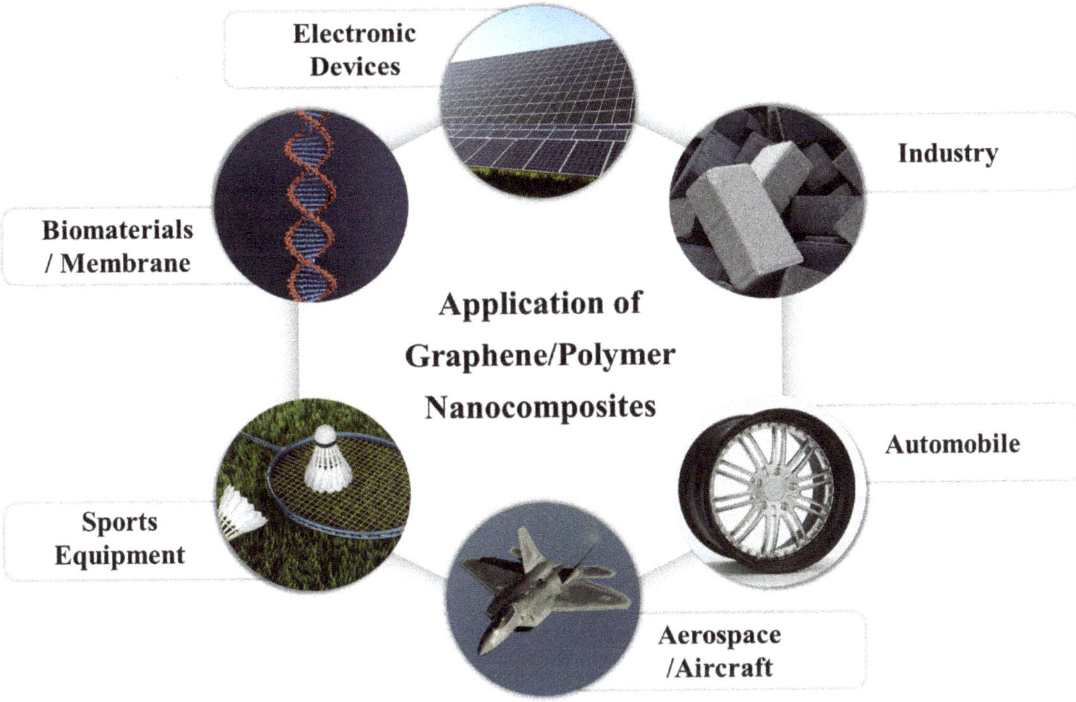

Figure 8. Applications of graphene/polymer nanocomposites.

Owing to the outstanding structural strength and conductivity of graphene, polymer nanocomposites have become a potential candidate for aerospace applications [150]. Thermosetting polymers are predominantly used as matrices for aerospace as they are thermally stable, chemically inactive, and have reasonable mechanical and electrical properties. However, most thermosetting polymers are fragile at low temperatures due to their stiff cross-linked structures, which mean that they crack easily according to thermal-fatigue loading [154]. Therefore, to resolve various issues, it is essential to improve the properties of the matrix or its interfacial bonding. Diverse types of graphene can be utilized for this purpose. This will be extremely helpful in satisfying the normal requirements of aerospace applications.

Graphene/polymer nanocomposites have advantages in diverse environmental applications including water treatment, energy production, and contaminant sensing [150,151,155]. Graphene/polymer nanocomposites preserve them against corrosion, thus, they were useful as adsorbents for the removal of metal ions, organic materials, and gases in the aquatic environment, and were also utilized to eliminate water contaminants.

Research is being conducted on the use of graphene as a new and renewable energy source. In solar cells, graphene either acts as the active medium or as a transparent/distributed electrode material [151,156]. It is a highly flexible and stretchable electrode that can be used in diverse electrolytes.

Graphene possesses properties that are distinct from those used in biotechnological and bio-based applications. Owing to its large surface area, chemical stability, and functionalization feasibility, it is a promising candidate for tissue engineering, drug delivery, and DNA sequencing [157]. Furthermore, ultrasensitive measurement equipment to sensitively detect diverse biological molecules, including glucose, hemoglobin, cholesterol, and DNA, can be prepared using chemically functionalized graphene. Graphene membranes with nanopores are ideal for achieving higher permeate flux, higher selectivity, and increased stability by controlling the pore size and aspect. Graphene paints were concatenated with conductive ink, antistatic, magnetic shielding, and gas barriers. Thus, to expand the application areas of graphene/polymer nanocomposites, new graphene chemical modifications to control the properties of products have been reported.

6. Future Perspectives

Among the various reinforcement materials, graphene has excellent mechanical, thermal, and electrical properties, and is regarded as a better alternative to general nanofillers. Thus, graphene/polymer nanocomposites have properties that are extraordinary for diverse applications. Several studies on the exceptional graphene properties have revealed that it plays an imperative role in enhancing the specific properties of polymer nanocomposites, which can be utilized in diverse applications [32,150,158]. Other than in the above-mentioned application, in order to utilize graphene/polymer nanocomposites, improvements are required in graphene dispersion in a matrix, interfacial interaction, mass production and remarkable quality, etc. The interfacial interaction between graphene and the polymer and the homogeneous dispersion of graphene generally enhances nanocomposite properties. A suitable interaction between graphene and polymer guarantees the technological advancement of graphene/polymer nanocomposites, but is still insufficient. Therefore, the challenge is to maintain the inherent properties of graphene in the nanocomposite as much as possible, many people continue to strive to overcome this issue.

7. Conclusions

Current polymer nanocomposites are useful in various aspects. Many studies have been reported demonstrating that graphene/polymer nanocomposites can be applied in various fields through the chemical functionalization of graphene used as fillers. The uniform dispersion of graphene is one of the factors that can improve performance. Additionally, the interaction between the polymer and the filler, the surface area, filler loading, and dispersion are various factors that determine the quality of the nanocomposite. Despite numerous challenges related to graphene, a suitable interaction between graphene and polymers is yet to be developed. However, the manufacturing industry cannot neglect the use of graphene (or graphene-like materials) in commercial products. It is expected that it will be transformed from a typical to a high-performance material that can offer the best solutions for industry and consumers. Therefore, the business influence of graphene/polymer nanocomposites is growing continuously.

Author Contributions: Conceptualization, I.-Y.J.; writing—original draft preparation, S.J.L., and S.J.Y.; writing—review and editing, I.-Y.J.; visualization, S.J.L., S.J.Y.; supervision, I.-Y.J.; project administration, I.-Y.J. All authors have read and agreed to the published version of the manuscript.

Funding: This research received no external funding.

Institutional Review Board Statement: Not applicable.

Data Availability Statement: Data presented in this study are available on request from the corresponding author.

Acknowledgments: This research was supported by Wonkwang University in 2022.

Conflicts of Interest: The authors declare no conflict of interest.

References

1. Natta, G.; Corradini, P. Structure and properties of isotactic polypropylene. In *Stereoregular Polymers and Stereospecific Polymerizations*, 1st ed.; Natta, G., Danusso, F., Eds.; Elsevier: Amsterdam, The Netherlands, 1967; Volume 2, pp. 743–746.
2. Chee, W.K.; Lim, H.N.; Huang, N.M.; Harrison, I. Nanocomposites of graphene/polymers: A review. *RSC Adv.* **2015**, *5*, 68014–68051. [CrossRef]
3. Fu, S.; Sun, Z.; Huang, P.; Li, Y.; Hu, N. Some basic aspects of polymer nanocomposites: A critical review. *Nano Mater. Sci.* **2019**, *1*, 2–30. [CrossRef]
4. Reynaud, E.; Gauthier, C.; Perez, J. Nanophases in polymers. *Metall. Res. Technol.* **1999**, *96*, 169–176. [CrossRef]

5. Young, R.J.; Kinloch, I.A.; Gong, L.; Novoselov, K.S. The mechanics of graphene nanocomposites: A review. *Compos. Sci. Technol.* **2012**, *72*, 1459–1476. [CrossRef]
6. Calvert, P. A recipe for strength. *Nature* **1999**, *399*, 210–211. [CrossRef]
7. Liu, M.; Jia, Z.; Jia, D.; Zhou, C. Recent advance in research on halloysite nanotubes-polymer nanocomposite. *Prog. Polym. Sci.* **2014**, *39*, 1498–1525. [CrossRef]
8. Mark, J.E. Ceramic-reinforced polymers and polymer-modified ceramics. *Polym. Eng. Sci.* **1996**, *36*, 2905–2920. [CrossRef]
9. Herron, N.; Thorn, D.L. Nanoparticles: Uses and relationships to molecular cluster compounds. *Adv. Mater.* **1998**, *10*, 1173–1184. [CrossRef]
10. Huang, P.; Shi, H.-Q.; Fu, S.-Y.; Xiao, H.-M.; Hu, N.; Li, Y.-Q. Greatly decreased redshift and largely enhanced refractive index of mono-dispersed ZnO-QD/silicone nanocomposites. *J. Mater. Chem. C* **2016**, *4*, 8663–8669. [CrossRef]
11. Albdiry, M.; Yousif, B.; Ku, H.; Lau, K. A critical review on the manufacturing processes in relation to the properties of nanoclay/polymer composites. *J. Compos. Mater.* **2013**, *47*, 1093–1115. [CrossRef]
12. Du, J.; Cheng, H.M. The fabrication, properties, and uses of graphene/polymer composites. *Macromol. Chem. Phys.* **2012**, *213*, 1060–1077. [CrossRef]
13. Li, X.; Wang, C.; Cao, Y.; Wang, G. Functional MXene materials: Progress of their applications. *Asian J. Chem.* **2018**, *13*, 2742–2757. [CrossRef] [PubMed]
14. Tu, S.; Jiang, Q.; Zhang, X.; Alshareef, H.N. Large dielectric constant enhancement in MXene percolative polymer composites. *ACS Nano* **2018**, *12*, 3369–3377. [CrossRef] [PubMed]
15. Sun, L.; Xiao, M.; Liu, J.; Gong, K. A study of the polymerization of styrene initiated by K–THF–GIC system. *Eur. Polym. J.* **2006**, *42*, 259–264. [CrossRef]
16. Xiao, M.; Sun, L.; Liu, J.; Li, Y.; Gong, K. Synthesis and properties of polystyrene/graphite nanocomposites. *Polymer* **2002**, *43*, 2245–2248. [CrossRef]
17. Li, Y.; Zhu, J.; Wei, S.; Ryu, J.; Wang, Q.; Sun, L.; Guo, Z. Poly(propylene) nanocomposites containing various carbon nanostructures. *Macromol. Chem. Phys.* **2011**, *212*, 2429–2438. [CrossRef]
18. Ha, H.W.; Choudhury, A.; Kamal, T.; Kim, D.-H.; Park, S.-Y. Effect of Chemical Modification of Graphene on Mechanical, Electrical, and Thermal Properties of Polyimide/Graphene Nanocomposites. *ACS Appl. Mater. Interfaces* **2012**, *4*, 4623–4630. [CrossRef]
19. Bian, J.; Lin, H.L.; He, F.X.; Wei, X.W.; Chang, I.-T.; Sancaktar, E. Fabrication of microwave exfoliated graphite oxide reinforced thermoplastic polyurethane nanocomposites: Effects of filler on morphology, mechanical, thermal and conductive properties. *Compos. Part A Appl. Sci. Manuf.* **2013**, *47*, 72–82. [CrossRef]
20. Ding, J.N.; Fan, Y.; Zhao, C.X.; Liu, Y.B.; Yu, C.T.; Yuan, N.Y. Electrical conductivity of waterborne polyurethane/graphene composites prepared by solution mixing. *J. Compos. Mater.* **2012**, *46*, 747–752. [CrossRef]
21. Fim, F.d.C.; Basso, N.R.S.; Graebin, A.P.; Azambuja, D.S.; Galland, G.B. Thermal, electrical, and mechanical properties of polyethylene–graphene nanocomposites obtained by in situ polymerization. *J. Appl. Polym. Sci.* **2013**, *128*, 2630–2637. [CrossRef]
22. Lim, Y.S.; Tan, Y.P.; Lim, H.N.; Huang, N.M.; Tan, W.T. Preparation and characterization of polypyrrole/graphene nanocomposite films and their electrochemical performance. *J. Polym. Res.* **2013**, *20*, 156. [CrossRef]
23. Tseng, I.-H.; Liao, Y.-F.; Chiang, J.-C.; Tsai, M.-H. Transparent polyimide/graphene oxide nanocomposite with improved moisture barrier property. *Mater. Chem. Phys.* **2012**, *136*, 247–253. [CrossRef]
24. Zhu, Y.; Murali, S.; Cai, W.; Li, X.; Suk, J.W.; Potts, J.R.; Ruoff, R.S. Graphene and Graphene Oxide: Synthesis, Properties, and Applications. *Adv. Mater.* **2010**, *22*, 3906–3924. [CrossRef] [PubMed]
25. Papageorgiou, D.G.; Kinloch, I.A.; Young, R.J. Mechanical properties of graphene and graphene-based nanocomposites. *Prog. Mater Sci.* **2017**, *90*, 75–127. [CrossRef]
26. Geim, A.K. Nobel Lecture: Random walk to graphene. *Rev. Mod. Phys.* **2011**, *83*, 851. [CrossRef]
27. Li, X.; Cai, W.; An, J.; Kim, S.; Nah, J.; Yang, D.; Piner, R.; Velamakanni, A.; Jung, I.; Tutuc, E.; et al. Large-Area Synthesis of High-Quality and Uniform Graphene Films on Copper Foils. *Science* **2009**, *324*, 1312–1314. [CrossRef]
28. Hernandez, Y.; Nicolosi, V.; Lotya, M.; Blighe, F.M.; Sun, Z.; De, S.; McGovern, I.T.; Holland, B.; Byrne, M.; Gun'Ko, Y.K.; et al. High-yield production of graphene by liquid-phase exfoliation of graphite. *Nat. Nanotechnol.* **2008**, *3*, 563–568. [CrossRef]
29. Niu, L.; Coleman, J.N.; Zhang, H.; Shin, H.; Chhowalla, M.; Zheng, Z. Production of Two-Dimensional Nanomaterials via Liquid-Based Direct Exfoliation. *Small* **2016**, *12*, 272–293. [CrossRef]
30. Abdelkader, A.M.; Cooper, A.J.; Dryfe, R.A.W.; Kinloch, I.A. How to get between the sheets: A review of recent works on the electrochemical exfoliation of graphene materials from bulk graphite. *Nanoscale* **2015**, *7*, 6944–6956. [CrossRef]
31. Jeon, I.-Y.; Shin, Y.-R.; Sohn, G.-J.; Choi, H.-J.; Bae, S.-Y.; Mahmood, J.; Jung, S.-M.; Seo, J.-M.; Kim, M.-J.; Wook Chang, D.; et al. Edge-carboxylated graphene nanosheets via ball milling. *Proc. Natl. Acad. Sci. USA* **2012**, *109*, 5588–5593. [CrossRef]
32. Phiri, J.; Gane, P.; Maloney, T.C. General overview of graphene: Production, properties and application in polymer composites. *J. Mater. Sci. Eng. B* **2017**, *215*, 9–28. [CrossRef]
33. Tripathi, S.N.; Rao, G.S.; Mathur, A.B.; Jasra, R. Polyolefin/graphene nanocomposites: A review. *RSC Adv.* **2017**, *7*, 23615–23632. [CrossRef]
34. Priyadarsini, S.; Mohanty, S.; Mukherjee, S.; Basu, S.; Mishra, M. Graphene and graphene oxide as nanomaterials for medicine and biology application. *J. Nanostruct. Chem.* **2018**, *8*, 123–137. [CrossRef]

35. Jeon, I.-Y.; Choi, H.-J.; Choi, M.; Seo, J.-M.; Jung, S.-M.; Kim, M.-J.; Zhang, S.; Zhang, L.; Xia, Z.; Dai, L.; et al. Facile, scalable synthesis of edge-halogenated graphene nanoplatelets as efficient metal-free eletrocatalysts for oxygen reduction reaction. *Sci. Rep.* **2013**, *3*, 1810. [CrossRef]
36. Jeon, I.-Y.; Ju, M.J.; Xu, J.; Choi, H.-J.; Seo, J.-M.; Kim, M.-J.; Choi, I.T.; Kim, H.M.; Kim, J.C.; Lee, J.-J.; et al. Edge-Fluorinated Graphene Nanoplatelets as High Performance Electrodes for Dye-Sensitized Solar Cells and Lithium Ion Batteries. *Adv. Funct. Mater.* **2015**, *25*, 1170–1179. [CrossRef]
37. Noh, H.-J.; Liu, S.; Yu, S.-Y.; Fan, Q.; Xiao, F.; Xu, J.; Jeon, I.-Y.; Baek, J.-B. Edge-NFx (x = 1 or 2) Protected Graphitic Nanoplatelets as a Stable Lithium Storage Material. *Batter. Supercaps* **2020**, *3*, 928–935. [CrossRef]
38. Kim, M.H.; Kang, Y.A.; Noh, H.-J.; Baek, J.-B.; Jeon, I.-Y. Direct preparation of edge-propylene graphitic nanoplatelets and its reinforcing effects in polypropylene. *Compos. Commun.* **2021**, *27*, 100896. [CrossRef]
39. Jeon, I.-Y.; Choi, H.-J.; Jung, S.-M.; Seo, J.-M.; Kim, M.-J.; Dai, L.; Baek, J.-B. Large-Scale Production of Edge-Selectively Functionalized Graphene Nanoplatelets via Ball Milling and Their Use as Metal-Free Electrocatalysts for Oxygen Reduction Reaction. *J. Am. Chem. Soc.* **2013**, *135*, 1386–1393. [CrossRef]
40. Song, H.D.; Im, Y.-K.; Baek, J.-B.; Jeon, I.-Y. Heptene-functionalized graphitic nanoplatelets for high-performance composites of linear low-density polyethylene. *Compos. Sci. Technol.* **2020**, *199*, 108380. [CrossRef]
41. Kang, Y.A.; Kim, M.H.; Noh, H.-J.; Baek, J.-B.; Jeon, I.-Y. Reinforcement of polystyrene using edge-styrene graphitic nanoplatelets. *J. Mater. Res. Technol.* **2021**, *10*, 662–670. [CrossRef]
42. Kim, M.-J.; Jeon, I.-Y.; Seo, J.-M.; Dai, L.; Baek, J.-B. Graphene Phosphonic Acid as an Efficient Flame Retardant. *ACS Nano* **2014**, *8*, 2820–2825. [CrossRef] [PubMed]
43. Jeon, I.-Y.; Zhang, S.; Zhang, L.; Choi, H.-J.; Seo, J.-M.; Xia, Z.; Dai, L.; Baek, J.-B. Edge-Selectively Sulfurized Graphene Nanoplatelets as Efficient Metal-Free Electrocatalysts for Oxygen Reduction Reaction: The Electron Spin Effect. *Adv. Mater.* **2013**, *25*, 6138–6145. [CrossRef]
44. Jeon, I.-Y.; Kim, H.M.; Kweon, D.H.; Jung, S.-M.; Seo, J.-M.; Shin, S.-H.; Choi, I.T.; Eom, Y.K.; Kang, S.H.; Kim, H.K.; et al. Metalloid tellurium-doped graphene nanoplatelets as ultimately stable electrocatalysts for cobalt reduction reaction in dye-sensitized solar cells. *Nano Energy* **2016**, *30*, 867–876. [CrossRef]
45. Jeon, I.-Y.; Choi, M.; Choi, H.-J.; Jung, S.-M.; Kim, M.-J.; Seo, J.-M.; Bae, S.-Y.; Yoo, S.; Kim, G.; Jeong, H.Y.; et al. Antimony-doped graphene nanoplatelets. *Nat. Commun.* **2015**, *6*, 7123. [CrossRef] [PubMed]
46. Baek, J.Y.; Jeon, I.-Y.; Baek, J.-B. Edge-iodine/sulfonic acid-functionalized graphene nanoplatelets as efficient electrocatalysts for oxygen reduction reaction. *J. Mater. Chem. A* **2014**, *2*, 8690–8695. [CrossRef]
47. Hong, J.; Bekyarova, E.; Liang, P.; de Heer, W.A.; Haddon, R.C.; Khizroev, S. Room-temperature Magnetic Ordering in Functionalized Graphene. *Sci. Rep.* **2012**, *2*, 624. [CrossRef] [PubMed]
48. Novoselov, K.S.; Geim, A.K.; Morozov, S.V.; Jiang, D.; Zhang, Y.; Dubonos, S.V.; Grigorieva, I.V.; Firsov, A.A. Electric Field Effect in Atomically Thin Carbon Films. *Science* **2004**, *306*, 666–669. [CrossRef]
49. Stoller, M.D.; Park, S.; Zhu, Y.; An, J.; Ruoff, R.S. Graphene-Based Ultracapacitors. *Nano Lett.* **2008**, *8*, 3498–3502. [CrossRef]
50. Johra, F.T.; Lee, J.-W.; Jung, W.-G. Facile and safe graphene preparation on solution based platform. *J. Ind. Eng. Chem.* **2014**, *20*, 2883–2887. [CrossRef]
51. McBride, J.R.; Lupini, A.R.; Schreuder, M.A.; Smith, N.J.; Pennycook, S.J.; Rosenthal, S.J. Few-Layer Graphene as a Support Film for Transmission Electron Microscopy Imaging of Nanoparticles. *ACS Appl. Mater. Interfaces* **2009**, *1*, 2886–2892. [CrossRef]
52. Sahoo, S.; Palai, R.; Katiyar, R.S. Polarized Raman scattering in monolayer, bilayer, and suspended bilayer graphene. *J. Appl. Phys.* **2011**, *110*, 44320. [CrossRef]
53. Novoselov, K.S.; Geim, A.K.; Morozov, S.V.; Jiang, D.; Katsnelson, M.I.; Grigorieva, I.V.; Dubonos, S.V.; Firsov, A.A. Two-dimensional gas of massless Dirac fermions in graphene. *Nature* **2005**, *438*, 197–200. [CrossRef]
54. Zhang, Y.; Tan, Y.-W.; Stormer, H.L.; Kim, P. Experimental observation of the quantum Hall effect and Berry's phase in graphene. *Nature* **2005**, *438*, 201–204. [CrossRef] [PubMed]
55. Geim, A.K.; Novoselov, K.S. The rise of graphene. *Nanosci. Technol.* **2007**, *6*, 11–19. [CrossRef] [PubMed]
56. Hsiao, M.-C.; Liao, S.-H.; Yen, M.-Y.; Liu, P.-I.; Pu, N.-W.; Wang, C.-A.; Ma, C.-C.M. Preparation of Covalently Functionalized Graphene Using Residual Oxygen-Containing Functional Groups. *ACS Appl. Mater. Interfaces.* **2010**, *2*, 3092–3099. [CrossRef] [PubMed]
57. Park, S.; Dikin, D.A.; Nguyen, S.T.; Ruoff, R.S. Graphene Oxide Sheets Chemically Cross-Linked by Polyallylamine. *J. Phys. Chem. C* **2009**, *113*, 15801–15804. [CrossRef]
58. Kumar, A.; Sharma, K.; Dixit, A.R. A review on the mechanical properties of polymer composites reinforced by carbon nanotubes and graphene. *Carbon Lett.* **2021**, *31*, 149–165. [CrossRef]
59. Bhunia, P.; Hwang, E.; Min, M.; Lee, J.; Seo, S.; Some, S.; Lee, H. A non-volatile memory device consisting of graphene oxide covalently functionalized with ionic liquid. *Chem. Commun.* **2012**, *48*, 913–915. [CrossRef]
60. Yu, W.; Sisi, L.; Haiyan, Y.; Jie, L. Progress in the functional modification of graphene/graphene oxide: A review. *RSC Adv.* **2020**, *10*, 15328–15345. [CrossRef]
61. Chua, C.K.; Pumera, M. Covalent chemistry on graphene. *Chem. Soc. Rev.* **2013**, *42*, 3222–3233. [CrossRef]
62. Sinitskii, A.; Dimiev, A.; Corley, D.A.; Fursina, A.A.; Kosynkin, D.V.; Tour, J.M. Kinetics of Diazonium Functionalization of Chemically Converted Graphene Nanoribbons. *ACS Nano* **2010**, *4*, 1949–1954. [CrossRef] [PubMed]

63. Xia, Z.; Leonardi, F.; Gobbi, M.; Liu, Y.; Bellani, V.; Liscio, A.; Kovtun, A.; Li, R.; Feng, X.; Orgiu, E.; et al. Electrochemical Functionalization of Graphene at the Nanoscale with Self-Assembling Diazonium Salts. *ACS Nano* **2016**, *10*, 7125–7134. [CrossRef] [PubMed]
64. Bouša, D.; Jankovský, O.; Sedmidubský, D.; Luxa, J.; Šturala, J.; Pumera, M.; Sofer, Z. Mesomeric Effects of Graphene Modified with Diazonium Salts: Substituent Type and Position Influence its Properties. *Eur. J. Chem.* **2015**, *21*, 17728–17738. [CrossRef] [PubMed]
65. Zhou, Z.; Yan, X.; Cook, T.R.; Saha, M.L.; Stang, P.J. Engineering Functionalization in a Supramolecular Polymer: Hierarchical Self-Organization of Triply Orthogonal Non-covalent Interactions on a Supramolecular Coordination Complex Platform. *J. Am. Chem. Soc.* **2016**, *138*, 806–809. [CrossRef]
66. Di Crescenzo, A.; Ettorre, V.; Fontana, A. Non-covalent and reversible functionalization of carbon nanotubes. *Beilstein J. Nanotechnol.* **2014**, *5*, 1675–1690. [CrossRef]
67. Lago, E.; Toth, P.S.; Pugliese, G.; Pellegrini, V.; Bonaccorso, F. Solution blending preparation of polycarbonate/graphene composite: Boosting the mechanical and electrical properties. *RSC Adv.* **2016**, *6*, 97931–97940. [CrossRef]
68. Park, S.; He, S.; Wang, J.; Stein, A.; Macosko, C.W. Graphene-polyethylene nanocomposites: Effect of graphene functionalization. *Polymer* **2016**, *104*, 1–9. [CrossRef]
69. Cheng, X.; Kumar, V.; Yokozeki, T.; Goto, T.; Takahashi, T.; Koyanagi, J.; Wu, L.; Wang, R. Highly conductive graphene oxide/polyaniline hybrid polymer nanocomposites with simultaneously improved mechanical properties. *Compos. Part A Appl. Sci. Manuf.* **2016**, *82*, 100–107. [CrossRef]
70. Liu, M.; Papageorgiou, D.G.; Li, S.; Lin, K.; Kinloch, I.A.; Young, R.J. Micromechanics of reinforcement of a graphene-based thermoplastic elastomer nanocomposite. *Compos. Part A Appl. Sci. Manuf.* **2018**, *110*, 84–92. [CrossRef]
71. Xu, Z.; Gao, C. In situ Polymerization Approach to Graphene-Reinforced Nylon-6 Composites. *Macromolecules* **2010**, *43*, 6716–6723. [CrossRef]
72. Wang, X.; Hu, Y.; Song, L.; Yang, H.; Xing, W.; Lu, H. In situ polymerization of graphene nanosheets and polyurethane with enhanced mechanical and thermal properties. *J. Mater. Chem.* **2011**, *21*, 4222–4227. [CrossRef]
73. Luong, N.D.; Hippi, U.; Korhonen, J.T.; Soininen, A.J.; Ruokolainen, J.; Johansson, L.-S.; Nam, J.-D.; Sinh, L.H.; Seppälä, J. Enhanced mechanical and electrical properties of polyimide film by graphene sheets via in situ polymerization. *Polymer* **2011**, *52*, 5237–5242. [CrossRef]
74. Milani, M.A.; González, D.; Quijada, R.; Basso, N.R.S.; Cerrada, M.L.; Azambuja, D.S.; Galland, G.B. Polypropylene/graphene nanosheet nanocomposites by in situ polymerization: Synthesis, characterization and fundamental properties. *Compos. Sci. Technol.* **2013**, *84*, 1–7. [CrossRef]
75. Park, H.; Lee, S.J.; Kim, S.; Ryu, H.W.; Lee, S.H.; Choi, H.H.; Cheong, I.W.; Kim, J.-H. Conducting polymer nanofiber mats via combination of electrospinning and oxidative polymerization. *Polymer* **2013**, *54*, 4155–4160. [CrossRef]
76. Greiner, A.; Wendorff, J.H. Electrospinning: A Fascinating Method for the Preparation of Ultrathin Fibers. *Angew. Chem.* **2007**, *46*, 5670–5703. [CrossRef] [PubMed]
77. Subbiah, T.; Bhat, G.S.; Tock, R.W.; Parameswaran, S.; Ramkumar, S.S. Electrospinning of nanofibers. *J. Appl. Polym. Sci.* **2005**, *96*, 557–569. [CrossRef]
78. Potts, J.R.; Dreyer, D.R.; Bielawski, C.W.; Ruoff, R.S. Graphene-based polymer nanocomposites. *Polymer* **2011**, *52*, 5–25. [CrossRef]
79. Shahryari, Z.; Yeganeh, M.; Gheisari, K.; Ramezanzadeh, B. A brief review of the graphene oxide-based polymer nanocomposite coatings: Preparation, characterization, and properties. *J. Coat. Technol. Res.* **2021**, *18*, 945–969. [CrossRef]
80. Lee, C.Y.; Bae, J.-H.; Kim, T.-Y.; Chang, S.-H.; Kim, S.Y. Using silane-functionalized graphene oxides for enhancing the interfacial bonding strength of carbon/epoxy composites. *Compos. Part A Appl. Sci. Manuf.* **2015**, *75*, 11–17. [CrossRef]
81. Son, D.-S.; Hong, J.-H.; Chang, S.-H. Determination of the autofrettage pressure and estimation of material failures of a Type III hydrogen pressure vessel by using finite element analysis. *Int. J. Hydrog. Energy* **2012**, *37*, 12771–12781. [CrossRef]
82. Ibrahim, A.; Klopocinska, A.; Horvat, K.; Abdel Hamid, Z. Graphene-Based Nanocomposites: Synthesis, Mechanical Properties, and Characterizations. *Polymers* **2021**, *13*, 2869. [CrossRef] [PubMed]
83. Lee, C.; Wei, X.; Kysar, J.W.; Hone, J. Measurement of the Elastic Properties and Intrinsic Strength of Monolayer Graphene. *Science* **2008**, *321*, 385–388. [CrossRef] [PubMed]
84. Pang, Y.; Yang, J.; Curtis, T.E.; Luo, S.; Huang, D.; Feng, Z.; Morales-Ferreiro, J.O.; Sapkota, P.; Lei, F.; Zhang, J.; et al. Exfoliated Graphene Leads to Exceptional Mechanical Properties of Polymer Composite Films. *ACS Nano* **2019**, *13*, 1097–1106. [CrossRef] [PubMed]
85. Kuila, T.; Bose, S.; Hong, C.E.; Uddin, M.E.; Khanra, P.; Kim, N.H.; Lee, J.H. Preparation of functionalized graphene/linear low density polyethylene composites by a solution mixing method. *Carbon* **2011**, *49*, 1033–1037. [CrossRef]
86. Batista, N.L.; Helal, E.; Kurusu, R.S.; Moghimian, N.; David, E.; Demarquette, N.R.; Hubert, P. Mass-produced graphene—HDPE nanocomposites: Thermal, rheological, electrical, and mechanical properties. *Polym. Eng. Sci.* **2019**, *59*, 675–682. [CrossRef]
87. Todd, A.D.; Bielawski, C.W. Thermally reduced graphite oxide reinforced polyethylene composites: A mild synthetic approach. *Polymer* **2013**, *54*, 4427–4430. [CrossRef]
88. Zhong, J.; Isayev, A.I.; Zhang, X. Ultrasonic twin screw compounding of polypropylene with carbon nanotubes, graphene nanoplates and carbon black. *Eur. Polym. J.* **2016**, *80*, 16–39. [CrossRef]

89. Ahmad, S.R.; Xue, C.; Young, R.J. The mechanisms of reinforcement of polypropylene by graphene nanoplatelets. *J. Mater. Sci. Eng. B* **2017**, *216*, 2–9. [CrossRef]
90. Abuoudah, C.K.; Greish, Y.E.; Abu-Jdayil, B.; El-said, E.M.; Iqbal, M.Z. Graphene/polypropylene nanocomposites with improved thermal and mechanical properties. *J. Appl. Polym. Sci.* **2021**, *138*, 50024. [CrossRef]
91. El Achaby, M.; Arrakhiz, F.-E.; Vaudreuil, S.; el Kacem Qaiss, A.; Bousmina, M.; Fassi-Fehri, O. Mechanical, thermal, and rheological properties of graphene-based polypropylene nanocomposites prepared by melt mixing. *Polym. Compos.* **2012**, *33*, 733–744. [CrossRef]
92. Md Said, N.H.; Liu, W.-W.; Khe, C.-S.; Lai, C.-W.; Zulkepli, N.N.; Aziz, A. Review of the past and recent developments in functionalization of graphene derivatives for reinforcement of polypropylene nanocomposites. *Polym. Compos.* **2021**, *42*, 1075–1108. [CrossRef]
93. Song, P.; Cao, Z.; Cai, Y.; Zhao, L.; Fang, Z.; Fu, S. Fabrication of exfoliated graphene-based polypropylene nanocomposites with enhanced mechanical and thermal properties. *Polymer* **2011**, *52*, 4001–4010. [CrossRef]
94. Han, S.; Chand, A.; Araby, S.; Cai, R.; Chen, S.; Kang, H.; Cheng, R.; Meng, Q. Thermally and electrically conductive multifunctional sensor based on epoxy/graphene composite. *Nanotechnology* **2019**, *31*, 75702. [CrossRef]
95. Berhanuddin, N.; Zaman, I.; Rozlan, S.; Karim, M.; Manshoor, B.; Khalid, A.; Chan, S.; Meng, Q. Enhancement of mechanical properties of epoxy/graphene nanocomposite. *J. Phys. Conf. Ser.* **2017**, *914*, 12036. [CrossRef]
96. Salom, C.; Prolongo, M.G.; Toribio, A.; Martínez-Martínez, A.J.; de Cárcer, I.A.; Prolongo, S.G. Mechanical properties and adhesive behavior of epoxy-graphene nanocomposites. *Int. J. Adhes. Adhes.* **2018**, *84*, 119–125. [CrossRef]
97. King, J.A.; Klimek, D.R.; Miskioglu, I.; Odegard, G.M. Mechanical properties of graphene nanoplatelet/epoxy composites. *J. Appl. Polym. Sci.* **2013**, *128*, 4217–4223. [CrossRef]
98. Klimek-McDonald, D.R.; King, J.A.; Miskioglu, I.; Pineda, E.J.; Odegard, G.M. Determination and Modeling of Mechanical Properties for Graphene Nanoplatelet/Epoxy Composites. *Polym. Compos.* **2018**, *39*, 1845–1851. [CrossRef]
99. Zhang, Y.; Wang, Y.; Yu, J.; Chen, L.; Zhu, J.; Hu, Z. Tuning the interface of graphene platelets/epoxy composites by the covalent grafting of polybenzimidazole. *Polymer* **2014**, *55*, 4990–5000. [CrossRef]
100. Wan, Y.-J.; Tang, L.-C.; Yan, D.; Zhao, L.; Li, Y.-B.; Wu, L.-B.; Jiang, J.-X.; Lai, G.-Q. Improved dispersion and interface in the graphene/epoxy composites via a facile surfactant-assisted process. *Compos. Sci. Technol.* **2013**, *82*, 60–68. [CrossRef]
101. Li, Y.; Sun, J.; Wang, J.; Qin, C.; Dai, L. Preparation of well-dispersed reduced graphene oxide and its mechanical reinforcement in polyvinyl alcohol fibre. *Polym. Int.* **2016**, *65*, 1054–1062. [CrossRef]
102. Kashyap, S.; Pratihar, S.K.; Behera, S.K. Strong and ductile graphene oxide reinforced PVA nanocomposites. *J. Alloys Compd.* **2016**, *684*, 254–260. [CrossRef]
103. Xu, Y.; Hong, W.; Bai, H.; Li, C.; Shi, G. Strong and ductile poly(vinyl alcohol)/graphene oxide composite films with a layered structure. *Carbon* **2009**, *47*, 3538–3543. [CrossRef]
104. Aslam, M.; Kalyar, M.A.; Raza, Z.A. Fabrication of reduced graphene oxide nanosheets doped PVA composite films for tailoring their opto-mechanical properties. *Appl. Phys. A* **2017**, *123*, 424. [CrossRef]
105. Yang, X.; Li, L.; Shang, S.; Tao, X.-M. Synthesis and characterization of layer-aligned poly(vinyl alcohol)/graphene nanocomposites. *Polymer* **2010**, *51*, 3431–3435. [CrossRef]
106. Cobos, M.; Fernández, M.J.; Fernández, M.D. Graphene Based Poly(Vinyl Alcohol) Nanocomposites Prepared by In Situ Green Reduction of Graphene Oxide by Ascorbic Acid: Influence of Graphene Content and Glycerol Plasticizer on Properties. *Nanomaterials* **2018**, *8*, 1013. [CrossRef]
107. Ma, H.-L.; Zhang, Y.; Hu, Q.-H.; He, S.; Li, X.; Zhai, M.; Yu, Z.-Z. Enhanced mechanical properties of poly(vinyl alcohol) nanocomposites with glucose-reduced graphene oxide. *Mater. Lett.* **2013**, *102–103*, 15–18. [CrossRef]
108. Zhao, X.; Zhang, Q.; Chen, D.; Lu, P. Enhanced Mechanical Properties of Graphene-Based Poly(vinyl alcohol) Composites. *Macromolecules* **2010**, *43*, 2357–2363. [CrossRef]
109. Liang, J.; Huang, Y.; Zhang, L.; Wang, Y.; Ma, Y.; Guo, T.; Chen, Y. Molecular-Level Dispersion of Graphene into Poly(vinyl alcohol) and Effective Reinforcement of their Nanocomposites. *Adv. Funct. Mater.* **2009**, *19*, 2297–2302. [CrossRef]
110. Loryuenyong, V.; Saewong, C.; Aranchaiya, C.; Buasri, A. The Improvement in Mechanical and Barrier Properties of Poly(Vinyl Alcohol)/Graphene Oxide Packaging Films. *Packag. Technol.* **2015**, *28*, 939–947. [CrossRef]
111. Fujimori, K.; Gopiraman, M.; Kim, H.-K.; Kim, B.-S.; Kim, I.-S. Mechanical and electromagnetic interference shielding properties of poly(vinyl alcohol)/graphene and poly(vinyl alcohol)/multi-walled carbon nanotube composite nanofiber mats and the effect of Cu top-layer coating. *J. Nanosci. Nanotechnol.* **2013**, *13*, 1759–1764. [CrossRef]
112. Ma, J.; Li, Y.; Yin, X.; Xu, Y.; Yue, J.; Bao, J.; Zhou, T. Poly(vinyl alcohol)/graphene oxide nanocomposites prepared by in situ polymerization with enhanced mechanical properties and water vapor barrier properties. *RSC Adv.* **2016**, *6*, 49448–49458. [CrossRef]
113. Bazzi, M.; Shabani, I.; Mohandesi, J.A. Enhanced mechanical properties and electrical conductivity of Chitosan/Polyvinyl Alcohol electrospun nanofibers by incorporation of graphene nanoplatelets. *J. Mech. Behav. Biomed. Mater.* **2022**, *125*, 104975. [CrossRef] [PubMed]
114. Ghobadi, S.; Sadighikia, S.; Papila, M.; Cebeci, F.Ç.; Gürsel, S.A. Graphene-reinforced poly(vinyl alcohol) electrospun fibers as building blocks for high performance nanocomposites. *RSC Adv.* **2015**, *5*, 85009–85018. [CrossRef]

115. Yadav, S.K.; Cho, J.W. Functionalized graphene nanoplatelets for enhanced mechanical and thermal properties of polyurethane nanocomposites. *Appl. Surf. Sci.* **2013**, *266*, 360–367. [CrossRef]
116. Jing, Q.; Liu, W.; Pan, Y.; Silberschmidt, V.V.; Li, L.; Dong, Z. Chemical functionalization of graphene oxide for improving mechanical and thermal properties of polyurethane composites. *Mater. Des.* **2015**, *85*, 808–814. [CrossRef]
117. Yoo, H.J.; Mahapatra, S.S.; Cho, J.W. High-Speed Actuation and Mechanical Properties of Graphene-Incorporated Shape Memory Polyurethane Nanofibers. *J. Phys. Chem. C* **2014**, *118*, 10408–10415. [CrossRef]
118. Lee, T.-H.; Yen, C.-T.; Hsu, S.-H. Preparation of Polyurethane-Graphene Nanocomposite and Evaluation of Neurovascular Regeneration. *ACS Biomater. Sci. Eng.* **2020**, *6*, 597–609. [CrossRef]
119. Lee, Y.R.; Raghu, A.V.; Jeong, H.M.; Kim, B.K. Properties of Waterborne Polyurethane/Functionalized Graphene Sheet Nanocomposites Prepared by an in situ Method. *Macromol. Chem. Phys.* **2009**, *210*, 1247–1254. [CrossRef]
120. Wang, X.; Xing, W.; Song, L.; Yang, H.; Hu, Y.; Yeoh, G.H. Fabrication and characterization of graphene-reinforced waterborne polyurethane nanocomposite coatings by the sol–gel method. *Surf. Coat. Technol.* **2012**, *206*, 4778–4784. [CrossRef]
121. Appel, A.-K.; Thomann, R.; Mülhaupt, R. Polyurethane nanocomposites prepared from solvent-free stable dispersions of functionalized graphene nanosheets in polyols. *Polymer* **2012**, *53*, 4931–4939. [CrossRef]
122. Wu, C.; Huang, X.; Wang, G.; Wu, X.; Yang, K.; Li, S.; Jiang, P. Hyperbranched-polymer functionalization of graphene sheets for enhanced mechanical and dielectric properties of polyurethane composites. *J. Mater. Chem.* **2012**, *22*, 7010–7019. [CrossRef]
123. Ma, W.-S.; Wu, L.; Yang, F.; Wang, S.-F. Non-covalently modified reduced graphene oxide/polyurethane nanocomposites with good mechanical and thermal properties. *J. Mater. Sci.* **2014**, *49*, 562–571. [CrossRef]
124. Raghu, A.V.; Lee, Y.R.; Jeong, H.M.; Shin, C.M. Preparation and Physical Properties of Waterborne Polyurethane/Functionalized Graphene Sheet Nanocomposites. *Macromol. Chem. Phys.* **2008**, *209*, 2487–2493. [CrossRef]
125. Fang, M.; Wang, K.; Lu, H.; Yang, Y.; Nutt, S. Covalent polymer functionalization of graphene nanosheets and mechanical properties of composites. *J. Mater. Chem.* **2009**, *19*, 7098–7105. [CrossRef]
126. Wan, C.; Chen, B. Reinforcement and interphase of polymer/graphene oxide nanocomposites. *J. Mater. Chem.* **2012**, *22*, 3637–3646. [CrossRef]
127. Aized, T.; Imran, M.; Raza, H.; Raza, M.R.; Gohar, G.A.; Iqbal, A. Effect of nano-filler graphene on nano-composite system of polystyrene-graphene. *Int. J. Adv. Manuf. Technol.* **2018**, *95*, 3707–3715. [CrossRef]
128. Raza, H.; Aized, T.; Khan, M.B.; Imran, M. Tensile testing of polystyrene graphene 2D nano composite membrane. *Int. J. Adv. Manuf. Technol.* **2018**, *94*, 4343–4349. [CrossRef]
129. Zhao, Z.; Cai, W.; Xu, Z.; Mu, X.; Ren, X.; Zou, B.; Gui, Z.; Hu, Y. Multi-role p-styrene sulfonate assisted electrochemical preparation of functionalized graphene nanosheets for improving fire safety and mechanical property of polystyrene composites. *Compos. B. Eng.* **2020**, *181*, 107544. [CrossRef]
130. Xu, H.; Li, X.; Li, P.; Ma, L.; Li, H.; Shi, L.; Wang, M.; Chen, H.; Song, G. Enhancing mechanical performances of polystyrene composites via constructing carbon nanotube/graphene oxide aerogel and hot pressing. *Compos. Sci. Technol.* **2020**, *195*, 108191. [CrossRef]
131. Wang, H.; Xie, G.; Fang, M.; Ying, Z.; Tong, Y.; Zeng, Y. Mechanical reinforcement of graphene/poly(vinyl chloride) composites prepared by combining the in-situ suspension polymerization and melt-mixing methods. *Compos. B. Eng.* **2017**, *113*, 278–284. [CrossRef]
132. Ahmed, R.M.; Ibrahiem, A.A.; El-Bayoumi, A.S.; Atta, M.M. Structural, mechanical, and dielectric properties of polyvinylchloride/graphene nano platelets composites. *Int. J. Polym. Anal. Charact.* **2020**, *26*, 68–83. [CrossRef]
133. Khaleghi, M.; Didehban, K.; Shabanian, M. Effect of new melamine-terephthaldehyde resin modified graphene oxide on thermal and mechanical properties of PVC. *Polym. Test.* **2017**, *63*, 382–391. [CrossRef]
134. Akhina, H.; Gopinathan Nair, M.R.; Kalarikkal, N.; Pramoda, K.P.; Hui Ru, T.; Kailas, L.; Thomas, S. Plasticized PVC graphene nanocomposites: Morphology, mechanical, and dynamic mechanical properties. *Polym. Eng. Sci.* **2018**, *58*, E104–E113. [CrossRef]
135. Wang, L.; Wei, X.; Wang, G.; Zhao, S.; Cui, J.; Gao, A.; Zhang, G.; Yan, Y. A facile and industrially feasible one-pot approach to prepare graphene-decorated PVC particles and their application in multifunctional PVC/graphene composites with segregated structure. *Compos. B. Eng.* **2020**, *185*, 107775. [CrossRef]
136. Gong, L.; Yin, B.; Li, L.-P.; Yang, M.-B. Nylon-6/Graphene composites modified through polymeric modification of graphene. *Compos. B. Eng.* **2015**, *73*, 49–56. [CrossRef]
137. Wang, C.; Hu, F.; Yang, K.; Hu, T.; Wang, W.; Deng, R.; Jiang, Q.; Zhang, H. Preparation and properties of nylon 6/sulfonated graphene composites by an in situ polymerization process. *RSC Adv.* **2016**, *6*, 45014–45022. [CrossRef]
138. Dixon, D.; Lemonine, P.; Hamilton, J.; Lubarsky, G.; Archer, E. Graphene oxide–polyamide 6 nanocomposites produced via in situ polymerization. *J. Thermoplast. Compos. Mater.* **2013**, *28*, 372–389. [CrossRef]
139. Liu, H.-H.; Peng, W.-W.; Hou, L.-C.; Wang, X.-C.; Zhang, X.-X. The production of a melt-spun functionalized graphene/poly(ε-caprolactam) nanocomposite fiber. *Compos. Sci. Technol.* **2013**, *81*, 61–68. [CrossRef]
140. Lan, Y.; Liu, H.; Cao, X.; Zhao, S.; Dai, K.; Yan, X.; Zheng, G.; Liu, C.; Shen, C.; Guo, Z. Electrically conductive thermoplastic polyurethane/polypropylene nanocomposites with selectively distributed graphene. *Polymer* **2016**, *97*, 11–19. [CrossRef]
141. Iniestra-Galindo, M.G.; Pérez-González, J.; Marín-Santibáñez, B.M.; Balmori-Ramírez, H. Preparation at large-scale of polypropylene nanocomposites with microwaves reduced graphene oxide. *Mater. Res. Express* **2019**, *6*, 105347. [CrossRef]

142. Chen, W.; Weimin, H.; Li, D.; Chen, S.; Dai, Z. A critical review on the development and performance of polymer/graphene nanocomposites. *Sci. Eng. Compos. Mater.* **2018**, *25*, 1059–1073. [CrossRef]
143. Xue, Q.; Lv, C.; Shan, M.; Zhang, H.; Ling, C.; Zhou, X.; Jiao, Z. Glass transition temperature of functionalized graphene–polymer composites. *Comput. Mater. Sci.* **2013**, *71*, 66–71. [CrossRef]
144. Jeong, J.S.; Moon, J.S.; Jeon, S.Y.; Park, J.H.; Alegaonkar, P.S.; Yoo, J.B. Mechanical properties of electrospun PVA/MWNTs composite nanofibers. *Thin Solid Film.* **2007**, *515*, 5136–5141. [CrossRef]
145. Kalaitzidou, K.; Fukushima, H.; Askeland, P.; Drzal, L.T. The nucleating effect of exfoliated graphite nanoplatelets and their influence on the crystal structure and electrical conductivity of polypropylene nanocomposites. *J. Mater. Sci.* **2008**, *43*, 2895–2907. [CrossRef]
146. An, J.-E.; Jeon, G.W.; Jeong, Y.G. Preparation and properties of polypropylene nanocomposites reinforced with exfoliated graphene. *Fibers Polym.* **2012**, *13*, 507–514. [CrossRef]
147. Liu, Y.; Fan, B.; Hamon, A.-L.; He, D.; Bai, J. Thickness effect on the tensile and dynamic mechanical properties of graphene nanoplatelets-reinforced polymer nanocomposites. *Graphene Technol.* **2017**, *2*, 21–27. [CrossRef]
148. Tang, Z.; Lei, Y.; Guo, B.; Zhang, L.; Jia, D. The use of rhodamine B-decorated graphene as a reinforcement in polyvinyl alcohol composites. *Polymer* **2012**, *53*, 673–680. [CrossRef]
149. Potts, J.R.; Lee, S.H.; Alam, T.M.; An, J.; Stoller, M.D.; Piner, R.D.; Ruoff, R.S. Thermomechanical properties of chemically modified graphene/poly(methyl methacrylate) composites made by in situ polymerization. *Carbon* **2011**, *49*, 2615–2623. [CrossRef]
150. Kumar, A.; Sharma, K.; Dixit, A.R. A review of the mechanical and thermal properties of graphene and its hybrid polymer nanocomposites for structural applications. *J. Mater. Sci.* **2019**, *54*, 5992–6026. [CrossRef]
151. Mohan, V.B.; Lau, K.-t.; Hui, D.; Bhattacharyya, D. Graphene-based materials and their composites: A review on production, applications and product limitations. *Compos. B. Eng.* **2018**, *142*, 200–220. [CrossRef]
152. Sun, X.; Liu, X.; Shen, X.; Wu, Y.; Wang, Z.; Kim, J.-K. Graphene foam/carbon nanotube/poly(dimethyl siloxane) composites for exceptional microwave shielding. *Compos. Part A Appl. Sci. Manuf.* **2016**, *85*, 199–206. [CrossRef]
153. Huang, C.J.; Fu, S.Y.; Zhang, Y.H.; Lauke, B.; Li, L.F.; Ye, L. Cryogenic properties of SiO_2/epoxy nanocomposites. *Cryogenics* **2005**, *45*, 450–454. [CrossRef]
154. Wu, Y.; Chen, M.; Chen, M.; Ran, Z.; Zhu, C.; Liao, H. The reinforcing effect of polydopamine functionalized graphene nanoplatelets on the mechanical properties of epoxy resins at cryogenic temperature. *Polym. Test.* **2017**, *58*, 262–269. [CrossRef]
155. Perreault, F.; Fonseca de Faria, A.; Elimelech, M. Environmental applications of graphene-based nanomaterials. *Chem. Soc. Rev.* **2015**, *44*, 5861–5896. [CrossRef] [PubMed]
156. Zhang, L.L.; Zhou, R.; Zhao, X.S. Graphene-based materials as supercapacitor electrodes. *J. Mater. Chem.* **2010**, *20*, 5983–5992. [CrossRef]
157. Wang, Y.; Li, Z.; Wang, J.; Li, J.; Lin, Y. Graphene and graphene oxide: Biofunctionalization and applications in biotechnology. *Trends Biotechnol.* **2011**, *29*, 205–212. [CrossRef]
158. Sanes, J.; Sánchez, C.; Pamies, R.; Avilés, M.-D.; Bermúdez, M.-D. Extrusion of Polymer Nanocomposites with Graphene and Graphene Derivative Nanofillers: An Overview of Recent Developments. *Materials* **2020**, *13*, 549. [CrossRef]

MDPI
St. Alban-Anlage 66
4052 Basel
Switzerland
Tel. +41 61 683 77 34
Fax +41 61 302 89 18
www.mdpi.com

Polymers Editorial Office
E-mail: polymers@mdpi.com
www.mdpi.com/journal/polymers